JEFF MOW
GEOLOGY 20
FALL TERM 1978
WATSON 213

Orogeny
Synclines

GEOMORPHOLOGY

GEOMORPHOLOGY

A Systematic Analysis
of Late Cenozoic Landforms

Arthur L. Bloom

Department of Geological Sciences
Cornell University

PRENTICE-HALL, INC., Englewood Cliffs, New Jersey 07632

Library of Congress Cataloging in Publication Data

BLOOM, ARTHUR LEROY, (date)
    Geomorphology.

    Bibliography: p.
    Includes index.
    1. Geomorphology.   2. Geology, Stratigraphic—
Quaternary.   I. Title.
GB401.5.B55   551.4   77–25816
ISBN   0–13–353086–8

Printed in the United States of America

10   9   8   7   6   5   4   3   2   1

PRENTICE-HALL INTERNATIONAL, INC., *London*
PRENTICE-HALL OF AUSTRALIA PTY. LIMITED, *Sydney*
PRENTICE-HALL OF CANADA, LTD., *Toronto*
PRENTICE-HALL OF INDIA PRIVATE LIMITED, *New Delhi*
PRENTICE-HALL OF JAPAN, INC., *Tokyo*
PRENTICE-HALL OF SOUTHEAST ASIA PTE. LTD., *Singapore*
WHITEHALL BOOKS LIMITED, *Wellington, New Zealand*

To Donna, with love.

Contents

PART III

Climatic Morphogenesis   307

Preface

Geomorphology is normally taught at the intermediate or advanced level in university and college curriculums. In writing this textbook, I have assumed that the student who uses it has had at least one introductory course in geology or physical geography. Elementary familiarity with climatic zones, planetary winds and ocean currents, the geologic time scale, common rocks and minerals, and topographic maps is assumed.

Geomorphology is studied by many more students than ever major in the subject or become practicing geomorphologists. The subject is part of the traditional undergraduate major in most geology and geography departments in North America and in physical geography departments elsewhere. Currently, course enrollments are swollen by nongeologists and nongeographers. Archeologists, pedologists, soils and hydraulics engineers, landscape architects, secondary-school science teachers, climatologists, limnologists, and conservationists are just a few of the specialists from outside the traditional fields of geology and geography who specifically seek geomorphic understanding. These students tend to draw the academic geomorphologist away from his or her central themes. They normally lack geologic background but contribute instead their own interests, skills, and needs. In the applied sciences, especially aerial photo interpretation, remote sensing, and land-use classification, geomorphic description must be pragmatic and useful.

At the same time, geomorphology is becoming more central to geological subjects. The new global tectonics emphasizes constructional landforms and also relates denudation history and explanatory description of landscapes to tectonic theories and structural analysis. The new general theories for the origin of surface relief and tectonism require an integration of geomorphology with structural geology, tectonics, geophysics, petrology, and stratigraphy.

The primary role of geomorphologists is to teach and demonstrate that landscapes change in response to energy supplied from a variety of sources. Among ourselves, we can engage in learned debates about whether the changes are sequential or random and whether the energy systems are best understood as open or closed, but we all accept the transient nature of the present landscape. Reduced to that level, geomorphology is still close to the natural philosophy from which it and other sciences originated. Geomorphology does not provide most of its students with a set of skills desired by prospective employers. Instead it provides a background, or reference framework, within which many specialists practice.

This book is organized largely along traditional lines, following an outline I have evolved by teaching a one-semester course in geomorphology. It considers first the constructional processes and landforms that result from tectonic and volcanic activity (Part I). Owing to the dynamic nature of our new plate-tectonic theory, we must consider constructional and erosional landforms on an equal level.

Next, as an introduction to destructional or erosional landforms (Part II), the energy environment at the surface of the earth is analyzed in some detail. It is my conviction that geomorphologists have been regrettably delinquent in considering the energy supply for the changes they so sweepingly describe. Geomorphic change, like all other physical processes, must be controlled ultimately by energy gradients and potentials. The concept of physical systems, with definable boundaries through which matter and energy move at definable rates, is especially well illustrated by the energy flow in geomorphic processes. The favorable reaction to the concept of a solar-driven "geomorphology machine" in a brief earlier book (Bloom, 1969) encouraged me to expand that topic considerably.

Erosional processes and the landscapes they produce from tectonic or structural blocks occupy 75 percent of the text. The logical scheme of explanatory description in terms of structure, process, and time is followed not out of blind loyalty to the spirit of W. M. Davis but from my studied conviction that the best way to describe landscapes is to explain their origins. For the practical needs of professional geographers, geologists, pedologists, and hydrographers, and for the cultural needs of intelligent tourists whose recreation is visiting areas of great scenic beauty, explaining how and when a landscape took its form must be the major emphasis in a textbook of systematic geomorphology.

Regional descriptions are included to demonstrate how certain kinds of landscapes formed, but this is not a textbook of regional physiography. Examples and illustrations have been chosen from as wide a range as I have been able to find in my own travel and in published descriptions in order to counter the natural tendency for provinciality that lessens the value of many geomorphology books. I regret that political and language barriers restrict my access to recent regional descriptions of the interior of the great Asian continent, which must contain some of the most spectacular arid, semiarid, and high-plateau landscapes on this planet. The more ordinary restriction of difficult travel conditions has previously limited the availability of studies from the tropical climatic zones, although the emergence of many new independent nations in the tropical regions of Africa and Asia has given impetus to geomorphic as well as economic and climatic studies there.

A notable development of recent years has been the accelerated study of landscapes in Australia and Papua New Guinea by government-sponsored teams of geomorphologists, pedologists and land-use experts. The initial outflow of technical reports is now being consolidated into major monographs and textbooks by Australian authors in a manner reminiscent of the exploration of the American west by the U.S. Geological Survey and its predecessors of the second half of the nineteenth century. We can anticipate that geomorphology will profit by the new work "down under" in the same way that the present science was shaped by the scientific pioneers of the American west.

Part III is a review of the geomorphic processes and forms that are distinctive of specific climatic regions. Arid and glacial landscapes have been regarded as "climatic accidents," but their extent during the late Cenozoic Era requires that they be treated as more than anomalies. Even so, they and their peripheral regions of semiarid and periglacial landscapes are best studied after the fluvial or "normal" processes are well understood. Much of climatic morphogenesis concerns the relative intensity, rather than the kind, of geomorphic processes. Chapter 18 concludes Part III of the book with a series of regional examples that demonstrate the complexity of the Quaternary climatic oscillations that have made our modern landscape so interesting and unique in geologic history.

Coasts deserve separate treatment (Part IV) because the energy input that shapes them is so different from the processes that shape subaerial landscapes. More importantly, the coastal zone is critical to human activities. Two-thirds of the world's people live in the 10 percent of the land defined as the coastal zone. The transience of sea level during the late Cenozoic Era has been a major factor in the development of the estuaries, lagoons, deltas, and reefs upon which so much modern commerce and food production depend. Future trends of sea-level movement are critical to coastal-zone activities.

The 71 percent of the terrestrial landscape on the floor of the ocean has barely been seen or photographed by scientists. Geomorphologists must not let their world end at low tide. The explanatory description of landforms in terms of structure, process, and time has been demonstrated to be just as useful for submarine geology as it is for subaerial studies on planet Earth.

I regret that the length of the manuscript and the continuing explosive growth of new information have precluded chapters on the morphology of the moon and the other terrestrial planets. It is only in the last 10 years that we have begun to rectify W. M. Davis's lament that "there are not enough kinds of observed facts on the small earth in the momentary present to match the long list of deduced elements of the scheme [of explanatory description]" (Davis, 1905, p. 156). The morphologists who describe the floor of the ocean and the face of the moon and planets must describe whole new lithologic materials and structures and the entirely exotic processes of weathering and erosion that occur under dense fluid "atmospheres" or in vacuum.

They require new, perhaps vastly expanded, time scales. The tectonic relief on the floor of the oceans is very young. By contrast, little has happened to the lunar surface in the last 25 million years, which we call the late Cenozoic Era here on earth. Yet I confidently predict that the schemes of explanatory description that are ultimately used for submarine and planetary landscapes will not be very different from the one developed for describing subaerial terrestrial scenery.

The references at the end of each chapter have been selected first for their relevance and second for their availability. Much excellent material is available in the form of technical memoranda of government agencies, offset-reproduced conference proceedings, and field excursion guides, but such "semipublished" or "near-print" material is not in general circulation and is doomed to early burial in archives. I have frequently paraphrased it, but it is rarely capable of being quoted or cited. I take this opportunity to thank the many friends and colleagues who have sent me offprints of their publications. These separate papers formed the nuclei of my chapter organization.

I thank many students and peers for suggestions, criticisms, and encouragement. Where I have disregarded their advice, I have done so at my own peril. V. E. Baker, J. H. Hartshorn, S. E. White, and D. F. Ritter read the entire manuscript at various stages and offered the finest constructive criticism. I thank them in the warmest personal way. Others, including R. W. Arnold, D. Hopley, S. S. Philbrick, D. A. Sangrey, P. C. Thomas, and D. L. Turcotte, I thank for their comments on chapters in their areas of special interest. Prentice-Hall editor W. H. Grimshaw saw me begin writing, and editor Logan Campbell saw me finish.

An unexpected pleasure of authorship was experiencing the generous and enthusiastic responses to my requests for illustrations. Many friends provided not only the photograph I requested, but several others from which I could choose. Seeing one's name in a caption is a trivial repayment, so I further acknowledge here my debt of gratitude to all contributors. Photographs not otherwise credited are my own. Most of my line drawings were redrafted by professional artists. Mindful of Sir Charles Cotton's gentle scorn (1955, p. 582) for "careless or delegated drafting," I can only extend my sinistral wrist for a deserved ethereal slap by my respected late teacher.

A final note on dimensions: The SI (metric) system is used almost throughout without conversions into English or obsolete metric dimensions. Geomorphologists in the United States have been slow to convert to the metric notation of all other sciences and most of the world's population. Here, as in the use of regional examples, our comfortable provinciality must be overcome.

ARTHUR L. BLOOM

REFERENCES

BLOOM, A. L., 1969, *The surface of the earth*: Prentice-Hall, Inc., Englewood Cliffs, N.J., 152 pp.

COTTON, C. A., 1955, Theory of secular marine planation: *Am. Jour. Sci.*, v. 253, pp. 580–89. (Reprinted 1974 in *Bold coasts*: A. H. & A. W. Reed, Wellington, N. Z., pp. 164–74).

DAVIS, W. M., 1905, *Complications of the geographical cycle*: Internat. Geog. Cong., 8th, Washington 1904, Repts., pp. 150–63. (Reprinted 1954 in *Geographical essays*: Dover Publications, Inc., New York, pp. 279–95).

CHAPTER 1

The Scope of Geomorphology

THE SYSTEMATIC STUDY OF LANDSCAPES

Every sensitive person, on viewing a mountain or river valley, instinctively wonders how the landscape assumed its form. Deeper thinkers sense that no scene is immutable but that all change with the passage of time. A few scientists of each generation choose *the systematic description and analysis of landscapes and the processes that change them* as an area of specialization. These scientists are geomorphologists.

The description and analysis of landforms (the component parts of landscapes) is a science only when it is done in a disciplined fashion. Many schemes of organized study can be conceived. **Regional description**, by district, nation, or natural provinces, is one logical approach (Thornbury, 1965; Hunt, 1974). The earliest geography books were largely regional descriptions of visible landmarks, passes, river valleys, and other landforms that helped guide the traveler. Regional, descriptive geomorphology remains as one of the accepted subdivisions of the science. A variant of growing importance, although not so much in North America, is the preparation of geomorphic maps that show landform components by distinctive colors and symbols.

At the beginning of the nineteenth century, the science of geology became distinguishable from "natural philosophy." As facts and theories about the origins of rocks and mountain ranges accumulated, they were absorbed into regional or descriptive geomorphology as an aid to understanding the origins of landforms. Thus **explanatory description** evolved, in which a landscape is not simply described by the heights of its hills and the declivities of its slopes but by the reconstructed geologic history of its evolution. Explanatory description is a higher form of analysis than regional description in that it requires inferences about past events. In a sense, a landscape

1

cannot be understood until the entire geologic history of the rocks and slopes is known. Fortunately, most landscapes are of comparatively recent geologic origin, and it is usually possible to begin an explanatory description with a phrase like: "As the area emerged from the Cretaceous sea . . ." or "Orogenic uplift in the Miocene Epoch produced a mountainous terrain . . . ." Whereas regional geomorphology, or physiography, is often regarded as part of the larger science of geography, explanatory or genetic geomorphology is mostly a geologic subject. Geomorphology is thus a part of two sciences, and geomorphologists may regard themselves as either geographers, or geologists, or preferably both (Dury, 1972).

The subject of landform analysis can also be organized on the basis of the processes that have operated and now operate to shape the land. Many recent shorter geomorphology books deal with specific geomorphic processes (for example, Leopold, et al., 1964; Ollier, 1969; Birkeland, 1974). Process-oriented geomorphology is closely related to climatology, because air temperature, precipitation, winds, and atmospheric humidity largely determine the response of rocks to subaerial exposure. Climatic morphogenesis is the approach to geomorphology that emphasizes the shaping of landforms in response to various climatic conditions (Birot, 1960, 1968; Tricart and Cailleux, 1965, 1972). This is an important topic, because many landscapes preserve the record of more than one set of climatic-controlled processes, and the correct explanatory description of such a landscape provides evidence for reconstructing ancient climates.

Much of the information about geomorphic processes is derived from other scientific fields. The elastic and plastic deformation of rocks under pressure and the thermal conductivity of rocks are studied by geophysicists, generally not with the primary intention of describing or analyzing landscapes. Seismologists are properly more concerned with the causes and prediction of earthquakes than with the landforms that earthquakes produce. Geochemists and mineralogists study the chemical reactions of rocks and minerals with hydrous solutions and thereby provide data about weathering processes. Pedologists study soil formation; hydrologists study the work of flowing water; sedimentologists study depositional processes and sediments; glaciologists study the physical properties of ice; oceanographers describe the submarine topography they see traced on their depth recorders: All of these scientists contribute to an understanding of the processes that shape the earth's surface. A geomorphologist must be at least conversant with these peripheral subjects.

How can regional description, historical reconstruction, and the details of the chemical and physical response of rocks to water and air be unified into a science of geomorphology? The best solution that has evolved is summarized by the trinity of *structure*, *process*, and *time*. Although generally attributed to the greatest geomorphologist, William Morris Davis (1850–1934), the organization of geomorphic information under the three headings is implicit in many earlier writings and is such a natural organization that it was probably unconsciously used many times before Davis codified it with such great force and clarity.

To illustrate the concept of geomorphic description in terms of structure, process, and time, consider the following descriptive paragraphs. The first is empirical; it states clearly *what* and *where*. The second is explanatory; it connotes *how* as well as *what* and *where*.

1. "The city of Ithaca, New York, is built on low ground at the south end of Cayuga Lake. Steep slopes surround the town, above which is a broad, rolling upland that becomes gradually more rugged southward into the low mountains of Pennsylvania.
2. "The city of Ithaca, New York, is built on a delta plain at the south end of the Cayuga Lake trough. Glacially oversteepened valley walls around the city are in contrast to the maturely dissected, cuestaform upland of the glaciated northern Appalachian Plateaus."

In the second paragraph the technical terms "delta plain," "cuestaform," and "plateaus" convey information about structure and genesis. Geomorphic processes are explicit in the terms "trough," "glacially oversteepened," "dissected," and "glaciated." Time is implied in the term "maturely dissected" in that a degree or stage of erosional dissection is specified. The first paragraph would be perfectly suitable for a tourist brochure or travel guide, but the second paragraph, in exactly the same number of words, conveys much more information to a trained reader. True, technical terms are used that must be understood by the reader, but technical terms are a kind of shorthand notation to convey concepts and increase understanding in an efficient manner.

If empirical description were the only goal of geomorphology, a regional landscape description like the two samples just given could be reduced (or expanded?) to a mathematical equation. Only three numerical terms are needed to describe the altitude above sea level of any point on the landscape and locate the point by its latitude and longitude coordinates. An integration of the changes in altitude with latitude and longitude will, in principle, give an equation that completely describes any specified area of land surface. But even if such an equation could be written, no one, not even the most astute mathematician, could envision the land surface that it describes. A radar-scanning computer on board an intercontinental rocket can recognize a landscape by mathematical terms, but human minds require descriptive words to convey images of landscapes. Significantly, the most satisfying way to describe a landscape seems to be by describing its origin, or genesis. Geomorphic description explains *how* and *when*, as well as *what* and *where*. Only the ultimate question of *why* is left to the philosophers.

THE LANDFORM AS THE UNIT OF SYSTEMATIC ANALYSIS

Landscapes are surfaces composed of an assemblage of subjectively defined, irregular, lesser surfaces. Each element of the landscape that can be observed in its entirety, and has consistence of form or regular change of form, is defined as a **landform**. The definition is necessarily subjective (Howard and Spock, 1940). To one geomorphologist, a hillside is a landform, continuous and homogeneous. Another observer sees gullies, ridges, and dozens of other small topographic features that either break the continuity of the hillside or establish its continuity, depending on the mental concept. There is no lower limit to the scale of observation. For instance, the simple question, "How long is the coast of Britain?" becomes an issue of great mathematical complexity when it is realized that at each closer scale of observation, new irregularities appear (Mandelbrot, 1967). In principle, to get the length of the British coast, the wetted perimeter of every pebble and sand grain on every beach at one instant would have to be summed! Therefore, the landform is simply defined as a continuous surface that can be viewed by an observer in its entirety, whether the observer is on hands and knees on the ground, or in an orbiting spacecraft.

Landforms may be *constructional,* as in the examples of fault scarps, volcanic cones, glacial moraines, or river deltas. However, because of the peculiar intensity of atmospheric weathering and erosion on our planet, most subaerial landforms are *destructional, or erosional.* Properly, the landscape is to be viewed as an assemblage of residual landforms, each in the process of being consumed or transformed by erosion. One of the fundamental goals of geomorphologic training is to be able to see landscapes not as an assemblage of slopes, ridges, and mountain peaks, but of erosional valleys, gullies, and hollows, between which the residual hills and mountains stand. The eye must be trained to see that the missing rock material is the significant feature, not the material that remains.

From these introductory comments, it can be appreciated how landforms came to be systematically evaluated and explained in terms of the Davisian trinity of *structure*, *process*, and *time* (or *stage*). Each landform can be visualized as enclosing or covering a mass of rock with specific physical and chemical properties and geometrically disposed discontinuities (bedding planes, joints, faults, etc.). All of these lithologic and mineralogic factors are collected in the geomorphic concept of *structure*.

We observe that a complex series of reactions takes place when rocks are exposed to water and air in our planet's gravitational field. Certain phenomena, such as downward movement of detached fragments, affect all matter. Other processes are uniquely effective only on specific structures. Thus landforms are the result of constructive or destructive *processes* acting on *structures*.

In closed systems, reactants tend toward stable equilibrium states. Some geomorphic processes, such as landslides or earthquakes, illustrate nearly instantaneous response by rocks to some overwhelming stress. The resulting piles of broken rock (and houses) represent a more stable configuration. Most geomorphic processes act much more slowly. A major topic of geomorphic study is whether all stream valleys are always *graded* (fully adjusted to the erosional processes that are actively shaping them) or whether the graded condition develops slowly and progressively. We see, then, that an account of the interaction of surface processes with geologic structures does not give a complete explanation of landforms unless we include in the explanation the length of time the process has been operating or the degree of equilibrium (*stage*) that has been reached. Relatively few landforms show purely constructional processes.

The systematic analysis of landforms in terms of structure, process, and time seems to be intuitive. Acknowledgment that landscapes are not immutable leads to the immediate conclusion that if change is in progress, something (structure) is being altered by something (process) to a definable extent (stage) or for a definite interval (time). Absolute time is difficult to measure on the geologic scale, and rates of geomorphic change are variable because of the wide range of processes and structures. For both of these reasons, the concept of a relative stage of development, rather than the absolute length of time a process has been acting, has become prevalent. It is probable that future trends of geomorphic research will be toward defining absolute rates of change and the true elapsed time since specified processes began to act on specified structures (Jennings, 1973, p. 123).

FURTHER CONSIDERATIONS OF STRUCTURE, PROCESS, AND TIME

Structure

All of the physical and chemical properties of rocks and minerals are loosely gathered into this general heading by geomorphologists. A useful distinction is made between the terrane, an area of a certain structure or rock type, as in the expressions "granitic terrane," or "limestone terrane," and the terrain, a landscape or, "the lay of the land." A military tactician must appreciate the terrain; he need not be concerned with the terrane. However, one of the first things a geomorphologist wants to know about a region is: "What is the structure?" By this question, he or she inquires about the minerals that form the rocks, the structural and stratigraphic arrangement of rock layers and masses, the previous tectonic displacement of the rocks, and the present state of deformation.

Structure is usually considered the passive result of previous depositional or deformational processes. Rocks are normally regarded as rigid solids, not likely to flow under their own weight to produce landforms of smooth, rounded domes. Yet it is well known that rocks have finite yield strengths, which when exceeded will result in permanent deformation. Under sufficient shear stress, rocks "flow" or "creep." The property of a substance that determines whether it will behave as a solid or fluid has been defined as *rheidity* (Carey, 1954) and is arbitrarily measured by the amount of time necessary for viscous flow to exceed by 1000 times the elastic (or recoverable) deformation, under specified conditions of temperature, pressure, and shear stress. If the rheidity of a material is measured by a short interval of time, the mass is likely to deform measurably during observation. For instance, glacier ice has a rheidity measured in weeks (10^6 s). It deforms readily to fill mountain valleys, and the profile of an ice sheet is an elliptical or parabolic curve, steepest at the edges and flatter inland (Fig. 16-9). It also fractures as a brittle substance to form open crevasses as deep as 50 m. Thus the surface morphology of an ice sheet is largely the result of plastic, or rheid, flow but also the result of fracturing and partial melting. Glacier ice is the second most common rock type at the surface of the earth (greatly exceeded by submarine basalt) forming a layer 1–2 km in thickness over one entire continent and the world's largest island.

Some other common rocks also have low rheidity under surface conditions on earth. Salt has a rheidity of about one year (10^7 s), and gypsum has a rheidity only about 10 times greater (10^8 s). Both rock types form spectacular extrusional landforms in Iran (Carey, 1954, pp. 80–81; Ala, 1974), where the arid climate prevents the "salt glaciers" and extruded sheets of gypsum from being dissolved away. Many volcanic landforms also have their shape determined by the rheidity of molten lava (Chap. 4). Even though the lava solidifies and becomes rigid on cooling, the landforms record the mobility of a former molten state.

Silicate crystalline rocks and nonevaporite sedimentary rocks have rheidities measured in tens of thousands of years (10^{11} s) and are properly considered as rigid solids when they are exposed at the surface of the earth. Yet within the cores of orogenic belts, these rocks flow, as proved by the intricate contortion of gneissic bands or the great folds exposed on eroding mountain sides (Fig. 11-12). On the largest scale of geomorphic units, the earth's surface is an approximate spheroid of rotation on which continents stand a few km higher than ocean floors, because the continents are composed of somewhat lower density rocks (Chap. 2). The average rheidity of rocks in the earth's interior must be very low in order for isostatic equilibrium between large areas of the surface to be so very nearly perfect. The shapes of other planetary masses are also fundamentally related to their masses and the strengths of their materials. Ten kilometers may be about the maximum relief possible on this planet (Weisskopf, 1975).

In summary, for most geomorphic considerations, rocks are to be regarded as rigid solids, brought near the surface of the earth by tectonic uplift but not likely to assume distinctive landforms as a result of rheid flow. Only in special cases, such as glaciers, salt domes, and lava flows, is the rheidity of rocks a factor in the description of landforms.

Process

A host of internal and external processes contribute to produce a landscape. Tectonic forces crumple rocks and push up continents and mountain ranges. Air and water react with rocks to cause the familiar results of weathering. Moving water, air, and ice erode rocks and transport the eroded debris to depositional sites. Organisms absorb mineral nutrients from the rocks on which they live. Gravity tends to pull down structures that rise above the general ground level. The rates of geomorphic processes vary enormously, from a few centimeters per thousand years for the surface weathering of ancient monuments to 50 m/s or more for an avalanche roaring down a mountainside.

Geomorphic processes also vary in intensity from one region to another, depending on climate, vegetation, and altitude above or depth beneath the ocean. A large portion of any geomorphology book must be devoted to analyzing the processes that act on geologic structures. The vital feature of geomorphic processes is that they, like all chemical and physical processes, are assumed to be time-independent. That is, we assume that under the same set of environmental conditions, processes have acted in the past and will act in the future exactly as they now can be observed to act. Hutton's great principle of uniformity of process, often called "Uniformitarianism," is a keystone of all scientific method.

Rigid uniformitarianism is not useful when studying pre-Quaternary landscapes or unconformities, however. Throughout the decipherable history of the earth, life has been evolving, and present living things are not the same as those that lived formerly. Many geomorphic processes are controlled by organisms, either directly, as when the steepness of hill slopes is determined by the holding capacity of tree or grass roots, or indirectly, as when soil chemical reactions are determined by the microorganisms in the soil water. It is hard to envision an early Cenozoic landscape, for example, on which no grasses had yet evolved to anchor the sod. Surely, some plants covered the pre-Miocene hills, but their fossil remains do not indicate the presence of grass. Modern landscapes devoid of grass are found in deserts and under dense forest, but the desert or forest climates are not necessarily analogous with all pre-Miocene climates.

Animals as well as plants affect landscapes. Even within the Quaternary Period, when most of our modern mammals evolved, faunal migrations have influenced landscape development. Grazing and browsing by domesticated animals may be responsible for huge areas of semiarid landscapes. *Homo sapiens*, a late-Cenozoic mammal, has reshaped many landscapes in the past 10,000 years.

These examples illustrate the dangers of regional or temporal extrapolations of the processes that are assumed to shape landforms. Although physical and chemical laws are indeed uniform through geologic time, the oscillatory climatic changes of the late Cenozoic Era have caused huge transpositions of climate-dominated processes on many regions (Chap. 18). Glaciated terrains in many middle-latitude regions, and loess and sand sheets that record former eolian transport processes in areas that are now humid and vegetated, remind us that few contemporary landforms can be assumed to have developed under any single set of processes.

Time

One of the fundamental principles of geomorphology is that the processes that act on geologic structures at the earth's surface change the appearance of landscapes with the passage of time. Anyone who has seen or felt the weather-beaten surface of an ancient stone monument understands that rocks react with water and atmosphere, and only a slight extension of reasoning is necessary to deduce that entire landscapes are shaped by the collection of processes that we call *weathering* and *erosion*. The key word in the relationship between structure and process is "progressive." If structures are affected by surface processes, are the changes progressive and sequential?

Do landscapes pass through predictable evolutionary stages with the passing of time? These questions are at the heart of the science, because in order to be a valid science, geomorphology must generate theories that have predictive value.

Practical or applied geomorphology needs prediction as much as theoretical geomorphology. Will an excavation wall remain stable, or will it collapse as it adjusts to subaerial exposure? Will the soil on a hillside remain in place when cultivated with row crops, or should it be seeded with grass and used as pasture? These are practical geomorphic questions in which the time scale is in years or tens of years.

In the theoretical analysis of landscapes, time is measured in thousands or even millions of years. How have landscapes evolved when exposed to the agents of weathering and erosion for a geologically significant interval of time such as the Quaternary Period? Traditionally, geomorphologists spoke of *relative* time only and compared the development of landscapes only in terms of the relative stage of erosional evolution. The basis for this convention is that some landscapes are on weak or easily eroded terranes, have strong degradational processes at work, and evolve rapidly. Others, forming on resistant structures or experiencing only weak processes, change very slowly. Little value was placed on specifying the rate of development of landscapes in terms of actual time.

With the perfection of methods for age determination by radioactive disintegration of natural isotopes, the determination of *real* time, and therefore real rates of change, has become possible. When applied to geomorphic problems, radiometric ages can answer such questions as when an ice sheet uncovered a region; when rivers began to dissect a volcanic ash bed; when sand dunes invaded a former river valley; or when a coral reef grew up to the ocean surface. Time is now specified to a precision that was not conceivable only a few decades ago, and the future of geomorphology lies in converting relative ages to real rates of change.

A corollary of the principle that processes of change alter structures through time is that *stages*, or degrees of completeness, of alteration can be defined. W. M. Davis conceived a geomorphic cycle,[1] in which specific stages could be recognized. His dangerously oversimplified terms, "youth," "maturity," and "old age" are familiar to every beginning student of geology and geography. They need careful reevaluation (Chap. 12).

Time has another special significance to geomorphology in that practically the entire subaerial landscape has evolved in the late Cenozoic Era, a time of great tectonic activity and climatic change on earth (Chap. 2). Probably the average geologic status of continental regions is one of low relief and slow erosion, or of shallow marine submergence and slow accumulation of sediment. Our present relief is fantastically diverse by comparison. Geomorphology is a necessary branch of geology precisely because in late Cenozoic time the continental blocks are unusually high above sea level and are undergoing rapid erosion. The latitudinal zonation of climates is extreme.

---

[1]His term was "geographic cycle," but the growth of geography into humanistic studies soon rendered the term inappropriate.

SUMMARY

At present, agents of geomorphic changes are retouching landscapes that are the result of a complex series of changing environments. Most subaerial landscapes are **palimpsests**, a word originally applied to parchment manuscripts that had been partially scraped clean and reused for later texts. As in a manuscript one or more older texts can sometimes be read beneath the latest words, so in the description of landscapes, previous processes and stages of development can be recognized beneath or within the landforms that are now being shaped. The highest skill of the geomorphologist is in deciphering those fragments of ancient development almost obscured by present processes. Sometimes the task is easy, as when the striking landforms of glacial erosion and deposition are found in regions of temperate climate, or when empty branching systems of river valleys are found in regions where no rain falls today. However, many landscapes do not record such obvious changes, and we must always apply the principle of uniformity of process with caution, because the present dominant processes may not be the key to past landform development. The science of geomorphology has two major goals. One is to organize and systematize the description of landscapes by intellectually acceptable schemes of classification, and the other is to recognize the deviations from these schemes that are evidence for changes in the environment that could not be inferred if normal conditions had not first been specified.

REFERENCES

ALA, M. A., 1974, Salt diapirism in southern Iran: *Am. Assoc. Petroleum Geologists Bull.*, v. 58, pp. 1758–70.

BIRKELAND, P. W., 1974, *Pedology, weathering, and geomorphological research:* Oxford Univ. Press, New York, 285 pp.

BIROT, P., 1960, *Le cycle d'érosion sous les différents climats:* Univ. do Brasil Centro de Pesquisas de Geografia do Brasil, Rio de Janeiro, 137 pp.

———, 1968, Cycle of erosion in different climates (Transl. by C. I. Jackson and K. M. Clayton): Univ. of California Press, Berkeley, 144 pp.

CAREY, S. W., 1954, The rheid concept in geotectonics: *Geol. Soc. Australia Jour.*, v. 1, pp. 67–117.

DURY, G. H., 1972, Some recent views on the nature, location, needs, and potential of geomorphology: *Professional Geographer*, v. 24, pp. 199–202.

HOWARD, A. D., and SPOCK, L. E., 1940, Classification of landforms: *Jour. Geomorph.*, v. 3, pp. 332–45.

HUNT, C. B., 1974, *Natural regions of the United States and Canada:* W. H. Freeman and Company, San Francisco, 725 pp.

JENNINGS, J. N., 1973, 'Any millenniums today, lady?' The geomorphic bandwaggon parade: *Australian Geog. Studies*, v. 11, pp. 115–33.

LEOPOLD, L. B., WOLMAN, M. G., and MILLER, J. P., 1964, *Fluvial processes in geomorphology:* W. H. Freeman and Company, San Francisco, 522 pp.

MANDELBROT, B., 1967, How long is the coast of Britain? Statistical self-similarity and fractional dimension: *Science*, v. 156, pp. 636–38.

OLLIER, C. D., 1969, *Weathering:* American Elsevier Publishing Company, Inc., New York, 304 pp.

THORNBURY, W. D., 1965, *Regional geomorphology of the United States:* John Wiley & Sons, Inc., New York, 609 pp.

TRICART, J., and CAILLEUX, A., 1965, *Introduction à la géomorphologie climatique:* Soc. d'edition d'enseignement supérieur, Paris, 306 pp.

———, 1972, *Introduction to climatic geomorphology* (transl. by C. J. K. de Jonge): Longman Group Ltd., London, 295 pp.

WEISSKOPF, V. F., 1975, Of atoms, mountains, and stars: A study in qualitative physics: *Science*, v. 187, pp. 605–12.

PART I

Constructional Processes
and Constructional Landforms

Only about 35 percent of the earth's lithosphere is older than 200 million years. The rest has been recycled, perhaps many times, by consumption and accretion at plate margins. The oceanic lithosphere is almost all younger than 200 million years. Continents include rocks with ages of over 3 billion years, because continental rocks tend to resist subduction and accumulate as thick crustal slabs. But the continental landscapes we now see are very young, mostly a product of processes that have been at work during late Cenozoic time.

The theory of plate tectonics is based on evidence that large, rigid lithospheric plates move over the surface of the earth at relative velocities of up to 10 cm/yr. Vertical movements measured in millimeters per year are also demonstrated, especially at plate margins. In the face of such evidence for a mobile lithosphere, geomorphologists must reevaluate the traditional concept that rock masses are passive structures from which landscapes are carved by erosion. The next three chapters establish the conceptual framework of landscape evolution during a time of great tectonic and volcanic activity.

Constructional landforms are here defined as all those subaerial terrains that progressively increase their mass or area above sea level. In Chapters 2, 3, and 4, the tectonic and volcanic constructional processes and landforms are systematically described. Constructional landforms that grow by biologic processes or the accretion of the waste products of erosion are not described until later chapters.

Tectonic and volcanic relief results from dynamic stresses and density changes, all of which are ultimately caused by thermal anomalies within the earth. Nothing is ever really "constructed" on any planet, because each total planetary mass remains very nearly constant. All constructional landforms are built by reprocessing or by redistributing some other mass.

CHAPTER 2

Cenozoic Diastrophism
and Constructional Processes

The Cenozoic Era, and especially that part of it defined as the Quaternary Period, has seen the modern continents generally emergent, with especially intense orogenic activity around the Pacific basin, across southern Europe and Asia in the Alpine-Himalayan zone, and around the Caribbean Sea. Our subaerial landscape is developed on relatively young tectonic structures, although some of the rocks exposed by recent uplift and erosion are among the oldest known. Epeirogenic and orogenic activity in the late Cenozoic Era have given us a landscape of exceptional diversity, complexity, and beauty. It may well be true that at any moment of geologic time orogenic belts have been actively rising on some part of the earth. However, if there have been other geologic intervals when extensive continental emergence, strong climatic gradients, and high mountain ranges were as notable as those of the late Cenozoic, they are not well documented in the stratigraphic record.

SCALE OF CONSTRUCTIONAL RELIEF FEATURES

A durable concept of both geomorphology and tectonics is that terrestrial relief features can be classified by size. Salisbury (1919) introduced the concept into systematic geomorphology with the hierarchy: (1) first-order relief features (continents and ocean basins); (2) second-order relief features (tectonic mountain belts, plateaus, and plains); and (3) third-order relief forms (erosion-carved mountains, hills, and valleys, as well as constructional volcanic mountains and equivalent forms). Although difficult to define and of little genetic value, the classification is useful, because each successive level includes features of successively smaller size, greater unity of origin, and more transient durability. Attention is drawn to successively greater detail in the analysis of a landscape. The larger orders of the earth's relief features are not usually analyzed in great detail, but a modernized size classification is introduced here as a prelude to the more complex details of landform analysis and description.

The Shape of the Earth

The best description for the shape of the whole earth is that it is an oblate, or slightly flattened, spheroid (Gaposchkin, 1974; King-Hele, 1976). This reference surface of gravity equipotential, the shape of the earth's mean sea level, and its extension beneath the land, is called the geoid. In addition to its fundamentally flattened spheroidal form, the geoid has a number of smaller, unexplained undulations (Fig. 2-1). A boat on the ocean east of Papua New Guinea is 192 m farther from the center of the earth than one south of Sri Lanka, yet by definition, both are at sea level.

The geoid is an important reference surface for geomorphic processes. Rivers, for instance, cease to flow when they reach sea level, unless they drain into depressions below sea level in arid regions (Chap. 13). Thus the geoid is the surface that would result if subaerial processes were ever to level all the continents completely to sea level. The geoidal projection of mean sea level under the subaerial landscape is the best definition of the term ultimate base level, which has been the subject of so much debate (Chap. 9).

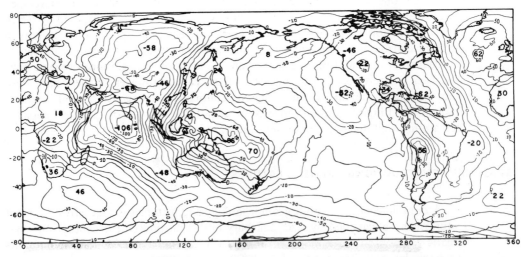

Figure 2-1. Geoid map of the earth. Contours show deviations from a spheroid of radius 6378.142 km with polar flattening of 1:298.255. Contour interval: 10 m. (D. L. Turcotte, based on Goddard Earth Model 8.)

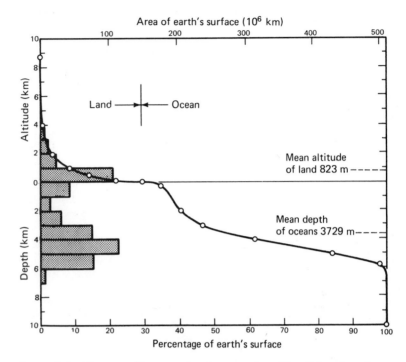

Figure 2-2. Hypsographic curve showing area of earth's solid surface above any given altitude or depth. Bar graph on left shows the distribution of area by 1-km intervals. (From Menard and Smith, 1966; Kossinna *in* Fairbridge, 1968.)

Continents and Ocean Basins

The largest units of the earth's rocky *relief* (altitude or depth above or below the geoid) are the continental blocks and the abyssal basins of the ocean floors. Continental structure covers about 35 percent of the earth's surface, but only 29 percent of the surface is now above sea level. On the continents, the average altitude is 823 m above sea level. The average depth of the oceans is about 3700 m. Most of the average depth is formed by abyssal basins, which cover about 30 percent of the earth's surface. Smaller regions of the continents and ocean floors rise much higher and plunge much deeper than the averages, but it is a striking and important characteristic of our planet's primary relief that the hypsographic curve is strongly bimodal (Fig 2-2). Tectonic theorists now recognize about 10 plates as the basic components of the earth's lithosphere. By this "new global tectonics," the continents are almost incidental masses that are being rafted along toward the zones where lithospheric plates are sinking back down into the mantle. Because of their low-density rocks, the continental slabs are not drawn down but accumulate on top of the mantle.

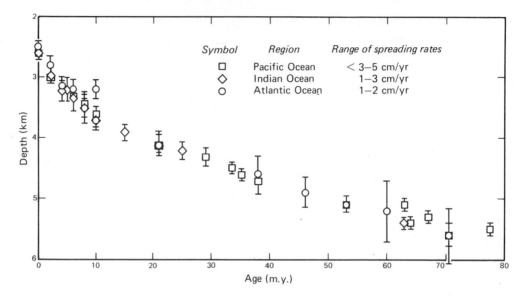

Figure 2-3. Average depth to smooth oceanic crust as a function of the age of the sea floor. The crust at slowly accreting plate margins is shallower than the average for the first 20 million years. Thereafter, depth is an accurate predictor of crustal age. (Data from Sclater, et al., 1971.)

Even if a few slabs or plates are the basic geophysical units of the earth's outer layers, the strong bimodality of the hypsographic curve justifies a primary geomorphic distinction between the continents and abyssal oceanic basins. Whatever the structural cause for their contrasting average altitude, their geomorphic differentiation is very real. How the continents have maintained their altitudes near or above sea level throughout at least the one-half billion years of Paleozoic and later geologic time is one of the most important topics of historical geology. In the face of evidence for erosion to depths of 20 km or more into continental structures (p. 106), the mechanism of long-term continental survival also assumes primary importance as a geomorphic topic.

The oceanic crust, experiencing geologically rapid creation and consumption, has a relatively simple exponential correlation of depth with age (Fig. 2-3). At an accreting plate margin along a mid-ocean ridge, the oceanic lithosphere is hot and has low density. Progressive cooling and thermal contraction of a lithospheric plate as it moves away from an accreting margin accounts well for both the observed geothermal heat flow and the oceanic depth, at least to a depth of about 6 km (Chap. 21).

Geomorphic (Physiographic) Provinces

Superimposed on the continents and ocean basins are major relief units that are recognizable by their size, continuity, uniformity of structure, and geologic durability. The familiar physiographic provinces of the United States (Fig. 2-4), as organized by Fenneman (1917, revised 1928), are examples of such relief features. Each province was defined as a geomorphic (or physiographic) entity that has resulted from a set of processes or a succession of processes acting on a specific structure (in the broad sense of a regional geologic terrane, as noted in Chapter 1) for a certain period of time or to a certain stage of completion. Fenneman's approach was thoroughly Davisian, as the previous sentence emphasizes. Nevertheless, the boundaries that he drew formally defined regions that had long been informally recognized as "provinces" in loosely defined terms as an aid in identifying the regional setting for local areas of study. Fenneman's provinces have been generally accepted as the geomorphic divisions of the United States. The provinces, subdivided into sections and grouped into larger divisions, are easily recognized as entities, even though their exact boundaries may be matters for scholarly debate. Words like "Colorado Plateau" and "Appalachian Valley and Ridge province" communicate immediate understanding to the informed reader.

The relief features commonly identified as geomorphic or physiographic provinces seem to be about the optimal size for regional or descriptive geomorphology. Books and maps of the regional geomorphology of the United States (Fenneman, 1931, 1938; Hammond, 1963; Thornbury, 1965), Canada (Bostock, 1964), Alaska (Wahrhaftig, 1965), and the floor of the Atlantic Ocean (Heezen, et al., 1959) are but a few examples of the use of geomorphic provinces to organize regional geomorphic description. One of the first tasks in regional description is the identification of the appropriate natural provinces on the basis of topographic coherence. Although they were originated to facilitate regional description of nations or continents, the definition of geomorphic provinces on the ocean floor (Chap. 21) nicely illustrates the fundamental use of structure, process, and time in the explanatory description of all landscapes.

Scenery

Von Engeln (1942) elegantly compared the first-order relief features to theaters: "permanent" structures "in which geologic dramas are enacted." The second-order relief features were compared to theatrical stages, which are rebuilt or altered from time to time. The third-order relief features are the scenery, the stage settings and backdrops that are mobile and temporary. Scenery is the assemblage of landforms that can be viewed from a single vantage point, the pleasure of the tourist and the raw material of the geomorphologist. The following chapters almost all concern landscapes at this scale. Most scenery, excluding such notable views as isolated volcanic cones (Fig. 4-10), is erosional rather than constructional.

RATES OF CONSTRUCTIONAL PROCESSES

Some landforms appear abruptly, literally overnight. In 1943, a Mexican farmer detected volcanic fumes escaping from cracks in the ground of his cornfield, and within a few days an active scoria cone had buried his farm, eventually to become

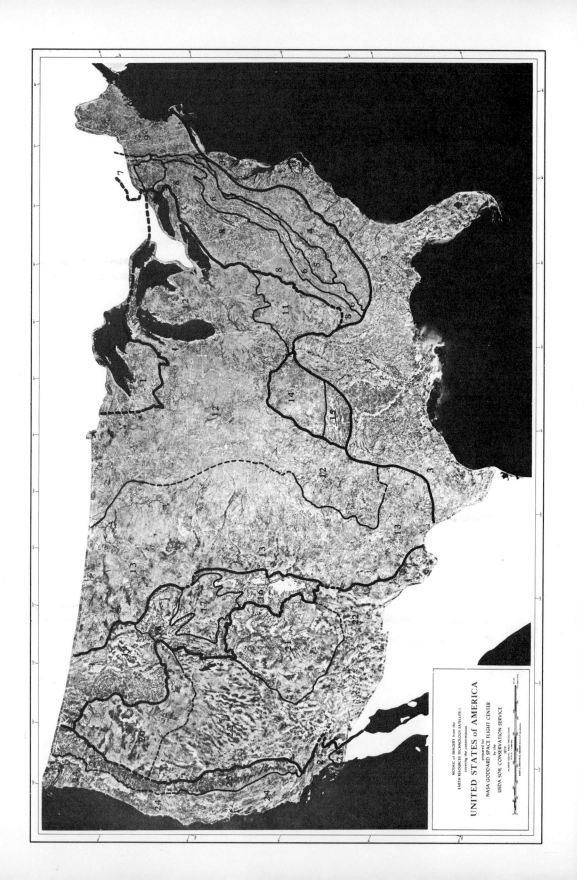

MOSAIC of IMAGERY from the
EARTH RESOURCES TECHNOLOGY SATELLITE-1
covering the conterminous

UNITED STATES of AMERICA

prepared for

NASA GODDARD SPACE FLIGHT CENTER

by the

USDA SOIL CONSERVATION SERVICE

Figure 2-4. (*Opposite page.*) Geomorphic (physiographic) provinces of the conterminous United States. (Base: mosaic of imagery from ERTS-1, prepared for NASA Goddard Space Flight Center by the U. S. Department of Agriculture-Soil Conservation Service, 1974.) Numbered provinces are listed below. (After Fenneman, 1928.)

Major Division	Province
Laurentian Upland	1. Superior Upland
Atlantic Plain	2. Continental Shelf (not shown on map)
	3. Coastal Plain
Appalachian Highlands	4. Piedmont province
	5. Blue Ridge province
	6. Valley and Ridge province
	7. St. Lawrence Valley
	8. Appalachian Plateaus
	9. New England province
	10. Adirondack province
Interior Plains	11. Interior Low Plateaus
	12. Central Lowland
	13. Great Plains province
Interior Highlands	14. Ozark Plateaus
	15. Ouachita province
Rocky Mountain System	16. Southern Rocky Mountains
	17. Wyoming Basin
	18. Middle Rocky Mountains
	19. Northern Rocky Mountains
Intermontane Plateaus	20. Columbia Plateaus
	21. Colorado Plateaus
	22. Basin and Range province
Pacific Mountain System	23. Cascade-Sierra Mountains
	24. Pacific Border province
	25. Lower Californian province

the new volcano, Parícutin. During the 1964 Alaskan earthquake, a sizable region of nearshore ocean floor was raised above tide level to become new land (Fig. 2-5). The duration of the movement was probably only a few minutes or even seconds, although no one witnessed the actual upheaval.

At somewhat slower rates, but still rapid by the ordinary scale of geologic change, the earth's surface is displaced along faults (Fig. 2-6) or is arched upward into anticlines and domes (Fig. 3-21). Areas that have been deglaciated within the last 10,000 years are still actively rising in isostatic response to the removed load of former ice sheets (Figs. 17-11, 17-13).

Rates of geologic processes cover many orders of magnitude, from less than 1 mm/1000 yr for deep-sea sedimentation rates to more than 50 m/s, for avalanches (Fig. 2-7). Regional denudation rates fall near the middle of the scale. Tectonic and volcanic landscapes can be built at rates at least an order of magnitude more rapidly than erosion can destroy them (Chap. 12). If intensities were otherwise, our terrestrial landscape would be very dull indeed.

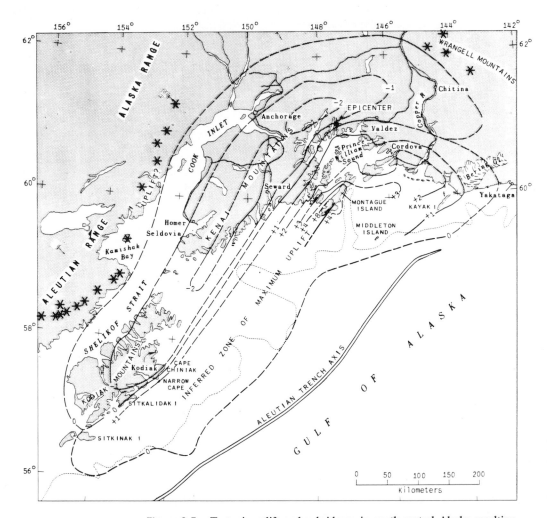

Figure 2-5. Tectonic uplift and subsidence in south-central Alaska resulting from the earthquake of March 27, 1964. Changes in land level are shown by contours (in meters) that are dashed where approximate or inferred. Edge of continental shelf (−200 m) is shown by dotted line. Active and dormant volcanoes are identified by asterisks. (Plafker, 1965; © 1965 by the American Association for the Advancement of Science.)

Figure 2-6. Emerged Pleistocene coral-reef terraces on the Huon Peninsula, Papua New Guinea. Terraces are on a fault block that has been rising at a rate of 1–3 m/1000 yr for at least 200,000 years. Figure 17–14 shows the terrace chronology for the last 140,000 years.

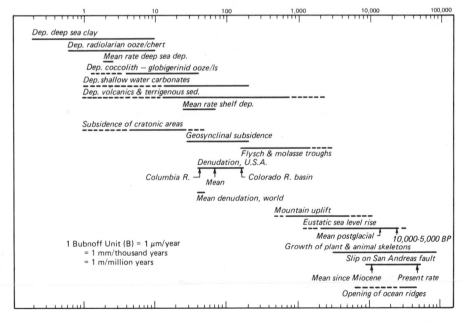

Figure 2-7. Comparative rates of some geologic processes on a scale of "Bubnoff Units" (Fischer, 1969). Regional denudation rates range from 30–150 B, whereas tectonic uplift rates range upward from 300 B.

Cenozoic Epeirogenic Uplift and Subsidence

Although 35 percent of the total earth's surface is underlain by continental structure, only about 29 percent of the total area is presently above sea level (Fig. 2-2). The remaining 6 percent of continental structure is submerged by no more than 200 m of sea water to form the continental shelves, the inner halves of which also were exposed during glacial intervals of lower sea level. Thus essentially all of the continental structure of the earth has a surface form determined by subaerial erosion or deposition, and most of it is presently exposed to view. Our present epoch is a time of extensive **epeirogenic, or continent-wide, uplift** (Hallam, 1963). No "lost continents" lie submerged beneath the oceans.

Epeirogeny is broad regional movement, with little folding or other local deformation. Except for mountain ranges, much of the continental landscape is eroded from sedimentary rocks that lie in their initial depositional attitude. Broad structural domes and basins with deformation-produced relief of a few hundred meters over distances of hundreds of kilometers are common in epeirogenic regions, but the resulting dips of the strata are usually less than 1 percent and are barely detectable in a single exposure. The cuestaform erosional landscape that develops on the broad domes and basins of epeirogenic regions is analyzed in detail in Chapter 11. Epeirogenic uplift is not considered likely to produce small-scale constructional landforms. More commonly, epeirogenic domes are breached by erosion to become topographic basins, and structural basins become low plateaus surrounded by outward-facing escarpments. The resulting *topographic inversion* by erosion on epeirogenic structures implies that the tectonic movements are either old or extremely slow relative to the rate of erosion.

Geologists in the Soviet Union and eastern Europe have evolved a new theory of epeirogenesis called **neotectonism** (the "newest" tectonics—the study of earth movements of the Quaternary Period and the later part of the Tertiary Period, including those in progress at the present time). According to this theory (Mescherikov, 1968), the ancient continental shields and the continental crust in general are subject to repeated late Cenozoic uplifts or depressions of 10–100 m. Contemporary wavelike motions with vertical components of as much as 10 mm/yr are reported. Most of the data for the contemporary motions are derived from repeated precise leveling surveys. Releveling of the geodetic network in England and Wales between 1912 and 1956 implied uplifts of up to 150 mm and subsidences of as much as 46 mm during the 44-year interval (Chorley, 1963, p. 964). Recent resurveys of the eastern United States show the Appalachians rising 6 mm/yr relative to the Atlantic coast. Other leveling lines show comparable rates of relative vertical motions (Brown and Oliver, 1976). Some of the differences between such successive surveys are probably due to unrecognized instrumental errors, but the possibility of large-scale but episodic or oscillatory warping as a continuing feature of epeirogenic tectonics must now be considered a factor in geomorphic analysis. Coastal landforms would be especially subject to vertical displacements of the scale suggested by the neotectonic theory, but drainage networks, river terraces, lake shorelines, and many other geomorphic features might also be subtly influenced. It should be noted that the geodesists who report the results of precise leveling surveys are generally more reluctant to claim earth movements as a cause of the discrepancies between successive surveys than are the geologists who quote the results.

The Greek root "epeiros" refers to a continent. Therefore, by literal translation, epeirogeny would not be the appropriate term to refer to movements of the ocean floor. However, the term can be extended to include all continent-sized warping, including thermal contraction of the oceanic lithosphere (Fig. 2-3). No such simple correlation of age and altitude can be demonstrated for continental landscapes. Some geomorphic provinces seem to have oscillated within a few hundred meters above and below sea level since early Paleozoic time. Precambrian rocks are exposed both in the cores of mountain ranges and in shield areas at sea level. Some tectonists suspect that contemporary compressive stress and seismicity in eastern North America may be related to Cenozoic plate motions (Sbar and Sykes, 1973). Possibly the moving plates ride over sublithospheric "bumps" or "hollows" that cause vertical epeirogenic movements. This is an area of active research, in which geomorphic evidence is as important as seismology, rock mechanics, and geodesy.

One special form of epeirogenic movement, of particular significance in the Quaternary Period, is glacial isostasy. The theory of isostasy holds that the pressure at the base of all vertical sections of the earth, above some deep level of compensation, is equal. Nature's "great experiments in isostasy" (Daly, 1940, p. 309; Cathles, 1975) were the alternate loading and unloading of large continental areas with ice sheets during late Cenozoic time (Chap. 17). Sea water must evaporate and accumulate on land as ice in order for ice sheets to grow. Thus each glaciation involved the mass transfer of a layer of water perhaps 100 m in thickness from the 71 percent of the earth's surface that is ocean, and the concentration of that water mass, as glacier ice averaging 2–3 km in thickness, on about 5 percent of the earth's surface. We assume but have not yet proved that the ocean floor responded to the removal of water load by isostatically rising. It is much easier to prove that the glaciated areas were basined beneath their loads of ice, because when the ice sheets were melting, lakes and seas often formed shorelines near the ice margins. The shorelines of the ice-marginal water bodies show regular tilting, always concentric upward toward the center of former maximum ice thickness. Because these shorelines must have been horizontal initially, their present slope is a measure of the differential postglacial isostatic uplift. Uplift of 300 m is known from areas on the eastern side of Hudson Bay, Canada, in a total time of less than 8000 years. Postglacial isostatic uplift at the northern end of the Baltic Sea totals 520 m, and an estimated 210 m of uplift yet remains to be accomplished (Flint, 1971, p. 352).

The rate of postglacial uplift is rapid at first, then progressively slower. It approximates an exponential decay curve with a *relaxation time* (the time required for the deformation to decrease to $1/e$, or 37 percent, of the original value) of from 1000–5000 years. Initial uplift rates of 9–10 m/century have been demonstrated for Greenland (Washburn and Stuiver, 1962; Ten Brink, 1974). Where uplift rates are so rapid, landforms such as storm beaches are raised in a few decades sufficiently high to be unreached by the next major storm, and wide, descending flights of beach ridges mark the margins of seas and lakes in former glaciated regions (Fig. 17-11).

In northern United States, Canada, and Scandinavia, the postglacial uplift is still in progress, although at a greatly decreased rate. Northern Sweden is presently emerging from the Baltic Sea at a rate of 9 mm/yr (Fig. 17-13); Finland annually gains 7 km² of new land by the uplift.

Cenozoic Orogeny

Orogeny clearly refers to mountain formation. Unfortunately, in geologic usage, the term has much more complex and contradictory implications. Gilbert (1890, p. 340) formalized the term: "displacements of the earth's crust which produce mountain ridges are called orogenic.—The process of mountain formation is orogeny, the process of continent formation is epeirogeny, and the two collectively are diastrophism." It was barely realized in Gilbert's time that mountainous relief is caused not by the tectonic processes that deformed the rocks in the orogenic belts but by later uplift of masses of these deformed rocks. In fact, many mountain ranges are the result of "postorogenic" or epeirogenic movements from which erosion has etched into mountainous relief the most resistant of the folded and metamorphosed rocks. Structural connotations have almost completely divorced the word from its obvious geomorphic implications, so that *orogeny* is now used for: "the processes by which rock structures within the mountain chains or foldbelts are created" (King, 1969, p. 45; but see Cebull, 1973, 1976). As such, it has little geomorphic significance, although for many general geomorphic descriptions the tectonic niceties of the term can safely be ignored.

In plate-tectonic theory, all orogenic belts are or were on plate margins. Orogenic movements are not necessarily more rapid than epeirogenic ones. However, they seem to be more continuous through time, less likely to stop or reverse, and restricted to relatively narrow, elongate regions.

Cenozoic mountain ranges are not randomly distributed but form distinctive linear or arcuate belts across the earth's surface. Not all earthquake-prone regions are mountainous, but a map of recent earthquakes (Fig. 2-8) outlines very well the most orogenic regions of the late Cenozoic Era, especially the circum-Pacific belt associated with active volcanoes and earthquakes along subduction zones. Many examples of rapid orogenic uplift around the Pacific are cited in Chapter 3. Areas in Japan have been uplifted as much as 2000 m during the Quaternary; the amount of tectonic uplift within the last 2 million years is commonly one-half to two-thirds of the highest altitude in any district (Research Group for Quaternary Tectonic Map, Tokyo, 1973).

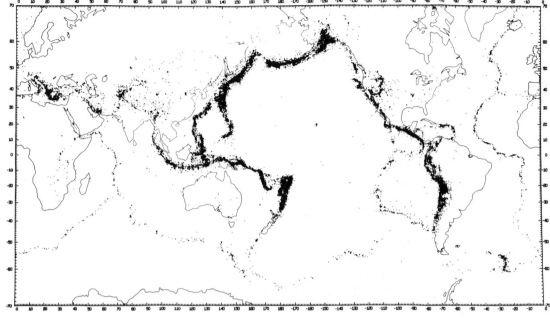

SEISMICITY OF THE EARTH, 1961-1967, ESSA, CGS EPICENTERS

DEPTHS 000-700 KM.

Figure 2-8. Seismicity of the earth, 1961–67. Each dot is the epicenter of a recorded earthquake. Depths 0–700 km. (Barazangi and Dorman, 1969.)

All the major landforms of New Zealand are tectonic except for depositional forms. Numerous pulished reports confirm that the late Cenozoic Kaikoura Orogeny is still in progress there. Regional vertical uplift and fault displacements of up to 11 m/1000 yr are described (Suggate, 1965; Sheppard, et al., 1975).

Similar deformational rates are demonstrated for many other tectonic regions. Emerged late Quaternary reefs in Papua New Guinea (Fig. 2-6) required uplift rates of 1–3 mm/yr to carry them above the range of glacial-controlled sea-level oscillations (Bloom, et al., 1974). Orogenic movements measured in millimeters per year, if continued for the nearly 2 million years of the Quaternary Period, would produce total uplift measured in kilometers, quite adequate to exceed erosion and create substantial mountain ranges. Along many converging plate margins, these movements are in progress today. They produce not only tectonic relief but also strong zonal climatic gradients, rain-shadow deserts, and other orographic climatic regions. They have also disrupted ocean currents, intensified latitudinal temperature gradients, and set the stage for the multiple Pleistocene glaciations that have shaped large areas of continents. A complete theory of Cenozoic plate tectonics will be a powerful tool in explaining present-day landforms, not only in orogenic zones but over the entire earth.

REFERENCES

BARAZANGI, M., and DORMAN, J., 1969, World seismicity maps compiled from ESSA, Coast and Geodetic Survey, epicenter data, 1961–1967: *Seismol. Soc. America Bull.*, v. 59, pp. 369–80.

BLOOM, A. L., BROECKER, W. S., CHAPPELL, J. M. A., MATTHEWS, R. K., and MESOLELLA, K. J., 1974, Quaternary sea level fluctuations on a tectonic coast: New ^{230}Th/^{234}U dates from the Huon Peninsula, New Guinea: *Quaternary Res.*, v. 4, pp. 185–205.

BOSTOCK, H. S., 1964, Provisional physiographic map of Canada: *Canada Geol. Survey Paper 64–35*, 24 pp.

BROWN, L. D., and OLIVER, J. E., 1976, Vertical crustal movements from leveling data and their relation to geologic structure in the eastern United States: *Rev. Geophys. and Space Phys.*, v. 14, pp. 13–35.

CATHLES, L. M., III, 1975, *Viscosity of the earth's mantle:* Princeton Univ. Press, Princeton, N.J., 386 pp.

CEBULL, S. E., 1973, Concept of orogeny: *Geology*, v. 1, pp. 101–2.

―――, 1976, Concept of orogeny: Reply: *Geology*, v. 4, pp. 388–89.

CHORLEY, R. J., 1963, Diastrophic background to twentieth-century geomorphological thought: *Geol. Soc. America Bull.*, v. 74, pp. 953–70.

DALY, R. A., 1940, *Strength and structure of the earth:* Prentice-Hall, Inc., Englewood Cliffs, N.J., 434 pp.

VON ENGELN, O. D., 1942, *Geomorphology:* The Macmillan Company, New York, 655 pp.

FAIRBRIDGE, R. W., 1968, Continents and oceans—statistics of area, volume and relief, *in* Fairbridge, R. W., ed., *Encyclopedia of geomorphology:* Reinhold Book Corporation, New York, pp. 177–86.

FENNEMAN, N. M., 1928, Physiographic divisions of the United States: *Assoc. Am. Geographers Annals*, v. 18, pp. 261–353.

―――, 1931, *Physiography of western United States:* McGraw-Hill Book Company, Inc., New York, 534 pp.

―――, 1938, *Physiography of eastern United States:* McGraw-Hill Book Company, Inc., New York, 714 pp.

FISCHER, A. G., 1969, Geologic time-distance rates: the Bubnoff unit: *Geol. Soc. America Bull.*, v. 80, pp. 549–51.

FLINT, R. F., 1971, *Glacial and Quaternary geology:* John Wiley & Sons, Inc., New York, 892 pp.

GAPOSCHKIN, E. M., 1974, Earth's gravity field to the eighteenth degree and geocentric coordinates for 104 stations from satellite and terrestrial data: *Jour. Geophys. Res.*, v. 79, pp. 5377–411.

GILBERT, G. K., 1890, Lake Bonneville: *U.S. Geol. Survey Mon. 1*, 438 pp.

HALLAM, A., 1963, Major epeirogenic and eustatic changes since the Cretaceous, and their possible relationship to crustal structure: *Am. Jour. Sci.*, v. 261, pp. 397–423.

HAMMOND, E. H., 1963, Classes of land-surface form in the forty eight states, U.S.A.: Map supp. no. 4, *Assoc. Am. Geographers Annals*, v. 54 [1964].

HEEZEN, B. C., THARP, M., and EWING, M., 1959, Floors of the oceans, I. The North Atlantic: *Geol. Soc. America Spec. Paper 65*, 122 pp.

KING, P. B., 1969, Tectonics of North America—a discussion to accompany the tectonic map of North America, Scale 1: 5,000,000: *U.S. Geol. Survey Prof. Paper 628*, 94 pp.

KING-HELE, D., 1976, The shape of the earth: *Science*, v. 192, pp. 1293–1300.

MENARD, H. W., and SMITH, S. M., 1966, Hypsometry of ocean basin provinces: *Jour. Geophys. Res.*, v. 71, pp. 4305–25.

MESCHERIKOV, Y. A., 1968, Neotectonics, *in* Fairbridge, R. W., ed., *Encyclopedia of geomorphology:* Reinhold Book Corp., New York, pp. 768–73 (see also pp. 223–27).

PLAFKER, G., 1965, Tectonic deformation associated with the 1964 Alaskan earthquake: *Science*, v. 148, pp. 1675–87.

Research Group for Quaternary Tectonic Map, Tokyo, 1973, *Explanatory text of the Quaternary tectonic map of Japan:* Natl. Research Center for Disaster Prevention, Tokyo, 167 pp.

SALISBURY, R. D., 1919, *Physiography*, 3d ed., rev.: Henry Holt & Co., New York, 676 pp.

SBAR, M. L., and SYKES, L. R., 1973, Contemporary compressive stress and seismicity in eastern North America: An example of intra-plate tectonics: *Geol. Soc. America Bull.*, v. 84, pp. 1861–81.

SCLATER, J. G., ANDERSON, R. N., and BELL, M. L., 1971, Elevation of ridges and evolution of the central eastern Pacific: *Jour. Geophys. Res.*, v. 76, pp. 7888–915.

SHEPPARD, D. S., ADAMS, C. J., and BIRD, G. W., 1975, Age of metamorphism and uplift in the Alpine schist belt, New Zealand: *Geol. Soc. America Bull.*, v. 86, pp. 1147–53.

SUGGATE, R. P., 1965, Tempo of events in New Zealand geological history: *New Zealand Jour. Geol. Geophys.*, v. 8, pp. 1139–48.

TEN BRINK, N. W., 1974, Glacio-isostasy: New data from West Greenland and geophysical implications: *Geol. Soc. America Bull.*, v. 85, pp. 219–28.

THORNBURY, W. D., 1965, *Regional geomorphology of the United States:* John Wiley & Sons, Inc., New York, 609 pp.

WAHRHAFTIG, C., 1965, Physiographic divisions of Alaska: *U.S. Geol. Survey Prof. Paper 482*, 52 pp.

WASHBURN, A. L., and STUIVER, M., 1962, Radiocarbon-dated postglacial delevelling in Northeast Greenland and its implications: *Arctic*, v. 15, pp. 66–73.

CHAPTER 3

Tectonic Landforms

It has become the convention to describe landforms that result from crustal movements as *tectonic*. Some geologists have used "structural" in the same sense, but as was noted in Chapter 1, "structure" in the geomorphic sense has a much broader connotation. Structurally controlled landforms, those that are shaped by differential erosion in response to rock structure, are analyzed in Chapter 11.

It could be argued that no subaerial relief can occur until crustal uplift has raised land above sea level and that therefore all subaerial landscapes are "tectonic" unless they are constructed by depositional (volcanic or sedimentational) processes. However, it is convenient to restrict the term to those landforms that are sufficiently undissected by erosion so that the shape of the fractured or deformed surface can be discerned. All degrees of transition are found between purely tectonic and totally erosional landscapes. The distinction is nevertheless one of the important topics of geomorphology, because if a landform is classed as tectonic, we assume that the deformation has been either recent, or rapid, or both, when compared to the erosion rate. The geomorphic recognition of tectonic landforms is of major value in working out the geologic history of a region. Tectonic landforms are the result of Quaternary, or at the oldest, late Cenozoic land movements. In most regions of tectonic landforms, movement continues today.

TECTONIC SCARPS

A **scarp** (shortened form of escarpment; any steep, abrupt slope or cliff along the margin of a plateau, terrace, or other topographic bench) may be formed by a variety of geologic causes, either tectonic, erosional, or depositional. It is well to retain a few purely descriptive topographic terms such as "scarp" and "bench" in the totally nongenetic sense and to add modifying adjectives to them as the need arises. Thus a **tectonic scarp** is a steep slope that results from differential movement of the earth's surface.

Fault Scarps

If a fault displaces the surface of the ground so that one side is higher or lower than the other, a **fault scarp** results. Very recent, high-angle faults in resistant rock may preserve the actual surface of faulting on the scarp complete with slickensides or fault breccia. Such fresh details of fault movement are rare, however. More commonly, fault scarps expose a surface layer of weathered, broken rock and soil. The resulting scarp quickly slumps or collapses to a slope of gravitational· stability and gives no clue to the attitude of the fault surface at depth. Even though the actual fault surface is normally not preserved, if the relief on a scarp is the direct result of faulting, the name *fault scarp* is appropriate. This is true even when subsequent erosion on the face of the raised block causes the scarp to retreat laterally some distance from the surface of fault movement, and when the scarp assumes a slope quite different from the dip angle or becomes gullied and dissected. The typical angle of repose (Fig 8-1) for scarps of all origins is between about 25° and 40°. Even historical fault scarps in humid regions quickly become grassy or brush-covered slopes (Fig. 3-1).

Figure 3-1. View south along the Wairarapa fault near Featherston, New Zealand. Parts of this scarp increased 2 m in height during the 1855 Wellington earthquake.

The more nearly vertical the dip of the fault surface, the more linear the resulting fault scarp will be across the preexisting landscape. In the case of a low-angle or nearly horizontal thrust fault, even without erosional dissection, the scarp will be along the contour at the average altitude of exposure; if the landscape is irregular, the scarp will be as sinuous or crenulate as the contour lines.

The absolute direction and amount of motion is not relevant to the formation of a fault scarp. Only the relative motion between adjacent blocks determines the way the scarp will face and its height. Many of the great scarps, or *Pali*, of the Hawaiian islands are arcuate fault scarps, facing seaward away from the crests of the volcanic domes (Fig. 3-2). Some authors suggest that the central volcanic areas have been raised up along these faults (Eaton and Murata, 1960; Macdonald, 1972, p. 278). Others hypothesize that flanks of the great volcanic shields are slumping seaward as large, arcuate slices (Cotton, 1944, p. 86). The fault scarp shown in Figure 3-2, like other normal faults, might be considered as a gigantic slump scar (Fig. 8-11).

A fault scarp may die out laterally, break into a series of *splinters*, or become a monoclinal scarp (Fig. 3-3). A major fault or narrow fault zone at depth may break the land surface as a series of closely spaced, *en echelon* scarps, none of which has any conspicuous continuity.

Fault scarps, like all other high, steep landforms, are prone to rapid erosional dissection. Streams that formerly flowed across the faulted landscape either will be

Figure 3-2.   View southwest along a coastal scarp of the Hilina fault system, south flank of Kilauea Volcano near Puu Kapukapu, Hawaii. The main Hilina Pali (scarp) forms the skyline. (Photo: R. I. Tilling, Hawaiian Volcano Observatory, U.S. Geological Survey.)

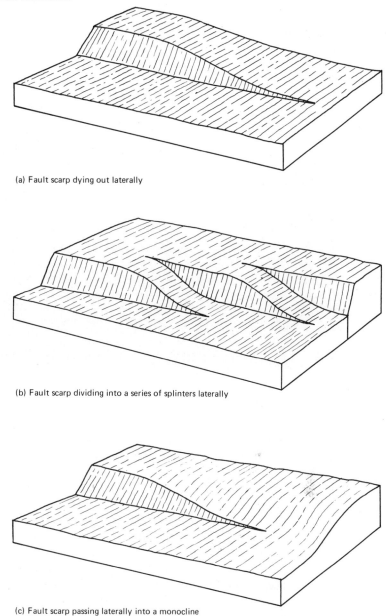

(a) Fault scarp dying out laterally

(b) Fault scarp dividing into a series of splinters laterally

(c) Fault scarp passing laterally into a monocline

Figure 3-3. Varieties of lateral termination of fault scarps.

rejuvenated if they flow from the upthrown to the downthrown block, or they will be impounded against the scarp if they flow in the opposite direction. As most dissected scarps represent repeated late Cenozoic uplift, the first case is much more common. Initially then, streams cascade down a fault scarp but quickly cut steep gorges that erode progressively headward from the scarp into the uplifted block (Figs. 3-4, 3-5).

Figure 3-4.   Steep, short stream valleys dissecting a fault scarp, north flank of the Cromwell Range, Huon Peninsula, Papua New Guinea.

The resulting valleys are narrow at the lower end and widen upstream into the up-thrown block (Fig. 3-5). These distinctively shaped scarp valleys have been called *bottle-necked,* or *wine-glass, valleys* (von Engeln, 1942, p. 377; Cotton, 1948, pp. 87, 396, 402). They are not restricted to fault scarps but form on any previously undis-sected steep slope or scarp.

Dissection by many gullies or valleys causes a scarp to be segmented into a series of **triangular facets**, planar surfaces with their bases aligned on or parallel to the fault trace (Fig. 3-6). Excessive debate has been generated about the significance of triangular facets on fault scarps. If misinterpreted as remnants of the original fault surface, the facets imply that the fault was a low-angle normal fault, because they rarely have inclinations greater than about 30°, the angle of repose. By contrast, low scarplets of recent movement at the base of the same facets may show by their straightness in plan or map view to be the trace of a high-angle fault. Thus *two* epi-sodes of faulting, on two faults that are unrelated but coincide at the ground surface, are incorrectly postulated.

In the advanced stage of erosional dissection, a fault scarp may look no different from the face of any other dissected hill or mountain. If the dissecting streams dis-

Figure 3-5.   Continuation of the view in Figure 3-4. Face-on view of a bottle-necked scarp valley.

charge their alluvium into a desert basin, the aggradation may extend headward across the fault trace into embayments in the uplifted block. If a river flows along the foot of the scarp, the alluvium is more likely to be carried away and the tectonic relief preserved, but stream erosion may undercut the fault scarp and make its recognition difficult.

Fault scarps may be buried by alluvium or other sedimentary cover. The tectonic relief of the scarp is then eliminated. If, at a later time, the covering sediments are eroded from the scarp, it is an *exhumed* landform. The relief then, even though identical to the original fault scarp, is no longer tectonic, but erosional. A scarp produced by structurally controlled erosion at an ancient fault is known as a **fault-line scarp** and is one of many kinds of structurally controlled erosional landforms (p. 257). The distinction between a fault scarp, of tectonic origin, and a fault-line scarp, formed by structurally controlled erosion, is not just a semantic matter. It is "the indispensable prerequisite to a correct interpretation of geological history" (Johnson, 1929, p. 356). The U. S. Nuclear Regulatory Commission stipulates that no nuclear power-plant site should by located in the vicinity of a "capable" fault. A "capable" fault is one that shows movement at or near the ground surface at least

Figure 3-6. Triangular facets aligned on the fault scarp of Maple Mountain, 15 km south of Provo, Utah. View east. (Photo: H. J. Bissell.)

once within the past 35,000 years or movement of a recurring nature within the past 500,000 years. Obviously, the correct interpretation of scarps in tectonic regions now has economic as well as theoretical significance.

Compound scarps can result from various combinations of tectonic and erosional processes. If a fault scarp were heightened by a river-cut bank at its base, the upper part of the scarp would be tectonic but the lower part erosional. Conversely, an exhumed fault-line scarp might be the locus of renewed faulting, and its lower part could thus become a true fault scarp (Cotton, 1950b).

Earthquake Scarps and Related Landforms

In seismic regions, many small-scale scarps or "scarplets" are the direct result of displacement during earthquakes (for example, Tchalenko and Berberian, 1975). They may be the actual fault scarp or secondary effects of differential compaction and slumping. Open *fissures* form during some earthquakes, probably due to consolidation of porous surface materials. Soil creep and erosion are likely to obliterate such features quickly. At the foot of recent fault scarps, closed basins or *sag ponds* may develop.

Figure 3-7. View south along an earthquake scarplet on the Wairarapa fault near Featherston, New Zealand (see also Fig. 3-1). Long-term movement is upward on the west side of the fault, but the local movement here was opposite during the 1855 earthquake.

Peculiarly oriented scarplets have been frequently observed along major fault scarps in New Zealand. They closely parallel the base of the scarp but face toward the main scarp so as to form a shallow trench (Fig. 3-7). They have been called *earthquake rents, reverse scarplets,* or *cicatrices* (Cotton, 1948, 1950b). They have been rarely reported from elsewhere. Although common along dip-slip fault scarps, their interpretation is not clear. Oscillatory or reversed motion is not likely to result from the stress fields that cause faulting. More likely, cicatrices on fault scarps represent local or shallow buckling of one of the fault blocks adjacent to the major fault surface.

Landforms Associated with Transcurrent Faulting

Transcurrent (strike-slip) faulting may produce no scarps in a landscape of low relief. However, if a series of gullies is offset by a transcurrent fault, many low scarps will result. Typically, if the fault crosses a group of ridges, all the spur ends or ridge crests will be displaced in the same direction (Fig. 3-8). Such half-displaced ridges

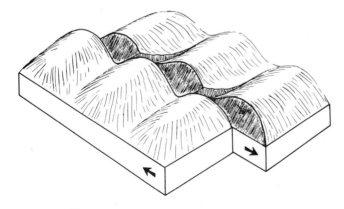

Figure 3-8. Shutter ridges caused by right-lateral movement on a transcurrent fault across ridge-and-gully topography. Streams are deflected along the fault trace or may be captured by the downstream segment of an adjacent gully.

were named *shutter ridges* by Buwalda (1936). Continued or intermittent transcurrent faulting, especially if it developed a zone of crushed, erodible rock, could be expected to encourage the erosion of a valley along the fault that would intersect and collect the drainage from a number of scarp gullies.

Along the San Andreas fault in California, minor stream displacement is obvious, but as yet no single stream follows the fault trace for any significant distance. Most of the drainage is across the fault. Perhaps the present system of stream valleys is too young to have been significantly displaced, or possibly the arid to subhumid climate in California has inhibited stream erosion. Measured right-lateral differential motion along the fault averages 2 cm/yr (Wallace, 1970). Some segments moved over 6 m during the 1906 San Francisco earthquake. Quaternary river terraces are displaced several kilometers; Pliocene strata are offset by 23 km in the Temblor Range; and Miocene and Eocene rocks are said to be displaced at least 32 km in the San Gabriel and San Bernardino Mountains. Stream channels are displaced by as much as 2.4 km in areas (Wallace, 1949; Sharp, 1954).

Some intermontane basins in New Zealand have been attributed to gaps that opened in the zone of tension where a major transcurrent fault makes an abrupt bend (Clayton, 1966). A similar explanation has been offered for the origin of the Dead Sea basin in the Middle East (Fig. 3-18; p. 47). Such tectonic depressions are similar to grabens, but their regional control is transcurrent, rather than dip-slip, faulting. Anticlines or horsts are forced up in comparable circumstances where the local stress is compressional rather than tensional.

Along the Alpine fault on the South Island of New Zealand, late Cenozoic transcurrent movement combined with high-angle thrust faulting has created a major mountain range on the eastern side of the fault. The downthrown west side of the fault is a coastal lowland with wavelike anticlines and synclines that strike normal to the main scarp. Cotton (1956) suggested that the western block has "buckled" under compressive stress into a series of wavelike folds (Fig. 3-9). In a land of lesser relief, such buckling of one of the walls of a transcurrent fault could produce alternate-facing scarps or cicatrices without the fault having more than incidental dip-slip motion.

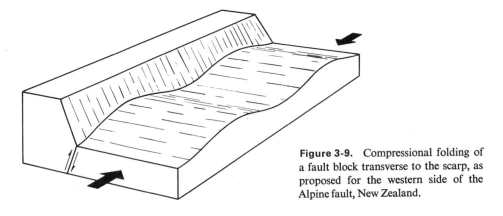

Figure 3-9. Compressional folding of a fault block transverse to the scarp, as proposed for the western side of the Alpine fault, New Zealand.

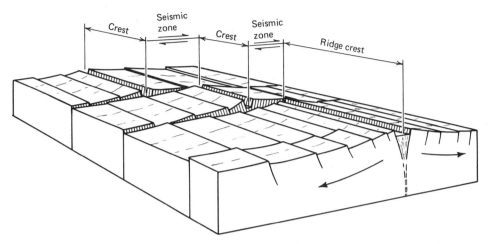

Figure 3-10. Transform faults that offset a spreading ridge in a left-lateral sense although the seismic portions of the faults are zones of right-lateral slip.

In 1965, a new type of transcurrent fault was devised to explain the offset of submarine topography and paleomagnetic anomalies (Wilson, 1965). The topographic offset of a spreading ridge across a *transform fault* is in the opposite sense from the lateral movement (Fig. 3-10). The transform fault was an entirely new concept and has been one of the major contributions to the theory of plate tectonics. Many geologists believe that all transcurrent faults on continents are segments of transform faults (Hamilton, 1971). Others (King, 1969) assert that the transform fault is a structure unique to the thin, basaltic oceanic crust, not to be found on the continents. Whichever view prevails, there seems no doubt that the ocean floor has been displaced hundreds of kilometers laterally along great transform faults, Some of them form submarine scarps several thousand kilometers long and 1–2 km high (Menard, 1964). From the viewpoint of an orbiting observer at heights of 100–1000 km above the earth's surface, large-scale tectonic patterns become obvious as "lineaments," or *linears*, (Hamilton, 1971; Lathram, 1972). Where surface geologic mapping has been done, many of the great linears coincide with major fracture or fault zones. Other linears have not yet been correlated with previously identified structures. Their relation to plate tectonics is assumed, but not yet proven.

FAULT VALLEYS AND FAULT-BLOCK MOUNTAINS

By a peculiar accident of geography and history, the earliest- and most-studied examples of block-faulted landscapes are in semiarid or arid climatic regions, where vegetation is sparse and geologic structures are boldly exposed. The regional assemblage of fault-block mountains in the Great Basin section of the Basin and Range province of the western United States (Fig. 2-4) led to the original name "basin ranges" for all the mountains of the region. Subsequently, "basin-range structure" was the general term for block faulting. Many deductions about the erosional evolution of landscapes on faulted structures are really based on climatic factors. Although it is true that few places in the world have such an array of well exposed fault-block landforms as the U.S. Basin and Range province, the systematic description of these tectonic landforms must not be overly conditioned by the climatically controlled erosional processes that act on them.

Horsts and Grabens

A **horst** is a fault block that has been uplifted relative to the blocks on either side; a **graben** is a depression produced by subsidence along faults. Most authors regard a graben as synonymous with a rift valley, but some (Willis, 1928) have attempted to define grabens as depressions between normal faults, and rift valleys as basins enclosed by escarpments along high-angle thrust faults. The distinction is not geomorphically useful, and it became unfortunately involved with the erroneous concept (p. 32) that facets on fault scarps are remnants of actual fault surfaces. Horsts do not seem to be particularly common landforms. Grabens, on the other hand, are well known spectacular landforms in many tectonic regions (Fig. 3-11).

Figure 3-11.   The central graben of Iceland. (Photo: S. Thorarinsson.)

Tilted-Block Mountains and Fault-Angle Depressions

Rather than being horsts, uplifted fault blocks are commonly tilted, and bounded on only one of their sides by fault scarps. A basin between two tilted elongate blocks is a **fault-angle depression** rather than a graben. An assembly of tilted blocks, grabens, and horsts is called simply a *fault-block landscape*.

Fault-block landscapes composed of originally flat-lying rocks are called the *Oregon type* (Davis, 1903; Cotton, 1950a, 1950b), named for the basalt area near Klamath Lake in southern Oregon. According to Davis (1903, p. 130): "There was good evidence for regarding the Oregon lava-block ranges as types of the youngest, most elementary mountain forms known to geographers." The faulting is so recent, the erosional dissection so slight, and the displaced flow sheets so strongly in control of surface slopes that photos of the region approach block diagrams in their simplicity (Donath, 1962). The relief is entirely tectonic. Small alluvial fans and lakes bury or drown some of the grabens and fault-angle depressions, but horsts and tilted-block mountains stand in bold profile.

Oregon-type landscapes are also found in Iceland (Fig. 3-11) and in the Afar region, near the Red Sea (Tazieff, 1970; Bannert, 1972). Part of the Afar triangle, as the region near Ethiopia is called, is below sea level and was an arm of the Red Sea until Pleistocene faulting isolated it and evaporation emptied the basin except for thick salt accumulations. Tilted basalt blocks (Fig. 3-12) are separated by fault-angle depressions that are empty or at most contain salt-encrusted playas or recent lava flows.

The *basin-range type* of fault-block landscape has replaced the Oregon type in most textbooks. The fault blocks of the Basin and Range province are not simply tilted, lifted, or dropped blocks of internally undeformed rock. Very early in the exploration of the region, Gilbert (1874, 1928) recognized that the faulting that produced the range-and-basin topography was younger than the orogeny that had intensely deformed and metamorphosed the rocks within the ranges. Louderback (1904, 1923) demonstrated that some tilted blocks are capped by lava flows that are neither folded nor thrusted but simply rest in sharp unconformity on erosional surfaces of low relief and are tilted because of the final episode of block faulting. This was the final proof needed to show that the Laramide structural deformation of late Cretaceous time was not the orogeny (p. 24) that created the fault-block ranges and their accompanying basins. In western Nevada, an interval of tectonic quiescence between 7.5 and 3.5 million years ago produced the widespread erosion surfaces across earlier structures, which were then covered with latest Pliocene and Pleistocene volcanic rocks and broken by Quaternary faulting into the modern topography (Gilbert and Reynolds, 1973).

The Ruby-East Humboldt Range of northeast Nevada is a typical basin range (Sharp, 1939, 1942). The complex internal structures are of pre-Miocene age; the less deformed Miocene alluvium and lacustrine beds either are faulted against or unconformably overlie the older structures; and the Quaternary alluvium unconformably overlies all older sediment and rock. An imposing fault scarp on the eastern

Figure 3-12. The Afar region of eastern Africa, a regional desert landscape of tilted fault blocks, horsts, and grabens in a late Cenozoic volcanic terrane. (Apollo 9 photograph AS-23-3539, NASA.)

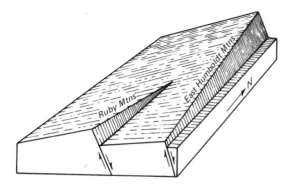

Figure 3-13. Block diagram of the Ruby and East Humboldt fault blocks. To show geometric relations, the pre-faulting surface is assumed flat and the post-faulting erosion of the two mountain ranges is ignored. The near edge of the diagram is about 25 km wide; vertical exaggeration is about twofold. (Sharp, 1939, Fig. 5.)

side is 1200–1500 m in height, but the westward slopes are more gentle. The western base of the range is linear, recent scarplets there are abundant, and the Miocene beds are displaced as much as 70 m, but the present tectonic form of the range is more like a pair of tilted blocks than a horst (Fig. 3-13). The blocks are bounded by normal faults that dip 60°–70°, where exposed in canyons.

The basin-range type of fault-block landscape has been described in some detail because of its significance to the historical development of American geology, and because drawings and photographs of basin ranges are featured in almost every American text. Lacking personal familiarity with fault-block landscapes, many geologists and geographers mentally associate them with deserts, forgetting that the arid or semiarid climate and internal drainage of most of the U.S. Basin and Range province impose special geomorphic processes on the tectonic landforms. In particular, almost all deductions about the sequential erosional development of fault-block mountains, fault-angle depressions, and grabens started from the assumption that alluvium eroded from the ranges would accumulate in the adjacent basins and progressively bury the base of the range (Fig. 13-3). For instance, Sharp (1942, p. 683) demonstrated that on the east scarp of the tilted fault block in the southern Ruby Mountains, almost one-third of the tectonic relief is buried.

To correct the climatic bias in geomorphic descriptions of fault-block mountains, we can follow Cotton (1948, 1950a, 1950b, 1958, 1968) and designate a third type of fault-block landscape as the *New Zealand type.* The Awatere fault scarp (Fig. 3-14), one of several northeast-trending fault scarps in eastern Marlborough, is representative of fault scarps in New Zealand and other humid regions, where through-flowing rivers dominate the erosional processes. The Awatere fault scarp "differs in one important respect from those of the basin-range type. A vigorous consequent river has been present parallel to the scarp, at or near its base, during much of its long erosional development—a river that has been capable of carrying away the debris furnished by smaller streams dissecting the scarp" (Cotton, 1950a, p. 196). "Distinction of the Awatere from the basin-range type [of fault scarp] depends on the absence of a bahada fringe, . . . all the debris derived from rapid early wasting has long ago been removed, and newer debris, though in abundant supply, continues to be carried off progressively by active river transportation" (Cotton, 1950b, p. 727).

Figure 3-14. View southwest along the Awatere fault scarp, South Island, New Zealand. Mountain block on the right has moved up and toward the camera relative to the valley along a high-angle reverse fault. (Photo: D. L. Homer, N. Z. Geological Survey.)

Figure 3-15. View northeast over Port Nicholson (Wellington Harbor), the drowned portion of a fault-angle depression. Wellington, New Zealand, in foreground. (Photo: Royal New Zealand Air Force. Crown Copyright reserved.)

New Zealand type fault-block landscapes are characterized, then, by fault-angle depressions and grabens that are river valleys or stream-dissected lowlands rather than alluvium-filled basins. Where near the sea, New Zealand type fault-angle depressions may be drowned to form excellent harbors, such as Port Nicholson, at Wellington, New Zealand (Fig. 3-15), Spencer and St. Vincent Gulfs near Adelaide, Australia, and the "Port Phillip Sunkland," the outer harbor of Melbourne, Australia (Twidale, 1968).

Early and extensive stream dissection of fault scarps and back slopes is the other diagnostic feature of the New Zealand type of fault-block landscapes. Instead of planar triangular facets, rounded or blunt-ended ridges descend to the fault trace, where they are sharply cut off (Figs. 3-14, 3-15). The dissecting valleys between the scarp segments are large and are occupied by vigorous perennial streams that have substantial catchment areas in the interior of the upthrown block.

The unique features of fault-block landscapes of the New Zealand type are not tectonic but erosional. Many major New Zealand faults are high-angle reverse faults, but as in the case of all fault scarps, the dip of the fault surface is not significant to the scarp angle. The fault scarp shown in Figures 3-4 and 3-5 is presumably of the New Zealand type, because it stands in a humid tropical climate, is heavily forested, and lacks alluvial fans at the mouths of the dissecting creeks.

Other fault-block landscapes in the humid tropics, such as the great graben that forms the Markham Valley westward from Lae, Papua New Guinea, also are of the New Zealand type. Wherever precipitation exceeds evaporation, the excess water must drain along the tectonic depressions to the sea, creating *consequent valleys* (p. 273) along the notches of fault-angle depressions or on the floors of grabens. Continued uplift on a scarp of the New Zealand type might create flights of uplifted river terraces on the scarp face in contrast to the multiple earthquake scarplets across alluvial fans at the base of a basin-range scarp. Considering that most of the world's landscapes are drained to the sea by rivers (Chap. 9), it is tempting to conclude that the New Zealand type of fault-block landscape is more common than the basin-range type. Perhaps we have been educated to expect the basin-range type of fault-block landforms and have therefore overlooked or ignored fault-block landforms in humid regions. Highly tectonic regions in southeast Asia, Japan, Alaska, and the Caribbean can be expected to provide many additional examples of the New Zealand type of tectonic landscape.

Rift Valleys

The great *rift valleys* of the Rhine, eastern Africa, Lake Baikal, and the Jordan Valley so completely dominate their tectonic and geomorphic settings that they deserve special treatment as tectonic landforms, although descriptively they are little more than large, branching, complex grabens.

Holmes (1965, pp. 1044–78) and most other authorities have emphasized the association of rift valleys with plateaus produced by broad epeirogenic updoming.

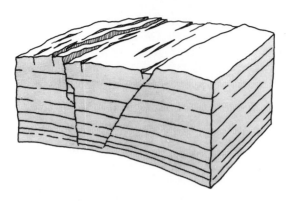

Figure 3-16. Experimental production of a rift valley by slowly arching layers of moist clay. (Redrawn from Holmes, 1965; experiments by Hans Cloos.)

Experiments (Fig. 3-16) show that excellent models of rift-valley morphology and tectonics, including the high plateaus that rise toward the rims of the depression and other details, result if moist clay layers are arched or updomed from beneath. Inward-facing normal faults form the boundary scarps, the model rift valley has a surface width approximating the thickness of the deformed layers, and no isostatic or mechanical inconsistencies develop. Instead of resulting from simple lateral tension, rift valleys are now interpreted as "keystones" subsiding along the crests of uparched plateaus.

The Rhine Graben is a small rift valley, compared to those of Africa and the Middle East. It is 30–40 km wide (Holmes, 1965, p. 1047) and, including several branching segments, about 300 km long. It cuts crystalline rocks or older folded structures, and Mesozoic and Cenozoic formations within the graben are strongly deformed and broken by secondary faults of Miocene and younger age. The fault scarps of the Rhine Graben are of the New Zealand type: dissected, rounded spurs ending abruptly at the alluvial plain of the Rhine River. Movement continued into the Pleistocene. Hot springs and health spas mark the boundary faults today.

The rift valleys of eastern Africa are a branching, *en echelon* series of grabens that divide into two major groups, the eastern and western rifts (Fig. 3-17). The total length of the system southward from the Red Sea approaches 3000 km. Rift valleys within the system are uniformly 30–50 km wide, approximately the thickness of the continental crust (Holmes, 1965, p. 1061). The boundary faults are dominantly normal. Recent volcanism and transverse buckling of the downdropped blocks has created divides within the rift valleys so that lakes in the southern part of the western rifts drain to the Congo, lakes in the northern part of the western rift drain to the Nile, Lake Nyasa drains south to the Indian Ocean via the Zambezi River, and most of the eastern rifts are areas of internal drainage into highly saline, alkaline lakes. Lake Tanganyika, part of the Congo drainage system, is 1400 m deep, its sediment bottom 650 m below sea level. As do many of the other rift lakes, it fills the floor of the rift valley from one fault scarp to the other.

In the midst of the western rift valley stands the Ruwenzori horst, the highest nonvolcanic mountain in Africa. Its summit at 5100 m is nearly 4 km higher than the plateau rims on each side of the rift valleys, which in turn are 400–500 m above the floors of the rifts. Clearly, the Ruwenzori massif is an uplifted horst, not simply a fault block that did not subside. It is said to be topped by the same mid-Cenozoic erosion surface that is arched upward toward the rift valleys across the adjacent plateaus (McConnell, 1972).

The plateau surface of Uganda, between the eastern and western rift valleys, has been strongly upwarped in the Quaternary (Bishop and Trendall, 1967). By Miocene time, a widespread erosion surface of low relief had beveled the metamorphic terrane of the plateau region. Miocene volcanic rocks buried this erosion surface, although much of it has subsequently been exhumed. Pre-Miocene rivers may have drained westward across Africa, perhaps via the Congo River, but by Miocene time the western rift had developed and the plateau drainage entered it. As the grabens progressively

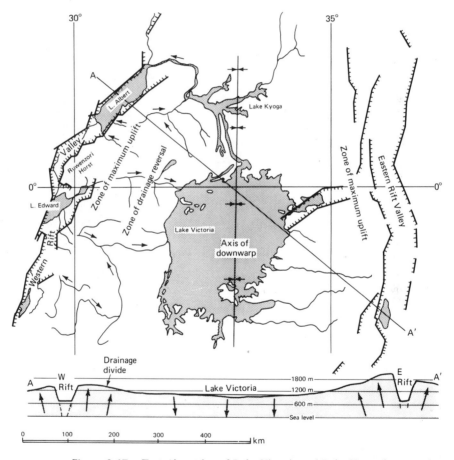

Figure 3-17. Tectonic setting of Lake Victoria and Lake Kyoga between the eastern and western rift valleys in eastern Africa. Former west-flowing streams were defeated by the uplifted shoulder of the western rift and reversed. Their drowned headwaters now form the lakes and drain north to the Nile. (After Holmes, 1965; Bishop and Trendall, 1967.)

Figure 3-18. Satellite photograph of the Sinai Peninsula viewed toward the northeast. The Dead Sea rift extends northward from the Gulf of Aqaba past the Dead Sea to Lake Tiberias. The eastern side of the rift has moved north relative to the western side. (Gemini XI photograph S66-54890, NASA.)

deepened in mid-Pleistocene time, the plateau erosion surface was lifted and flexed into a sharp western anticline and a broad central basin. An uplifted "shoulder" just east of the rim of the western rift valley defeated and reversed streams that formerly crossed it. The headwaters of these streams ponded to form Lakes Victoria and Kyoga (Fig. 3-17), which deepened against the uplifted plateau rim until they spilled around it to the north. The present Victoria Nile River resulted, draining into Lake Albert over an exhumed fault-line scarp at Murchison Falls (Bishop, 1965, p. 317). Lake Kyoga is a drowned dendritic drainage system on the warped early Pleistocene surface. The warping continues: 90 m of relative uplift across the defeated drainage systems may have occurred between 50,000 and 20,000 years ago. The implied tectonic differential uplift rate of 3 mm/yr approaches that of strongly orogenic regions (p. 25) although the plateau movements are certainly of epeirogenic scale.

The Dead Sea occupies a basin segment of a rift valley 600 km long and only 10–20 km wide that extends northward from the Red Sea to Lebanon (Neev and Emery, 1967) (Fig. 3-18). The eastern scarp is straight and continuous along a single border fault. The western side of the graben is more complex, with several step faults or fault splinters involved. Fault scarps along the Dead Sea rift valley are predominantly of the basin-range type. The region is one of extreme aridity.

The Dead Sea rift has a more complex structure than the Rhine Graben and the eastern African rift systems (Quennell, 1958). Left-lateral movement along a crooked transcurrent fault is believed to have opened a quadrilateral gap in the vicinity of the Dead Sea (Fig. 3-18). Pliocene and Pleistocene vertical movements along the boundary faults and several oblique transverse faults have created the present basins. The shore of the Dead Sea is the lowest subaerial place on the earth's surface.

Lake Baikal, the world's deepest freshwater lake (1600 m), lies in a rift valley north of Mongolia along the crest of the Baikal arch. The tectonic setting is similar to that of the eastern African rift lakes. Fault scarps plunge directly into the lake, with only small deltas. Most of the regional drainage is away from the lake along radii of the uplifted Baikal arch.

A *median rift valley* is a characteristic feature of many submarine midocean ridges or rises (Figs. 3-10, 21-7). The rift valleys of midocean ridges are bounded by a series of parallel normal faults, each with a throw of several hundred meters. The fault blocks that parallel the central rift valley are commonly lifted and rotated so that their upper surfaces slope away from the rift axis. The morphology of median rift valleys is consistent with their tectonic setting on the crests of expanding ridges at accreting plate margins (Ballard and van Andel, 1977).

LANDFORMS MADE BY FOLDING

Fault scarps are bold, obvious landforms that attract the eyes of geomorphologists and sightseers. Much less has been written about tectonic landforms that result from warping or flexing of the earth's surface without fracture and displacement on faults. A few examples can be cited from regions where earth movements are obvious.

Perhaps by calling attention to known examples, this review will stimulate readers to consider other possibilities, especially in regions of humid climate, heavy vegetation cover, active soil-forming processes, and mild diastrophism.

A major difficulty in recognizing tectonic landforms produced by folding is that a visible reference surface is not generally available. Whereas every structure and the landscape itself is sharply displaced across a fault scarp, warping or flexing of an irregular landscape or homogeneous structure is extremely difficult to recognize. Only where approximately planar erosional and depositional surfaces, such as shore platforms or river terraces are exposed, is warping likely to be visible. Even then, surveyed profiles may need to be taken along the surfaces to prove that they are warped.

Monoclinal Scarps

Tectonic scarps with height, steepness, and lateral continuity comparable to fault scarps are formed by steep monoclines (Fig. 3-19). As is the case with fault scarps, monoclinal scarps are regarded as tectonic landforms only if their relief can be shown to be due to tectonic movement and if substantial parts of the original surface of deformation can be recognized. Monoclines, as other folds, are described by the attitude of the deformed strata, not by the shape of the displaced landscape.

Monoclines are the most distinctive structure of the Colorado Plateau (Kelley, 1955). The longest is over 400 km in length; strata are displaced vertically as much as 4300 m. In the Colorado Plateau many land surfaces coincide with particularly resistant formations, so structural benches are flexed sharply along the monoclines to produce imposing scarps. The major period of deformation in the region was the late Cretaceous–Eocene Laramide orogeny (Kelley, 1955, p. 797). It is not known whether monoclinal flexing has continued to the present.

No systematic distinction of monoclinal landforms has ever been made analogous to the distinction between fault scarps and fault-line (erosional) scarps. It would

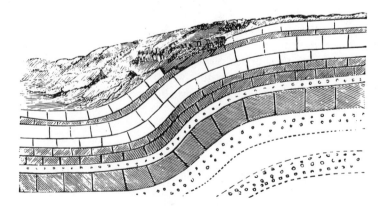

Figure 67.—A monoclinal fold.

Figure 3-19.   First illustration of a monoclinal fold. (Powell, 1875, Fig. 67.)

seem that the great monoclinal scarps of the Colorado Plateau were folded at some depth. Tertiary strata postdate the folding. Perhaps such scarps would better be included in Chapter 11 as examples of structurally controlled erosion rather than as tectonic landforms.

Topographic Domes

A variety of processes can produce domal tectonic landforms. Shallow intrusions of ice, salt, mud, or molten rock can uparch the overlying rock or sediment. Shallow intrusions may also extrude to create new landforms. Active anticlines in rock structures normally produce topographic domes in the landscape above. All these landforms are tectonic in the sense that they are built by earth movements, even if the movements that are involved are at a very shallow depth.

Salt Tectonics.   More than 400 domal or piercement structures with salt in their cores are known from the northern part of the Gulf of Mexico (Halbouty, 1967). The salt is presumed to have risen hydrostatically or under tectonic pressure from source beds of Jurassic or older age that are now buried under many kilometers of rapidly accumulated overburden. A thousand meters is regarded as the minimum thickness of a buried salt bed that will flow into domes (Murray, 1961, p. 260). Some of the U.S. Gulf Coast salt domes have risen through 7500 m or more of covering strata. They are circular or ridgelike in plan view and may be up to 10 km in diameter. Similar structures are known or suspected in the Colorado Plateau, the Canadian Arctic, many parts of the Continental Shelves, Saudi Arabia, France, Germany, Iran, and the Soviet Union. Because of their potential for petroleum production, salt domes are actively sought by exploration geologists.

Only five salt domes on the U.S. Gulf Coast have positive topographic relief (Gould and Morgan, 1962, p. 314). Although the domes penetrate Pleistocene and even Holocene sediments, the region has been one of deltaic progradation and fluviatile deposition. Most of the surface relief on the Texas–Louisiana coastal plain consists of meander scrolls, point bars, and low river terraces (Fig. 10-3). Lateral migration of coastal-plain rivers has obliterated most of the salt-dome relief. Avery Island, Louisiana, has a highest point of 46 m and forms a dome about 2.3 km in diameter. Salt is found at a depth of only about 4 m below the surface of Avery Island. Several salt springs and ponds drain radially off the dissected dome.

Many Gulf Coast salt domes are marked by central depressions ringed by low cuestas of the upwarped overlying rocks. *Annular* drainage patterns (Fig. 11-16) are common (Muehlberger, et al., 1962, p. 266). Such morphology is *structurally controlled*, but not *tectonic*, and is considered further in Chapter 11.

In the Canadian Arctic archipelago, Tertiary or older folding forced Paleozoic anhydrite and gypsum into anticlinal cores and domes. In contrast to the rapid solution of evaporites in humid regions, in the polar deserts these massive, unjointed, monomineralic rocks are resistant ridge- and dome-formers (St. Onge, 1959). Whether the highlands of gypsum in the Sverdrup Islands are tectonic or structural has not been determined (Stott, 1969, p. 37), but aerial reconnaissance shows that many of the higher hills are underlain by massive white gypsum, intrusive into darker sedimentary rocks that form surrounding lowlands.

Mud Lumps.   A minor but interesting form of tectonic domal landform has been described from the Mississippi delta region. Rapid forward growth of distributary channels may cause local loading of unstable prodelta clay by 100 m of deltaic sand, mud, and organic sediment. Diapiric intrusions through the overlying sandy sediment by plastic clay causes updoming or even extrusion as low islands near the mouths of distributaries (Morgan, et al., 1968). Near the mouth of South Pass, 45–50 different mudlump islands have been observed and mapped since 1867. Some reappear at intervals after being eroded off. Surface exposures show muds with marine faunas that must have been initially deposited in water depths of 100–120 m.

Laccoliths.   Shallow igneous sills may swell to a planar-convex lens shape and arch overlying sedimentary strata to produce topographic domes. The name *laccolith* (lens-shaped intrusion) was given for such forms in the Henry Mountains of Utah in one of the classic works of early American geology (Gilbert, 1877). Some of the larger type examples subsequently have been shown to be igneous stocks (Hunt, et al., 1953) and not laccoliths.

The laccoliths of the Colorado Plateau were intruded in late Eocene or Miocene time, probably at a depth of at least 1600 m (Hunt, et al., 1953, p. 147). Because some of the laccoliths are thicker than that, surface arching must have resulted. However, it is doubtful that any intrusive relief remains over the laccoliths in the Henry Mountains. The regional drainage pattern suggests that major rivers, flowing on an erosion surface a thousand meters or more above the present landscape, were diverted by the rising laccoliths and migrated laterally off the domes (Hunt, et al., 1953, p. 212). The subsequent geomorphic history of the Henry Mountains laccoliths, as is true of the monoclines of the region, is one of structurally controlled erosion (Chap. 11).

A laccolith has been referred to as "an unsuccessful attempt to form a volcano" (Hawkes and Hawkes, 1933, p. 396; attributed to G. K. Gilbert). In the San Francisco Mountains, near Tucson, Arizona, viscous siliceous lava produced both volcanic cones and laccoliths. At least one mountain, Mount Elden, has a form that is transitional between a viscous extrusive plug and a laccolith. The volcanic peaks have lost about 10 percent of their volume by erosional dissection since early Pleistocene time, whereas laccoliths of comparable age have lost fully 50 percent (Robinson, 1913, pp. 50, 80). The calculations, based on direct measurement of landforms of relatively simple geometry, help us to understand the comparative rarity of undissected tectonic domes over shallow intrusives. The overlying strata are probably stretched and fractured, making them highly susceptible to erosion.

Rising or Live Anticlines.   Orogenic regions such as the margin of the Pacific Ocean have domes or arches in the landscape over the crests of structural anticlines. Warped coastal and river terraces (Fig. 3-20) are a common form of evidence. In New Zealand, several active anticlines warp a coastal-plain landscape (Te Punga, 1957). One of them, the Pohangina anticline (Fig. 3-21), can be traced for 30 km. At its low

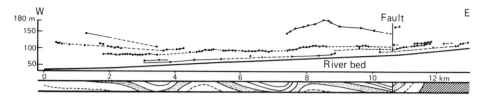

Figure 3-20. Longitudinal profile of the Oguni River and its terraces, Yamagata Prefecture, northeastern Japan. Vertical scale is exaggerated tenfold. Quaternary terraces are progressively upwarped over the anticlines in the Miocene and Pliocene strata shown on the geologic cross section beneath the profile. A fault near the eastern end of the profile displaces older terraces. (Sugimura, 1967, Fig. 2.)

southern end, the arched coastal-plain surface is only slightly dissected; farther north, the anticline crest is at an altitude of 350 m, and dissection of both flanks is complete. In between, the steeper eastern flank is the more dissected. Recent alluvium and Pleistocene coastal-plain formations dip 2°–4° off the west flank of the anticline, but on the steeper eastern flank, beds dip as steeply as 60°. On this rising anticline in a humid climate, vigorous gullies migrate rapidly headward on the steeper flank, and stream divides have shifted laterally beyond the crest of the structure onto the less-steep dip slope. Many drainage diversions have been noticed in the century since the land was cleared and fenced for agriculture. One of the anticlines in the New Zealand group may have risen 170 m in the last 20,000 years, at an average rate of 8.5 mm/yr. Seismic surveys and the regional geologic setting suggest that active anticlines in both New Zealand and Japan (Ota, 1975) are surface manifestations of rising fault blocks of indurated basement rock. Each anticline is thought to drape over the highest edge of an upfaulted block beneath.

Figure 3-21. View north along the nearly undissected crest of the Pohangina anticline, 6 km west of Ashhurst, Manawatu district, New Zealand. The anticline crest forms the divide near the left margin. East-flowing consequent streams are beginning to dissect the anticline. (Photo: C. C. Rich.)

Numerous examples of rising anticlines can be cited from other orogenic regions. Lees (1955) described a Persian *Qanat*, or horizontal tube well, that had been arched by a rise of 18 m in the middle part of its 4-km course across an anticline in the Khuzistan Plain at Shaur, Iran. The tunnel was dug 1700 years ago, giving a differential uplift rate across the structure of 10 mm/yr, even more rapid than the New Zealand examples previously cited. In both Japan and California, Holocene deformation of river terraces corresponds well with continuing deformation documented by geodetic releveling (Sugimura, 1967; Castle, et al., 1974).

REFERENCES

BALLARD, R. D., and VAN ANDEL, T. H., 1977, Morphology and tectonics of the inner rift valley at lat. 36°50′ N on the Mid-Atlantic Ridge: *Geol. Soc. America Bull.*, v. 88, pp. 507–30.

BANNERT, D., 1972, Afar tectonics from space photographs: *Am. Assoc. Petroleum Geologists Bull.*, v. 56, pp. 903–15.

BISHOP, W. W., 1965, Quaternary geology and geomorphology in the Albertine Rift Valley, Uganda: *Geol. Soc. America Spec. Paper 84*, pp. 293–321.

BISHOP, W. W., and TRENDALL, A. F., 1967, Erosion-surfaces, tectonics and volcanic activity in Uganda: *Geol. Soc. London Quart. Jour.*, v. 122, pp. 385–420.

BUWALDA, J. P., 1936, Shutterridges, characteristic physiographic features of active faults [abs.]: *Geol. Soc. America, Proc. 1936*, p. 307, [1937].

CASTLE, R. O., ALT, J. N., SAVAGE, J. C., and BALAZS, E. I., 1974, Elevation changes preceding the San Fernando earthquake of February 9, 1971: *Geology*, v. 2, pp. 61–66.

CLAYTON, L., 1966, Tectonic depressions along the Hope fault, a transcurrent fault in North Canterbury, New Zealand: *New Zealand Jour. Geol. Geophys.*, v. 9, pp. 95–104.

COTTON, C. A., 1944, *Volcanoes as landscape forms:* Whitcombe and Tombs, Ltd., Christchurch, N. Z., 416 pp. (Republished with minor corrections, 1952.)

——, 1948, *Landscape as developed by the processes of normal erosion*, 2d ed.: Whitcombe and Tombs Ltd., Wellington, N. Z., 509 pp.

——, 1950a, Tectonic scarps and fault valleys: *Internat. Geog. Cong., 16th, Lisbon 1949, Comptes rendus*, pp. 191–200.

——, 1950b, Tectonic scarps and fault valleys: *Geol. Soc. America Bull.*, v. 81, pp. 717–58.

——, 1956, Geomechanics of New Zealand mountain-building: *New Zealand Jour. Sci. and Technology*, sec. B, v. 38, pp. 187–200.

——, 1958, *Geomorphology*, 7th ed.: Whitcombe and Tombs Ltd., Christchurch, N. Z., 505 pp.

——, 1968, Tectonic landscapes, *in* Fairbridge, R. W., ed., *Encyclopedia of geomorphology:* Reinhold Book Corporation, New York, pp. 1109–16.

DAVIS, W. M., 1903, Mountain ranges of the Great Basin: *Mus. Comp. Zool. Harvard Bull. 42*, pp. 129–77 (reprinted 1954 in *Geographical essays:* Dover Publications, Inc., New York, pp. 725–72).

DONATH, F. A., 1962, Analysis of basin-range structure, south-central Oregon: *Geol. Soc. America Bull.*, v. 73, pp. 1–15.

EATON, J. P., and MURATA, K. J., 1960, How volcanoes grow: *Science*, v. 132, pp. 925–38.

VON ENGELN, O. D., 1942, *Geomorphology:* The Macmillan Company, New York, 655 pp.

GILBERT, C. M., and REYNOLDS, M. W., 1973, Character and chronology of basin development, western margin of the Basin and Range province: *Geol. Soc. America Bull.*, v. 84, pp. 2489–509.

GILBERT, G. K., 1874, Preliminary geological report, expedition of 1872: *U.S. Geog. Geol. Survey W. 100th Mer.* (Wheeler), *Progress Rept.*, pp. 48–52.

———, 1877, Report on the geology of the Henry Mountains [Utah]: *U.S. Geog. Geol. Survey Rocky Mtn. Region* (Powell), pp. 18–98.

———, 1928, Studies of basin-range structure: *U.S. Geol. Survey Prof. Paper 153*, 92 pp.

GOULD, H. R., and MORGAN, J. P., 1962, Coastal Lousiana swamps and marshlands, Field Trip No. 9, *in* Rainwater, E. H., and Zingula, R. P., eds., *Geology of the Gulf Coast and central Texas and guidebook of excursions:* Houston Geological Soc., Houston, Tex., pp. 287–341.

HALBOUTY, M. T., 1967, *Salt domes, Gulf region, United States and Mexico:* Gulf Publishing Co., Houston, Tex., 425 pp.

HAMILTON, W., 1971, Recognition on space photographs of structural elements of Baja California: *U.S. Geol. Survey Prof. Paper 718*, 26 pp.

HAWKES, L., and HAWKES, H. K., 1933, Sandfell laccolith and dome of elevation (Iceland): *Geol. Soc. London Quart. Jour.*, v. 89, pp. 379–98.

HOLMES, A., 1965, *Principles of physical geology*, 2d ed.: Ronald Press, New York, 1288 pp.

HUNT, C .B., AVERITT, P., and MILLER, R. L,., 1953, Geology and geography of the Henry Mountains region, Utah: *U.S. Geol. Survey Prof. Paper 228*, 234 pp.

JOHNSON, D. W., 1929, Geomorphic aspects of rift valleys: *Internat. Geol. Cong., 15th, South Africa, 1929, Comptes rendus*, v. 2, pp. 354–73.

KELLEY, V. C., 1955, Monoclines of the Colorado Plateau: *Geol. Soc. America Bull.*, v. 66, pp. 789–804.

KING, P. B., 1969, Tectonics of North America—a discussion to accompany the tectonic map of North America, scale 1:5,000,000: *U.S. Geol. Survey Prof. Paper 628*, 95 pp.

LATHRAM, E. H., 1972, Nimbus IV view of the major structural features of Alaska: *Science*, v. 175, pp. 1423–27.

LEES, G. M., 1955, Recent earth movements in the Middle East: *Geol. Rundschau*, v. 43, pp. 221–26.

LOUDERBACK, G. D., 1904, Basin Range structure of the Humboldt region: *Geol. Soc. America Bull.*, v. 15, pp. 289–346.

———, 1923, Basin Range structure in the Great Basin: *Calif. Univ. Pub. Geol. Sci.*, v. 14, pp. 329–76.

MACDONALD, G. A., 1972, *Volcanoes:* Prentice-Hall, Inc., Englewood Cliffs, N.J., 510 pp.

MCCONNELL, R. B., 1972, Geological development of the rift system of east Africa: *Geol. Soc. America Bull.*, v. 83, pp. 2549–72.

MENARD, H. W., 1964, *Marine geology of the Pacific:* McGraw-Hill Book Company, New York, 271 pp.

MORGAN, J. P., COLEMAN, J. M., and GAGLIANO, S. M., 1968, Mudlumps: diapiric structures in Mississippi delta sediments: *Am. Assoc. Petroleum Geologists Mem.*, no. 8, pp. 145–61.

MUEHLBERGER, W. R., CLABOUGH, P. S., and HIGHTOWER, M. L., 1962, Palestine and Grand Saline salt domes, eastern Texas, Field Trip No. 6, *in* Rainwater, E. H., and Zingula, R. P., eds., *Geology of the Gulf Coast and central Texas and guidebook of excursions:* Houston Geological Soc., Houston, Tex., pp. 266–77.

MURRAY, G. E., 1961, *Geology of the Atlantic and Gulf Coastal Province of North America:* Harper and Brothers, New York, 692 pp.

NEEV, D., and EMERY, K. O., 1967, The Dead Sea: *Israel Geol. Survey Bull. 41*, 147 pp.

OTA, Y., 1975, Late Quaternary vertical movement in Japan estimated from deformed shorelines: *Royal Soc. New Zealand Bull. 13*, pp. 231–39.

POWELL, J. W., 1875, *Exploration of the Colorado River of the west and its tributaries:* U.S. Govt. Printing Office, Washington, D.C., 291 pp.

QUENNELL, A. M., 1958, Structural and geomorphic evolution of the Dead Sea Rift: *Geol. Soc. London Quart. Jour.*, v. 114, pp. 1–24.

ROBINSON, H. H., 1913, San Franciscan volcanic field, Arizona: *U.S. Geol. Survey Prof. Paper 76*, 213 pp.

ST. ONGE, D. A., 1959, Note sur l'érosion du gypse en climat périglaciaire: *Rev. Canadienne de Geog.*, v. 13, pp. 155–62.

SHARP, R. P., 1939, Basin-range structure of the Ruby-East Humboldt Range, northeastern Nevada: *Geol. Soc. America Bull.*, v. 50, pp. 881–920.

———, 1942, Stratigraphy and structure of the southern Ruby Mountains, Nevada: *Geol. Soc. America Bull.*, v. 53, pp. 647–90.

———, 1954, Physiographic features of faulting in southern California, *in* Jahns, R. H., ed., Geology of Southern California: *California Dept. of Natural Resources, Div. of Mines Bull. 170*, Chap. V, Contr. 3, pp. 21–28.

STOTT, D. F., 1969, Ellef Ringnes Island, Canadian Arctic Archipelago: *Canada Geol. Survey Paper 68–16*, 44 pp.

SUGIMURA, A., 1967, Uniform rates and duration period of Quaternary earth movements in Japan: *Osaka City Univ. Jour. Geosci.*, v. 10, pp. 25–35.

TAZIEFF, H., 1970, The Afar triangle: *Sci. American*, v. 222, no. 2, pp. 32–40.

TCHALENKO, J. S., and BERBERIAN, M., 1975, Dasht-e Bayaz fault, Iran: Earthquake and earlier related structures in bed rock: *Geol. Soc. America Bull.*, v. 86, pp. 703–9.

TE PUNGA, M. T., 1957, Live anticlines in western Wellington: *New Zealand Jour. Sci. and Technology*, sec. B., v. 38, pp. 433–46.

TWIDALE, C. R., 1968, *Geomorphology: with special reference to Australia:* Thomas Nelson (Australia) Ltd., Melbourne, 406 pp.

WALLACE, R. E., 1949, Structure of a portion of the San Andreas rift in southern California: *Geol. Soc. America Bull.*, v. 60, pp. 781–806.

———, 1970, Earthquake recurrence intervals on the San Andreas fault: *Geol. Soc. America Bull.*, v. 81, pp. 2875–90.

WILLIS, B., 1928, The Dead Sea problem: Rift valley or ramp valley: *Geol. Soc. America Bull.*, v. 39, pp. 490–542.

WILSON, J. T., 1965, A new class of faults and their bearing on continental drift: *Nature*, v. 207, p. 343–47.

CHAPTER 4

Volcanoes

The emphasis of this chapter is on volcanoes as constructional landforms: parts of the modern landscape that either are being built or were built recently enough for their constructional form to be recognized in spite of dissection. Volcanic activity ("eruptions," "volcanicity," and "volcanism" are synonyms) must be described, as must the products of eruptions, in order to understand the structure and genesis of volcanic landforms. As with all landforms, the structures and genesis are vital to an understanding of the scenery. Erosional development on volcanic structures is deferred and considered with other forms of structurally controlled erosion (Chap. 11). By this arrangement, full attention can here be given to volcanoes as a constructional assemblage of dramatic, often strikingly beautiful, landforms, built by processes that have had legendary impact on human senses.

Volcanoes have had an uncertain place in systematic geomorphology. W. M. Davis (1905) treated volcanism as an "accident" that occurs so arbitrarily in time and place and is so disruptive to the erosional development of landscapes that the landforms cannot be treated in a systematic manner. A conventional arrangement of geomorphology textbooks is to review both constructional and erosional landforms on volcanic structures in a single chapter late in the book. Some books define volcanoes as the vents or fissures from which hot subsurface materials are transferred to above ground level. Others define volcanoes as the mountains or hills of erupted rock around the vents. Literally, then, there is no general agreement as to whether volcanoes should be defined as holes or hills. Cotton (1944, preface) summarized the issue: "An approach to volcanism may be made through either petrology or geomorphology, a volcano and its mechanism being regarded either as a geological crucible or as a builder of landforms." Information on volcanoes, their activity, and their products can be found in petrology books (Bayly, 1968) as well as geomorphology books and descriptive monographs (Cotton, 1944; Rittmann, 1962; Ollier, 1969; Green and Short, 1971; Macdonald, 1972; Bullard, 1976).

It is generally understood that volcanoes are characterized by eruptive activity. That is, hot gases, liquids, molten rock, and shattered rock fragments are forcibly or rapidly ejected from openings in the earth's surface. Classification of volcanoes beyond that level of understanding has been based on a wide variety of criteria, such as: (1) chemical composition and temperature of the ejecta; (2) physical state of the ejecta—whether a gas, liquid, or solid phase is dominant; (3) historic record of volcanicity—active, dormant, or extinct; (4) shape of the aperture—whether a central vent or linear fissure, for instance; (5) nature of the volcanic activity—exhalative gas discharge, effusive lava flows, explosive ejection of fragments, frothing of gas-charged liquid, fire-fountains, etc.; or (6) shape of the resulting landforms—mounds, rings, craters, domes, cones, shields, or plateaus.

GEOGRAPHIC AND GEOLOGIC SIGNIFICANCE OF VOLCANOES

Entire geomorphic provinces are volcanic in origin. The great Cenozoic basalt plateaus of Iceland, India, and the northwestern United States are among the largest constructional relief features of the subaerial landscape (although exceeded greatly by the ice caps of Greenland and Antarctica). An estimated 2 million km² of land surface have been covered to depths of 500–3000 m by lava flows in geologically recent time (Rittmann, 1962, p. xiii). Much larger areas of ocean floor have constructional topography of lava flows and shallow intrusives (Chap. 21). Probably the entire earth has been covered by layers of fragmental volcanic ejecta (ash and dust) that range in thickness from tens of meters to thin layers of dust. More than 500 volcanoes are classed as active in historic times, and tens of thousands of extinct volcanoes can be recognized by their form or structure.

Volcanoes are not accidentally distributed on the surface of the earth (Fig. 4-1). About 62 percent of the active volcanoes are on the rim of the Pacific basin, in the circum-Pacific "ring of fire." Another 22 percent are in Indonesia, 10 percent are in the Atlantic Ocean (including the Caribbean region), and a few percent are in each of Africa and the Mediterranean–Middle East. The remaining few form the Hawaiian Islands and other midocean islands.

Even along the lines of greatest concentration, as on the Pacific margins, volcanoes are not uniformly distributed. Most are in the arcs of the Aleutian and Kurile islands, the Kamchatka Peninsula, and Japan; other large groups are in the island arcs of Melanesia, and the New Zealand–Tonga volcanic belt. Alaska, Central America, and Mexico have many active volcanoes, but on the entire Pacific margin of the conterminous United States and Canada, only seven volcanoes have been

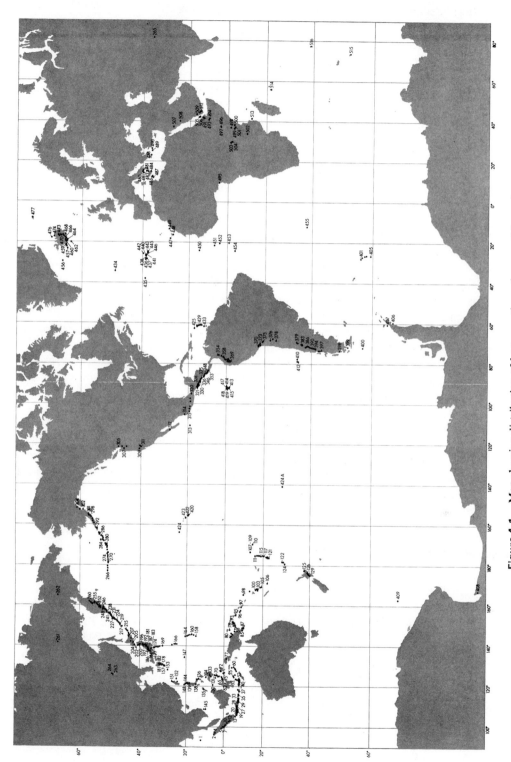

Figure 4-1. Map showing distribution of known active volcanoes. Other sub-marine eruptions must greatly outnumber the known volcanoes, but their distribution is unknown. (Macdonald, 1972, Fig. 14-1.)

active in historic times (Macdonald, 1972, App. 1). Converging plate margins are the loci of most known volcanic activity, as well as most earthquakes (compare Figs. 4-1 and 2-8). Submarine eruptions along midocean rises (Fig. 21-6) must be a major factor in generating new lithosphere along accreting plate margins, but the process has been observed only a few times (Chap. 21). Other regions of volcanic activity, including archipelagoes of midocean volcanic islands, may be related to "hot spots" or the rising limbs of convection cells within the earth's mantle.

Volcanoes have attracted renewed geomorphic interest in recent years because new techniques of geochronometry permit precise dating of eruptive rocks. It is now possible, by measuring the volume of rock eroded from a dated volcanic landscape, to measure the absolute rates of landscape modification (Chap. 12). Because many volcanic landforms have symmetrical geometry, their original constructional shape can be restored with much greater confidence than is possible with tectonic constructional landforms.

In addition to dating whole volcanic landscapes, key marker beds of volcanic ash of known age and place of origin can blanket millions of square kilometers of landscapes formed by other processes. It is hard to overstate the scientific value of a distinctively colored or otherwise identifiable ash bed inserted at an instant of geologic time into a stratigraphic sequence or onto an evolving landscape. When that layer can be dated precisely, it is even more valuable (p. 407).

A final point about the geomorphic importance of volcanoes concerns their wide latitudinal and climatic distribution. Volcanic landforms are constructed independently of any climatically controlled processes. Volcanic edifices are built in or on the Antarctic ice cap, in the tropical forests of Melanesia and Indonesia, in deserts, and in every other geomorphically significant climate. In each instance, the structure and form of the constructed landforms are similar. A great opportunity is thus available for geomorphologists to study contrasting erosional forms and rates in many climatic regions on constructed volcanic landforms that are not only similar but datable. This aspect of geomorphic volcanology alone justifies an extended systematic treatment of volcanic constructional landforms.

VOLCANIC ACTIVITY AND PRODUCTS

The products of eruption can be described as the results of certain types of activity, such as *exhalative* (gas), *effusive* (lava), or *explosive* (tephra). Alternately, characteristic modes of activity and products can be defined by describing active volcanoes that behave in certain ways. Both devices are used in the following pages.

Exhalative Activity and Mud Volcanoes

Hot springs, geysers, fumaroles, solfateras, and many other terms are used to describe vents that continuously or intermittently discharge hot water, steam, and other gases but rarely any solid or molten rock. Such vents are not generally considered volcanoes even though their activity is similar to volcanic eruption. They build minor landforms such as *sinter mounds* and *cones* of precipitated minerals.

Mud volcanoes make impressive small landforms over exhalative vents. Several in the Copper River Basin of Alaska were described by Nichols and Yehle (1961). They are between 45 m and 95 m in height and discharge mineralized warm water and gas, including light hydrocarbons probably derived from the decay of buried peat beds or coal. One of the largest cones contains abundant fragments of lava and volcanic glass that are unlike the glacial drift of the region but typical of shallow hydrothermal eruptions. Travertine crusts intermingle with the hardened mud and angular rock fragments of the crater rims. Such cones are probably the minimum accepted "volcanic" landform to most geomorphologists.

In 1951 a spectacular mud volcano in northeastern California suddenly erupted steam, gases, and mud to an estimated height of 1000 m (White, 1955). Fine debris, including pellets of mud, fell on a town and farms 7 km downwind from the eruption. The entire episode ended within four days. After a few months of quiescence, probably with some ground subsidence, the area was flooded with boiling water and mud springs. This kind of mud volcano draws its energy from volcanic heat, but most of the water seems to be of near-surface origin. If pressure is suddenly reduced at the top, by chance overflow of a pool, for instance, the confining pressure is decreased throughout the entire water column. A chain reaction of steam generation ensues exactly as in a geyser eruption, except that overlying fine-grained sediment is liquefied and erupted as hot mud, along with blocks of indurated bedrock.

Identical mud-cone landforms are built by an entirely nonvolcanic process. In orogenic regions, where methane-rich, plastic, organic muds accumulate rapidly and are rapidly buried, gas and water may liquefy the mud and cause it to flow under tectonic pressure from vents or fissures along faults (Fig. 4-2). The best known examples are on the Makran coast of West Pakistan, 100 km west of Karachi (Snead,

Figure 4-2.   A dormant mud volcano, about 50 m high, on the south flank of the larger Chandragup mud volcano on the Arabian Sea coast of Pakistan. (Photo: R. E. Snead.)

1964). These cool-mud volcanoes, and others in comparable areas of hydrocarbon-rich sediment accumulation, are genetically more related to the mud lumps of the Mississippi delta (p. 50) than to volcanic landforms.

Effusive Activity: Lava

Lava is the name given to erupted molten rock and to the subsequently cooled, solid equivalent, provided that it has an obvious flowlike surface form. More often, solidified lava is simply called "volcanic rock," along with solidified ejecta, because the fine grain sizes or glassy textures of eruptive rocks may not permit a clear conclusion about origin. **Effusive** (as contrasted to explosive) activity refers to copious outpourings of lava from a vent or fissure.

Most lavas are molten silicates, although silica-free carbonate melts are known, especially in eastern Africa (Dawson, 1962). A lava flow of elemental sulfur in Japan has been commercially mined. The ordinary range of SiO_2 in volcanic rocks is from about 35 percent to about 75 percent. At any given temperature, acidic (silica-rich) lavas are more viscous than basic (silica-poor) lavas (Bayly, 1968, p. 33). The composition of the molten phase is a major factor in landform development because of the control it exerts on viscosity. Low-silica Hawaiian basalt freely flows down slopes of only a few degrees. The resulting landforms are likely to be low compared to their lateral extent, with smooth, domal form. On the other extreme, silica-rich viscous lava may not flow from a vent but form a protruding plug or spine.

Temperature and dissolved gases are the two other factors that determine the viscosity of lava. Confining pressure alone normally elevates the melting temperature of solids, but when water is dissolved in a silicate melt, the melting temperature is lowered. Although acidic magmas can dissolve more water than basic magmas, and thereby be highly mobile at depth, they become highly viscous at surface pressure, when the volatiles have boiled off. In the process of degassing, they become violently explosive.

Surface pools and flows of basic lava are usually at temperatures of 1100°C–1200°C. Their viscosity, calculated from field measurements of laminar flow rates, is about 4.5×10^4 poises (Nichols, 1939). This viscosity is comparable to melted sucrose (sugar) at a temperature of 109°C, when it will flow as a pasty liquid from a test tube. On cooling to a temperature of 1160°C, some of the minerals in a silica-poor melt begin to crystallize, and the viscosity increases rapidly and by many orders of magnitude with only slightly lower temperature. The heat of fusion is probably a factor that keeps most lava flows in the relatively narrow temperature range noted. These considerations of temperature, volatile pressure, and composition help to clarify why effusive activity is almost totally confined to silica-poor, or basic, melts. By far the most characteristic effusive lava is *basalt*, so much so that other mineralogical compositions can be ignored for geomorphic interest.

Lava emerges from vents or fissures as white-hot sheets or tongues that flow downhill (Fig. 4-3). Each lobe or sheet is called a lava *flow*; a remarkably long single

Figure 4-3.   Active lava flows, Mauna Loa, Hawaii. (Photo: Hawaiian Volcano Observatory, U.S. Geological Survey.)

flow from the 1950 Mauna Loa eruption traveled 24 km downhill to the sea from a fissure at about a 3050-m elevation at an average rate of 9.5 km/hr (Bullard, 1976, p. 278). The 1881 lava flow from Mauna Loa lasted 11 months and traveled 80 km northeastward to within a few hundred meters of the town of Hilo, where it stopped (Dutton, 1884, p. 154). At its terminus, the flow is less than 6 m thick.

Lava flows follow preexisting topography downhill and may completely fill river valleys. On more open slopes, they can be diverted by walls or dikes no more than 10 m high (Macdonald, 1972, p. 422). One lobe may congeal and stop when a lateral lobe breaks out higher upslope and follows an easier path.

The upper surface of a basalt lava flow is usually highly *vesicular* or porous due to gas bubbles that escape during the final stages of cooling. If sufficiently spongelike in texture, the rock is called *scoria*. Two contrasting surface textures develop on flows, both commonly known by the traditional Hawaiian terms: *aa* for an angular, blocky, scoriaceous surface, and *pahoehoe* (or ropy) for smoothly twisted, convolute surfaces (Fig. 4-4). Pahoehoe surfaces develop on hotter, more fluid, flows. An extreme variant of pahoehoe or ropy lava occurs when hot, fluid lava flows into or erupts under water. Blobs or lobes of lava up to a meter in diameter form steam-jacketed, tough, flexible skins and pile up like pillows or sandbags while still molten in their interiors. The result is *pillow lava* (Fig. 21-6), a diagnostic feature of subaqueous eruption and perhaps the most abundant rock type on earth (Moore, 1975).

Figure 4-4. Pahoehoe lava draped over a former sea cliff south of Kilauea Volcano, Hawaii. (Photo: D. A. Swanson, U.S. Geological Survey.)

Repeated observations confirm that pahoehoe flows remain hot and fluid inside even though their top and bottom surfaces and lateral margins have solidified. One result of fluid lava becoming encased in cooler margins is the formation of *lava tubes*. These caves, many of which are several km in length and large enough to be entered, mark the draining out of the molten lava from within a solidified surface. As they drain, air may enter at their lower end and, escaping somewhere higher upstream, set up a convective "blast furnace" condition that glazes the interior walls and creates stalactites of melted rock (Greeley and Hyde, 1972).

Rough aa surfaces on flows result from the cooled and broken surface layer being carried forward by laminar flow and then being "plowed under" at the front of the advancing lava. The motion is analogous to that of the endless tread on a tracked vehicle. *Blocky lava* is sometimes distinguished from aa surfaces on the basis of a slablike or blocky character rather than the extremely scoriaceous, spinose, or "clinkery" aa texture (Macdonald, 1972, p. 91).

Lava flows, in common with intrusive dikes and sills and thick sheets of fused tephra, develop distinct sets of intersecting contraction joints upon cooling that result in the familiar *hexagonal* or *columnar* jointing (Fig. 11-1). These primary structures exert powerful controls on the landforms eroded from volcanic terranes (Chap. 11).

Explosive Activity: Tephra

Eruptive rock may be fragmented by a sudden, foaming loss of volatiles, by abrupt contact with groundwater or surface water, by grinding abrasion in the volcanic neck, or by liquid dispersion in a violent explosion. Volcanic ejecta that settle out of air or water are sometimes called *pyroclastic* or *volcaniclastic* sediments or rocks. **Tephra** is a less cumbersome collective term for all fragmental ejecta from volcanoes (Macdonald, 1972, p. 23). It is derived from the Greek word for ash and is intended to be functionally equivalent to *lava* and *magma* (Thorarinsson, 1951, p. 5). No connotation of grain size is intended or implied, although the related subscience of **tephrochronology** is often thought of as the use of ash beds in stratigraphic correlation and dating, ignoring the coarser grades of ejecta.

Tephra consists of fragments with a wide range of grain sizes and shapes. *Ash* is the most common term for sand-size and finer tephra. Larger fragments or aggregates are called *lapilli* (gravel size; ejected either molten or solid), *blocks* (cobble or boulder-sized solid ejecta), or *bombs* (twisted, air-cooled masses ejected in a molten state) (Macdonald, 1972, pp. 122–41). The terminology is elaborate and extensive, because explosive volcanic eruptions are among the most feared and discussed phenomena on earth.

Tephra is strongly size-sorted during air transport. Bombs and blocks fall only as far from the vent or fissure as the explosive force is able to hurl them. Lapilli and ash are air-borne for many kilometers, the grain size and thickness of the resulting tephra layer decreasing with increasing distance from the source (Fisher, 1964; Porter, 1972).

A **nuée ardente,** or glowing cloud, is the most terrifying form of explosive volcanic activity. In May, 1902, a nuée ardente from Mt. Pelée, Martinique, killed 30,000 people in the city of St. Pierre. The eruption, the first of its kind to be scientifically studied, gave the name Peléan to such peculiarly explosive activity (p. 66). The phenomenon can be produced either by a lateral, or directed, eruption of superheated steam and glowing hot tephra or by "boiling over" of foaming lava. The leading edge of a nuée ardente may speed downhill at 50 m/s. The fluidized mass is heavier than air, so it remains in contact with the ground and flows in a turbulent fashion.

Volcanologists emphasize that nuées ardentes are soundless, evidence that in spite of their forward velocity comparable to the winds of a hurricane, the glowing fragments in them are cushioned by hot gas and are not abraded by the rapid, turbulent motion. The temperature in the nuée ardente that devastated St. Pierre is estimated to have been between 650° and 1060°C.

Most of the volume of a nuée ardente is hot gas and superheated steam. The tephra accumulation over the ruins of St. Pierre was less than a meter in thickness, generally only about 30 cm. A nuée ardente in the Philippines charred trees to a height of 4.5 m, clearly marking the thickness of the flowing layer, yet left only 13 mm of ash on the ground (Macdonald, 1972, p. 5). On cooling, the steam condenses to make scalding hot mud of the tephra. A cloud of air-borne tephra that towers above the fast-moving ground layer is technically not part of the nuée ardente but is usually associated with it (Sparks, et al., 1973).

The nuée ardente form of explosive activity is believed to be responsible for some of the largest constructional volcanic landforms. These are the *ignimbrite* or *welded tuff* sheets that cover thousands of square kilometers in New Zealand, the Yellowstone Plateau, Katmai National Monument in Alaska, the Lake Toba region of northern Sumatra, and probably elsewhere (p. 79). The ignimbrite may be hundreds of meters thick.

Less thick than ignimbrite sheets but even more widespread are *ash showers* that result from some explosive eruptions. Explosive force cannot propel tephra more than a few kilometers, so wind must be the transporting force for widespread ash showers. They are cold when they reach the ground. When a tephra sheet is mapped in detail (Fig. 4-5), the direction of prevalent wind during the eruption is obvious.

Widespread ash showers are the great tools of tephrochronology. Individual ash beds can be identified by the refractive index of the glass shards (if they are not altered by hydration, which occurs rapidly in humid regions) or by trace-element chemical analysis (Borchardt, et al., 1971). They can sometimes be dated directly by fission-track density on the glassy shards, or they can be traced laterally to a vent and correlated to a lava flow or welded tuff dated by the potassium/argon method. Late Quaternary ash beds also may be dated by the radiocarbon ages of underlying or overlying peat beds or other organic sediments (Borchardt, et al., 1973).

Tephrochronology is becoming an important tool for measuring rates of constructional and erosional landscape development, especially in the volcanic terranes of Japan, New Zealand, Iceland, and western North America. One example from

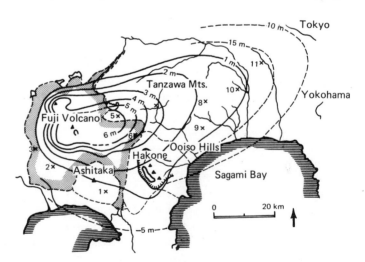

Figure 4-5. Isopach map of the Younger Fuji Tephra Formation (solid contours) and the Older Fuji Tephra Formation (dashed contours). Locations of measured sections are shown by numbered crosses. (Machida, 1967.)

Japan suggests the potential of the method. West of Tokyo, beds of lapilli and ash totaling 100 m in maximum thickness comprise the Older Fuji Tephra, which was deposited by hundreds of ash falls over a long period of time, probably between 25,000 and 10,000 years ago (Machida, 1967). It contains preceramic artifacts in its upper part. The Younger Fuji Tephra, which also accumulated during multiple ash falls, is 6 m thick in the vicinity of Mt. Fuji but less than 1 m thick near Tokyo, 80 km to the east (Fig. 4-5). It overlies archeologic sites with pottery, indicating a young age. The uppermost tephra layer is known to date from historical eruptions since 781 A.D. The weathered zone, fine ash, and soil between the Older and Younger Fuji Tephras shows that the latest historic activity of Fuji began after a long period of quiescence. These widespread tephra layers are used to measure tectonic displacement on faults, warping of coastal terraces, subsidence in the metropolitan Tokyo area, and many other tectonic movements of geomorphic interest. They have also been used to study rates of soil formation and for dating archeological sites. In Arctic Canada, floating pumice 5000 years old marks a distinctive emerged beach from which postglacial isostatic uplift can be accurately measured.

Subaqueous Volcanism: Hyaloclastite

Pillow lavas, an extreme form of pahoehoe surface texture, have been noted as one diagnostic feature of subaqueous eruptions of fluid lava. They are especially common on the deep ocean floor, where basalt is the only eruptive rock, and where the high hydrostatic pressure prevents degassing of the lava and minimizes fracturing.

More viscous lavas, and those erupted at lesser depths, develop shattered glassy margins on pillows and flow surfaces (p. 491). The related volcanic product is **hyaloclastite** (literally: glassy-fragment rock). Most hyaloclastites thus far identified are in Iceland, where basalt eruptions under glaciers or through glaciers have been frequent. Ancient hyaloclastites are common in association with pillow lavas and deep-sea sedimentary rocks and are presumed common on the sea floor today. In Marie Byrd Land of Antarctica, hyaloclastites make up significant proportions of several volcanic peaks that protrude through the ice sheet. The volcanic rocks range in age from 36 million to 0.5 million years (Oligocene to Pleistocene). The abundance of hyaloclastites in the volcanic rocks suggests that a thick ice sheet has been present in West Antarctica during most of the late Cenozoic Era (LeMasurier, 1972).

Characteristic Eruptive Types

It has been the tradition of volcanologists to use the names of certain well-known volcanoes to classify characteristic modes of eruptive activity. As no two volcanoes are alike, such a classification could become nothing less than a description of all known eruptions. A comprehensive summary of eruptive types, based on ten named volcanoes or other widely recognized modes of activity, has been provided (Macdonald, 1972, pp. 210–54). Only the four most traditional "typical" eruptive types and a few of their major variants are listed here in order to illustrate this alternative method of classifying volcanic activity. None of the named volcanoes erupt with only the activity described, however.

Hawaiian Eruption.   The simplest, most predictable, and safest type of eruption to study is the effusive outpouring of basalt lava from craters, lava lakes, or fissures. Individual flows are on the order of 10 m thick, and either spread widely over open slopes or pour down valleys as lava rivers. Little gas or tephra is produced. The great basalt plateaus of the Columbia Plateau and Iceland are large-scale examples of Hawaiian activity.

Strombolian Eruption.   The persistent lava-fountain activity of Stromboli, reflected at night as a red glow on the underside of a towering steam plume, has caused the volcano to be known for millenniums as the "lighthouse of the Mediterranean." Strombolian activity is similar to Hawaiian, except that the somewhat more viscous lava is ejected upward in fountains from a lava lake in the crater at regular, rhythmic intervals of about 15 minutes, and flows are unusual. The continued source of heat for maintaining the liquid pool of lava seems to be bubbles of burning gas that rise freely through the melt from several vents that have remained in about the same position for at least 350 years (Bullard, 1976, p. 248). The vapor cloud over the volcano is white, showing that no tephra is generated.

Vulcanian Eruption.   Vulcano, an island north of Sicily, was named for the Roman god of fire and smelting and in turn gave its name (with modified spelling) to the entire range of volcanic eruptive activity and landforms. Vulcanian activity, based specifically on observations of the 1888–90 eruption there, is explosive. The crater fills with viscous lava, but rather than remaining molten as in nearby Stromboli, the lava solidifies and is explosively ejected as a great cauliflower cloud of dark tephra. Bombs, blocks, and lapilli shower the surrounding area. Pumice is a common component of the ejecta. Only minor lava flows result. Vulcano is known to have erupted explosively at least 12 times between the fifth century B.C. and the 1888 eruption. After each eruptive cycle, the volcano is dormant for decades or centuries.

An awesome feature of Vulcanian eruptions is the brilliant lightning flashes that play in the cloud or between it and the ground. The friction of tephra particles generates great electrical discharges. Thunderstorms are generated by the heat, and torrential muddy rainfall is an added geomorphic feature of this type of activity. Some volcanologists consider Vesuvius a better example of Vulcanian activity than Vulcano. Narrative accounts of the Vesuvius eruption of 79 A.D. are widely quoted (Bullard, 1976, pp. 192–98). The popular image of a volcanic eruption is surely the Vulcanian type.

Peléan Eruption.   The nuées ardentes of Mount Pelée, Martinique, were described in a preceding section. The activity is violently explosive, the result of very viscous, gas-rich, acidic lava plugging the vent and either frothing violently over the crater rim or breaking out laterally. An explosion of the Peléan type differs from a Vulcanian eruption in that the very hot gas and lava mixture is not carried skyward by the updraft to become cold tephra but spreads downslope as a nuée ardente, continuing to evolve gas that cushions the flowing fragments. In a closely related process, a stiff lava plug may choke the vent and force a lateral, or directed, explosion of the gas and lava emulsion.

Another variant of Peléan activity was displayed by a 1912 ash eruption near Katmai volcano in Alaska, which buried the Valley of Ten Thousand Smokes. In that eruption, glowing ash flows welled out of numerous fissures nearly simultaneously. The fluidized ash filled the valley floor, burying water-saturated alluvium. The numerous fumaroles and hot springs that subsequently formed gave the valley its new name. Some authors give a unique name, *Katmaian,* to such massive explosion of tephra. The widespread occurrence of ignimbrite sheets testifies to the importance of Peléan, or Katmaian, activity.

It will be noted that the fourfold classification above can be consolidated to a twofold classification of effusive, or lava, volcanoes, and explosive, or tephra, volcanoes (Rittmann, 1962, p. 117). Not included in the fourfold classification are violent gas or steam explosions that erupt only shattered old rock, with little or no inclusion of new lava or tephra. These explosions are called *ultravulcanian*, because their violence exceeds the Vulcanian type, or *phreatic*, because they are caused by ground water or sea water entering the magma chamber below a vent. The enormous 1883 eruption of Krakatoa, in Indonesia, was a major phreatic eruption.

Some authors correlate an increase in explosive volcanism in the last two million years to Quaternary glaciations (Kennett and Thunell, 1975). If there is a correlation, it is a complex one. Glaciation might cause an increase in volcanicity by isostatic strain in the lithosphere, or volcanic activity might trigger an ice age by reducing incoming solar energy with a veil of fine tephra.

VOLCANIC LANDFORMS

Having extensively described and classified the types of volcanic activity and the products of volcanic eruptions, we now turn to the explanatory description of constructional volcanic landforms. The viscosity of the magma is the major factor that determines the type of eruptive activity and the products that are erupted. (Viscosity, of course, is a parameter that is determined by the several more fundamental properties of composition, temperature, and dissolved-gas content of the magma rising within the feeder conduits.) Viscosity and the size of the various edifices can be combined as the basis for a classification of volcanic landforms (Table 4-1; Fig. 4-6).

The vertical component of Table 4-1 is the viscosity of the magma, or as an equivalent, the type of activity (effusive, mixed, or explosive). For a horizontal com-

TABLE 4-1. CLASSIFICATION OF VOLCANIC LANDFORMS

Quality of Magma	Type of Activity	Quantity of Eruptive Material Small ← → Great			
Fluid, very hot, basic	Effusive	Lava flows	Exogenous domes	Basalt plateaus and shield volcanoes	
				Icelandic Hawaiian	
Increasing viscosity, gas content, and silica percentage	Mixed	Scoria cones with flows	Composite cones or Strato-volcanoes	Volcanic fields with multiple cones	
		Loose tephra cones and thick flows			
		Endogenous domes (plug domes, tholoids, spines)	Ruptured endogenous domes with thick lava flows		
Viscous, relatively cool, acidic	Explosive	Maars of tephra	Maars with ramparts	Collapse and explosion calderas	Ignimbrite sheets
Extremely viscous, abundant crystals	Explosive, mostly gas	Gas maars	Explosion craters		

SOURCE: Simplified from Rittmann, 1962, Tables 4 and 5

ponent, the table uses the volume of erupted matter. Increasing quantities of lava and tephra do not simply produce larger constructional landforms of the same shape. The intriguing concept of *morphological capacity* was introduced by Rittmann (1962, p. 113) to account for changes in landform with increasing size. *Scoria cones* (Fig. 4-7), for instance, can only grow to a certain size before they can no longer support the column of lava rising in the throat of the volcano. Then, flows break out on the flank or even from beneath the base of the cone (Fig. 4-8) and the subsequent growth of the cone follows a different trend.

In the following paragraphs, the forms that are more common and more significant to a general explanatory-descriptive scheme are described in a sequence beginning with the forms built by the least viscous, effusive lava eruptions, and continuing to consideration of the forms built by extremely viscous lava and explosive tephra eruptions. Within each eruptive type, a sequence of forms from small to large is reviewed.

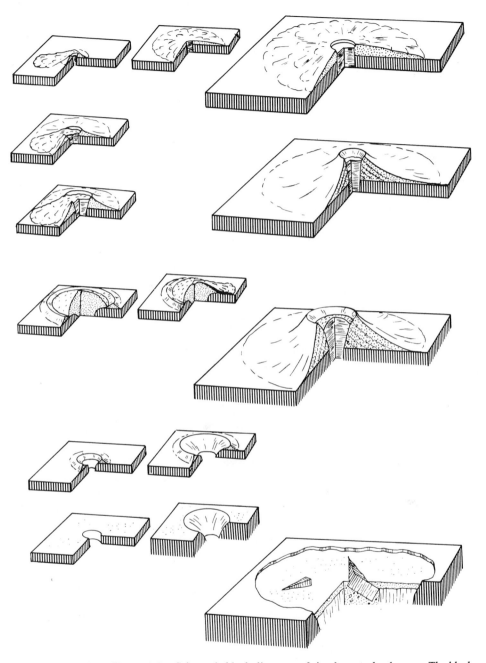

Figure 4-6. Schematic block diagrams of simple central volcanoes. The blocks are arranged according to the classification scheme of Table 4-1 except that the largest volcanic structures are not represented. (From Rittmann, 1962, Fig. 57.)

Figure 4-7. Scoria cones partly burying a lava flow, Sheeptrail Butte, Craters
of the Moon National Monument, Idaho. (Photo: U.S. National Park Service.)

Figure 4-8 Small spatter cone (hornito) over a partly collapsed lava tube,
breached by a lava flow that emerged from its base. North flank of Mauna Loa
Volcano, Hawaii. (Photo: J. P. Lockwood, Hawaiian Volcano Observatory,
U.S. Geological Survey.)

Basalt Flows, Domes, and Shield Volcanoes

The basic geomorphic unit of a lava eruption is the *flow*, previously described. Flows may reduce the preexisting relief by filling valleys or may increase it by building elongate, flat-topped lobate sheets and tongues of lava on top of the previous landscape (Figs. 4-3, 4-8). Their edges and upper surfaces may have pahoehoe or aa texture or may be a jumble of lava blocks. On the flow, a variety of small landforms may be found, such as *tumuli*, formed by small-scale lacolithic swellings on a pahoehoe surface, *pressure ridges, trenches* from collapsed lava tubes, and *spatter cones* or *hornitos*, built where secondary gas eruptions from within or beneath the flow throws up small cones of lava clots (Fig. 4-8). *Lava blisters* are created by steam pockets beneath thin flows that have spread over marshy ground. Thorarinsson (1951, pp. 63–73) reviewed the origins of these minor landforms and introduced a new term, *pseudocraters*, for lines or clusters of explosion craters along valley-filling lava flows. He demonstrated that the crater fields and lines of Iceland are not vents along the fissures that fed the flows but result from steam eruptions when flows spread over water-saturated alluvium.

Effusive flows from a central vent in flat terrain may build a low dome over the vent. A dome will develop only if slopes are nearly flat because of the low viscosity of basalt flows. These domes built of successive flows piled around eruptive vents are called **exogenous domes,** to distinguish them from **endogenous domes**, or cumulo-domes of viscous lava that swell within a skin or crumbly encrustation of obsidian and pumice by slow, continued extrusion of highly viscous lava (Cotton, 1944, pp. 158–60). The latter forms are described on p. 77.

Exogenous domes grade upward in size to become **shield volcanoes.** No definitive size criteria have been suggested. Shield volcanoes of the Icelandic type are between 100 m and 1000 m in elevation and have a base diameter about twenty times their height. Surface slopes on shield volcanoes average only a few degrees, although slopes of 20° may surround the summit crater. Late-stage caldera subsidence may flatten the summit profile. A genetic sequence for the Icelandic-type shield volcanoes in the Galápagos Islands has been deduced by Nordlie (1973).

Shield volcanoes of the Hawaiian type rank with basalt plateaus and ignimbrite sheets as the largest volcanic landforms. They are not simply scaled-up Icelandic shield volcanoes or exogenous domes; again, the morphologic capacity of the domal form is involved. When a lava dome has grown to some critical size by central eruption, it is no longer strong enough to support a central lava lake or carry a central conduit up to the summit crater. Flank eruptions break out, sometimes along the normal faults that encircle the growing mass (Fig. 3-2), and the dome grows peripherally with only slow increase in height. Slopes around Mauna Loa, for instance, range from only 3°–11° (Wentworth and Macdonald, 1953, p. 7). The crest of the dome becomes a broad, convex-upward plateau around a rimless *pit crater*, or a *summit caldera* (Fig. 4-9). Because of hydrostatic balance in the internal magma chambers, a pit crater may temporarily hold a lava lake. When the lake drains down

Figure 4-9.   View eastward across the summit of Mauna Loa shield volcano, including Mokuaweoweo Caldera and South Pit (right). (Photo: U.S. Navy, contributed by D. P. Cruikshank.)

due to a flank eruption, the crater rim subsides as giant slump blocks along concentric normal faults.

The size of Hawaiian-type shield volcanoes is enormous (Fig. 21-8). Mauna Kea and Mauna Loa, the two major shields of the five that merge to form the island of Hawaii, have summit elevations of 4206 m and 4175 m above sea level. More significantly, they rise smoothly from a submarine base about 400 km in diameter, and the total height of the multiple shield volcano is about 10 km. Including its submerged portion, Mauna Kea is the highest mountain on earth.

Basalt Plateaus

At various intervals of geologic time, regions measuring more than 10^4 km² have been engulfed by massive eruptions of fluid basalt. Central vents are not responsible; rather, *fissure eruptions* of regional extent provide the lava. Each fissure eruption is monogenetic. Dikes can be seen that cut individual flows but rarely are continuous through the entire thickness of multiple flows. The former land surface bows down beneath the accumulating sheets of basalt, either because of isostatic compensation or because of the enormous volume of magma that is erupted.

The result of such massive lava effusions are plateaus of considerable size and monotonous structure. The largest such area in the United States is the 130,000 km² of the Columbia Plateau of the states of Washington and Oregon, where the eruptions are of Miocene age. Adjacent to the south is the smaller Snake River Plain with an area of 52,000 km² extending across southern Idaho and merging into the Columbia Plateau. The basalt flows of the Snake River Plain are of Quaternary age, although no historic eruptions are known. These basalts overlie 1000 m of Pliocene ignimbrite along an elongate structural depression called the Snake River downwarp (Malde and Powers, 1962; Eaton, et al., 1975). The largest known basalt plateau is the Deccan district of northwestern India, which now covers 260,000 km² and originally may have been twice that large (Macdonald, 1972, p. 256). The basalts there are of Cretaceous and Eocene age and are 2 km thick in places. Other similar basalt plateaus of recent enough age to form constructional landscapes are in southern Brazil, Manchuria, and central Siberia, and in fragments, around the North Atlantic basin on Greenland, Iceland, Ireland, the Faeroe Islands, and Jan Mayen.

Except for exogenous domes or minor scoria mounds, the constructional morphology of a basalt plateau is extraordinarily flat. Flows have initial slopes of 1° or less, barely detectable to the eye. The impression is created that the entire regional landscape is the surface of a single lava flow, which is, of course, not the case. The extent of individual flows within the Deccan Plateau is a matter of vigorous debate (Choubey, 1973, 1974; Gupte, et al., 1974). In central Oregon, some individual lava flows have volumes two or three times as great as large andesitic cones (McBirney, et al., 1974).

Erosional dissection of basalt plateaus is strongly controlled by the sheetlike continuity and columnar jointing of the flows. Flat benches and vertical cliffs dominate the landscape (Fig. 11-1). Hills are commonly flat-topped. The Germanic name for the steplike landscape (*treppen*) has survived in English not as a geomorphic term but as the petrographic term "trap" or "traprock" for all dense, dark igneous rocks that look like basalt.

Tephra Cones and Composite Cones (or Strato-Volcanoes)

Most historic volcanoes are characterized by explosive rather than effusive activity. Rittmann (1962, p. 153) defined the percentage of tephra in the total ejecta (tephra plus lava) as the *explosion index* of volcanicity and showed that for the circum-Pacific, Indonesian, and Caribbean volcanoes, the explosion index generally has exceeded 90 percent. Others have noted that the entire Quaternary Period has been a time of great explosive volcanism (McBirney, et al., 1974; Kennett and Thunell, 1975). None of these estimates include the enormous amount of quiet extrusive basalt volcanism along the midocean ridges, however (Chap. 21).

Mixtures of tephra and lava, erupted by the explosive Vulcanian or Peléan activity, build cones with radial slopes at or less than the angle of repose of the erupted material. All proportions of tephra and lava are found in the **composite cones** that form the earth's most perfectly symmetrical landforms (Fig. 4-10).

Figure 4-10.   Mount Mayon (height: 2400 m) in southern Luzon, Philippines; reputedly the most perfectly symmetric volcanic cone. (Photo: Philippine Department of Tourism.)

Small *tephra cones* have straight, steep sides at the angle of repose of tephra and scoria (Porter, 1972) (Fig. 4-11). They are permeable enough to resist erosional dissection by streams, but straight gullies may descend radially from their rims. Larger cones cannot be built entirely of scoria or tephra, because the material is not strong enough to support continued summit eruptions. Lava flows that mantle the slopes help to support the cone. More importantly, several sets of systematic intrusive dikes and sills (Fig. 4-12) are integral parts of large volcanic edifices, and these, more than the eruptive lava flows, provide a strong framework to the cone. The intermingling of tephra, lava flows, intrusive pipes, dikes, and sills makes the structure of large volcanic cones too complex for any simple designation of lava or tephra cone, so they are called *composite cones*, or less commonly, *strato-volcanoes*. Their complex internal structure plays an important controlling role in their subsequent erosional dissection (Chap. 11).

The magnificent symmetry of composite cones has been the subject of much scientific and esthetic inquiry. Their smooth, concave-up profiles (Fig. 4-10) are usually called *logarithmic* or *exponential curves* and are attributed to summit lava flows, slumping, or angle-of-repose conditions (Cotton, 1944, pp. 235–36; Macdonald, 1972, pp. 281–82). However, an initial, constructional concave-up profile can be predicted from the pattern of fallout from air-borne tephra (Fig. 4-5). If a tephra cloud of uniform-size ash spreads with a prevailing wind over a wedge-shaped sector of a cone, the ratio of ash accumulation at any two places beneath the cloud will be the reciprocal of their distances from the vent. The resulting curve is hyperbolic, not

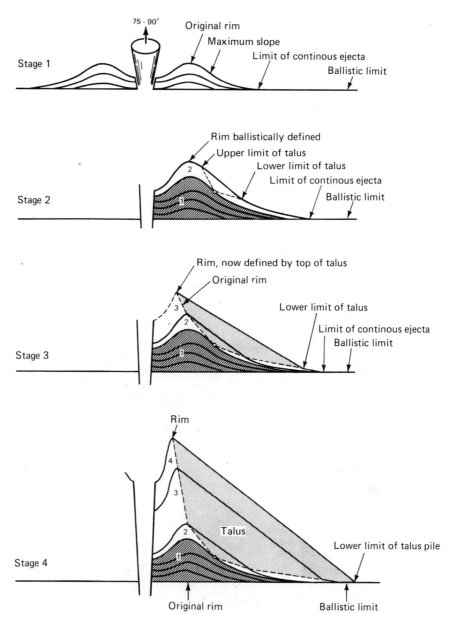

Figure 4-11. Four major stages in the development of tephra cones. Stage 1: Tephra ring with shape controlled by ballistic trajectories of particles. Stage 2: Angle-of-repose talus forms a straight midslope segment. Stage 3: As cone grows, talus extends upward to crater rim. Stage 4: Talus extends from rim to beyond ballistic limit. From this stage until the morphological capacity of the cone is exceeded, the slope angle remains constant. (Redrawn from McGetchin, et. al., 1974, Fig. 14; © 1974 by the American Geophysical Union.)

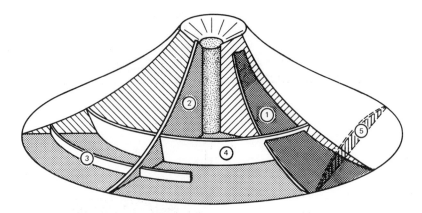

Figure 4-12. Systems of dikes and sills combine with lava flows to strengthen large composite cones. (1) Mantle sill or buried flow, (2) radial dike, (3) ring dike, (4) cone sheet, (5) peripheral dike. (Redrawn from Rittmann, 1962, Fig. 73.)

Figure 4-13. Wilson Butte, a tholoid or plug dome at Mono Craters, California. Diameter of base: 1 km. Surface features include explosion craters, spines, and an encircling talus. (Photo: E. I. Smith.)

logarithmic. Complexities of size-sorting, fallout rates, and shifting wind patterns complicate the pattern (Fisher, 1964), but the flanks of many composite cones rise toward the vent in smooth hyperbolic curves until the angle of repose is reached, then become straight. The genetic implication is that the shape of composite cones is established by tephra fallout but is preserved by the lava flows and various intrusive structures that support the edifice.

Plug Domes, Cumulo-Domes, Tholoids, and Spines

Very viscous lava of acidic composition may surge upward into a crater but be too stiff to flow and too degassed to explode. The result is an **endogenous dome** (Table 4-1), a "skinned-over," bulbous protrusion of lava from the vent. These domes vary greatly in size and shape from the awesome "shark-fin" spine that gradually extruded from Mount Pelée for nine months after the St. Pierre disaster and eventually reached a height of more than 300 m above the crater, to the mass of Mount Lassen, which is a dome about 1.6 km in diameter and over 600 m high. **Tholoid** (dome-shaped) is the term generally used for a flattened hemispheroidal dome within a crater rim (Fig. 4-13). **Cumulo-domes** or **plug domes** are more irregular or cylindrical masses of rigid lava that form hills or low domal mountains (Smith, 1973). Sometimes their surfaces are cratered, or covered with small flows and spinelike viscous extrusions, proving that they are not entirely endogenous.

The surface of an endogenous dome is constantly stretched by continued inflation, so blocks or breccia accumulate around its base or on its flanks and may nearly bury the dome. The breccia is usually of obsidian glass and forms straight taluses radiating from the sides of the dome. Rarely, the striated, polished surface of an extruded plug will be visible above its talus, and may solidify, preserving the marks of extrusion on its flanks.

Plug domes should not be confused with **volcanic necks,** which are the result of long-continued erosion of a composite cone to expose the solidified lava-filled conduit or neck. A plug dome is a constructional landform; a volcanic neck is a structurally controlled erosional remnant. Both may be present in the same volcanic region, as in the Auvergne region of France. There, the same term, *Puy,* is given to domes and eroded necks.

Maars, Pumice Rings, and Tuff Rings

Extremely explosive monogenetic eruptions blast out broad, shallow craters below general ground level with pumice or tuff rings around them (Ollier, 1967). The Eifel district of Germany, at the northern end of the Rhine graben (p. 44), is the type locality for these landforms. The name **maar** (not to be confused with the lunar *mare*) refers to the shallow lake that fills the crater in a humid climate such as western Europe's. Maars (Germanic plural: *maare*) and their enclosing ring walls are surrounded by widespread thin sheets of tephra, especially in the prevailing downwind direction. If there is no central lake, such an explosive eruption results in no

Figure 4-14.   Crater Elegante, a maar in the Pinacate volcanic field, Sonora, Mexico (Arvidson and Mutch, 1974). Pre-explosion tephra and lava beds are exposed on the crater walls, overlain by a tephra ring. (Photo: R. E. Arvidson.)

geomorphic expression other than a low rampart or ring wall, grading radially outward into thin tephra, and inward into a pumice-covered shallow crater underlain by shattered country rock (Fig. 4-14).

Purely gas explosions produce craters surrounded by little or no ejecta. A lake may form in the resulting crater (a *gas maar*), or the geologic evidence for the eruption may be nothing more than an area of shattered rock. If volcanic breccia can be identified, the dike or stocklike intrusive mass is called a *diatreme*. Some large circular areas of shattered sedimentary rocks in the central United States were once thought to have resulted from gas eruptions, but because of the total lack of volcanic rocks, they were called *cryptovolcanoes*. Most of them are now interpreted as ancient meteorite impact scars.

Large Explosion Craters and Calderas: Ignimbrite Sheets

Vulcanian or Katmaian types of explosive activity, or the even more explosive ultravulcanian or phreatic type, may create a "constructional" landscape of catastrophically destructive appearance. A number of large composite cones are known to have had their entire summits blown off or subside inward subsequent to a major explosion. The subsidence hypothesis is favored, because the mass of ejecta is substantially less than the volume of the resulting explosion crater or caldera, and also because as much as 95 percent of the tephra traceable to a source in an explosion caldera is derived from new magma rather than older volcanic or other kinds of rock (Macdonald, 1972, p. 305).

The distinction between craters and calderas was arbitrarily set at a diameter of 1 mile (1.6 km) by Macdonald (1972, p. 290). Although arbitrary, this dimensional distinction is related to the morphological capacity concept mentioned earlier in this chapter, in that summit craters up to a mile in diameter are normal components of the constructional morphology of composite cones, whereas calderas larger than that are probably always due to subsidence during or following a Katmaian-type terminal eruption.

Large explosion craters and calderas should be distinguished from the pit craters and summit calderas of basalt shields (Fig. 4-9) that are the site of nonexplosive lava lakes or lava fountains. Explosion craters are surrounded by tephra sheets that thicken and become coarser toward the crater. The inward-facing walls of explosion craters are often slump scars or fault scarps of blocks that subsided into the funnel-shaped opening prepared by an explosion, so the resulting crater is substantially broader than the actual explosive vent.

Calderas are huge basins, ranging in diameter from the defined minimum of 1.6 km to the largest one yet recognized, 70 km long and 45 km wide, largely buried beneath the thick ignimbrite sheets of the Yellowstone Plateau (Eaton, et al., 1975). They may be on the summits of composite cones. For example, the site of Crater Lake, Oregon, is a caldera 8–10 km in diameter and 1100 m deep that was formed by the Mount Mazama tephra eruption 6700 years ago. Other calderas are so large that the area of subsidence includes whole clusters of volcanic vents, or grabenlike elongate trenches across ignimbrite plateaus. Calderas are bounded by inward-facing fault scarps, and tilted fault blocks or horsts may rise above the basin floors. Late-stage scoria cones or basalt flows may build minor volcanic landforms on the floor of a caldera. In humid regions, calderas are frequently the sites of deep lakes.

The largest calderas are those associated with massive ignimbrite sheets (Marshall, 1932). The best hypothesis for this relationship is that a large subsurface reservoir of magma persists long enough to become compositionally stratified, with the most acidic, viscous, and gas-rich fraction near the top. When the magma begins to subside within the chamber (by a cause and to a place unknown), the pressure drop causes a massive effusion of pumice and glowing hot tephra along fissures that are associated with the subsidence of the caprock over the magma chamber. In a geyserlike self-perpetuating sequence, the explosive eruption lowers the pressure still more and further eruptions occur. In some of the great ignimbrite sheets, the later eruptives are more basic than the earlier, showing that deeper levels of more basic magmas were progressively expelled. The quantities are staggering. The Mazama tephra, for instance, totals at least 41 km^3, although the resulting caldera volume is 50 percent greater than that (Macdonald, 1972, p. 307). Even larger are some of the calderas and ignimbrite sheets of Colorado and New Mexico, where late Cenozoic explosive eruptions were common. The Valles Caldera in the Jemez Mountains of New Mexico erupted at least two welded tuffs of about 200 km^3 each during early Pleistocene time, then subsided to form a basin 20–25 km in diameter and 600 m deep, with a central mountain complex 12–16 km in diameter and over 900 m high (Smith and Bailey, 1968). The Timber Mountain Caldera in Nevada, of Pliocene age, erupted an estimated 10^3 km^3 of ignimbrite. Commonly, these large calderas had later dome mountains form in their central basins, with minor volcanism along a graben on the crest of the central dome. Such forms were named *resurgent caldrons* (or caldera) by Smith and Bailey (1969).

The Yellowstone Plateau was built during three great cycles of Katmaian volcanism, each of which culminated in the eruption of massive ignimbrite sheets. They occurred approximately 1.9, 1.2, and 0.6 million years ago. The huge caldera was filled by several thousand meters of tephra and rhyolite flows (Eaton, et al., 1975). Individual ash beds totaling hundreds of meters in thickness cooled as single units, with columnar joints as well developed as in lava flows. Tephra from the multiple Yellowstone eruptions is well known in the older Pleistocene deposits of the Great Plains from Texas to Kansas (p. 407).

REFERENCES

ARVIDSON, R. E., and MUTCH, T. A., 1974, Sedimentary patterns in and around craters from the Pinacate volcanic field, Sonora, Mexico: Some comparisons with Mars: *Geol. Soc. America Bull.*, v. 85, pp. 99–104.

BAYLY, B., 1968, *Introduction to petrology:* Prentice-Hall, Inc., Englewood Cliffs, N.J., 371 pp.

BORCHARDT, G. A., HARWARD, M. E., and SCHMITT, R. A., 1971, Correlation of volcanic ash deposits by activation analysis of glass separates: *Quaternary Res.*, v. 1, pp. 247–60.

BORCHARDT, G. A., NORGREN, J. A., and HARWARD, M. E., 1973, Correlation of ash layers in peat bogs of eastern Oregon: *Geol. Soc. America Bull.*, v. 84, pp. 3101–8.

BULLARD, F. M., 1976, *Volcanoes of the earth:* Univ. Texas Press, Austin, Tex., 579 pp.

CHOUBEY, V. D., 1973, Long-distance correlation of Deccan basalt flows, central India: *Geol. Soc. America Bull.*, v. 84, pp. 2785–90.

———, 1974, Long-distance correlation of Deccan basalt flows, central India: Reply: *Geol. Soc. America Bull.*, v. 85, pp. 1008–10.

COTTON, C. A., 1944, *Volcanoes as landscape forms:* Whitcombe and Tombs Ltd., Christchurch, N. Z., 416 pp. (Republished with minor corrections, 1952.)

DAVIS, W. M., 1905, Complications of the geographical cycle: *Internat. Geog. Cong., 8th, Washington, 1904, Repts.*, pp. 150–63. (Reprinted 1954 in *Geographical essays:* Dover Publications, Inc., New York, pp. 279–95.)

DAWSON, J. B., 1962, Geology of Oldoinyo Lengai: *Bull. volcanol.*, v. 24, pp. 349–87.

DUTTON, C. E., 1884, Hawaiian volcanoes: *U.S. Geol. Survey 4th Ann. Rept.*, pp. 81–219.

EATON, G. P., CHRISTIANSEN, R. L., IYER, H. M., PITT, A. M., MABEY, D. R., BLANK, H. R., ZIETZ, I., and GETTINGS, M. E., 1975, Magma beneath Yellowstone National Park: *Science*, v. 188, pp. 787–96.

FISHER, R. V., 1964, Maximum size, median diameter, and sorting of tephra: *Jour. Geophys. Res.*, v. 69, pp. 341–55.

GREELEY, R., and HYDE, J. H., 1972, Lava tubes of the cave basalt, Mount St. Helens, Washington: *Geol. Soc. America Bull.*, v. 83, pp. 2397–418.

GREEN, J., and SHORT, N. M., 1971, *Volcanic landforms and surface features: A photographic atlas and glossary:* Springer-Verlag New York, Inc., New York, 522 pp.

GUPTE, R. B., KARMARKAR, B. M., KULKARNI, S. R., and MARATHE, S. S., 1974, Long-distance correlation of Deccan basalt flows, central India: Discussion: *Geol. Soc. America Bull.*, v. 85, pp. 1007–8.

KENNETT, J. P., and THUNELL, R. C., 1975, Global increase in Quaternary explosive volcanism: *Science*, v. 187, pp. 497–503.

LEMASURIER, W. E., 1972, Volcanic record of Antarctic glacial history: Implications with regard to Cenozoic sea levels, *in* Price, R. J., and Sugden, D. E., eds., *Polar geomorphology:* Instit. Brit. Geographers Spec. Pub. 4, pp. 59–74.

MACDONALD, G. A., 1972, *Volcanoes:* Prentice-Hall, Inc., Englewood Cliffs, N.J., 510 pp.

MACHIDA, H., 1967, The recent development of Fuji Volcano, Japan: *Tokyo Metropolitan Univ. Geog. Repts.*, no. 2, pp. 11–20.

MALDE, H. E., and POWERS, H. A., 1962, Upper Cenozoic stratigraphy of western Snake River plain, Idaho: *Geol. Soc. America Bull.*, v. 73, pp. 1197–220.

MARSHALL, P., 1932, Notes on some volcanic rocks of the North Island of New Zealand: *New Zealand Jour. Sci. and Technology*, v. 13, pp. 198–200.

MCBIRNEY, A. R., SUTTER, J. F., NASLUND, H. R., SUTTON, K. G., and WHITE, C. M., 1974, Episodic volcanism in the central Oregon Cascade Range: *Geology*, v. 2, pp. 585–89.

MCGETCHIN, T. R., SETTLE, M., and CHOUET, B. A., 1974, Cinder cone growth modeled after northeast crater, Mount Etna, Sicily: *Jour. Geophys. Res.*, v. 79, pp. 3257–72.

MOORE, J. G., 1975, Mechanism of formation of pillow lava: *Am. Scientist*, v. 63, pp. 269–77.

NICHOLS, D. R., and YEHLE, L. A., 1961, Mud volcanoes in the Copper River Basin, Alaska, *in* Raasch, G. O., ed., *Geology of the Arctic:* Univ. of Toronto Press, Toronto, pp. 1063–87.

NICHOLS, R. L., 1939, Viscosity of lava: *Jour. Geology*, v. 47, pp. 290–302.

NORDLIE, B. E., 1973, Morphology and structure of the western Galápagos volcanoes and a model for their origin: *Geol. Soc. America Bull*, v. 84, pp. 2931–55.

OLLIER, C. D., 1967, Maars, their characteristics, varieties, and definition: *Bull. volcanol.*, v. 31, pp. 45–73.

————, 1969, *Volcanoes*: Australian National Univ. Press, Canberra, 177 pp.

PORTER, S. C., 1972, Distribution, morphology, and size frequency of cinder cones on Mauna Kea volcano, Hawaii: *Geol. Soc. America Bull.*, v. 83, pp. 3607–12.

RITTMANN, A., 1962, *Volcanoes and their activity* (transl. by E. A. Vincent): Wiley-Interscience, New York, 305 pp.

SMITH, E. I., 1973, Mono Craters, California: A new interpretation of the eruptive sequence: *Geol. Soc. America Bull.*, v. 84, pp. 2685–90.

SMITH, R. L., and BAILEY, R. A., 1969, Resurgent cauldrons, *in* Coats, R. L., Hay, R. L., and Anderson, C. A., eds., Studies in volcanology: *Geol. Soc. America Mem. 116*, pp. 613–62.

SNEAD, R. E., 1964, Active mud volcanoes of Baluchistan, West Pakistan: *Geog. Rev.*, v. 54, pp. 546–60.

SPARKS, R. S. J., SELF, S., and WALKER, G. P. L., 1973, Products of ignimbrite eruptions: *Geology*, v. 1, pp. 115–18.

THORARINSSON, S., 1951, Laxárgljúfur and Laxárhraun: A tephrochronological study: *Geograf. Annaler*, v. 33, pp. 1–88.

WENTWORTH, C. K., and MACDONALD, G. A., 1953, Structures and forms of basaltic rocks in Hawaii: *U.S. Geol. Survey Bull. 994*, 98 pp.

WHITE, D. E., 1955, Violent mud-volcano eruption of Lake City Hot Springs, northeastern California: *Geol. Soc. America Bull.*, v. 66, pp. 1109–30.

PART II

Subaerial Destructional (Erosional) Processes and Erosional Landforms

However they may be constructed, subaerial landforms cannot stand immutable. The rocks that make them were formed in physical and chemical environments that are grossly out of equilibrium with the oxygen-rich, hydrous atmosphere into which they are thrust or erupted by constructional processes, and they are sure to crumble or dissolve. Landscapes are carved from structures by agents of chemical and physical weathering and erosion. Water, which occurs naturally as both a chemically reactive fluid and an erosive solid at earthly temperatures and pressures, is the primary agent. The energy drive is provided by solar radiation, converted into heat at the earth's surface and eventually reradiated into space. Gravity and internal heat are secondary sources of energy for geomorphic change. We call these processes *destructional, degradational,* or *erosional,* because they tend to destroy the edifices that are constructed by diastrophism and volcanism. The eroded debris accumulates in tectonic basins on the continents or is washed to the sea by rivers. Distinctive landforms are built by the accumulated waste, but the far greater proportion of the subaerial landscape consists of erosional remnants, still in the process of reduction.

Present-day processes conform to universal thermodynamic laws but are also influenced by the steady evolution of the earth's lithosphere, hydrosphere, atmosphere, and biosphere. We can see a geologic record of about one billion years of weathering and erosional processes similar to those now operating, although different to the degree that different forms of life were involved in the processes.

The certainty of geomorphic change by subaerial processes acting on geologic structures has been the subject of philosophical comment since the beginning of written history, at least. A Chinese scholar, in 240 B.C., observed: "Nothing under heaven is softer or more yielding than water; but when it attacks things hard and resistant there is not one of them that can prevail [Waley, 1935, p. 238]."[1] One of the oldest geomorphic deductions is: "Every valley shall be exalted and every mountain and hill shall be made low; and the crooked shall be made straight, and the rough places plain [Isaiah, XL, 4]."

This philosophical tradition of skepticism about the dubious durability of the "everlasting" hills was summarized in a single statement by J. W. Powell, one of the great geomorphologists who learned his trade while exploring the mountains of the western United States. *Powell's dictum* is: "We may now conclude that the higher the mountain, the more rapid its degradation; that high mountains cannot live much longer than low mountains, and that mountains cannot remain long as mountains: they are ephemeral topographic forms. Geologically all existing mountains are recent; the ancient mountains are gone [Powell, 1876, p. 193]."[2]

New knowledge of the absolute rates of erosion make it clear how true Powell's dictum really is. All mountains except volcanoes are carved from either late Cenozoic orogenic belts or areas of more ancient terranes that were uplifted by late Cenozoic epeirogeny. The sediment loads carried in modern rivers demonstrate that the water-sheds ought to be reduced to near sea level in 10–12 million years. The great issue before geomorphologists today is not whether erosional processes can accomplish the work they are alleged to have done but how continents have managed to survive in spite of such enormous denudation. With this outlook, geomorphologists have joined sedimentologists, volcanologists, and tectonists in a reexamination of the Cenozoic history of continents and ocean basins, using the inferred motions of lithospheric plates as a guiding theory. The explanatory description of major landforms becomes an account of complexly interrelated or alternating episodes of erosion and diastro-phism, with continental ice sheets and other climate-controlled processes adding fascinating complexity.

The following eight chapters consider the processes and landforms of subaerial weathering and fluvial erosion and make deductions about sequential changes in these forms through time. Although climatic influence on fluvial landforms must be considered, the distinctively climate-controlled processes and landforms are reserved for Part III.

[1]See References, Chapter 5.
[2]See note 1 above.

CHAPTER 5

Energy Flow
in Geomorphic Systems

THE SYSTEMS CONCEPT IN GEOMORPHOLOGY

All physical phenomena may be conceived as being organized into **systems**, a system being defined as a collection of objects and the relationships between those objects. **Closed systems** are those that have boundaries across which no energy or matter moves; **open systems** have a flow of energy and matter through their boundaries. A closed system must change toward a time-independent equilibrium state, according to the second law of thermodynamics, in which the ratios of various physical phases remain constant, the amount of free energy (such as chemical potential, potential energy of position, etc.) is minimal, and the entropy (unavailable energy, often visualized as the diffuse heat of random molecular motion) is maximal. An open system, on the other hand, may achieve a **steady state**, wherein the system and its phases remain in a constant condition even though matter and energy enter and leave it (von Bertalanffy, 1950).

Only the most simple geomorphic processes may be studied as closed systems. Certain weathering reactions, such as solution of limestone by acidified rain water, can be duplicated by adding measured quantities of the reactants to a test tube and observing that after a period of time, predicted amounts of new compounds have formed and are in stable equilibrium with each other. Hot, newly erupted lava radiates and conducts heat energy to its surroundings until the temperature gradient is reduced to zero and the mass has solidified. Then it ceases to give off heat. Even in these simple examples of physical and thermal systems, it is obvious that the natural phenomena are much more complex than these brief statements imply.

The entire subaerial portion of the earth's surface can be profitably regarded as an open system (Chorley, 1962). Matter is supplied by diastrophism or volcanism, with an additional trivial net increase of mass from meteoritic infall. Energy is derived from solar radiation, gravitation, rotational inertia, and internal heat. Surface processes erode the rock structures, and the eroded debris is deposited elsewhere, usually below sea level. It is conceivable that a steady state could be reached in such a complex open system so that the form of the landscape would not change even though the mass that composes it was constantly changing and energy was being constantly expended. One of the great issues of geomorphology is whether landforms change sequentially through time or whether they achieve steady states and then change only slowly or not at all.

Compare a landscape (a "geomorphic system") with a rotary kiln used to manufacture Portland cement (Fig. 5-1). Simple raw materials, primarily limestone and shale, are fed into the upper end of a rotating, heated drum. The fragments tumble and are pulverized within the drum as they move toward the lower end, but in the presence of heat they chemically react to form the desired end product, a powder that reacts with water to form calcium silicate minerals of great adhesive strength. The heat that is not stored in the newly created chemical compounds is liberated as wasted energy, along with the mechanical frictional losses. Mechanical energy is provided by a gravity potential and also by the kinetics of rotation. It can be seen that if the supply of raw materials is adjusted to the supply of mechanical and thermal energy, a desired product is produced at a desired rate. If material is supplied at a more rapid rate, the energy supply also has to be increased to maintain the flow of the desired product. A steady state at various levels of net flow can be obtained by mutually adjusting the several variables of material and energy supply in this simple open system.

A landscape above sea level (or above the geoid) is similar to the rotary kiln model. Rainfall sweeps weathered rock fragments downhill to the sea in rivers. The rate of flow is a function of the climatic variables of temperature and precipitation, the gradient is established primarily by diastrophism, and the intensity of chemical and physical change of the transported debris is a function of the time of travel through the system. As long as the energy and material flow are adjusted, the system stays in a steady state.

Suppose, however, that the intake end of a rotary kiln had been unwisely built on the supply dump, which is finite. As the supply of raw material (the constructional landform) is consumed, the longitudinal gradient of the kiln is reduced. In order to maintain quality control during a longer residence time in the kiln, the heat must be reduced. The rate of cement production gradually declines. By proper control, even such a poorly designed system would appear to be in a steady state over any brief period of observation, although on an extended time scale the potential energy of the system would be declining. This is analogous with erosional landscapes that are not being renewed by constructional processes. Although rivers may appear to be *graded*, that is, nicely adjusted open systems in steady states doing just the work they are required to do, on a longer time scale they are seen to be open systems of declining energy, only apparently in a steady state during short periods of observation.

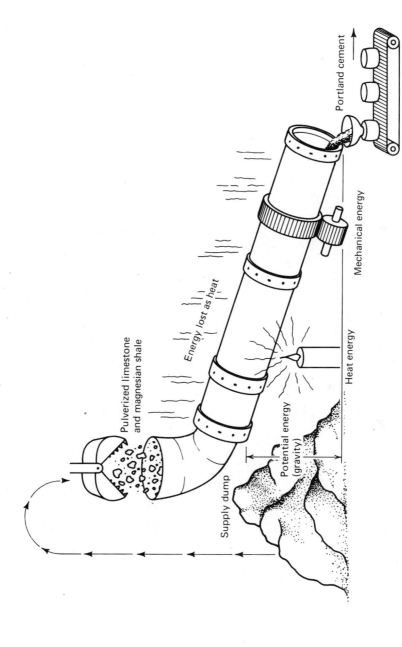

Figure 5-1. The rotary kiln analog to an open geomorphic system.

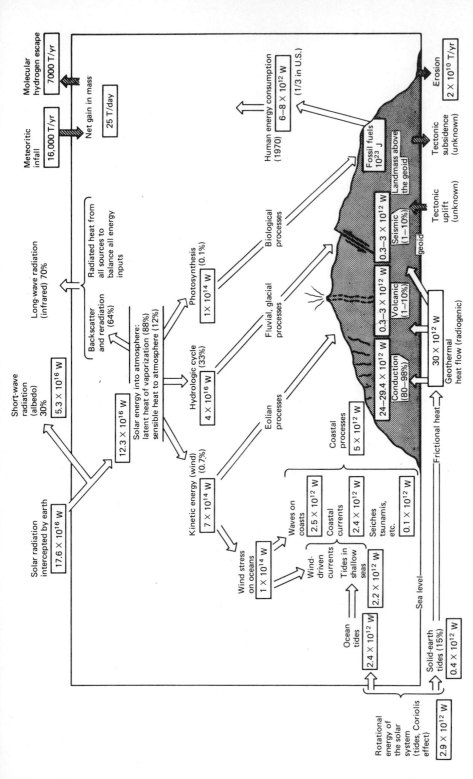

Figure 5-2. Energy and matter flux in the geomorphic system. System boundaries are the geoid and the air and water interfaces. Energy enters the system as solar radiation, rotational energy, and radiogenic heat. The rate of outgoing infrared radiation balances all incoming energy. Matter enters the system by tectonic uplift above the geoid and leaves by tectonic subsidence and erosion. Trivial amounts of matter enter as meteorites and escape as molecular hydrogen. (Flohn, 1969; Hubbert, 1971; Inman and Brush, 1973; other sources cited in this chapter.)

We assume that the mass of the earth is constant, ignoring the continuous small increment of meteoritic dust and the extremely rare impact of large extraterrestrial fragments. In principle, it should be possible to define the geoid (Chap. 2) as the boundary of an open system and to monitor the movement of rock material upward through it by diastrophism and volcanism and back downward through it by diastrophism and erosion (Fig. 5-2). Because the geoid is in equilibrium, the masses transported through this reference surface must balance, and lateral transport systems must also operate above and below the geoid to maintain isostatic equilibrium. The study of the geomorphic system has not yet progressed to the point where such mass balances can be calculated, however.

The energy balance in the geomorphic system is becoming quite well known owing to related research in biology, meteorology, astronomy, and geophysics. The earth's surface has not grown significantly colder or warmer in geologic time, so it must dissipate to space as much heat as it receives from various sources. In the universe as a whole, gravity is the dominant form of energy (Dyson, 1971), greatly exceeding heat, light, and nuclear energy. Within our geomorphic system, solar radiation (in itself a result of the gravitational force that sustains thermonuclear fusion in the sun) dominates all other energy sources. A diagram of the energy flow at the surface of the earth (Fig. 5-2) shows the major sources of energy for geomorphic change.

It is important for the next generation of geomorphologists to understand better the energy potentials available for any system they wish to study. Many poor generalizations have been made in the past that were not consistent with energy considerations. Trends toward quantification require that energy as well as material rates and balances be calculated. The remainder of this chapter is an evaluation of the energy flow in the geomorphic system, essentially a detailed review and discussion of Figure 5-2. A few rare or trivial forms of energy that might produce landforms, such as meteorite impact, lightning discharge, cosmic radiation, and magnetic fields, are not considered further. Humans, who have become significant geomorphic agents during the last few thousand years, are now using energy on a scale comparable to that released by wind, tides, and geothermal heat. The impact of humans is unique, because they can concentrate power so intensely.

THE POWER AVAILABLE FOR GEOMORPHIC CHANGE

Figure 5-2 outlines the rate of energy conversion (power) near the surface of the earth. Solar radiation, mostly in the visible and near-infrared wavelengths, supplies about 99.98 percent of the energy entering the geomorphic system. About 30 percent of the solar radiation is directly reflected, mostly by clouds (the planetary **albedo**, or brightness, is the measure of this reflected energy). Most of the nonreflected energy is converted to thermal vibration of air, water, and mineral molecules as sensible heat, activating chemical and physical reactions and possibly even heating rocks until they fracture. A very significant fraction is absorbed as latent heat of vaporization by the abundant water on the surface of the earth and is transported by water vapor in the atmosphere until it is released by condensation in clouds. Much smaller fractions of the energy are converted directly to kinetic motion of air and water masses, driving convective and advective currents in the atmosphere and oceans, or are absorbed by photosynthetic plants. A very small amount of the energy drives geomorphic processes. All of this energy is ultimately returned to outer space by reradiation from the moderately warmed earth.

Solar radiation provides energy for geomorphic change at a rate 4000 times faster than all other energy sources combined, yet these other sources cannot be ignored. They include the gravitational and inertial forces associated with the mass and motion of the earth, moon, sun, and other objects of the solar system, expressed at the earth's surface by tides in both the lithosphere and the hydrosphere. Also included is the heat that flows outward through the surface from the interior of the earth. Only about 1–10 percent of the earth's internal heat is released by volcanic activity. A comparable amount is released during earthquakes. The rest is conducted to the surface along the geothermal gradient from the hotter interior to the cooler surface rocks and is a small but worldwide phenomenon. In ways not yet understood, heat energy and geothermal gradients in the interior of the earth, aided by kinetic rotational forces and gravity, drive all of the diastrophic activity in the lithosphere. Any comparison of internal and external energy sources for geomorphic processes must consider that the power of the latter is 4000 times greater than the former.

One difficulty with evaluating the power of the subaerial geomorphic system is that there is no simple way to measure the *power density*, or localized concentration of energy expenditure. For instance, the small proportion of geothermal heat flow that is concentrated at volcanoes creates spectacular landforms, whereas the other 90–99 percent drives tectonic processes but makes no specific landforms. Sunlight, when fixed by photosynthesis, accumulated as fossil fuel, burned in an engine, and focused at the tip of a drill bit, can alter the landscape along a highway right-of-way at a rate that millions of years of normal incident radiant energy could not accomplish. Biological systems, especially human beings, have an exceptional ability to store and concentrate diffuse energy at the earth's surface (Häfele, 1974; Holdren and Ehrlich, 1974).

SOLAR RADIATION

A major goal of geomorphology is to understand how the immense but diffuse energy of solar radiation is converted into mechanical work that shapes landscapes. We can visualize the process in terms of a "geomorphology machine" (Fig. 5-3) in which a steam engine fired by the sun powers a series of fans, saws, files, grinders, and hydraulic jets to reduce the landscape. The analogy with a steam engine is a good one, because it is the phase changes in the abundant water on the earth's surface that convert solar energy into geomorphically significant mechanical work. Of even greater significance is the fact that solar heat keeps the surface of our planet in the temperature range in which water naturally occurs in solid, liquid, and vapor phases. In a sense,

Figure 5-3.   The "geomorphology machine." (Bloom, 1969.)

we are inside the boiler and condenser of the machine rather than witnessing the final mechanical conversion of the solar energy.

   The sun radiates energy over an electromagnetic spectrum of wavelengths that is closely approximated by a perfectly radiating black body at an absolute temperature of 5785°K (Goody and Walker, 1972). The radiant energy intercepted by a surface at the earth's mean orbital radius from the sun is about 1.4 kW/m². This value is called the **solar constant**. The earth intercepts solar energy at a rate equal to the product of the solar constant and its cross-sectional area. However, the earth is not a disk but approximately a rotating sphere, with a surface area of $4\pi r^2$ that is four times as large as the equivalent disk cross section. Thus the average input of solar radiation on each square meter of the upper atmosphere is only one-quarter of the solar constant, or 349 W/m² (Flohn, 1969, p. 10). The total power of solar radiation shown in Figure 5-2 is equal to this average energy expenditure on the total area of the earth (510 $\times$ 10¹² m²).

In addition to reflecting nearly one-third of the total incoming energy, the atmosphere acts as a selective filter to further restrict the wavelengths of solar radiation that reach the earth's surface. In the lower atmosphere, water vapor (H_2O) and carbon dioxide (CO_2) strongly absorb incoming solar radiation in specific longer infrared wavelengths. It is one of the remarkable balances of nature that the earth is warmed by the sun to just the right temperature so that it reradiates energy out through the atmosphere into space at wavelengths that are in spectral "windows" between the incoming wavelengths that are so strongly absorbed by water and by carbon dioxide. The mean terrestrial surface temperature of 288°K (15°C) is a result of the balance between both the wavelengths and the intensity of incoming and outgoing radiation. As long as the earth has had its present atmospheric composition, the mean surface temperature cannot have fluctuated more than a few degrees.

Terrestrial Thermal Gradients

Solar energy is absorbed and reradiated unequally by the lower atmosphere, hydrosphere, and lithosphere. The result is a complex of thermal gradients that cause atmospheric and oceanic circulation and produce climatic regions on the earth's surface. Because climate is such an important aspect of geomorphic processes, the major thermal energy gradients need specific discussion.

Latitudinal Gradients.   For several reasons, most of the solar energy is received by the earth within the tropics. First, the sun is high in the tropical sky at all seasons, and the reflectance is least when the incoming radiation is at a high angle to the surface. Second, the tropics contain large oceanic areas, favoring absorption of heat by water. Third, the atmosphere is cloudy and humid, absorbing reradiated earth heat. Fourth,

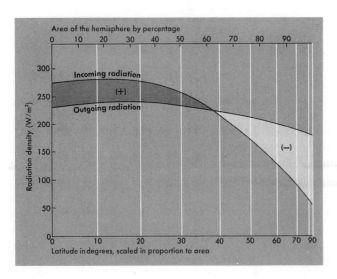

Figure 5-4. Radiation balance of the northern hemisphere. Within 38° north of the equator, energy is received at a faster rate per unit area than it is reradiated; north of 38° each unit area radiates at a faster rate than it receives energy from the sun. The southern hemisphere has a similar, but not identical, balance. (After Houghton, 1954.)

the surface area between successive parallels of latitude is larger near the equator than near the poles. Almost 40 percent of the earth's surface is within the tropics. For all these reasons, in the lower latitudes more energy is received from the sun than is reradiated back into space (Vonder Haar and Suomi, 1969).

The polar regions receive very little of the total solar radiation. The low incident angle of the sunlight, even during the time of 24-hour daylight in high latitudes, promotes reflection. Both the Arctic Ocean and the Antarctic Continent are covered with ice and snow, which further aid in reflecting the incoming radiation. It is true that for more than a month of midsummer when the sun shines continuously, each polar region in turn receives more radiation per day than is ever received per day in the tropics, but the season is very brief, and for the rest of the year the energy input at the poles is far lower than the output. Figure 5-4 summarizes the latitudinal distribution of incoming and outgoing radiation in the northern hemisphere and shows that at a latitude of about 38° north, the annual heat budget of the earth is balanced. The southern hemisphere is assumed to have a similar budget. The excess heat received in the tropics is carried poleward by winds and ocean currents, and it balances the net deficit between incoming and outgoing radiation at high latitudes.

Altitudinal Gradients. The atmosphere is reasonably transparent to incoming solar radiation of visible and near-infrared wavelengths but is strongly absorbent in certain wavelengths of the outgoing longer wave infrared radiation that leaves the solid earth. Thus the atmosphere is heated from beneath, and a *normal temperature lapse rate* of about −6.4°C/1000 m is recorded during ascent from sea level. With increasing altitude, the mean annual temperature on a mountainside steadily decreases until it is below freezing. If precipitation is adequate, a permanent snow line is reached. The altitude of the regional snow line varies from nearly 6 km above sea level in the subtropical dry belts to sea level in the polar regions. Near the equator, the snow line dips to about 5000 m because of the greater cloudiness and precipitation there than in the adjacent clear, dry subtropical belts.

If a mountain is high enough to intersect the regional snow line, it will carry a year-round snow cap and possibly glaciers (Chaps. 16–18). The sediment loads in streams draining glaciated and nonglaciated mountain peaks suggest that the presence of glaciers in its headwaters increases a stream's erosion rate by a factor of 4 to 25 (Table 12-3). Thus altitude alone can be a powerful factor in determining the effectiveness and even the kind of geomorphic processes acting on a landform. Powell's dictum (p. 84) easily comes to mind.

Although mountain peaks are cold, they may nevertheless be subject to a very high energy flux. Unimpeded by most clouds, haze, and atmospheric dust, the incoming radiation intensity on a mountain greatly exceeds the average at sea level. Reradiation is also rapid. Sunlit rock surfaces at high altitudes may be much warmer than the air only a few centimeters away, but as soon as the sun sets, the rocks reradiate into cold, dry space and cool rapidly. Freeze and thaw cycles are frequent and intense (Chap. 15).

Seasonal Gradients; Land–Sea Contrasts. Because the earth's plane of daily rotation is inclined $23\frac{1}{2}°$ to the orbital plane, the sun is vertically overhead across a wide intratropical belt during some part of the year. The well-known phenomena of seasonality result. In extratropical regions, the names "winter" and "summer" express the dominant role of temperature change. Within the tropics, where temperatures are generally high, seasonality is defined by precipitation maxima and minima, so that rainy seasons, or wet monsoons, alternate annually with dry seasons of several months duration.

Many geomorphic processes are directly associated with seasonality. The annual alternation of torrential rain and prolonged drought creates a distinctive savanna landscape. Subtropical desert landforms are shaped by flash floods and wind. Tundra regions have long smooth slopes formed by soil creep and earth flow in a seasonally thawed, water-saturated surface layer that overlies perennially frozen ground. Climate-controlled geomorphic processes give such a distinctive shape to landscapes that climatic regions can be used to organize an alternative approach to systematic geomorphology (Part III).

Associated with seasonal climatic changes are the temperature contrasts and gradients between land and sea. Maritime climates are characterized by moderate daily and seasonal temperature contrasts, because the abundance of water and water vapor provides a "buffer" of latent heat. The continental interiors, on the other hand, have extreme seasonal temperature contrasts. The greatest seasonal contrasts are not in the subtropical deserts but in high-latitude continental interiors. The greatest range of mean annual air temperature on earth is 62°C, in Siberia. In contrast, both the annual and daily temperature range on small tropical islands is commonly less than 3°C. There, temperature change is totally irrelevant to geomorphic processes. In spite of their moderate temperatures, the general high humidity of maritime climates promotes chemical weathering, and in cold maritime regions, glaciers are likely to be significant geomorphic features of mountains and plateaus.

Daily Temperature Fluctuations. Daily temperature changes are especially marked in tropical highlands and in subtropical deserts. In the Sahara, for instance, the daily temperature fluctuation in a shady location is over 60°C. The daily temperature range of exposed rock surfaces exceeds 120°C. These fluctuations are reliably reported to shatter rocks (p. 109). In nondesert regions, night also is normally cooler than day, but reradiation back to earth from the humid atmosphere keeps the diurnal range small. Geomorphic processes related to condensation of dew on rocks, or freezing, may be dominated by a daily cycle.

The Hydrologic Cycle

A very large proportion of the solar radiation intercepted by the earth is isothermally absorbed as *latent heat of vaporization* by water (Fig. 5-2). Ours is truly the hydrous planet, and water is as anomalous a compound in its thermal properties as in its chemical activity. About 2.4 kJ are required to evaporate 1 gram of water at

20°C. The large area of oceans in tropical regions ensures that much of the incoming radiation is absorbed not as an increase in sensible temperature but as latent, or insensible, heat by the change in state of water from liquid to vapor. When the water vapor condenses to form clouds of droplets, the latent heat is liberated to the surrounding atmosphere. The largest single component in the horizontal transfer of heat over the surface of the earth from equatorial to polar regions is undoubtedly this latent heat of vaporization and condensation.

On the most awesome scale, local intense evaporation and precipitation over tropical oceans spawn the funnel-like spiraling convective cells variously called *cyclones, hurricanes,* or *typhoons.* These storms draw huge quantities of heat from the surface waters of the ocean and transfer it to higher altitudes and latitudes. In the process, they generate winds and waves that cause as much geomorphic change and human tragedy as any other form of catastrophe. Tropical cyclones are an extreme example of highly concentrated solar energy.

The hydrologic cycle, or interchange of water between oceans, atmosphere, land, glaciers, and organisms, is summarized in Figure 5-5. Two kinds of information are shown: The first is an estimate of the *amounts* of water that are present in various parts

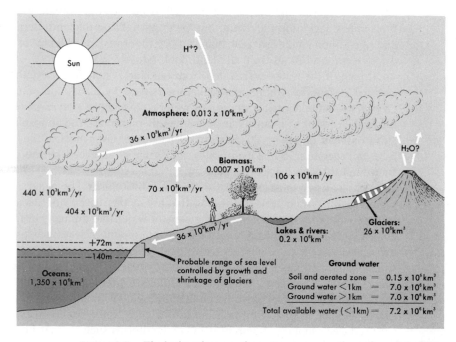

Figure 5-5. The hydrosphere, or the water on or near the surface of the lithosphere. For easy comparisons, all the amounts of water in the reservoirs are given in millions of cubic kilometers (10^6 km³); all the annual exchanges among the reservoirs (the hydrologic cycle) are given in thousands of cubic kilometers per year (10^3 km³/yr). The accuracy of the various estimates varies widely. (Compiled from various sources, including Nace, 1969; Penman, 1970; Lvovitch, 1973.)

of the outer layers of the earth; the second is the calculated *annual exchange* of water between the various reservoirs. The figure ideally portrays a steady-state open system. The energy inflow equals the outflow, and, with negligible exceptions, the amount of water in the system is constant.

First, examine the amounts of water in the system. In their order of importance, the reservoirs of water are: (1) oceans, (2) glaciers, (3) ground water, (4) lakes and rivers, (5) atmosphere, and (6) biomass (all living matter). Actually, over 97 percent of all the water near the surface of the earth is in the oceans, and most of the remainder is in glaciers. The interior of the earth must contain some water, chemically combined or dissolved in solid or molten rock, but for at least the last billion years, the amount of water at or near the surface of the earth has been approximately constant. Only a small amount of new water might be added annually by condensation of volcanic gases, and an equally small amount of water is probably lost from the earth annually by the photochemical dissociation of water vapor by solar radiation.

The hydrologic cycle is usually shown as a rate of transfer, commonly as cubic kilometers or cubic miles of water per year. It can be converted into energy or power units, too. No high degree of accuracy is claimed for any of them.

We see from Figure 5-5 that 440,000 km³ of water are evaporated from the sea annually, representing a layer 122 cm thick from the entire ocean. Most of the water precipitates back onto the ocean, but the excess falls on the land. Because more snow and rain fall on land every year than can be evaporated, 36,000 km³ of water annually drips, seeps, and flows from the land back into the sea.

Much of the water evaporated from the land is not simply evaporated from the exposed surfaces of lakes and streams but is used by plants and animals. A field of crops may annually use an amount of water equivalent to a layer 45–60 cm deep over the field. A forest of Douglas fir may annually pump into the atmosphere the equivalent of a layer of water 120 cm deep over its area. Over most of the land, rain water falls on vegetation. Some evaporates, and some is absorbed by the plant roots and transpired through leaves. Because the variables cannot be clearly distinguished, the term *evapotranspiration* is used for the collective effect. The water requirements of typical plants on the parts of the land that are neither desert nor glacier could evapotranspire all of the 70,000 cubic kilometers of water that annually evaporate from the land, so it is likely that each year a significant amount of water passes through a biologic cycle as a part of the larger hydrologic cycle. Only a small part of the evapotranspired water, perhaps 1 percent, is held at any time as living matter, and the turnover is continuous. The energy required for biological productivity is only about 0.1 percent of the total received from the sun (Woodwell, 1970), but by combining water and carbon dioxide in photosynthesis, an estimated 2×10^{11} tons of dry organic matter are produced annually. Some of it builds landforms such as peat bogs.

Glaciers store large amounts of water on land, temporarily removing it from the hydrologic cycle. If the present glaciers were to melt, sea level would rise about 72 m and submerge the most heavily populated areas of the earth. In the last two million years, continental ice caps have repeatedly developed and melted again, each time temporarily upsetting the hydrologic cycle. During maximum glaciation (Chap. 17), the amount of glacier ice on land may have been three times as great as the present ice volume, and sea level was lowered as much as 140 m, exposing most of the continental shelves (Fig. 17-14).

All of these aspects of the hydrologic cycle relate to geomorphology, but one aspect in particular must be emphasized. The average continental height is 823 meters above sea level (Fig. 2-2). If we assume that the 36,000 cubic kilometers of annual runoff flow downhill an average of 823 meters, the potential mechanical power of the system can be calculated. Potentially, the runoff from all lands would continuously generate almost 9×10^9 kW. If all this power were used to erode the land, it would be comparable to having (in the English units of power and area) one horse-drawn scraper or scoop at work on each 3-acre piece of land, day and night, year around. Of course, a large part of the potential energy of the runoff is wasted as frictional heat by the turbulent flow and splashing of water. The "geomorphology machine" has a low but definable efficiency, eroding and transporting rock debris down to the oceans (Caine, 1976).

ROTATIONAL ENERGY OF THE EARTH-MOON SYSTEM

Terrestrial Gravity

Every mass has an inherent attraction for every other mass. This weak force, called *gravitation*, is an intrinsic property of mass. Although its nature is yet to be defined, we know something about its operation. We can assume, for one thing, that the gravitational force between two large objects operates as though the entire mass of each object were concentrated at the center of the mass, or the center of gravity. Because all objects at the surface have small masses as compared to that of the planet, we usually regard the force of gravitation as a tendency for an object at the surface to fall or accelerate toward the center of the earth.

The gravitational force is reduced by the tendency of objects on this rotating spheroid to maintain straight-line motion and hence to be lifted from the surface (the centrifugal effect), especially near the equator. Objects moving horizontally appear to deviate to the right in the northern hemisphere and to the left in the southern hemisphere when viewed by an observer who is being carried in a circular path by the earth's rotation. This *Coriolis effect* is proportional to the sine of latitude and the velocity of the object. It has been claimed to cause northern-hemisphere rivers to erode their right banks preferentially, but the effect is actually much too weak. The Coriolis effect is of great importance to atmospheric and oceanic circulation, however.

Gravity, as we commonly use the word, refers to the net force of gravitation as reduced by the centrifugal and other lesser effects. The geoid is controlled purely by terrestrial gravitation and rotational forces (Fig. 2-1). It is worth noting that the word "level" as in "sea level" does not connote flatness, or planarity, on earth but rather a line or surface parallel to the geoid. A billiard table must be level, but if it were truly flat or planar, all the balls would, in principle, roll to the center and stop. We commonly diagram sea level as a straight line, as in showing the concave long profile of rivers (Fig. 9-6). In reality, of course, the bed of a major river has a convex-skyward longitudinal profile, just as has any line on the geoid. The Mississippi River, for instance, originates in Lake Itasca, Minnesota, at an altitude of 450 m above sea level. In flowing 3800 km southward to sea level, it follows a convex-skyward meridianal arc over 15° in length, more than $\frac{1}{6}$ of a quadrant. No reasonably sized scale drawing could show the difference in altitude of head and mouth of the river along the convex arc of its southward course. Furthermore, because of the oblate spheroidal shape of the earth, the mouth of the Mississippi is actually several kilometers "higher" (with respect to the center of the earth) than its source. These amusing but real considerations are easily overlooked in most geomorphic thinking but should at least be mentioned once, so that words like "up," "down," "horizontal," and "level" can be seen in their true sense.

All water that falls as rain or snow on the 29 percent of the earth's surface that projects above the sea tends to move downhill by gravity, back to the ocean. Thus every raindrop that strikes the ground has potential energy proportional to the product of its mass and the altitude above sea level at which it strikes. The few localities where dry land extends well below sea level (Death Valley, California, at -86 m; the Dead Sea at -390 m; and others) are exceptions to the rule that sea level is the limit for downhill flow of water. However, each of the basins that extends below sea level is in an arid region where few raindrops ever fall. In fact, such basins never last very long in humid regions, because they are soon filled with water to overflowing, and sea level again becomes the ultimate downward limit of water flow from the basins. The tendency of water to flow downhill is the cause of most erosional landforms.

Rock particles as well as raindrops are attracted toward the center of the earth by the force of gravity. When rock material moves downslope under the influence of gravity, but without a transporting agent such as flowing water or glacier ice, the process is called *mass-wasting* (Chap. 8). Mass-wasting includes not only such spectacular events as landslides and avalanches but also the slow, barely perceptible process of soil creep.

The small contribution of rotational forces to the massive gravity field of the earth may in some way be related to diastrophism. A correlation between earthquake frequency, volcanism, and a "wobble" of the rotational axis of the earth has been reported (Myerson, 1970; Anderson, 1974).

Extraterrestrial Gravity: Tides

The gravitational forces of the moon, sun, and nearby planets are expended as frictional energy in the earth's hydrosphere and lithosphere. The force of extraterrestrial gravity causes variations in the surface gravity field of only about 1 part in 5 million. Most of the energy generated by extraterrestrial gravitation is expended in turbulent friction of ebbing and flowing tides in shallow marginal seas (Hendershott, 1973). Because of the orbital motions of the planets and our moon, tides vary cyclically, but complexly. According to the estimates used in compiling Figure 5-2, the tidal energy expended on the lithosphere contributes about 15 percent of the geothermal heat flow. It is hard to account for the 2.9 terawatts (1 TW = 10^{12} W) of power consumed by tidal drag. Perhaps internal waves on boundary layers within the oceans contribute to the consumption. A related problem is that shallow oceanic basins have varied greatly in size and distribution, even within the late Cenozoic Era. We cannot assume that the dissipation of tidal energy has been uniform throughout time.

Tidal energy is an important factor in coastal processes (Chap. 19), and for most purposes, tides are best studied in the context of the marine environment. However, *earth tides*, with a regular semidiurnal rhythm similar in period to ocean tides but smaller in amplitude, can be measured by sensitive strain meters and gravity meters. The lithosphere apparently has two high and two low tides, with an amplitude of about 30 cm, that follow the moon in its orbit. As the earth rotates daily through the tidal "bulges," rocks are constantly but minutely flexed. The combined effect of oceanic and lithospheric tidal friction is to decrease the number of daily rotations of the earth per year by about 1 day every 10 million years (Wells, 1963). Our moon has long since "locked on" the earth with one rotation per month for the same reason, so that we never see the far side of the moon (Goldreich, 1972).

The role of lunar gravity as a geomorphic process (excluding the important tides in the hydrosphere) is not well known. Tidal friction causes the earth's axis to precess and cyclically changes the intensity of seasonal climates (p. 410). Some seismologists suspect that earthquakes occur with greater frequency during the full and new phases of the moon, when the sun's gravitational force is additive to the moon's. The hypothesis is strengthened by the discovery that "moonquakes" are recorded with greater frequency every 28.5 days when the moon is at perigee, its closest orbital approach to the earth (Latham, et al., 1971).

One of the longer cyclic components of tides is the 18.6-year regression of the lunar nodes, a shift of the lunar orbit that causes a 10 percent variation in the average tide-producing force. Geyser activity in Yellowstone Park has been correlated with the 18.6-year cycle (Rhinehart, 1972). Attempts also have been made to correlate volcanic activity and diastrophism with tide-producing forces (Shaw, et al., 1971; Hamilton, 1973). On a very long time scale, it is possible that the slight flexing of crustal rocks by earth tides could lead to "fatigue" or brittle fracture. Joint patterns in young rocks may be propagated upward from older patterns beneath by such weak, but persistent, forces.

INTERNAL HEAT

Geothermal Gradient

Direct observation proves that the interior of the earth is hotter than the exterior. The increase of temperature with depth is the geothermal gradient, a typical near-surface gradient being about 20°C/km. The deepest mines and wells penetrate less than 10 km into the earth, and beyond that depth, the gradient can only be inferred from indirect evidence. Clearly, the near-surface geothermal gradients must decrease sharply at depth, otherwise the mantle could not behave as a solid and transmit seismic shear waves (Clark, 1971, pp. 118–26).

A steep geothermal gradient is obviously important for volcanic processes, including exhalative activity. However, any rock that solidifies or lithifies at depth and is later exposed by erosion has moved upward through the geothermal gradient. Cooler near-surface temperatures should cause thermal contraction of rocks. Fractures caused by elastic stresses in rocks are at least in part the result of their passage upward along the geothermal gradient. A complicating factor is the pressure gradient in the earth that counteracts the geothermal gradient (p. 106).

Heat Flow

The energetics of geomorphic processes are better understood in terms of the energy that flows to the surface from the interior of the earth rather than the temperature gradient within the earth. Heat flow is the product of the geothermal gradient and the thermal conductivity of the rocks through which the gradient was measured. The rate of energy receipt at the earth's surface from internal heat is about 30×10^{12} watts, calculated from the measured average heat flow on continents and in ocean sediments, and with an additional estimate for high heat flow at accreting plate margins (Williams and Von Herzen, 1974; Chapman and Pollack, 1975). According to Figure 5-2, volcanoes and hot springs carry about 1–10 percent of the total internal heat flow to the surface by convection, the remainder being conducted to the surface of ocean floors and continents, especially along midocean rises. High heat flow is observed in volcanic regions, but cause and effect are not clear. Other sources estimate that as much as 1–10 percent of the total heat flow is dissipated by seismicity.

The geothermal heat flow is mostly generated by radioactive decay of natural isotopes of uranium, thorium, and potassium, with a contribution from rotational or tidal friction. Although small in comparison to the solar energy received by the earth, this internal heat must provide nearly the entire energy drive for diastrophism and volcanism. It is odd that the heat flow in volcanic and orogenic areas is rarely more than two or three times greater than the global average. One of the puzzles of geophysics is the small range of heat-flow values.

Lattice Energy of Minerals

A difficult minor factor to evaluate in the energetics of geomorphic processes is the highly ordered atomic lattices of mineral phases that crystallize at the high temperatures within the earth. When exposed to weathering reactions, especially oxidation and hydrolysis, these minerals normally react exothermally and produce new phases that are more stable under surface conditions. In a fashion better explained by a physical chemist than by a geomorphologist, the free energy of the original crystal lattice is degraded to lost heat at the earth's surface. When the debris of weathering and erosion is buried at great depths in geosynclines and recrystallized, internal heat is again absorbed. The process has been repeated countless times in earth history.

Degradation of Geothermal Energy

At some future distant time, the radiogenic energy source for diastrophism will be expended, and when the surface is worn down, no internal force will be able to renew it again. The half-lives of the radioactive isotopes that supply the heat are on the order of 10^9 or 10^{10} years, however, so in the total geologic history of the earth, only a part of the internal energy supply has been consumed. This is another grand example of an energy system that is slowly declining, although it appears essentially in a steady state even when measured by the scale of geologic time.

REFERENCES

ANDERSON, D. L., 1974, Earthquakes and the rotation of the earth: *Science*, v. 186, pp. 49–50.

VON BERTALANFFY, L., 1950, Theory of open systems in physics and biology: *Science*, v. 111, pp. 23–29.

BLOOM, A. L., 1969, *The surface of the earth:* Prentice-Hall, Inc., Englewood Cliffs, N.J., 152 pp.

CAINE, N., 1976, A uniform measure of subaerial erosion: *Geol. Soc. America Bull.*, v. 87, pp. 137–40.

CHAPMAN, D. S., and POLLACK, H. N., 1975, Global heat flow: A new look: *Earth and Planetary Sci. Letters*, v. 28, pp. 23–32.

CHORLEY, R. J., 1962, Geomorphology and general systems theory: *U.S. Geol. Survey Prof. Paper 500-B*, 10 pp.

CLARK, S. P., 1971, *Structure of the earth:* Prentice-Hall, Inc., Englewood Cliffs, N.J., 126 pp.

DYSON, F. J., 1971, Energy in the universe: *Sci. American*, v. 225, no. 3, pp. 50–59.

FLOHN, H., 1969, *Climate and weather* (transl. by B. V. de G. Walden): McGraw-Hill Book Co., Inc., New York, 253 pp.

GOLDREICH, P., 1972, Tides and the earth–moon system: *Sci. American*, v. 226, no. 4, pp. 42–52.

GOODY, R. M., and WALKER, J. C. G., 1972, *Atmospheres:* Prentice-Hall, Inc., Englewood Cliffs, N.J., 150 pp.

HÄFELE, W., 1974, A systems approach to energy: *Am. Scientist*, v. 62, pp. 438–47.

HAMILTON, W. L., 1973, Tidal cycles of volcanic eruptions: Fortnightly to 19 yearly periods: *Jour. Geophys. Res.*, v. 78, pp. 3363–75.

HENDERSHOTT, M. C., 1973, Ocean tides: *EOS (Am. Geophys. Un., Trans.)*, v. 54, pp. 76–86.

HOLDREN, J. P., and EHRLICH, P. R., 1974, Human population and the global environment: *Am. Scientist*, v. 62, pp. 282–92.

HOUGHTON, H. G., 1954, On the annual heat balance of the northern hemisphere: *Jour. Meteorol.*, v. 11, pp. 1–9.

HUBBERT, M. K., 1971, The energy resources of the earth: *Sci. American*, v. 225, no. 3, pp. 60–70.

INMAN, D. L., and BRUSH, B. M., 1973, The coastal challenge: *Science*, v. 181, pp. 20–32.

LATHAM, G., EWING, M., DORMAN, J., LAMMLEIN, D., PRESS, F., TOKSOZ, N., SUTTON, G., DUENNEBIER, F., and NAKAMURA, Y., 1971, Moonquakes: *Science*, v. 174, pp. 687–92.

LVOVITCH, M. I., 1973, Global water balance: *EOS (Am. Geophys. Un., Trans.)*, v. 54, no. 1, pp. 28–42.

MYERSON, R. J., 1970, Long-term evidence for the association of earthquakes with the excitation of the Chandler wobble: *Jour. Geophys. Res.*, v. 75, pp. 6612–17.

NACE, R. L., 1969, World water inventory and control, *in* Chorley, R. J., ed., *Water, earth, and man:* Methuen & Co. Ltd., London, pp. 31–42.

PENMAN, H. L., 1970, The water cycle: *Sci. American*, v. 223, no. 3, pp. 98–108.

POWELL, J. W., 1876, Report on the geology of the eastern portion of the Uinta Mountains: *U.S. Geol. and Geog. Survey Terr.* (Powell), Washington, D.C., 218 pp.

RHINEHART, J. S., 1972, 18.6-year earth tide regulates geyser activity: *Science*, v. 177, pp. 346–47.

SHAW, H. R., KISTLER, R. W., and EVERNDEN, J. F., 1971, Sierra Nevada plutonic cycle: Part II, tidal energy and a hypothesis for orogenic-epeirogenic periodicities: *Geol. Soc. America Bull.*, v. 82, pp. 869–96.

VONDER HAAR, T. H., and SUOMI, V. E., 1969, Satellite observations of the earth's radiation budget: *Science*, v. 163, pp. 667–69.

WALEY, A., 1935, *The way and its power:* Houghton Mifflin Company, Boston, 262 pp.

WELLS, J. W., 1963, Coral growth and geochronometry: *Nature*, v. 197, pp. 948–50.

WILLIAMS, D. L., and VON HERZEN, R. P., 1974, Heat loss from the earth: New estimate: *Geology*, v. 2, pp. 327–28.

WOODWELL, G. M., 1970, Energy cycle of the biosphere: *Sci. American*, v. 223, no. 3, pp. 64–74.

CHAPTER 6

Rock Weathering

The processes that alter the physical or chemical state of rocks at or near the surface of the earth, without necessarily eroding or transporting the products of alteration, are collectively called **rock weathering**. Weathering does not make landforms but only altered or broken rock from which landforms are shaped. Weathering must always be studied in detail as a set of mineralogical or geochemical processes, because minerals are the chemical species of which the earth is made, and every mineral has specific chemical and physical responses to the near-surface environment. However, for many geomorphic generalizations, it is sufficient to speak of weathering at the scale in which an entire mass of rock (an assemblage of minerals) is the unit of alteration.

Weathering is the assemblage of rock-altering processes that are powered by an exogenic, essentially solar, energy supply. The depth of weathering is thereby restricted by the depth to which exogenic-powered processes can operate. A limit of 1 km on our planet can be suggested, based on the maximum reported depth of perennially frozen ground (Büdel, 1968), the maximum known depths of chemical weathering in the tropics (Ollier, 1965), and the deepest circulating ground water (Schneider, 1964; Ollier, 1965, p. 301). Most weathering is much shallower, and its intensity is inversely proportional to depth.

Weathering is distinguished from other destructive processes by the inclusion in its definition of the concept of *in situ*, or nontransported, alteration. Mass-wasting and erosion, considered in subsequent chapters, always involve translocation or transportation of material. Of course, when a mineral grain reacts with water, the soluble products are carried away in solution. Yet the basic structures of the rock mass, including bedding and other primary layered structures, granularity, and intrusive dikes or veins, are not obliterated until extremely advanced stages of weathering (Fig. 6-1). Weathering is the precursor of mass-wasting and erosion, although weathering continues even after a rock fragment has become dislodged from a hillside ledge or pried loose from a stream bed. Indeed, weathering processes do not end until a rock fragment has been finally buried or submerged beyond contact with the atmosphere or circulating ground water.

A rational classification of weathering processes must be arbitrary. Basically, if artificially, weathering processes are subdivided into a mechanical, or physical, group and a chemical group (Reiche, 1950). Mechanical weathering is also called **disintegration**, implying that pieces of rock are taken apart or disaggregated without alteration. In contrast, chemical weathering is also called **decomposition**, emphasizing the breakdown of the chemical composition of the mineral grains that make a rock. The more that is learned about weathering processes, the less sharp becomes the distinction between the mechanical and chemical groups, because the same thermodynamic principles govern both. The most rational geomorphic approach is to consider mechanical weathering processes first, because mechanical fracturing of rocks usually is necessary before air, water, and organisms can begin their largely chemical attack.

Figure 6-1. Deeply weathered syenite under glacial drift in southwestern Maine. A narrow basalt dike cuts the syenite vertically, left of the shovel. Horizontal foliation and the dike have not been disrupted, although the entire exposure can be dug with a shovel.

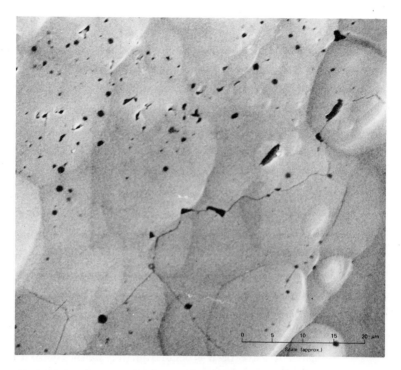

Figure 6-2. Photomicrograph of the boundary between grains of feldspar (upper left) and quartz (lower right) in Rutland quartzite. Feldspar grain is full of pores or tubes about 1 μm in diameter. On the grain boundary is a nearly continuous crack with larger triangular cavities aligned on it. (Photo: W. F. Brace.)

MECHANICAL WEATHERING PROCESSES

Whenever a brittle mass such as rock is fractured, whether by internal or external stresses, the resulting fragments have sizes with a statistical distribution that follows *Rosin's law* (Bennett, 1936; Blatt, et al., 1972, pp. 37, 58). Rosin distribution is not Gaussian, or Normal; instead, fracturing produces an excess of fine fragments. The theory of crushing has important applications to mining and quarrying operations, but it is also instructive as a rigorous mathematical explanation of why rocks broken by any mechanical process yield many small fragments with a large total surface area.

Mechanical fracturing is controlled by pre-existing cracks in or between mineral grains. Cracks form from thermal, mechanical, and chemical stresses that few rocks can escape. Microfractures and micropores as small as 1 to 10 μm have been photographed (Brace, et al., 1972) (Fig. 6-2). The fundamental elastic properties of rocks and minerals are probably controlled by cracks, so a consideration of mechanical weathering can be built upon an assumption of their presence (Stout, 1974; Simmons, et al., 1975).

The most important processes by which rocks are mechanically broken or dis-
integrated are: (1) differential expansion with pressure release; (2) thermal expansion
and contraction, including fire damage; and (3) growth of foreign crystals in cracks
and pores. The mechanical pressures generated by growing and moving organisms
might be important enough to rate a place on this list, as well. Each of these processes
affects different rock types in different ways, and the second and third processes are
strongly dependent on climatic conditions.

A common effect of weathering is the detachment of slabs, sheets, spalls, or chips
from rock surfaces. **Exfoliation** is the general term for the loosening or separation of
concentric surface shells or layers of rock. Exfoliation is caused by chemical as well
as thermal and physical processes, but only the physical processes and results are
described in the following section. Other examples of exfoliation are described in
subsequent sections.

Pressure Release on Unloading; Sheeting

Several lines of evidence can be used to prove that rocks now at the earth's surface
were formerly buried at depths of 20 km or more. The reconstructed geometry of
structures in the Valley and Ridge province of the Appalachian Mountains requires
that at least 8 km of folded sedimentary rocks have been removed since late Paleozoic
time. From the metamorphic grades of certain mineral assemblages, 24 km of rock are
estimated to have been eroded from the southeastern Adirondacks (Whitney and
McLelland, 1973). Post-Jurassic erosion to a depth of 8–16 km has been required to
expose the igneous rocks of the Sierra Nevada (Bateman and Eaton, 1967); 16–24 km
of post-Triassic erosion has exposed the tightly folded sedimentary rocks of the
Wellington District, New Zealand (Stevens, 1974, p. 26). The Alps in Switzerland have
suffered 30 km of denudation in the last 30 million years, one of the largest values and
most rapid rates to be found in the literature (Clark and Jager, 1969).

The confining pressure that is typically released by erosion of the magnitudes
just listed is on the order of 1.5 to 8×10^5 kPa (1.5–8 kilobars). The coefficients of
elastic expansion of typical rocks are such that expansion of 0.1–0.8 percent must
result (Birch, 1966). Thermal contraction due to cooling largely compensates for the
decompressive expansion (Skinner, 1966; Haxby and Turcotte, 1976), but in the upper
few kilometers of the crust, a net expansive (or compressive) stress results. If a massive
rock is confined on all sides but one, corresponding to the land surface, the expansion
must be toward the free face. In massive rocks, fractures parallel to the land surface,
known as **sheeting joints** or simply **sheeting**, result (Fig. 6-3).

Sheeting is a near-surface phenomenon, most common in massive rocks such as
igneous intrusives, quartzites, and thickly bedded sandstones. Rocks with other kinds
of fractures, such as columnar joints or bedding planes, are able to expand upon
unloading without developing sheeting, even though they may develop minor sets of
relief joints parallel to gorge walls and excavations. Quarries in granite near Boston,
Massachusetts, have an average thickness of joint-bounded sheets of 10 cm to 1 m
near the surface, but the average sheet thickness increases rapidly to about 5 m at a

Figure 6-3.   Sheeting joints on the southeast flank of Stone Mountain, a granite dome near Atlanta, Georgia. An estimated 12–15 km of overlying rock have been removed from Stone Mountain in 300 million years (Whitney, et al., 1976). (Photo: C. A. Hopson.)

depth of 20 m, and to more than 10 m at depths of 30–40 m. At greater depths, the sheeting joints become so widely spaced as to be insignificant (Jahns, 1943, p. 97). Sheeting joints on granitic monoliths in Yosemite National Park, California, are spaced from a few centimeters apart at the surface to a few meters apart at depth. The aggregate thickness of the sheets rarely exceeds 30 m (Matthes, 1930, p. 116). In tunnels at the Glen Canyon dam site, all joints parallel to the canyon wall were encountered within 10 m of the surface (Bradley, 1963, p. 522).

Few rock structures have been attributed to so many causes as sheeting. Jahns (1943, p. 78) reviewed seven processes that deserve serious consideration as the specific cause of sheeting:

1. Tensional or contractional strains set up during cooling of an igneous mass.
2. Local or regional compressional stresses (tectonic).
3. Insolation, with attendant daily and secular temperature changes.
4. Progressive hydration and formation of chemical alteration products in susceptible minerals.
5. Mechanical action of fire, frost, and vegetation.
6. Diminution of primary confining pressure by removal of superincumbent load.
7. Combinations of the above causes.

Of the various proposed causes of sheeting, only number 6 generally satisfies the observed field relations. A minority opinion favors number 2, however (Twidale, 1973). Sheeting cuts across primary structures and intrusive contacts within igneous rocks; the curvature and orientation are much too local to be related to tectonic stresses; it is similar in climatic regions of great insolation variation, such as polar and tropical, wet and dry regions; the minerals within the sheets show negligible chemical alteration; and sheeting invariably conforms to the present or a previous land surface (Dale, 1923; Hopson, 1958).

Sheeting joints develop only near the land surface probably because the geothermal gradient decreases to much less than normal in the outer 200 m of the crust (Clark, 1971, p. 120), and the temperature of the rock becomes nearly constant. The overburden pressure, however, continues to decrease at a constant rate as overlying rock is peeled off by erosion. No compensating contraction due to decreased temperature is possible at shallow depths, so the expanding outer layers of rock are bowed upward or outward toward the land surface, and sheeting joints develop. Pre-existing microcracks determine the final fracturing, and because they are so abundant in rocks, slight directional stresses can cause large, continuous joints to propagate.

Massive igneous rocks and thick-bedded sandstones typically have low geothermal gradients. Their "tight" structure permits efficient heat conduction. These are the same rocks that exhibit the best sheeting joints, not only because no other fractures are available to absorb the expansion or release of pressure, but because the low geothermal gradient provides a minimal compensatory contraction upon cooling.

The size of monolithic domes is largely determined by the spacing of faults and joints other than sheeting, but their shape is determined by sheeting. If a rock mass is irregularly fractured at depth, the most fractured portions adjust readily as the superincumbent load is removed, but portions with few or widely spaced fractures store the stress of expansion until the overburden pressure is so low that failure occurs. Such failure is parallel to the land surface at a depth of 100 m or less, so the sheeting joints conform closely to local topography. In glaciated mountains, sheeting on cirque walls and floors is concave skyward, parallel to the bowl shape of the cirques, whereas the sheeting joints are convex on the adjacent convex summits (Ollier, 1969, p. 6). At the rims of cirques, the upland and cirque-wall joint sets intersect.

A combination of relief joints and expansion toward a free face upon unloading produces structures that are controlled by erosion but that subsequently control erosion. A notable correspondence of valleys with small-scale anticlines in Northampton, England, was explained by a process called *cambering*, whereby brittle sandstone beds fractured along vertical joints and moved apart, while underlying plastic claystones flowed or bulged upward between the separated blocks (Hollingsworth, et al., 1944). A sinuous anticline in the Muav Limestone along the floor of the Grand Canyon has been similarly attributed to elastic rebound or plastic flow caused by erosion-induced pressure release (Sturgal and Grinshpan, 1975). Large-scale circular or arcuate patterns in the crystalline terrane of North Carolina have been attributed to the structural control of sheeting (Hack, 1966).

Thermal Expansion and Contraction

For rocks buried at shallow depth, thermal contraction or expansion is minor except during cooling of igneous rocks. However, a dark-colored rock in bright sunlight on a desert or mountain top may heat from below 0°C to 50°C each day (Roth, 1965) and cool a comparable amount each night. Rock is a poor conductor of heat, so the diurnal thermal effects are shallow, perhaps extending only a few centimeters deep. Shattered, angular rock fragments litter the stony plateaus in arid parts of Australia. The fragments can be reassembled into cobbles and boulders up to 30 cm in diameter. Ollier (1963, 1969) suggested that these "gibbers" are heat-cracked, because there is no other plausible weathering process. Thermal "crazing" is also thought by Ollier to shatter the surfaces of quartzite boulders to a depth of 10 cm or less and to produce blocky detritus.

Strangely, no one has ever verified experimentally that solar heating is intense enough to break stones. Specimens of granite have been heated dry to 105°C and cooled to −10°C for 2000 cycles with no evidence of deterioration (Kessler, et al., 1940). In another often-quoted study, Griggs (1936) alternately heated the surface of a 3-inch cube of dry granite for 5 minutes to about 140°C, then cooled it for 10 minutes with an electric fan to about 30°C, for 89,400 cycles, equal to 244 years of diurnal heating. His experiment lasted over 3 years, but no change in the rock specimens could be detected, even by microscopic examination. Larger specimens may be more subject to thermal stresses (Rice, 1976). There are also stories of rocks shattering in extreme cold, their fragments strewn about over the snow (Washburn, 1969, pp. 32–33). Even if the process is insignificant on earth, thermal expansion and contraction is likely to be one of the major processes of lunar weathering.

Fire has a notably destructive effect on rocks, especially those with granitic texture. Fire, alternating with sudden water quenching, was probably the principle quarrying tool of primitive people. Generally, limestone (including marble) and sandstone are more resistant than granite to fire damage, and coarse-grained rocks are less resistant than fine-grained rocks to fire damage (Tarr, 1915, p. 349). Unequal thermal expansion of the dominant minerals is the controlling factor. For instance, the thermal volumetric expansion of quartz is three times that of feldspar. In addition, quartz, feldspar, and many other common minerals expand under heat in a highly anisotropic fashion, with expansion along certain crystallographic axes being as much as 20 times greater than along other axes.

Growth of Foreign Crystals in Cracks and Pores

Certain minerals, such as pyrite (FeS_2) in shale or slate, oxidize readily on exposure to oxygen-bearing ground water or moist atmosphere. The newly formed iron oxides are of lower density and larger volume than the original minerals, and the volume increase can be sufficient to split weaker rocks or cause spalls or "pop-outs" in stronger rocks.

Of greater significance to weathering processes are the numerous hydrated salts that are water-soluble (Evans, 1970). When ground water percolates upward into a rock and evaporates, any dissolved salts precipitate. Rainwater may also accumulate salts, especially when it flows over soot-stained city buildings and through masonry joints. The force of crystallization around nuclei of precipitating salts is substantial. Dendritic salt crystals, or even crusts, may form in cracks, along mortar joints, or on grain boundaries within a permeable but otherwise coherent rock.

A particularly troublesome salt is hydrous calcium sulfate (the mineral *gypsum*). In industrial regions, where the air is polluted by sulfurous smoke from fossil fuels, rainwater becomes a dilute sulfuric acid (Likens and Bormann, 1974) that corrodes limestone and marble buildings. The reaction product of the corrosion is gypsum, a relatively insoluble salt that crystallizes in rock cracks, flakes off thin pieces of rock, and accelerates chemical attack. The formation of salt crystals in rock openings is regarded as a mechanical weathering process, but the close relationship with chemical processes is obvious.

Many water-soluble salts precipitate as hydrates or invert to hydrated compounds within the normal ranges of atmospheric temperature and humidity. Hydration pressure is more effective than simple crystal growth, because as temperature and humidity change seasonally or daily, the salts may hydrate and dehydrate repeatedly (Winkler and Wilhelm, 1970). The hydration pressures of common salts range from 1×10^4 kPa to over 2×10^5 kPa (100 to over 2000 bars). In theory, the pressure should be highest at low temperature and high relative humidity.

Figure 6-4.   Detail of the entrance vault at Cathedral St. Coretin, Quimper, France. Original carvings of the late fifteenth century are in Caen Stone, a soft Jurassic limestone widely used in western European architecture. Three inner archivaults have been restored recently, with similar stone. Outer carvings have been nearly obliterated by 500 years of weathering.

Mirabilite (Glauber's salt, $Na_2 SO_4 \cdot 10 H_2O$) dehydrates to thenardite (Na_2SO_4) readily, quickly, and reversibly. At 39°C, the dehydration is complete in about 20 minutes (Winkler and Wilhelm, 1970, p. 570). The standard test for building-stone durability makes use of this hydrated salt. Blocks of construction-grade granite that can be subjected to 5000 cycles of alternate freezing (6 hours at −12°C) and thawing (1 hour in water at 20°C) with barely visible disintegration, crumble after an average of only 42 cycles when soaked in a saturated solution of sodium sulfate (17 hours at room temperature) and dried (7 hours at 105°C) (Kessler, et al., 1940; Evans, 1970).

Until the last century, salt precipitation and hydration were weathering phenomena of concern only in desert regions with a high water table, as for instance along the Nile River, or the Rio Grande. However, many of the same salts are leached out of bricks and mortar and attack stone and concrete masonry in ancient and modern cities. The physical force of hydrating crystals is causing growing concern over ways to maintain and preserve historical monuments in heavily populated, industrialized regions (Fig. 6-4) (Winkler, 1973).

Ice and Hydrofracturing by Freezing

A foreign crystal of peculiar abundance in surface rocks is ice. When water freezes under atmospheric conditions, its molecules organize into a rigid hexagonal crystalline network, and it increases 9 percent in specific volume (volume per unit mass). Intuition and oversimplified explanations of the freezing process have led to the wide use of the term *frost wedging* to describe the fracturing of rocks by freezing and thawing. The actual process is much more complicated than formerly thought, and frost wedging is abandoned here as a geomorphic term.

A very high confining pressure is required to depress the freezing point of water only a few degrees. Conversely, if confinement prevents water from beginning to freeze until a below-zero temperature is reached, very large pressures are generated on the walls of the confining chamber (Fig. 6-5). As more ice freezes and expands, the pressure on the remaining water increases, and the temperature must be lowered further in order to continue the process. Pressure continues to build up until the crucial tempera-

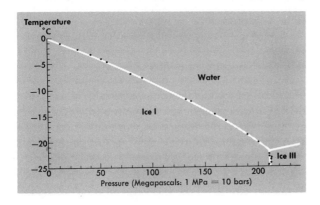

Figure 6-5. The pressure increase in confined pure water with lowered freezing temperature. At confining pressures greater than 216 MPa, more dense phases of ice crystallize. (Data from Bridgman, 1911.)

ture of $-22°C$ is reached. Below $-22°C$, the pressure remains about constant, because a more dense kind of ice (Ice III) begins to crystallize.

Few rocks can withstand the huge pressure generated by freezing confined water. However, total confinement is required, and there is no natural way for water to be totally confined in rock. Ice is not strong enough to seal water into a fracture and produce the maximum possible pressure. The plastic deformation of ice becomes so rapid at stresses in excess of only 100 kPa (1 bar) that it is quickly extruded (Fig. 16-3). Furthermore, water in very thin cracks will not freeze, even at very low temperatures, because of the strong capillary adhesion of the water to rock. Unfrozen pore water in cold rocks maintains electrical, and therefore presumably hydraulic, properties down to a temperature of $-125°C$, although most of the water seems to have frozen by $-60°C$ (Mellor, 1970, p. 50).

Water in shallow soil, or standing on the surface of the ground, freezes at $0°C$ and freely expands 9 percent, because it is unconfined. Small pebbles, garden soil, and grassy turf are "heaved" by unconfined freezing, but no great pressure is involved. Most of the rock fractures attributed to freezing are probably due to capillary or monomolecular films of unfrozen water that spread along pre-existing cracks in rocks. Thin films of water form a semicrystalline "ordered" molecular structure on mineral surfaces (especially on clays) that can pry apart microfractures and propagate cracks (Dunn and Hudec, 1966; Winkler, 1968). The powerful molecular forces in films of pure water and in saline solutions are analogous, drawing more water to the region where the ordered structure is developing (Evans, 1970, p. 167). Studies of "ordered films" of water emphasize the disappearing distinction between mechanical and chemical weathering processes.

Shallow freezing may aid in forcing capillary films of water deeper along tight joints and pre-existing microfractures, thereby disaggregating rocks well below the depth of actual freezing. Engineering studies of dam sites in the northern Allegheny Mountains show that this *hydrofracturing* effect extends to depths of 12–15 m, even though the ground is rarely frozen beyond a depth of 1 m (Philbrick, 1974). Similar processes may operate in any humid climate with periodic freezing. Many mountain slopes above the tree line are wastelands of shattered rock rubble that is gradually being broken down to particle sizes that can be attacked by other weathering and erosional agents. In humid subpolar regions of perennially frozen ground, freezing and hydrofracturing become the dominant weathering processes (Chap. 15).

Plants and Animals as Agents of Mechanical Weathering

The prying or wedging action of growing plant roots, especially of trees, is often described as mechanical weathering. Two-dimensional networks or sheets of interlaced roots can be followed for many meters along bedding planes or joints, deep into fresh rock. It has been supposed that the growing roots exert a pressure on the rock and force cracks to open. However, the efficacy of roots as agents of mechanical weathering probably has been overestimated, just as their importance as agents of chemical weathering has been underestimated. Roots follow the paths of least resistance and conform to each little irregularity of a crack, but they do not exert much force on the rock. Cracks opened by other processes can be maintained by roots, however, and decaying vegetable matter and washed-in dirt can keep rock surfaces wet and chemically active. In addition, when trees sway in a strong wind, their roots pry apart rocks.

The "strangler fig" (*Ficus* sp.) of tropical regions has the ability to envelop other trees, masonry, and bedrock with snakelike, anastomosing stems (Fig. 6-6). As the stems grow and cross each other, they fuse, encasing a host tree with a wooden latticework sufficiently strong to prevent its further growth and cause its eventual death (Argo, 1964). They also pry apart masonry and natural joint blocks, but it is doubtful that they could crush or pry apart sound rock. The mature *Ficus* may grow to huge size, however, and when such a tree is toppled during a storm, large masses of rock may be displaced. Under temperate and boreal forests, soils usually include many mounds and pits formed by tree roots that were torn up when trees fell (Denny and Goodlett, 1956, p. 59). Because translocation is so obviously involved, *root throw*, along with animal burrows and human engineering works, is better treated as a category of erosion than as weathering.

Figure 6-6. Stems of *Ficus* encasing prehistoric stone blocks, Kusaie Island, Micronesia. Blocks are 30–50 cm high.

CHEMICAL WEATHERING PROCESSES

The general trends of chemical weathering can be predicted from the conditions under which rock-forming minerals crystallize or precipitate. Having formed in an environment of high thermal energy and high pressure, the minerals in rocks at the surface of the earth tend to weather by exothermic chemical reactions that produce new compounds of greater volume and lower density. **Oxidation** is one of the most typical exothermic, volume-increasing reactions between minerals and the wet atmosphere; especially common is the reaction of iron-bearing minerals with oxygen dissolved in water. Other typical weathering reactions are **carbonation**, the reaction of minerals with dissolved CO_2 in water; **hydrolysis**, the decomposition and reaction with water; **hydration**, the addition of water to the molecular structure of a mineral; **base-exchange**, the exchange of one cation for another between a solution and a mineral solid; and **chelation**, the incorporation of cations from the mineral into organic compounds. Hydration has been discussed in a preceding section as a primarily mechanical process, as demonstrated by the volumetric expansion of hydrated salts that enter cracks and pores, yet it is certainly a chemical process when the water is absorbed or reacts with the minerals that make the rock.

Water in Chemical Weathering

All chemical weathering reactions on this planet involve water either as a reactant or as the carrier of the reaction products. It is not necessary, therefore, to list *solution* as a unique chemical weathering process, because solutions in water are always a part of chemical weathering.

Water is by far the most abundant material near the surface of the earth. In the outer five kilometers of the earth, water is about three times as abundant as the next most common single substance, the feldspar group of minerals (Kuenen, 1955). Water is the only compound that occurs naturally at the earth's surface in gaseous, liquid, and solid states. Its general solvent power and its surface tension are greater than those of any other fluid. Its heat of vaporization is the highest of all substances; recall the significance of this property in the conversion of solar energy to geomorphic work (Chap. 5). Its maximum density at $4°C$ in the liquid phase, with expansion toward the freezing temperature as well as away from it, is a strange and almost unique property. A recitation of the properties of water could go on at length, but this summary is enough to suggest that the surface of the earth is generally saturated with an abundant and chemically active compound, made available in unending supply by the hydrologic cycle, that readily attacks and reacts with rock-forming minerals. The chemical potential of rain is largely expended by various reactions with minerals soon after the drops strike the ground. River water has just about the dissolved chemical load that could be predicted from analyzing the rocks through and over which it has flowed.

Oxidation: Weathering of Iron-Bearing Minerals

Iron-bearing minerals commonly contain iron in the ferrous state. This is true for the major iron sulfide (pyrite, FeS_2), iron carbonate (siderite, $FeCO_3$), and various iron silicates. On contact with water that has oxygen in solution, the ferrous iron oxidizes to the ferric state and forms nearly totally insoluble ferric oxides or hydroxides. Only oxygen-free, chemically reducing, alkaline water permits iron to remain in the more soluble ferrous state. Thus, when aerated water reacts with iron-bearing minerals, the reaction is more complex than simple solution. *Oxidation–reduction reactions* are inevitable, with the result that instead of the mineral going into hydrous solution as ionic species, new compounds are formed, some of which are nearly insoluble.

Weathering by oxidation takes place with water as the intermediary. Unprotected iron surfaces stay clean and bright in dry air, as is proved by the condition of tools and equipment abandoned and later found in the polar regions or of aircraft that have been found in the Libyan desert decades after they were abandoned. But iron rusts, or oxidizes, quickly beneath even a dew-thick film of moisture. There is always enough dissolved oxygen in rain water and circulating ground water to oxidize metallic iron and to change the ferrous iron in mineral compounds to the more oxidized ferric state. As long as water is in contact with atmospheric molecular oxygen on the one hand and incompletely oxidized iron on the other hand, oxygen dissolves from the air, diffuses through the water, and combines with the iron. When the iron or other elements from a mineral grain combine with oxygen, the original mineral structure is destroyed, and the remaining mineral components are free to participate in other chemical reactions. The oxidation-reduction reactions of water with compounds of aluminum, magnesium, manganese, chromium, and some less common metals are similar to the reactions with iron compounds. The several oxides often occur together.

Iron oxides have intense yellow or red colors. Rocks and soil are stained various shades of red or brown by mere traces of iron oxide. The reactions are so rapid that a brown surface layer is the first indication of weathering on most rocks. Black iron-sulfide muds from the anaerobic conditions of a swamp, for instance, oxidize and develop a brown color in a few minutes after exposure to air. Iron-rich rock-forming minerals such as pyroxenes, amphiboles, and biotite are usually rust-colored on all but the most recent natural exposures.

Iron oxides, and the oxides of related metals, are exceptionally stable chemical compounds, even on the geologic time scale. *Ironstone*, or *ferricrete* (hardened iron oxide in lateritic soil; see p. 131), on an ancient land surface in Australia has been inferred to be at least 13 million years old (pre-Middle Miocene) from potassium/argon dates on lava flows that overlie the ancient, deep, weathering profile (Dury, et al., 1969). Precambrian iron ores are widely distributed on the continents. Some have been buried and later exhumed. It has been suggested that every geologic system contains "ironstones," or iron-oxide-rich sedimentary formations, that formed during contemporaneous weathering (Degens, 1965, p. 87).

Hydrolysis and Base-Exchange: Reactions of
Carbonated Water With Silicates

Hydrolysis, like oxidation, is more than just a mineral dissolving in water; both the mineral and water molecules decompose and react to form new compounds. It is assumed that hydrolysis takes place in a great excess of water, that is, in an aqueous solution. It is the most important chemical weathering reaction of silicate minerals.

Pure water ionizes only slightly, but it does react with some easily weathered silicate minerals in the following fashion:

$$Mg_2SiO_4 + 4\,H^+ + 4\,OH^- \longrightarrow 2\,Mg^{++} + 4\,OH^- + H_4SiO_4$$

| olivine (forsterite) | + | 4 ionized water molecules | = | ions in solution | + | silicic acid in solution |

The result of such complete hydrolysis is that the mineral is entirely dissolved, assuming that a great excess of water is available to carry the ions in solution. Silicic acid, one of the reaction products, is such a weak acid that we can disregard its name and simply think of it as silica (SiO_2) dissolved in water.

The foregoing reaction is deceptively simple, because it implies that water is an ionized compound. Such is not the case. Pure water is known to be a very poor H^+ donor and electrolyte. Its electrical conductivity is extremely low, proving that it has few free ions. Of great importance to weathering processes, therefore, is the fact that carbon dioxide gas dissolves readily in water to form a weak acid, or H^+ donor. Even though the terrestrial atmosphere contains only 0.03 percent CO_2 (by weight), raindrops absorb CO_2 as they fall, so that rainwater is usually slightly acidic, with a pH of 5.7 or less. The reversible reactions between water and carbon dioxide are as follows:

$$CO_2 + H_2O \rightleftharpoons H_2CO_3 \rightleftharpoons H^+ + HCO_3^-$$

| gas | + | water | = | carbonic acid | = | hydrogen ion | + | bicarbonate ion |

Cold water dissolves more carbon dioxide gas than warm water; water under pressure also dissolves more gas. For every condition of temperature and pressure, an equilibrium is established among the several compounds and ions so that an increase or decrease of any one of them is offset by shifts among the others to evenly distribute the change. For instance, an increase in available carbon dioxide gas quickly causes an increase in acidity of the equilibrium aqueous solution. Addition of a strong mineral acid such as H_2SO_4 from atmospheric pollution drives the equilibria to the left and decreases the amount of bicarbonate ion in solution (Johnson, et al., 1972).

Soil air is greatly enriched in carbon dioxide by the decay of humus. Soil air drawn from the biologically active upper layer of a soil profile may have from 10–1000 times more CO_2 than the free atmosphere; that is, CO_2 may constitute as much as 30 percent of soil air as compared to the 0.03 percent of dry normal atmosphere (p. 138). Biogenic carbon dioxide in soil air is surely the major source of carbonated ground water.

Any reaction that increases the H^+ ion concentration in water also increases the effectiveness of hydrolysis. Carbon dioxide dissolving in soil water is the most common and important way that water is provided with H^+ ions, or acidified, for hydrolysis. We can elaborate the chemical equation previously given for the reaction between water and olivine as follows:

$$Mg_2SiO_4 + 4\,CO_2 + 4\,H_2O \longrightarrow 2\,Mg^{++} + 4\,HCO_3^- + H_4SiO_4$$

| olivine (forsterite) | + | carbon dioxide | + | water | = | magnesium and bicarbonate ions in solution | + | silicic acid in solution |

This hydrolysis reaction is much more common than the one with pure water shown previously. Note that the H^+ ions that react to form silicic acid and bicarbonate are derived from water molecules through the carbonation process and the accompanying formation of bicarbonate ions in solution. Carbonic acid is consumed by silicate weathering, and the resulting solutions become more alkaline because of the dissolved bicarbonate. The common cations freed by silicate weathering are quite soluble in solution with bicarbonate ions, and most of the dissolved load of rivers is in the form of bicarbonates.

Quartz (SiO_2) is an extremely common silicate mineral, but it is only slightly soluble in water, and solution reaction is very slow. Silica glass and other polymorphs of quartz are significantly more soluble in water, but they are not common. The world average for the dissolved silica concentration in river water is 13 parts per million, about twice the calculated solubility of quartz (Livingston, 1963). Therefore, reactions such as those given previously for olivine and those that follow for feldspars are the most likely sources for silica in solution. Quartz (the oxide of silicon) is chemically inert at the earth's surface, and can be classed with the oxides of iron and aluminum as stable compounds.

Probably the most common weathering reaction on earth is the hydrolysis of feldspars by carbonic acid in water. A typical, although simplified, weathering reaction between potassium feldspar (orthoclase) and carbonated water is as follows:

$$2\,KAlSi_3O_8 + 2\,H_2CO_3 + 9\,H_2O$$

| orthoclase | + | carbonic acid | + | water |

$$\longrightarrow Al_2Si_2O_5(OH)_4 + 4\,H_4SiO_4 + 2\,K^+ + 2\,HCO_3^-$$

| = | kaolinite— a clay mineral | + | silicic acid in solution | + | potassium and bicarbonate ions in solution |

Calcium and sodium feldspars (plagioclase) hydrolyze even more readily than orthoclase in carbonated water. In the foregoing example, as in many others, the initial mineral and acid are consumed, and the end products are: (1) a clay mineral, (2) silica in solution, and (3) bicarbonate of potassium, sodium, or calcium in solution. The process is essentially a surface reaction between the solution and the mineral grain (Berner and Holdren, 1977). Clay minerals are stable residual solids under all but the most humid and tropical climatic conditions, where they may be desilicified to hydrated aluminum oxides (bauxite).

Soil scientists in recent years have recognized the importance of another chemical weathering reaction, *base-exchange*. Base-exchange involves a mutual transfer of cations such as Ca^{++}, Mg^{++}, Na^+, or K^+ between an aqueous solution rich in one cation and a mineral rich in another. The rate of exchange depends on the chemical activity and the abundance of the various cations as well as on the acidity, temperature, and other properties of the solution. In a sense, hydrolysis is a special kind of base-exchange, in which hydrogen ions (H^+) in solution displace the cations in silicate mineral lattices (Frederickson, 1951; Deju, 1971). The exchange of cations between minerals and ground water may expand or collapse the mineral structure and free other chemical components. As in other chemical reactions, if one mineral grain in a rock is so destroyed, adjacent grains are detached and exposed to the same and other weathering processes.

The residual clay minerals of hydrolysis are hydrated compounds. The closely related hydration and hydrolysis of feldspars to form clays are the major processes in the weathering of granite and related coarse-grained polymineralic rocks. Weathering feldspar grains expand and, combined with the oxidized and hydrolyzed iron-rich minerals such as biotite, cause the granite to crumble into a mass of disaggregated grains called *grus* (Fig. 6-1).

Chelation: Chemical Weathering by Plants

Chelation is a complex organic process by which metallic cations are incorporated into hydrocarbon molecules (Lehman, 1963). The word "chelate," which means "clawlike," refers to the tight chemical bonds that hydrocarbons may impose on metallic cations. Many organic processes, in order to function, require metallic-organic chelates. The role of iron in hemoglobin to carry oxygen from the lungs is a good example. Synthetic chelating agents are widely used in analytical chemistry laboratories.

Soils in which plants are growing and decaying are presumed to contain some organic compounds that can chelate metallic cations, although only a few of the complex compounds have been identified with certainty (Loughnan, 1969, pp. 47–49). By analogy with laboratory procedures, if natural organic molecules in soil can remove metallic cations directly from mineral grains by chelation, the mineral lattices are disrupted, and possibly fragments of crystal lattices, including otherwise insoluble elements such as aluminum, can be mobilized.

Vascular plants promote chemical weathering by mutually reinforcing interactions among chelation, hydrolysis, and base-exchange reactions. Each plant rootlet maintains a negative electrical charge on its surface (Keller and Frederickson, 1952) and a surrounding field of H^+ ions. Hydrolysis and base-exchange are strongly promoted by the availability of H^+ ions in the vicinity of the root hair, and organic compounds in the roots absorb and chelate the essential metallic cations as they hydrolyze. Through vascular plants, the energy derived from sunlight falling on green leaves is used to weather minerals many meters underground.

An impressive demonstration of the ability of plants to weather rock was provided by Lovering and Engel (1967). In controlled-environment greenhouses, they grew *Equisetum* (scouring rush) and three grasses in pots of freshly crushed, sterilized rock samples. The silica uptake rate by the plants, all of which are known to be silica accumulators, was equivalent to removing all the silica from a basalt substrate 30 cm thick in only 5350 years. This remarkable rate of withdrawal of silica from igneous rocks is exclusive of any chemical weathering or leaching by percolating water, although some of the silica taken up by the plants might have been first dissolved by other weathering reactions.

Less complex, nonphotosynthetic plants also have the ability to dissolve elements from minerals. Fungi (*Penicillum* sp.) grown in a nutrient solution mixed with finely crushed rock greatly increased the quantity of Si, Al, Fe, and Mg entering solution, as compared to the amount of elements entering solution in similar, but uninoculated, control solutions (Silverman and Munoz, 1970). During the incubation period of the experiment, the pH of the nutrient solution dropped from 6.8 to below 3.5 in 7 days. As the acidity increased, the rate of mineral solution correspondingly increased. Other experiments were conducted with a variety of organic and mineral acid solutions of equivalent concentrations to the fungal growth medium, with essentially similar effects on rocks. Thus fungal weathering was proved to be due not to some special biogenic process but simply to chemical weathering in acidified water.

Laboratory experiments such as those just cited prove that plants are capable of putting into solution or absorbing substantial fractions of the total mass of rock on which they grow, in intervals from a few thousand years to as short as one week. Even though the rocks in the experiments had been mechanically crushed in advance, and the plants were growing under optimal conditions, the amounts of translocated mineral matter are impressive. Field study supports the experimental work. Some parts of Iceland today are barren basalt plains, essentially devoid of macroscopic plants even though the climate is humid. Chemical weathering on these barren plains is one-half to one-third as fast as on similar areas that have plant cover, as judged by the dissolved loads in streams (Cawley, et al., 1969). Even in regions devoid of macroscopic plants, algae and fungi may contribute to weathering processes. Blue-green algae have been discovered living in near-surface pores and cracks in sandstone and quartzite in both the Sahara and Antarctica, with no known source of liquid-phase water (Friedmann and Ocampo, 1976).

Weatherability of Silicate Minerals

By comparing the rates of chemical reactions between water and various powdered minerals in controlled laboratory experiments, it is possible to predict which minerals will weather most rapidly under natural conditions. Extensive studies of this sort, especially on the silicate minerals that form most of the earth's crust, have led to the recognition of a mineral stability series or **weathering series**, a list of common silicate minerals arranged in order of relative susceptibility to chemical weathering (Fig. 6-7).

Forsterite (29,789)   Anorthite (31,935)

Augite (30,728)

Hornblende (31,883)

Biotite (30, 475)

Albite (34,335)

Orthoclase (34,266)

Muscovite (32,494)

Quartz (37,320)

Figure 6-7. Common minerals of igneous rocks arranged as a weathering series, with the least resistant minerals at the top and more resistant minerals progressively lower in the series. Energies of formation in parentheses (kcal/mol) are calculated from the bond strengths between oxygen and the various cations. This same arrangement is more commonly known as Bowen's Reaction Series for the sequence of formation of minerals from a silicate melt. (After Blatt, et al., 1972, p. 234.)

The silicate minerals that crystallize at the highest temperatures and that have the lowest silicon-oxygen ratios weather most rapidly. The strength of the silicon-oxygen bond is so much greater than the bonding energies of oxygen with metallic cations that it determines the total bond energy for each mineral molecule. The higher the Si: O ratio, and therefore the higher the bond energy, the more resistant the mineral, with few exceptions (Blatt, et al., 1972, pp. 233–35). Biotite weathers faster than might be expected from its bond energy, because the iron oxidizes and the layered mica structure expands greatly on hydrolysis. Biotite weathering is a large contributor to the granular disaggregation of granite to grus (Birkeland, 1974, pp. 74, 138; Isherwood and Street, 1976).

Another way to study relative chemical weathering rates is to make a ratio of the concentration of each cation in solution in ground water or rivers with the abundance of that cation in the regional terrane. Feth, et al. (1964) established a **cation mobility series** in this way for a granitic terrane in California; Na, Ca, and Mg are the most mobile elements, K and Si are less mobile, and Al and Fe are by far the least mobile. Goldich (1938) obtained comparable results by comparing fresh and weathered samples of a quartz-feldspar-biotite gneiss from Minnesota and a calcic plagioclase-amphibole-olivine diabase from Massachusetts.

In general, the cations that are lost to solution in greatest proportion are those that are derived from easily weathered plagioclase feldspars and mafic minerals. An exception is iron, which is most abundant in the dark mafic minerals that weather most easily, but which oxidizes to an insoluble oxide that does not move away in solution. Another anomalous result of weathering involves potassium, which is of comparable crustal abundance and chemical behavior to sodium, yet which has a chemical mobility of a much lower order, comparable to silica. Potassium is "fixed" in the crystal lattice of many clay minerals and tends to stay behind in residual clays, whereas the other alkalic ions are carried away in solution (Loughnan, 1969, pp. 45–47).

ROCK WEATHERING AND SOIL FORMATION

Definitions

The term **soil** is used to describe the rock detritus at the surface of the earth that has been sufficiently weathered by physical, chemical, and biological processes so that it supports the growth of rooted plants. This is an agricultural definition, emphasizing that soil is a biologic as well as a geologic material. Engineers are less specific about their definition of a soil. To them, all the loose, unconsolidated, or broken rock material at the surface of the earth, whether residual from weathering at that place, or transported by rivers, glaciers, or wind, is soil. However, for most of the engineers' "soils," genetic terms such as "alluvium" or "glacial drift" are available. For a general term, *regolith* is useful. In most contexts, the engineering or agricultural implication of the term is clear enough so that tedious definitions are unnecessary (Legget, 1967).

Soils are characterized by **horizons**: distinctive weathered zones, approximately parallel to the surface of the ground, that are produced by soil-forming processes. A **soil profile** is a vertical cross section through these horizons (Fig. 6-8). The horizons in a soil profile illustrate the principle that the many mechanical, chemical, and biological weathering processes operate simultaneously near the surface of the earth but at different rates and to various depths. By color, chemical analyses, grain size, and other diagnostic criteria, soil profiles are divided into horizons and subhorizons that clearly record the intensity and duration of the various soil-forming, or weathering, processes. In spite of the strong genetic implications of horizons, they are intended to be defined by observable facts, not genetic inferences.

The five principal horizons, from the surface down into unaltered rock, are conventionally given the capital-letter symbols O, A, B, C, and R (Fig. 6-8). The **O horizon** is the upper organic horizon, dominated by fresh or partly decomposed organic material. The other four horizons are dominantly mineral in nature, defined by the manner and degree in which they differ from the parent material. The **A horizon** is a zone in which organic matter has accumulated and mixed with mineral detritus and from which clay, iron, or aluminum has been transferred downward to underlying horizons. The **B horizon** is characterized by the secondary accumulation or enrichment of clay, iron, or aluminum and by development of distinctive structures not present in the parent material, such as granular, blocky, or prismatic aggregates of partly decomposed minerals, loosely or firmly cemented by the clay or oxides that have moved into the horizon. The B horizon typically has more intense colors than the overlying and underlying horizons. The **C horizon** is below the zone of major biologic activity. It retains some characters of the parent material but may be mechanically fractured, stained by oxidation, or loosely recemented by calcium carbonate, gypsum, iron oxide, silica, or the more soluble salts. The **R horizon** is the underlying **parent material**, such as bedrock, alluvium, or other material from which the soil has formed. Although not specified in the definition of the soils scientists, most geologists would observe the effects of mechanical weathering on bedrock in the upper part of the R horizon. Its lower limit is not specified by pedologists.

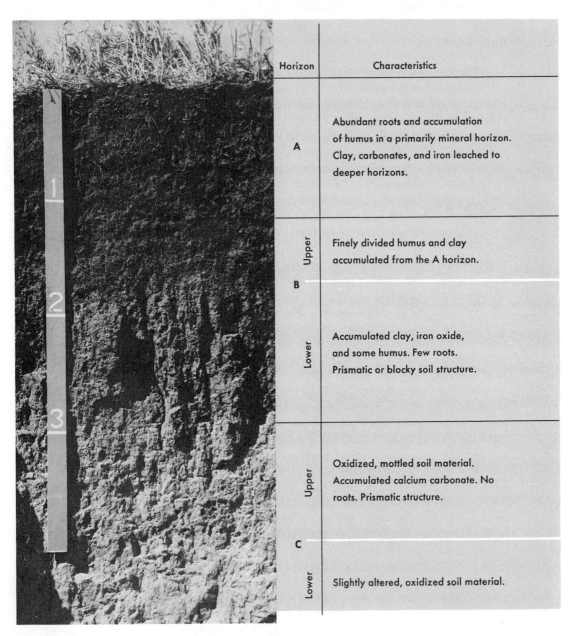

Horizon		Characteristics
A		Abundant roots and accumulation of humus in a primarily mineral horizon. Clay, carbonates, and iron leached to deeper horizons.
B	Upper	Finely divided humus and clay accumulated from the A horizon.
	Lower	Accumulated clay, iron oxide, and some humus. Few roots. Prismatic or blocky soil structure.
C	Upper	Oxidized, mottled soil material. Accumulated calcium carbonate. No roots. Prismatic structure.
	Lower	Slightly altered, oxidized soil material.

Figure 6-8. A Mollisol soil profile in Nebraska showing horizons A, B, and C and the notable characteristics of each. Depth scale is in feet. The R horizon is not exposed in this profile. (Photo: W. M. Johnson, U.S. Department of Agriculture-Soil Conservation Service.)

Soil Genesis

Five key factors of soil formation have been long recognized by pedologists. They are: *parent material, climate, organisms, slope,* and *time.* The kind of soil that forms is the result of chemical and mechanical weathering of some rock type ("structure," in the geomorphic sense defined on p. 5) under a certain climatic regimen of temperature and precipitation in the presence of plants, animals, and microorganisms (that are at least in part controlled by geologic and climatic factors) on some kind of landform through an interval of time. Under certain conditions, each of the five variables can exert the controlling influence. Three of the factors—climate, organisms, and slope —may change during the progress of soil genesis.

Geologists are inclined to assume that parent material is the dominant soil-forming factor. A residual soil cannot contain minerals that are not present in, or cannot be made by chemical weathering of, the parent material. We assume that a soil on weathered limestone will be calcareous and that one on quartz sand will be siliceous. However, the chemical weathering processes that have been reviewed earlier in this chapter should demonstrate that the residual or new compounds may be in very different proportions from the primary minerals. For instance, the Paleozoic Beekmantown limestone in Tennessee is composed of 90 percent calcium and magnesium carbonate, 7 percent quartz sand, and 3 percent detrital clay. A simple calculation shows that solution of 5 m of this limestone produces a residual parent material 1 m thick if the residuum has one-half the bulk density of the original rock. The residuum is 70 percent sand and 30 percent clay and hydrated iron oxides. Unless we comprehend the relative solubilities of limestone, quartz, and clay minerals, it is hard to believe that a typical red, sandy, clay loam soil on the Beekmantown limestone can result from *in situ* weathering.

An even more extreme example of the contrast between bedrock and residual soil is the great surficial deposit of lateritic bauxite (hydrated aluminum and iron oxides) at Weipa, Queensland, Australia (Loughnan and Bayliss, 1961). The bedrock is sandstone consisting of 90 percent quartz and 10 percent kaolinite (Fig. 6-9). Under a hot monsoon climate, with strongly seasonal rainfall averaging over 1500 mm/yr, most of the quartz has been leached to a depth of over 8 m, and the upper 5 m is mostly hydrated aluminum oxides stained by iron oxides. Only 5 percent of the thick surface layer is quartz, yet the bedrock is dominantly quartz, which is a relatively insoluble mineral even under the intense heat and rainfall of northern Queensland. A long interval of weathering must be assumed in order to explain the Weipa bauxite ore, because the smooth curves of the weathering profile shown in Figure 6-9 are clear evidence that the surface bauxite ore is the residuum of the quartz sandstone at depth.

The two examples just cited demonstrate that residual soils can be very different from their parent material, given the right combination of the other four soil-forming factors. Nevertheless, "structural" control of weathering is as important in soil genesis as it is in landscape development.

Climate must compete with parent material for the status of "first among equals" in the list of soil-forming factors. Pedologists, like geomorphologists, probably can

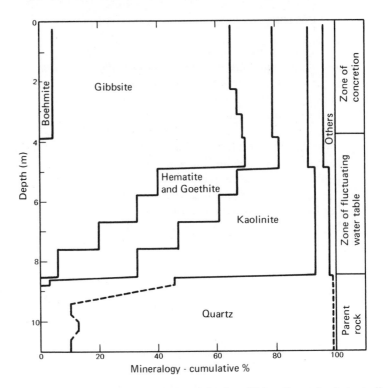

Figure 6-9.   Mineralogy in relation to depth at Weipa, Queensland, Australia. (Loughnan and Bayliss, 1961.) See also pp. 130–131.

be divided into two groups: those who believe that weathering processes (primarily a function of climate) give the distinctive character to a soil or a landscape in spite of lithologic or structural differences, and those who believe that structure always shows through in spite of climate-controlled differences in the intensity of soil-forming and weathering processes. Readers of this book should compare the approach of Part I on constructional landforms with the approach of Part III on geomorphic processes unique to certain climates. Either structure or process can be emphasized, depending on the interest and experience of the writer or reader.

Organisms as a factor in soil formation have been reviewed in their role as chemical weathering agents. Some pedologists view plants as a secondary response to climate and parent material. Others emphasize the role of soil microorganisms and give reduced emphasis to the rooted macroflora. And others see organisms as little more than the creators of a certain chemical soil environment in which essentially chemical weathering reactions form the various horizons of the profile.

Slope, or topographic influence on soil formation, is emphasized more by pedologists than by geologists. The geologist sees a slope as the result of processes acting on structures through time (Chap. 8). The pedologist sees slope as a pre-existing factor in the formation of soil. If the slope is steep, then runoff is rapid, erosion removes the soil as fast as it forms, little water enters the soil, and profiles are thin and poorly developed. On more level terrain, runoff is inhibited and more water enters the ground to weather minerals and translocate clay and other mobile components of the profile. Yet while a soil is forming, so is the slope around it. Pedologists need the geomorphic appreciation of the origin of slopes, lest they assume that the primary condition of a region is of soil-less, but otherwise fully shaped, landforms. This cannot be true, because weathering, and therefore soil formation, is as much a factor in slope development as slope is a factor in soil development. No simple cause-and-effect relationship is possible.

The role of the fifth soil-forming factor—time—is at the heart of both geomorphology and pedology. Simply and directly asked, do soils and landscapes continue to evolve through time at rates appropriate to the energy expended on them? Do soils and landscapes reach some end form in a closed system and cease to evolve; or do they reach some dynamic equilibrium between processes and form and maintain thereafter a steady-state morphology, even though energy and matter continue to move through their open physical systems? These are the underlying questions of explanatory description, and if soils are to be described in terms of the interaction of five soil-forming factors, pedologists must be geomorphologists, just as geomorphologists must be pedologists.

Soil Classification

The very definition of soil as the abode of rooted plants emphasizes the practical reason for studying and classifying soils. Each developed nation has a scheme of soil mapping that enables its agronomists and economists to plan the effective use of available land. Although genetic assumptions are useful and unavoidable, modern soil classifications attempt to classify according to definable, observable chemical and physical properties of the horizons in soil profiles.

In 1965, the United States Department of Agriculture adopted a totally new classification of soils to replace the classification that had been in use since 1938 (Soil Survey Staff, 1975). The new system had been under development since 1951 as a series of "approximations" until it was formally adopted for use by the U.S. Soil Conservation Service and state cooperative programs. Although created to meet the needs of describing and mapping soils in the United States, the new U.S. soil taxonomy was designed for worldwide applicability. Special attention was given to tropical soils, which received woefully inadequate treatment in previous soils classifications. The classification is outlined here in the confident belief that it is the best and most comprehensive classification of soils and soil-forming processes yet devised. Students of geomorphology should become familiar with the new, sometimes curiously compounded, terminology. It is regrettable that some recent textbooks of geology and geography have persisted in using the highly useful, but now obsolete, terminology of the 1938 U.S. classification.

The nomenclature of the U.S. soil taxonomy (Soil Survey Staff, 1975) is systematic. The hierarchy of terms, in descending order of rank, is: order, suborder, great group, subgroup, family, and series. Ten soil orders are defined, based on the presence or degree of development of the various horizons in soil profiles. The name of each soil order (Table 6-1) ends in *sol* (L. *solum*, soil) and contains a formative element that is used as the final syllable in the names of taxa in suborders, great groups, and subgroups. Thus the name of each of the 47 suborders consists of two syllables, a formative element from a list of terms (Table 6-2) prefixed to the formative element of an order. The name of each great group, now about 185 in number, consists of three or four syllables, created by prefixing a suitable element from a list (Table 6-3) to the

TABLE 6-1. NAMES OF SOIL ORDERS,
WITH SIMPLIFIED DEFINITIONS

ALFISOL	Soil with gray to brown surface horizon, medium to high base supply, and a subsurface horizon of clay accumulation. Formative element: *alf*.
ARIDISOL	Soil with pedogenic horizons, low in organic matter, usually dry. Formative element: *id*.
ENTISOL	Soil without pedogenic horizons. Formative element: *ent*.
HISTOSOL	Organic (peat and muck) soil. Formative element: *ist*.
INCEPTISOL	Soil with weakly differentiated horizons showing alteration of parent materials. Formative element: *ept*.
MOLLISOL	Soil with a nearly black, organic-rich surface horizon and high base supply. Formative element: *oll*.
OXISOL	Soil that is a mixture principally of kaolin, hydrated oxides, and quartz. Formative element: *ox*.
SPODOSOL	Soil that has an accumulation of amorphous materials in subsurface horizons. Formative element: *od*.
ULTISOL	Soil with a horizon of clay accumulation and low base supply. Formative element: *ult*.
VERTISOL	Cracking clay soil. Formative element: *ert*.

TABLE 6-2. PRINCIPAL FORMATIVE ELEMENTS TO BE USED
IN NAMING SUBORDERS

Names of suborders consist of two syllables, for example, *Aqualf*. Formative elements are as follows:

alb	L. *albus*, white; soils from which clay and iron oxides have been removed.
aqu	L. *aqua*, water; soils that are wet for long periods.
arg	modified from L. *argilla*, clay; soils with a horizon of clay accumulation.
bor	Gr. *boreas*, northern; cool.
fluv	L. *fluvius*, river; soils formed in alluvium.
ochr	Gr. base of *ochros*, pale; soils with little organic matter.
orth	Gr. *orthos*, true; the common or typical.
psamm	Gr. *psammos*, sand; sandy soils.
ud	L. *udus*, humid; of humid climates.
umbr	L. *umbra*, shade; dark colors reflecting much organic matter.
ust	L. *ustus*, burnt; of dry climates with summer rains.
xer	Gr. *xeros*, dry; of dry climates with winter rains.

TABLE 6-3. PRINCIPAL FORMATIVE ELEMENTS TO BE USED
IN NAMING GREAT GROUPS

Names of great groups consist of more than two syllables and are formed by adding a prefix to the suborder name, for example, *Cryoboralf*. Some of the formative elements are as follows:

cry	Gr. *kryos*, icy cold; cold soils.
dystr, dys	modified from Gr. *dys*, ill; infertile.
eutr, eu	modified from Gr. *eu*, good; fertile.
frag	modified from L. *fragilis*, brittle; a brittle pan.
gloss	Gr. *glossa*, tongue; deep, wide tongues of albic materials into the argillic horizon.
hapl	Gr. *haplous*, simple; the least advanced horizons.
quartz	Ger. *quarz*, quartz; soils with very high content of quartz.
torr	L. *torridus*, hot and dry; soils of very dry climates.
trop	modified from Gr. *tropikos*, of the solstice; humid and continually warm.

---

name of a suborder. The names of the 970 subgroups are created by an adjective modifying the name of a great group. At first glance, the resulting names are totally exotic, but all can be decoded quickly and precisely. All the formative elements are derived from Greek or Latin roots, are as short as possible, and are designed for use in any modern language that uses the Latin alphabet. By compounding the several ranks of formative elements, any soil type on earth can be classified, and in the definition of the formative elements, the soil is also described.

For example, the Dunkirk Series of central New York State is a well drained soil formed in silty, calcareous lacustrine deposits of late Pleistocene age. The Dunkirk Series is classified at the subgroup level as a *Glossoboric Hapludalf*. Decoded, using the formative elements from Tables 6-1, 6-2, and 6-3, the Dunkirk Series belongs to a subgroup of the great group *Hapludalf*, of the suborder *Udalf*, and of the order *Alfisol*. Translating the formative elements of the name, the soil has a gray to brown surface horizon, moderate to high base (cation) saturation, and clay accumulation in a subsurface horizon (formative element *alf*). It is of the suborder of Alfisol that forms in humid regions (formative element *ud*), specifically, one of the alfisols of temperate and tropic regions that are usually moist but may be intermittently dry in some horizons during summer. It is of the great group that has the least complex horizons of the suborder (formative element *hapl*); specifically, it is a brownish or reddish Udalf with only a slight loss of clay from the A horizon and a moderate accumulation of clay in the B horizon. It is without a dense, clay-rich crust (fragipan) in the profile. It is in an area characterized by mean summer and mean winter soil temperatures that differ by 5°C or more and a mean annual temperature of 8°C or more (temperate climate). The great group Hapludalf is equivalent to the Gray-Brown Podzolic soils of the 1938 classification. To classify the Dunkirk Series to the subgroup level, the compound adjective *glossoboric* precedes the great group name. *Glossic* refers to minor but characteristic tongues of the clay-depleted A horizon that extend down into the clay-enriched B horizon; if this tonguing were more pronounced, the soil would be a *Glossudalf* instead of a *Hapludalf*. *Boric* is from the root *boreal*; as defined for this subgroup, the mean annual soil temperature is less than 10°C.

The decoding exercise of the preceding paragraph is tedious and complex for a novice, but it illustrates that approximately a full page of written description about a soil and its climatic environment is contained in the name of the soil at the subgroup level. If additional modifiers concerning texture, mineralogy, grain size, permeability, and soil temperature are prefixed to describe the soil at the family level, a comprehensive impression of the soil, its genesis, and its agricultural potential can be gained just from the name alone. A map of the principal orders, suborders, and great groups of soils found in the United States is published in the U.S. National Atlas (U.S. Geol. Survey, 1970, Plate 38) with a useful accompanying descriptive text. Other than by consulting the definitive document (Soil Survey Staff, 1975) or appropriate textbooks of soil classification and morphology (Buol, et al., 1973; Birkeland, 1974; and Steila, 1976), the Atlas map and text are recommended as a good introduction to the U.S. soil taxonomy.

STRUCTURE, CLIMATE, AND ROCK WEATHERING

To what extent does structure influence or control rock weathering? To what degree are landforms shaped by climate-controlled weathering processes? The remainder of this chapter considers these two questions, in review and in anticipation of expanded treatment in later chapters where structural and climatic control of mass-wasting and erosion are also considered.

Structural Influences in Rock Weathering

Lithology, the physical character of a rock, exerts an obvious structural control on weathering. Some rocks are massive; others have strong *fabrics* that preferentially control mechanical disaggregation and chemical disintegration. Some are permeable; others are nearly impermeable. Related to lithology are the *secondary fractures* that have resulted from tectonic deformation or other stresses. Fractures of many kinds, from microscopic pores and cracks to major faults, break rocks and create surfaces that air and water can reach.

Mineralogic factors are part of the structure that controls weathering. Chemical composition is obviously important, as noted by the fact that each example of chemical weathering cited earlier in the chapter used a particular chemical compound (mineral) in the equation. Two other less obvious mineralogic factors also influence rock weathering. One is grain size, and the other can be called "crystallinity," or the amount of crystalline, as opposed to amorphous or glassy, minerals in a rock. For example, different textural types of compositionally identical granite have varying resistance to weathering and erosion. In the humid tropical climate of Hong Kong, coarse porphyritic granite is least resistant, coarse-grained granite is stronger, and medium-grained granite, fine-grained granite, and finally granite porphyry dikes are the most resistant (Ruxton and Berry, 1957, p. 1273). Obsidian, basalt glass, and amorphous silica (chert and chalcedony) are notably more soluble in water than is crystalline quartz. Hydrated rinds have formed on the surface of obsidian and basalt-glass artifacts only a few thousand years old. Their thickness can be calibrated as an archeologic dating method.

Almost every rock type has a distinctive weathering character. Limestones are often smooth, gray, and lichen-covered, or deeply pitted by solution (Chap. 7). Shales split and crumble. Granites form sandy grus. The list could be extended at length. The weathered appearance of rock is so useful that it is included among the diagnostic criteria for defining *formations*, the basic units of geologic mapping (Dunbar and Rodgers, 1957, pp. 267, 273).

Climatic Influence on Rock Weathering

Climate is the integration of weather over a period of years. Weather is related to weathering by more than the identical origin of the terms; the same moisture, temperature, and seasonality factors that define a regional climate also define the weathering processes that operate there.

Strakhov (1967) drew an instructive, if idealized, meridianal cross section of the earth's weathered layer (Fig. 6-10). From polar to equatorial latitudes, his figure not

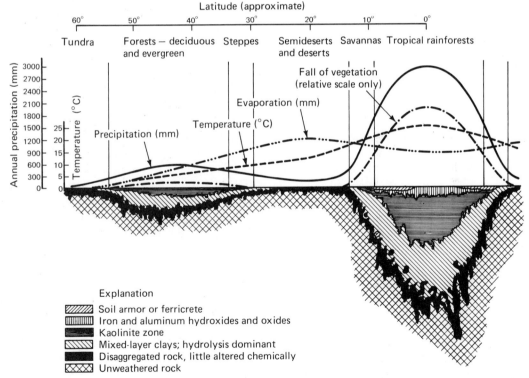

Figure 6-10. Depth and character of weathering in relation to latitude. (Adapted from Strakhov, 1967, Fig. 2.)

only shows the varying depths of weathering but illustrates that certain sets of weathering and soil-forming processes follow each other in prescribed order. Mechanically fractured rocks, but little altered chemically, are at the surface in polar and subtropical deserts. Either because of extreme cold or lack of precipitation, or both, these are latitudinal belts of minimal chemical and biological weathering.

Desert weathering in warm climates proceeds with a lack of precipitation, as it does in cold deserts, except that ground water is present, even if at great depth. If the water table is near the surface, evaporation draws salt-laden ground water toward the surface, where the salts precipitate and cause exfoliation. Surface *flaking* of granular rock under projecting ledges in arid parts of Australia (Dragovich, 1967) occurs because dew and other atmospheric moisture is uniformly absorbed to a depth of 15 mm or less, and small chips or flakes of slightly hydrated but otherwise fresh rock are spalled off.

In a humid climate with seasonal freezing, cold weather brings a dormant period for both vegetation and soil microorganisms. Trees shed their leaves or needles, and the humus layer at the surface of the ground accumulates faster than microorganisms, worms, and insects can consume it. Rainwater or snowmelt that soaks through the rotting humus collects organic compounds and carbon dioxide that chelate and hydrolyze metallic cations from the underlying minerals and generally leave a silica-rich clay residue. Iron oxides and clay minerals that are washed from the surface layer accumulate in the upper meter of soil. Chemical weathering processes penetrate only a few meters into rock, but mechanical weathering, especially hydrofracturing, may extend considerably deeper (Fig. 6-10). Clay minerals in the soil prevent water from freely penetrating the ground and encourage overland runoff. The resulting landscape develops broad, gentle, soil-covered slopes shaped by soil creep and stream erosion. Ridge crests commonly have thin soils or exposed bedrock ledges.

In the tropics, hydrolysis and residual clay-mineral formation reach to depths of 100 m or more (Fig. 6-10). The great depth of tropical weathering is due in part to temperature, but abundant precipitation is an even greater climatic factor. In spite of the enormous biological productivity of a tropical rainforest, little humus accumulates because of continuous microbial and fungal attack and rapid recycling of nutrients. The heavy rainfall literally flushes the soil free of all soluble compounds.

Chemical weathering in the tropics is so intense that clay minerals, the stable residuals of weathering in cooler latitudes, are desilicified. A typical reaction of kaolinite in a great excess of water is:

$$Al_2Si_2O_5(OH)_4 + 5\,H_2O \longrightarrow 2\,Al(OH)_3 + 2\,H_4SiO_4$$

$$\text{(kaolinite)} \quad + \text{(water)} \quad = \quad \text{(gibbsite)} + \text{(silicic acid in solution)}$$

The silica that is removed by leaching is shown as silicic acid in solution, as in earlier illustrations of hydrolysis (p. 117). The actual form of the leached silica is not certain. Some is in amorphous or colloidal form and may reprecipitate at the water table in seasonally dry tropical regions to form nodules, crusts, and layers of chalcedony or opal. Gibbsite, the aluminum hydroxide residual, recrystallizes slowly as insoluble hydrated alumina:

$$2 \, Al(OH)_3 \longrightarrow Al_2O_3 \cdot 3 \, H_2O$$

Gibbsite is the major mineral in the heterogeneous earthy mixture of hydrated aluminum oxides called *bauxite*, the ore of aluminum. Tropical weathering of iron-poor aluminosilicates produces a surface layer of earthy or nodular bauxite (Fig. 6-9) that can be handled by ordinary earth-moving equipment.

In latitudinal belts peripheral to the tropical rainforests are the regions of tropical savanna climate (Chap. 13). During the wet season, normally when the sun is overhead, the savannas are flooded, but during the dry half of the year, almost no rain may fall. The seasonal wetting promotes deep weathering, but the dry season accelerates generally irreversible oxidation of aluminum and iron hydroxides. Massive *laterite* crusts may form cap rock that exerts a strong structural control on the landscape (Chap. 11). In some regions, the layer is called *soil armor*, or *ferricrete*.

The chemistry of tropical weathering is complex and as yet relatively unstudied. At least some of the soils stay soft and clay-rich as long as they are under forest cover and remain wet throughout the year. If they are allowed to dry, either seasonally or by exposure through agricultural use or road construction, they irreversibly harden, apparently within a few years. These are the true **laterites**, defined by their tendency to harden. Large areas of laterite in southeast Asia are believed to have been created by the traditional slash-and-burn, or *swidden*, agricultural pattern. After gardening, a patch of burned forest floor becomes hard and infertile, and a new patch must be cleared.

Tropical weathering has been discussed at length because of the dramatic intensity and depth of chemical activity. The contrasts with weathering in cooler and less humid climates are those of degree and not of kind. The latitudinal temperature gradient of the earth is really not very great, especially when considered on the Kelvin scale of absolute temperature, which controls the kinetics of chemical reactions. The intensity of tropical weathering is caused by the continuous biological activity and massive quantities of rainfall that flush through the weathered layer.

In general, the intensity of tropical weathering minimizes the effects of structural control. The depth of the weathered zone is equal to the local relief on hills, escarpments, and other landforms, so that entire landscapes may be within the weathered zone, rather than the weathered zone being a rind on the surface of the landforms. Geomorphologists trained in the cool, humid climates of Europe and North America have always been profoundly impressed with the scale of climate influence in tropical landscapes. As the developing nations of the tropics have become more influential, and as tropical mineral resources such as bauxite and laterite have become the primary raw materials of industrial nations, tropical weathering has been given much greater attention.

REFERENCES

ARGO, V. N., 1964, Strangler fig: Native epiphyte: *Nat. Hist.*, v. 73, no. 9, pp. 26–31.

BATEMAN, P. C., and EATON, J. P., 1967, Sierra Nevada batholith: *Science*, v. 158, pp. 1407–17.

BENNETT, J. G., 1936, Broken coal: *Jour. Inst. Fuel*, v. 10, no. 49, pp. 22–39.

BERNER, R. A., and HOLDREN, G. R., JR., 1977, Mechanism of feldspar weathering: Some observational evidence: *Geology*, v. 5, pp. 369–72.

BIRCH, F., 1966, Compressibility; elastic constants, *in* Clark, S. P., Jr., ed., Handbook of physical constants: *Geol. Soc. America Mem. 97*, pp. 97–173.

BIRKELAND, P. W., 1974, *Pedology, weathering, and geomorphological research:* Oxford Univ. Press, New York, 285 pp.

BLATT, H., MIDDLETON, G. V., and MURRAY, R. C., 1972, *Origin of sedimentary rocks:* Prentice-Hall, Inc., Englewood Cliffs, N.J., 634 pp.

BRACE, W. F., SILVER, E., HADLEY, H., and GOETZE, C., 1972, Cracks and pores: A closer look: *Science*, v. 178, pp. 162–64.

BRADLEY, W. C., 1963, Large-scale exfoliation in massive sandstones of the Colorado Plateau: *Geol. Soc. America Bull.*, v. 74, pp. 519–28.

BRIDGMAN, P. W., 1911, Water, in the liquid and five solid forms, under pressure: *Am. Acad. Arts Sci. Proc. (Daedalus)*, v. 47, pp. 439–558.

BÜDEL, J., 1968, Geomorphology—principles, *in* Fairbridge, R. W., ed., *Encyclopedia of geomorphology:* Reinhold Book Corp., New York, pp. 416–22.

BUOL, S. W., HOLE, F. D., and MCCRACKEN, R. J., 1973, *Soil genesis and classification:* Iowa State Univ. Press, Ames, 360 pp.

CAWLEY, J. L., BURRUSS, R. C., and HOLLAND, H. D., 1969, Chemical weathering in central Iceland: An analog of pre-Silurian weathering: *Science*, v. 165, pp. 391–92.

CLARK, S. P., JR., 1971, *Structure of the earth:* Prentice-Hall, Inc., Englewood Cliffs, N.J., 132 pp.

CLARK, S. P., Jr., and JAGER E., 1969, Denudation rate in the Alps from geochronology and heat flow: *Am. Jour. Sci.*, v. 267, pp. 1143–60.

DALE, T. N., 1923, Commercial granites of New England: *U.S. Geol. Survey Bull. 738*, 488 pp.

DEGENS, E. T., 1965, *Geochemistry of sediments:* Prentice-Hall, Inc., Englewood Cliffs, N.J., 342 pp.

DEJU, R. A., 1971, A model of chemical weathering of silicate minerals: *Geol. Soc. America Bull.*, v. 82, pp. 1055–62.

DENNY, C. S., and GOODLETT, J. C., 1956, Microrelief resulting from fallen trees, *in* Denny, C. S., Surficial geology and geomorphology of Potter County, Pa.: *U.S. Geol. Survey Prof. Paper 288*, pp. 59–66.

DRAGOVICH, D., 1967, Flaking, a weathering process operating on cavernous rock surfaces: *Geol. Soc. America Bull.*, v. 78, pp. 801–4.

DUNBAR, C. O., and RODGERS, J., 1957, *Principles of stratigraphy:* John Wiley & Sons, Inc., New York, 356 pp.

DUNN, J. R., and HUDEC, P. P., 1966, Water, clay and rock soundness: *Ohio Jour. Sci.*, v. 66, no. 2, pp. 153–67.

DURY, G. H., LANGFORD-SMITH, T., and MCDOUGALL, I., 1969, Minimum age for the duricrust: *Australian Jour. Sci.*, v. 31, pp. 362–63.

EVANS, I. S., 1970, Salt crystallization and rock weathering: A review: *Rev. Géomorph. Dynamique*, v. 19, no. 4, pp. 153–77.

FETH, J. H., ROBERSON, C. E., and POLZER, W. L., 1964, Sources of mineral constituents in water from granitic rocks, Sierra Nevada, California and Nevada: *U.S. Geol. Survey Water-Supply Paper 1535-I*, 70 pp.

FREDERICKSON, A. F., 1951, Mechanism of weathering: *Geol. Soc. America Bull.*, v. 62, pp. 221–32.

FRIEDMANN, E. I., and OCAMPO, R., 1976, Endolithic blue-green algae in the dry valleys: Primary producers in the Antarctic desert ecosystem: *Science*, v. 193, pp. 1247–49.

GOLDICH, S. S., 1938, Study in rock weathering: *Jour. Geology*, v. 46, pp. 17–58.

GRIGGS, D. T., 1936, The factor of fatigue in rock exfoliation: *Jour. Geology*, v. 44, pp. 783–96.

HACK, J. T., 1966, Circular patterns and exfoliation in crystalline terrane, Grandfather Mountain, North Carolina: *Geol. Soc. America Bull.*, v. 77, pp. 975–86.

HAXBY, W. F., and TURCOTTE, D. L., 1976, Stresses induced by the addition or removal of overburden and associated thermal effects: *Geology*, v. 4, pp. 181–84.

HOLLINGWORTH, S. E., TAYLOR, J. H., and KELLAWAY, G. A., 1944, Large-scale superficial structures in the Northampton ironstone field: *Geol. Soc. London Quart. Jour.*, v. 100, pp. 1–44.

HOPSON, C. A., 1958, Exfoliation and weathering at Stone Mountain, Georgia, and their bearing on disfigurement of the Confederate Memorial: *Georgia Mineral Newsletter*, v. 11, pp. 65–79.

ISHERWOOD, D., and STREET, F. A., 1976, Biotite-induced grussification of the Boulder Creek Granodiorite, Boulder County, Colorado: *Geol. Soc. America Bull.*, v. 87, pp. 366–70.

JAHNS, R. H., 1943, Sheet structure in granite: Its origin and use as a measure of glacial erosion in New England: *Jour. Geology*, v. 51, pp. 71–98.

JOHNSON, N. M., REYNOLDS, R. C., and LIKENS, G. E., 1972, Atmospheric sulfur: Its effect on the chemical weathering of New England: *Science*, v. 177, pp. 514–16.

KELLER, W. D., and FREDERICKSON, A. F., 1952, Role of plants and colloidal acids in the mechanism of weathering: *Am. Jour. Sci.*, v. 250, pp. 594–608.

KESSLER, D. W., INSLEY, H., and SLIGH, W. H., 1940, Physical, mineralogical, and durability studies on the building and monumental granites of the United States: *U.S. Nat. Bureau of Standards Jour. Res.*, v. 25, pp. 161–206.

KUENEN, PH. H., 1955, *Realms of water:* John Wiley & Sons, Inc., New York, 327 pp.

LEGGET, R. F., 1967, Soil: Its geology and use: *Geol. Soc. America Bull.*, v. 78, pp. 1433–59.

LEHMAN, D. S., 1963, Some principles of chelation chemistry: *Soil Sci. Soc. America Proc.*, v. 27, pp. 167–70.

LIKENS, G. E., and BORMANN, F. H., 1974, Acid rain: A serious regional environmental problem: *Science*, v. 184, pp. 1176–79.

LIVINGSTONE, D. A., 1963, Data of geochemistry, Chap. G. Chemical composition of rivers and lakes: *U.S. Geol. Survey Prof. Paper 440-G*, 64 pp.

LOUGHNAN, F. C., 1969, *Chemical weathering of the silicate minerals:* American Elsevier Publishing Co., Inc., New York, 154 pp.

LOUGHNAN, F. C., and BAYLISS, P., 1961, Mineralogy of the bauxite deposits near Weipa, Queensland: *Am. Mineralogist*, v. 46, pp. 209–17.

LOVERING, T. S., and ENGEL, C., 1967, Translocation of silica and other elements from rock into *Equisetum* and three grasses: *U.S. Geol. Survey Prof. Paper 594-B*, 16 pp.

MATTHES, F. E., 1930, Geologic history of the Yosemite Valley: *U.S. Geol. Survey Prof. Paper 160*, 137 pp.

MELLOR, M., 1970, Phase composition of pore water in cold rocks: *U.S. Army Corps of Engineers Cold Regions Res. Eng. Lab. Res. Rept. 292*, 61 pp.

OLLIER C. D., 1963, Insolation weathering: Examples from central Australia: *Am. Jour. Sci.*, v. 261, pp. 378–81.

———, 1965, Some features of granite weathering in Australia: *Zeitschr. für Geomorph.*, v. 9, pp. 285–304.

———, 1969, *Weathering:* American Elsevier Publishing Co., Inc., New York, 304 pp.

PHILBRICK, S. S., October 1, 1974, Personal communication.

REICHE, P., 1950, *Survey of weathering processes and products*, 2d ed.: New Mexico Univ. Press, Albuquerque, N.M., 95 pp.

RICE, ALAN, 1976, Insolation warmed over: *Geology*, v. 4, pp. 61–62.

ROTH, E. S., 1965, Temperature and water content as factors in desert weathering: *Jour. Geology*, v. 73, pp. 454–68.

RUXTON, B. P., and BERRY, L., 1957, Weathering of granite and associated erosional features in Hong Kong: *Geol. Soc. America Bull.*, v. 68, pp. 1263–92.

SCHNEIDER, R., 1964, Relation of temperature distribution to ground-water movement in carbonate rocks of central Israel: *Geol. Soc. America Bull.*, v. 75, pp. 209–15.

SILVERMAN, M. P., and MUNOZ, E. F., 1970, Fungal attack on rock: Solubilization and altered infrared spectra: *Science*, v. 169, pp. 985–87.

SIMMONS, G., TODD, T., and BALDRIDGE, W. S., 1975, Toward a quantitative relationship between elastic properties and cracks in low porosity rocks: *Am. Jour. Sci.*, v. 275, pp. 318–45.

SKINNER, B. J., 1966, Thermal expansion, *in* Clark, S. P., Jr., ed., Handbook of physical constants: *Geol. Soc. America Mem. 97*, pp. 75–96.

Soil Survey Staff, 1975, Soil taxonomy: A basic system of soil classification for making and interpreting soil surveys: *U.S. Dept. Agr. Handbook 436*, 754 pp.

STEILA, D., 1976, *Geography of soils:* Prentice-Hall, Inc., Englewood Cliffs, N.J., 222 pp.

STEVENS, G. R., 1974, *Rugged landscape: The geology of central New Zealand:* A. H. & A. W. Reed, Wellington, N. Z., 286 pp.

STOUT, J. H., 1974, Helicoidal crack propagation in aluminous orthoamphiboles: *Science*, v. 185, pp. 251–53.

STRAKHOV, N. M., 1967, *Principles of lithogenesis*, v. 1 (transl. by J. P. Fitzsimmons): Oliver and Boyd, Edinburgh, 245 pp.

STURGUL, J. R., and GRINSHPAN, Z., 1975, Finite-element model for possible isostatic rebound in the Grand Canyon: *Geology*, v. 3, pp. 169–71.

TARR, W. A., 1915, A study of some heating tests, and of the light they throw on the cause of the disaggregation of granite: *Econ. Geology*, v. 10, pp. 348–67.

TWIDALE, C. R., 1973, On the origin of sheet jointing: *Rock Mechanics*, v. 5., pp. 163–87.

U.S. Geological Survey, 1970, *National atlas of the United States of America:* U.S. Dept. of Interior, Geol. Surv., Washington, D.C., 417 pp.

WASHBURN, A. L., 1969, Weathering, frost action, and patterned ground in the Mesters Vig district, northeast Greenland: *Medd. om Grønland*, v. 176, no. 4, 303 pp.

WHITNEY, J. A., JONES, L. M., and WALKER, R. L., 1976, Age and origin of the Stone Mountain Granite, Lithonia district, Georgia: *Geol. Soc. America Bull.*, v. 87, pp. 1067–77.

WHITNEY, P. R., and MCLELLAND, J. M., 1973, Origin of coronas in metagabbros of the Adirondack Mts., N.Y.: *Contr. Mineral. and Petrol.*, v. 39, pp. 81–98.

WINKLER, E. M., 1968, Frost damage to stone and concrete: Geological considerations: *Eng. Geology*, v. 2, pp. 315–23.

——, 1973, *Stone: Properties, durability in man's environment:* Springer-Verlag, New York, 230 pp.

WINKLER, E. M., and WILHELM, E. J., 1970, Salt burst by hydration pressures in architectural stone in urban atmosphere: *Geol. Soc. America Bull.*, v. 81, pp. 567–72.

CHAPTER 7

Karst

In the previous chapter, solution was excluded as a specific process of chemical weathering, on the argument that all chemical processes of rock weathering involve aqueous solutions. In one special case, however, solution of massive quantities of abnormally soluble rock gives rise to a terrain so distinctive that almost every dialect has a special set of descriptive terms for the landforms. The Germanized form of an ancient Slovenian term for such a region of solution topography, now part of Yugoslavia, is *karst*. The term connotes both an assemblage of landforms and a set of processes. It has become a truly international term, and although an enormous and provincial vocabulary has accumulated to describe karst processes and forms, including caves and their deposits, the one word offers a central and unifying concept for the special landforms on exceptionally soluble rocks.

DEFINITIONS AND HISTORICAL CONCEPTS

Karst is a *"terrain with distinctive characteristics of relief and drainage arising primarily from a higher degree of rock solubility in natural waters than is found else-where* [Jennings, 1971, p. 1]." This sensible and practical definition stresses geomorphic *form*, but emphasizes *process* in stressing the role of solution in natural waters, and emphasizes *structural control* in stressing that abnormally soluble rocks are involved. Solution becomes a process of both weathering and erosion inasmuch as the weathered products are removed from the surface and carried away in solution.

Attention to the definition is important. For some students, karst is synonymous with only limestone landscapes. However, other soluble-rock terranes may be karst, and limestone weathering in some regions is not characterized by solution. In both hot and polar deserts, massive limestone formations become prominent cliff-formers. Without water, karst does not form. Also, equating karst with limestone eliminates many interesting karst landscapes on gypsum, salt, dolomite, glacier ice, and perennially frozen ground.

Karst also has been defined as a dry landscape, characterized by subterranean drainage rather than by surface streams. It is true that subsurface drainage is common in karst regions, but it must be considered the effect, not the cause, of excessive solubility and solution. The landscape of southeast England and northwest France, for instance, is underlain by thick beds of Cretaceous chalk, a soluble, fine-grained, soft, and massively permeable limestone. Almost all of the chalk terrain lacks surface streams except during peak runoff intervals, yet the chalk does not form karst. The uniform permeability precludes it.

No branch of geomorphology is so plagued by terminology as "karstology." Part of the problem is psychological in that every language and dialect has special, often highly imaginative, terms to describe caverns and other unusual karst landforms. The other part of the problem is historical. The first systematic studies of karst were carried out in the Dinaric Alps of a Yugoslavian region traditionally known as The Karst. Albrecht Penck (the author of the first textbook of geomorphology) and his students popularized the study of karst landforms (Cvijić, 1893; Penck, 1894). They translated local Slovenian terms for the forms first into German, then into French, then Italian, and subsequently into other languages. In each country, the words were given slightly different equivalents. Only since World War II has there been international integration of terms and theories of hydrogeologists, speleologists, and geomorphologists to create a body of karst literature.

Even before Penck and his students began their systematic study of the Yugoslavian karst, descriptions of tropical karst had been published that emphasized different landforms than were known from the Mediterranean or other temperate karst regions. Sawkins (1869) described the "cockpits" of Jamaican karst, a maze of solution depressions separated by steep ridges. Climatic control of karst processes received greater attention in the decades of 1920 and 1930 with studies of tropical karst in south China and in Java (Roglić, 1972). The *tower karst* landscape (p. 154) with isolated, steep- or vertical-sided hills rising from flat plains in the humid tropics, was proposed to be a fundamentally different, climate-controlled karst landscape. The barrel-shaped or conical isolated peaks were judged to be the residuals of marginal corrosion by acid swamp water, a process unique to warm, humid regions.

Hydrologists have long been aware of the unique problems of ground-water movement in soluble rocks. Theories of ground-water movement have had to be modified from models of uniform motion along regional hydrostatic gradients to models more related to flow in pipes or networks of pipes. The very assumption of a water table in karst has been questioned by some hydrologists, even though geomorphologists have debated whether caves form above or below the water table.

The introduction of hydrologic ideas about subsurface water movement leads to a consideration of the role of caves in karst geomorphology. Topologically, a cave open to the atmosphere is part of the landscape, yet caves are considered "subterranean," unfamiliar, and the objects of fearful superstition by most people. Cave exploration has developed its own heroes and terminology. The literature on cave morphology and genesis must be sought in many journals not normally read by geomorphologists.

THE KARST PROCESS

Solution, as the term is used in the context of karst, is essentially the hydrolysis of $CaCO_3$ (p. 116). Solution of calcite in ground water made acid by dissolved CO_2 is represented by the following reaction:

$$CaCO_3 + H_2O + CO_2 \rightleftharpoons Ca^{+2} + 2\,HCO_3^-$$

As simplified, the reaction omits the several steps by which atmospheric CO_2 becomes dissolved in water to form carbonic acid, which in turn ionizes to free H^+ ions that convert CO_3^{-2} to the more soluble form of bicarbonate ion HCO_3^-. The Ca^{+2} ion does not form a hydroxide but remains in ionic equilibrium with the bicarbonate. It should be noted that the dissolved ions are those that also enter solution when a calcium-rich silicate rock weathers by hydrolysis. The reversible reaction is forced to the right if carbon dioxide in solution is increased. If carbon dioxide is removed from the left side of the reaction, a saturated bicarbonate solution will become cloudy with finely divided $CaCO_3$ precipitate, which is much less soluble than the bicarbonate. Limestone solubility is controlled primarily by addition or loss of CO_2 gas to a great excess of water. Changes of temperature or pressure, mixing of unlike water masses, and biologic process, can all promote solution or deposition.

The solubility of $CaCO_3$ in water at various partial pressures of CO_2 is graphed in Figure 7-1. The partial pressure of CO_2 in an aerated aqueous solution is only 30 pascals and the related solubility of $CaCO_3$ is very low, only about 63 mg/1. However, under anaerobic conditions in a soil the partial pressure of CO_2 may increase to 30 kPa and result in an increase of $CaCO_3$ in solution to 700 mg/1, more than a tenfold increase over the amount in aqueous solution at equilibrium with

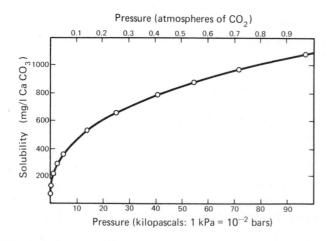

Figure 7-1. Solubility of $CaCO_3$ in water at various partial pressures of CO_2 ($t = 16.1°C$). (After Adams and Swinnerton, 1937.)

normal atmosphere. Most ground water contains dissolved $CaCO_3$ in a concentration of between 200 and 400 mg/1, indicating CO_2 partial pressures well within the range of observed composition of soil air. Limestone solution, and therefore karst, is clearly accentuated by biologically generated CO_2 in decaying humus.

Biological Effects

In addition to providing the decaying humus that enriches soil air in CO_2, plants and animals may corrode limestone directly. Folk, et al. (1973) designated a distinctive minor karst landform on Grand Cayman Island as **phytokarst**. The surface is intricately pitted, with pits intersecting to form a sharp-edged, spongy network of pinnacles over 2 m in height. Algae cover the surface and penetrate the karst surfaces to a depth of 0.1–0.2 mm. Presumably, the algae growing in the limestone generate acidic solvents much like the fungi described previously (p. 119).

Organic wastes accumulate on the floor of caves occupied by bats. The noxious phosphatic mixture, sometimes mined commercially as *guano*, strongly corrodes limestone. A very similar process occurs on coral islands, where flocks of sea birds roost. The reaction of bird guano and limestone produces commercial deposits of rock phosphate, which has been mined extensively, especially on Nauru and Ocean Island in the Pacific (Fig. 7-2). The rock phosphate fills karst cavities in limestone to a depth of 20 m (Power, 1925). On many atolls, the location of former groves of trees in which sea birds roosted can be inferred from areas of phosphatized reef limestone. The acid humus of trees also corrodes peat-filled depressions at their bases.

Figure 7-2. Karst pinnacles on Nauru Island exposed by removing the overlying phosphate rock. (Photo: British Phosphate Commissioners.)

Temperature Effect

Water at 10°C dissolves about twice as much atmospheric CO_2 as water at 30°C; water at 0°C dissolves almost three times as much as at 30°C. Cold water saturated with CO_2 should be more "aggressive" than warm water in dissolving soluble rocks. A logical conclusion is that karst should be well developed in cold, humid climates, especially in snow climates. Snow air is enriched in CO_2 because atmospheric nitrogen and oxygen molecules diffuse out of snow more readily than the larger CO_2 molecules. However, the theoretical saturation values are not the determining factors in karst solution. Even though cold water can hold more CO_2, if the cold landscape is barren of vegetation and the cold soils are poor in humus and microbial activity, there is no source of dissolved CO_2 except the atmosphere. Karst is not a conspicuous feature of cold regions.

Geothermal gradients may be perturbed by ground-water flow, intensified by downward water movement in recharge areas and attenuated in discharge areas (Domenico and Palciauskas, 1973). The solution effects, if any, of the slight thermal anomalies are unknown.

Pressure Effect

In theory, ground water moving downward under hydrostatic pressure will absorb additional CO_2 in proportion to the partial pressure graph (Fig. 7-1) and will dissolve more $CaCO_3$. The theory is excellent, but once ground water has moved deeper than the A horizon of the soil profile, little additional CO_2 is available for absorption, so pressure has little additive effect on solution. The release of pressure when percolating ground water under hydrostatic pressure enters the free atmosphere of a cave may be an important factor in dripstone deposition, however.

Related to the pressure effect is turbulence. Dissolved CO_2 gas may be exsolved by splashing or turbulent flow in an underground stream, and $CaCO_3$ may be deposited. For most purposes, however, the movement of ground water through granular materials is not believed to be turbulent. Models of turbulent, laminar, or diffuse flow, even in limestone aquifers, do not give significantly different predictions of karst solution (Thrailkill, 1968, p. 28).

Mixing Effect

The nonlinear solubility of calcium carbonate in carbonated water gives rise to an interesting phenomenon when two unlike water masses mix. For clarity, a portion of Figure 7-1 is enlarged (Fig. 7-3) to show the *mixing effect*. Suppose that deep ground water, saturated with $CaCO_3$ at a low CO_2 partial pressure, and water from the aerated zone, also saturated but at a higher CO_2 partial pressure, meet and mix at the water table. Because of the curved solubility graph, mixing of any two saturated but unlike water masses invariably results in an undersaturated solution. This mixing effect is probably the reason that most caves develop just below the water table (Moore and Sullivan, 1977).

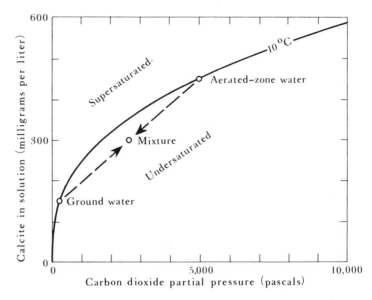

Figure 7-3.   The result of mixing unlike water masses that are each saturated with respect to calcite is a water mass that is undersaturated. (Diagram courtesy of G. W. Moore.)

Flow-Rate Effect

Somewhat related to the mixing effect are the solubility changes related to variations in rate of flow of surface and underground water. If water is moving slowly over or through soluble rock, it should approach saturation equilibrium. However, during rapid runoff periods, river water is notably undersaturated, because the water is not in contact with mineral soil long enough. If such undersaturated water, normally with a low dissolved CO_2 partial pressure, enters a karst terrain and mixes with saturated ground water, the extreme difference in their composition greatly enhances the mixing effect. This process may be especially important in enlarging caves that normally drain outward to a surface river but become backflooded when the river is in flood (Thrailkill, 1968, p. 36).

Climatic Control of Karst Processes

Temperature, precipitation, and biological activity are definitive of climatic regions. It should not be surprising, then, that karst processes should vary in intensity among climatic zones. The individual effects on solution of temperature, available water, and organisms have been considered in preceding paragraphs, but they can be integrated into a climatic model of the karst process, reviewed by latitudinal belts as in the manner of Figure 6-10.

The lack of conspicuous karst in polar regions may be due primarily to the low rate of chemical reactions in cold water, the negligible bacterial decay of humus, the short season of runoff, and permafrost that limits infiltration. Corbel (1959a, b) emphasized the importance of karst solution in cold humid regions, but few other authors have supported him (Jennings, 1971, p. 186). Ford (1971) made extensive measurements of dissolved components in limestone regions of the Canadian Rockies. He found that glacial meltwater above the tree line was saturated with respect to calcite at only 50–90 mg/1 $CaCO_3$, indicating that little CO_2 was in solution. Below the tree line, ground water was not saturated with respect to calcite until it contained 100 to 265 mg/1 $CaCO_3$, and even creeks and lake water had enough dissolved CO_2 to carry 100–140 mg/1 $CaCO_3$ in solution at saturation. Ford concluded that the ground water on forested mountains in a subarctic climate zone is as capable as the ground water in temperate or tropical regions of dissolving limestone but that there is no particular "aggressiveness" associated with glacier or snowbank meltwater.

The cool-humid midlatitude zone of increased weathering (Fig. 6-10) has well-developed karst. Precipitation is greater than in polar regions, more of the precipitation falls as rain, infiltration is not inhibited by frozen ground, and the seasonal leaf-fall and dormant period promote humus accumulation. Ground water is generally abundant. This is the climate for *doline karst*, a maze of surface sinkholes. Caves develop just beneath the water table, probably controlled by the level of nearby rivers.

In the subhumid and semiarid steppe and savanna grasslands, karst is again inhibited by the lack of abundant moisture. During the hot or dry seasons, ground water moves upward in the grassland soils, depositing, rather than dissolving, carbonates. Mollisols, the rich black grassland soils, are characteristically high in cations and do not acidify downward-percolating water during the wet season.

From the emphasis on the necessity for water in karst processes, it is obvious that deserts are not karst-prone. Minor solution features develop, but limestone characteristically assumes a role of ridge- and cliff-former, perhaps rivaling granite in structural control of landforms. Thin soils, negligible plant cover, and lack of effective precipitation mark the subtropical desert region as an extreme of nonkarst development. Caves in deserts are known, but at least some are relict from former more humid climatic conditions.

In the tropical rainforest climate, karst develops to a degree unequaled elsewhere. The CO_2-rich atmosphere beneath the rainforest canopy is well known. Organic acids, including nitrous compounds, add to the aggressiveness of carbonated ground water. Above all, copious quantities of water are constantly moving through the ground.

Structural Control of Karst Processes

The most obvious structural factor in karst is lithology. Most karst is on limestone, but dolomite, gypsum, salt, glacier ice, and perennially frozen ground (permafrost) all show karst landforms in some regions. The term has been deplorably extended (with modifying prefixes as in "thermokarst" and "pseudokarst") to a variety of landforms, such as those formed by compaction of fine-grained volcanic tephra, subsidence of desert soils by a lowering of the water table, and even unusual chemical weathering of silicate rocks.

Limestone is an extremely diverse rock type. To be defined as a limestone for industrial purposes, a rock should contain more than 80 percent calcium and magnesium carbonate, but many limestones are shaly or sandy. Karst limestones, in general, are quite pure. For example, the Dinaric karst is in limestones that range from 80–98 percent $CaCO_3$, with many of the analyses in the upper part of the range (Herak, 1972, p. 28). The Annville Limestone of Ordovician age that underlies a karst area near Hershey, Pennsylvania, is more than 95 percent $CaCO_3$ (Foose, 1953, p. 624). The lower Tertiary White Limestone of the famous "cockpit country" of Jamaica rarely has more than 2 percent noncarbonate minerals and generally has less than 0.5 percent (Versey, 1972, p. 447). By contrast, the Luray Caverns of Virginia are in coarsely crystalline Ordovician dolomite that contains as much as 10 percent quartz, chert, feldspar, and clay, but no calcite (Hack and Durloo, 1962, p. 8).

Next to lithology, the crystallinity, stratification, and fractures of a soluble rock are the structural factors that most control karst processes. Highly permeable limestone such as chalk, or reef limestones that have massive primary voids through which water can freely drain, do not develop good karst. Dense, fine-grained limestone with close-spaced bedding planes and many joints is considered ideal for karst development (Thornbury, 1969, p. 305). The intersections of vertical joint sets are conducive to concentrating the downward movement of ground water and are dissolved to make open *shafts*. Joints and bedding planes in flat-lying rocks localize lateral movement and dissolve lateral *galleries* along their intersections (Deike, 1969). Flow lines in ground water probably determine the depth and orientation of caves, but most minor features are controlled by structures.

KARST HYDROGEOLOGY

The science of underground water movement in karst is closely related to the process of solution and to the formation of caves. Unfortunately, ground-water movement in karst is highly irregular and complex. Open solution shafts and galleries are so much larger than the intergranular pore spaces through which water normally moves that some authors question the applicability of the fundamental concept of a water table beneath karst terrain (Jennings, 1971, p. 90).

In the classical theories of hydrogeology, precipitation percolates into soil and rock and infiltrates downward through a zone of variable and discontinuous intergranular water and air. This zone is known as the **zone of aeration**, or *vadose zone*. At some depth, ranging from zero to hundreds of meters, pore spaces are full of water. This is the **zone of saturation**, or *phreatic zone*. The surface of the zone of saturation is the **water table** (Fig. 7-4). Ground water moves so slowly that it establishes local hydrodynamic gradients, and in humid regions the water table is a subdued replica of the topography, rising beneath hills and intersecting the landscape at the surface of rivers, lakes, or springs.

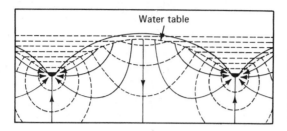

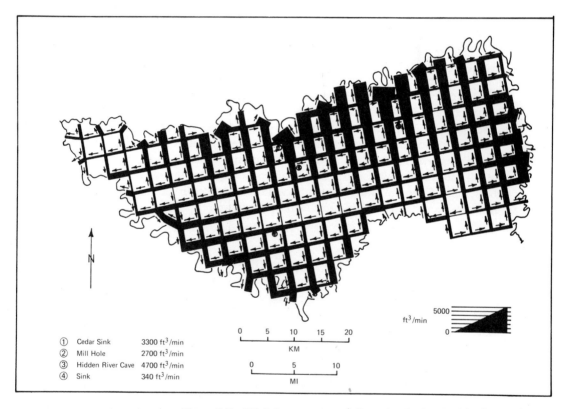

	Cedar Sink	3300 ft³/min
①	Cedar Sink	3300 ft³/min
②	Mill Hole	2700 ft³/min
③	Hidden River Cave	4700 ft³/min
④	Sink	340 ft³/min

Figure 7-5. Digital computer two-dimensional pipe model of ground-water movement in the vicinity of Mammoth Cave, Kentucky. (Thrailkill, 1972, Fig. 4.)

Water beneath the water table moves in accordance with laws of hydraulics along *flow lines*, perpendicular to surfaces of pressure equipotential (Fig. 7-4). From each point on the water table, there is but one unique flow path to an outlet. The general hydrogeologic equation, known as *Darcy's law*, is fully analogous to Ohm's law for electricity, because the amount of flow (amperage) per unit area of cross section is proportional to the pressure-potential difference (voltage) and inversely proportional to the resistance (ohmage) and the length of the conduit.

The fundamental problem of karst hydrogeology is scale. Models of classical flow (Fig. 7-4) are large compared to the grain size or pore size of the granular medium through which water moves. Darcy's law would not necessarily hold if the model were so small as to contain only a few unrepresentative channels. In karst flow, initially small planar openings such as bedding planes, faults, and joints are enlarged by solution, especially along intersections between them. The openings then become large as compared to the scale of the model. Karst flow is more similar to flow in branching irregular pipes than through a uniformly permeable, granular medium.

Computers can model the flow of karst ground water as a network of pipes of specified shape and internal friction, with water added at the intersections and draining out at the margins of the network. Figure 7-5 is a map model of the karst plain in the vicinity of Mammoth Cave, Kentucky (Thrailkill, 1972). Water is assumed to move downward through the zone of aeration through open shafts beneath sinkholes. The plain is surrounded by intrenched rivers, into which ground water emerges as springs and seeps. The computer model reproduces the flow of the large springs and also predicts discharge at other points on the periphery of the network. It correctly models the large flow through the underground river in the lower level of Mammoth Cave. It could be used to predict the sites of wells that would intercept the maximum flows in the zone of saturation. With further refinement and incorporation of solubility factors, a computer model could be used to predict future cavern enlargement.

KARST LANDFORMS

Regional and Temporal Significance of Karst

Limestones and dolomites constitute between 5 and 15 percent by weight of the total sedimentary rock mass (Garrells and Mackenzie, 1971, p. 40); evaporites constitute an additional 5 percent. About 15 percent of the area of the conterminous United States has karst-prone rock at or near the surface (Davies and LeGrande, 1972, p. 469). Karst landforms are worthy of study not only because of their dramatic form and unusual origin but because of their abundance.

The late Cenozoic Era has been a time of unusually extensive karst landscapes. Both the Indonesian-Philippines-New Guinea orogenic belt and the West Indies-Caribbean belt are in climatic zones of submarine carbonate deposition and coral-reef growth. Cenozoic tectonism has uplifted coral limestone and other carbonate rocks, and abundant karst landforms have resulted.

Quaternary glacier-controlled fluctuations of sea level (p. 406) have alternately exposed and drowned karst landscapes on all carbonate-rock coasts. When sea level is lowered by expanding glaciers, coastal regions develop deep ground-water circulation, and karst landforms and caves are actively developed. When, as at the present time, the level of the sea is at a high interglacial level, coastal karst terranes are drowned. Dripstone, formed in aerated caves, has been collected by divers from caves 45 m below sea level in the Bahama Islands (Kohout, et al., 1972), and from a depth of 125 m in a "blue hole" in a British Honduras reef (Dill, 1971). Hundreds or thousands of "blue holes" below sea level today are almost certainly dolines that formed during glacial-age low sea levels. The water supply for coastal cities on drowned karst, as in the state of Florida and in Greece, is drawn from karst aquifers below sea level. The shape of atolls also depends to a large extent on karst solution during Pleistocene low sea levels (p. 456).

Minor Karst Landforms[1]

Limestone and dolomite, and to a lesser degree other soluble rocks, commonly develop distinctive surface weathering features. Surfaces may be smoothly rounded forms (Figs. 7-6, 7-7), or hemispherical pits that intersect as razor-sharp points and ridges. In the extreme, a spongelike framework of rock is all that remains (Fig. 7-8). Phytokarst (p. 139) is independent of gravity, with surface pits corroding equidimensionally beneath an algal film, whereas solution forms tend to elongate vertically or downslope.

Figure 7-6. Minor karst landforms on Devonian limestone near Syracuse, New York.

[1]Because of the multilingual terminology of karst forms, a single boldface term is defined. At the first use, synonyms follow in parentheses but are not defined. The reader is urged to consult glossaries or expanded texts for the detailed shades of meaning (see Monroe, 1970; Jennings, 1971; Sweeting, 1972).

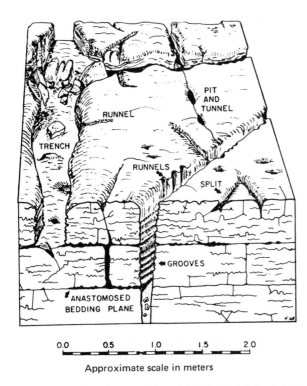

0.0   0.5   1.0   1.5   2.0

Approximate scale in meters

Figure 7-7. Labeled sketch of minor karst forms on Silurian dolomite near Hamilton, Ontario. (After Pluhar and Ford, 1970, Fig. 2.)

Figure 7-8. Spongelike karst weathering on Pleistocene coral limestone, Huon Peninsula, Papua New Guinea. One-meter axe handle is nearly concealed.

Solution Pits, Facets, Flutes, and Runnels.   On exposed rock (bare karst), horizontal surfaces develop shallow **solution pits** (rainpits, makatea). Larger features have been named **solution pans** (tinajitas) (Smith and Albritton, 1941). On planar surfaces inclined a few degrees from the horizontal, sheet runoff produces flat **solution facets** (solution bevels). On steeply inclined karst surfaces, elongate **solution flutes** (Rillenkarren, lapies) are common landforms. These are straight, parallel-sided grooves with sharp crests separating them. They follow the steepest slope down a rock face. **Solution runnels** (Rinnenkarren) are branching channels in networks on gently inclined bare surfaces. Many runnels probably first form under soil cover but, having formed, are perpetuated by runoff on bare rock surfaces.

Covered Karst: Grikes, Solution Pipes, and Solution Notches.   If soil or an organic mat covers a limestone surface, the effectiveness of solution increases manyfold. Solution pits are even more common under soil cover than on bare ledges, but the divides between pits are likely to be more rounded and smooth than the knife-edged ridges between intersecting pits on bare karst.

Vertical openings such as joints become widened by solution under soil cover (Figs. 7-6, 7-7). The British term is **grike** (Kluftkarren, solution slot). Grikes intersect as circular, vertical **solution pipes** (karst wells, potholes). At the level where bare rock projects above the soil level, a **solution notch** may develop. A distinct undercut measures the greater effectiveness of karst processes below ground level.

An extreme example of a solution notch has been named a **swamp slot** (Wilford and Wall, 1965). It is probably restricted to the humid tropics. Acid swamp water corrodes a narrow slit a meter or more horizontally into the base of limestone cliffs. Residual clay or an organic detrital layer may prevent downward solution, especially if the water table is at the swamp surface and if percolation is slow. The swamp slot is a basic component of tropical *tower karst* (p. 154), characterized by marginal corrosion around swampy plains at the base of steep-sided residual karst hills.

Minor Karst on Dolomite.   Minor surface karst on dolomite is comparable to but usually less dramatic than on limestone. An exception is the roughly pitted, unpleasantly sharp-edged weathered surface of the Bighorn Dolomite, a massive cliff-forming rock of Ordovician age in the northern Rocky Mountains. The rough surface on the Bighorn Dolomite has been shown to be caused by variable amounts of quartz and amorphous silica (opal) in the rock. The differential surface weathering is largely due to the greater resistance of the silica, which forms as much as 15 percent of the rock (Johnson and Biggs, 1955). Some of the differential weathering and staining is also due to oxidation of scattered pyrite grains.

The Niagaran dolomite of Silurian age is a cuesta-forming massive rock along its outcrop belt from central New York and southern Ontario (where it is the cap rock of Niagara Falls), northward and westward around Lakes Huron and Michigan, and into eastern Wisconsin. It is a coarsely recrystallized, permeable dolomite. The minor karst features are controlled by joints and bedding planes and only to a minor extent by slight permeability differences (Fig. 7-7) (Pluhar and Ford, 1970).

Major Karst Landforms

Karst is essentially a surface process, especially in vegetated and soil-covered landscapes. Most of the solution takes place in or just beneath the soil profile, where ground water has its maximum solubility. Thus there is a continuum of landforms from the minor features just described to distinct landforms comparable to the hills and valleys of fluvial landscapes. The forms described in this section are of a scale that can be studied on aerial photographs and topographic maps. It should be noted that portraying karst topography on contour maps poses special problems for the cartographer. Hachured depression contours, contour lines that cross to show overhanging cliffs, and dotted contours in natural bridges are some of the unusual mapping conventions that are required to contour karst landforms.

Dolines (Sinkholes, Sinks, Swallow Holes, Cockpits, Blue Holes, Cenotes). The fundamental genetic component of karst topography is the **doline**, or limestone sink (Figs. 7-9, 7-10). Whether due to joint control (the most common cause), differential solubility, or random events, localized areas of karst terrain are lowered more rapidly

Figure 7-9. Doline about 120 m in diameter and 45 m deep, which abruptly collapsed on the night of December 2, 1972, near Montevallo, Alabama. (Photo: U.S. Geological Survey.)

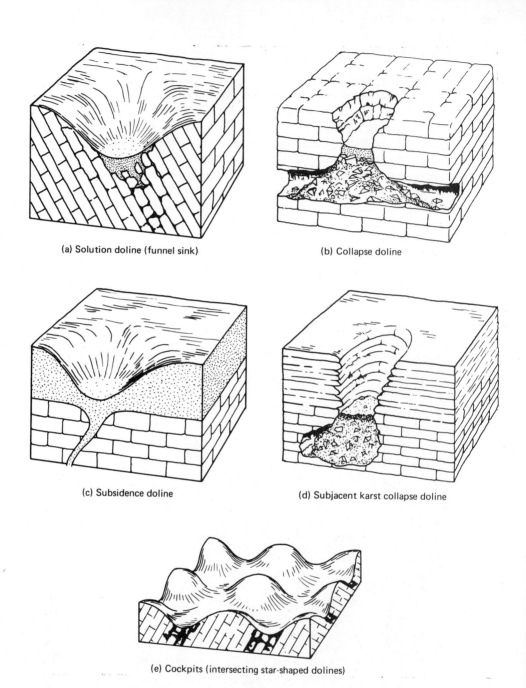

(a) Solution doline (funnel sink)

(b) Collapse doline

(c) Subsidence doline

(d) Subjacent karst collapse doline

(e) Cockpits (intersecting star-shaped dolines)

Figure 7-10. Five major classes of dolines. (From Jennings, 1971, Figs. 36, 9, and 58.)

than the surrounding area and form closed depressions. Their sides slope inward at the angle of repose of the adjacent material, typically 20°–30°. Some have much more gentle side slopes, and others, especially those forms known as *collapse sinks*, have vertical or overhanging cliffs. They are usually circular in plan view and less commonly elongate or oval. They range in size from shallow soil depressions a few meters in diameter and a meter deep to major landforms several kilometers in diameter and hundreds of meters deep.

Doline and its synonyms are genetic terms. They describe a landform produced by the karst process. Identically shaped closed depressions can form by subsidence, vulcanism, wind deflation, or glaciation, in fact, by any process that selectively translocates a mass of rock and permits the surrounding material to slump into the excavation. In areas such as southern Indiana, where glacial moraines cross a region of karst topography, real judgment and careful investigation are required to distinguish dolines from glacial kettles.

Dolines are the karst analog of river valleys. They are the fundamental unit of erosion as well as weathering. As the dissolving water goes underground, the eroding landmass literally disappears "down the hole," and the resulting landform has much in common with the surface forms on dry sand above a hole in the floor of a container (Fig. 7-9).

The bottom of a doline may open into a subterranean passage, down which water pours during a rainstorm. More commonly, the subsurface passage is partly blocked by rock rubble, soil, or vegetation, and temporary lakes form until the water percolates away. Some karst lakes are notorious for irregularly flushing out a clay plug at depth and draining, only to unexpectedly fill at some later date. The karst lake district of north-central Florida is a real estate lawyer's paradise, full of "shorefront" property that suddenly becomes property at the edge of a saw grass "prairie" or a mud flat.

Five major classes of dolines are recognized (Fig. 7-10). The two most contrasting types are the funnel-shaped *solution doline* and the steep or cliffed *collapse doline*. In the field, distinction between the two varieties of doline is rarely as obvious as the sketches of Figure 7-10 suggest. *Subsidence dolines* and *subjacent karst collapse dolines* are surface forms in nonsoluble rock, caused by solution of a buried karst. A thick residual soil, alluvium, or loess may mantle an active karst landscape, and the surface material will gradually settle or be carried into a buried doline. Frequently, small dolines (*swallow holes*, *ponors*) form in a stream bed, and although alluvium completely conceals the buried karst relief, the stream abruptly disappears and abandons the downstream segment of its former channel.

Dolines in tropical regions develop rapidly and grow very large. Adjacent dolines intersect, fuse, or engulf smaller ones to form compound, reticulate, honeycomb, or even star-shaped depressions (Williams, 1972; Monroe, 1976). The floors are frequently swamps at the water table. The *cockpit country* of Jamaica is the classic locality for tropical dolines (Versey, 1972). One doline or cockpit in Puerto Rico is so nearly a perfect hemisphere in cross profile that it has been surfaced with metal mesh as a giant radio telescope (Fig. 7-11).

Figure 7-11.   The radio telescope at the Arecibo Observatory, Puerto Rico. The reflector surface is a doline about 300 m in diameter and 70 m deep. The floor of the doline probably drains through a cave into the Tanama River, seen entering a natural bridge in the right foreground. (Photo: R. C. Hamilton, Cornell University Office of Public Information.)

***Uvalas* (*Compound Sinkholes, Karst Windows, Blind Valleys*).**   A genetic progression can be visualized in which dolines progressively abstract surface runoff into groundwater circulation and leave networks of *dry valleys* as relict surface forms. The Serbo-Croatian term **uvala** was given to a compound doline, or a chain of intersecting dolines (Cvijić, 1960). A **karst window** is similar, in that an unroofed segment of an underground stream channel becomes a surface valley in the window, and passes underground again downstream. If the exposed segment has normal stream valley form and an alluvial floor but disappears downstream into a swallow hole, the term **blind valley** is often used. A *natural bridge* (Fig. 7-11) may mark a segment of unroofed cave or may form by progressive underground abstraction of surface flow, either across the neck of an intrenched stream meander or from a high-level stream to a lower adjacent stream.

A **karst valley** is a special form of blind valley that occurs where a surface drainage pattern first developed on nonsoluble rock and later eroded into soluble rock, when subsurface drainage totally defeated the surface flow. It has all the properties of a stream valley except that it lacks a stream. Karst valleys are common in areas of nearly flat-lying, interbedded limestone, shale, and sandstone. Many good examples are shown on topographic maps of the Mammoth Cave area in Kentucky.

Poljes (Plans, Wangs, Hojos).   The Dinaric region of Yugoslavia is in a belt of Alpine folding (Fig. 20-3). The karst topography is on carbonate rocks that are folded and faulted, often in close proximity to insoluble rocks. Large blind valleys or karst valleys, enclosed by either soluble or insoluble rocks, usually elongated along tectonic axes, are called **poljes** there. Because of their orientation on tectonic trends, they have been interpreted as grabens or fault-angle depressions. Some are, but others are simply large-scale examples of structurally controlled karst processes. Their floors truncate folded limestone formations at a local base level determined by a narrow gorge through insoluble rock or a series of swallow holes (ponors) at the downstream end of the polje. Poljes are large valleys; the largest in Yugoslavia is 40 km long and 6–8 km wide. In general, uvalas are measured in hundreds of meters; poljes in kilometers (Herak, 1972, p. 37).

Poljes have broad, flat floors that abut sharply against steep enclosing walls. On occasion, during floods, the swallow holes by which some valley floors are drained become choked with debris, and the entire polje floor becomes a shallow lake. Some Yugoslavian poljes flood annually when the water table is high and are fertile agricultural land during drier seasons. Their hydrology is the subject of sophisticated Yugoslavian research (Herak, 1972).

Residual Karst

There is a tendency among many geomorphologists to consider solution landforms as a genetic sequence: doline → uvala → polje. Dolines grow and merge into uvalas through time in a genetic sequence. However, poljes are more than large, complex uvalas. They have a structural control that does not lend itself to a genetic sequence following the much smaller, process-controlled forms.

The concept of a karst "cycle of erosion," beginning with the genetic sequence of doline and uvala and ending with a plain of solution dotted with residual limestone hills, can be traced to the historical impact of the visit with Albrecht Penck to the Dinaric karst by W. M. Davis in 1899 (Davis, 1901). At the time, the Davisian concept of a cycle of erosion was achieving great popularity. Cvijić (1893) had previously published a distinguished monograph on karst under the supervision of Penck. Penck and Davis inspired him to rework his observations into a cyclic concept, which he published in France while he taught there during World War I. The paper was not well known because of the wartime communication problems in Europe, but a review was published in English (Sanders, 1921) reproducing Cvijić's original block diagrams. A major manuscript by Cvijić (1960) was lost by his publisher and not rediscovered and published until 33 years after his death.

The deduced residual landforms of the karst cycle were not well illustrated by the Dinaric karst. Small limestone hills (hums) rise above the floors of Yugoslavian poljes, but the structural complexities of the region made it unsuitable as an area for a model of a karst erosion cycle. Later authors used the tropical karst of Jamaica and southeast Asia as models for the more advanced stages of a deduced karst cycle, neglecting the obvious differences in climatic controls of processes. As a result, the deduced karst "cycle of erosion" starts in a Mediterranean climate but ends in the tropics. Rather than persist in following the cyclic concept, it is more productive to review residual karst landforms simply as the products of extreme karst processes, especially common in tropical humid climates.

Lehman (1936) introduced the term **kegelkarst** (cone karst) for a landscape in Java where the terrain is approximately equally shared by conical dolines and conical residual hills. In fact, both in Jamaica and in Java, the typical kegelkarst residual hills are more nearly hemispheric than conic, and the dolines are star-shaped or valleylike among the hills.

Tower karst (turmkarst, mogote, pepino hill) is distinguished from kegelkarst in that the residual hills are steep-sided or vertical and are separated from each other by swamps or alluvial plains (Fig. 7-12). Marginal solution at the edge of the swampy plains may be largely responsible for the steep walls of tower karst. The residuals are riddled with caves; a swamp slot may deeply notch the base of the towers.

The tower karst of northern Vietnam extends northwestward from near Haiphong harbor, where it is drowned by the postglacial rise of sea level, along the Red River to the Chinese frontier. It continues into southern China, especially in Kwangsi Province. Many art historians believe that the strange, needlelike mountains of classical Chinese art originated in the subtropical karst landscape of the southern provinces of China (Fig. 7-13). The scenes seem exotic and dreamlike to Western

Figure 7-12. Tower karst in Kweilin, Kwangsi Province, China. Compare with Figure 7-13. (Photo: J. H. Ferger, M.D.)

青綠關山迥
嶀嵋道路長
客人多結束
李白周詳繪
高名和利郎
發芳興忙年
陳失姓氏北宋
近平唐
甲午新秋
尚題

Figure 7-13. A south China landscape as painted by an unknown artist of the T'ang dynasty (A.D. 618–907). Compare with Figure 7-12. (Collection of the National Palace Museum, Taipei, Taiwan, Republic of China.)

eyes, but they are actually reasonable geomorphic sketches. It should be noted that later classical Chinese art idealized the forms of mountains so that all look equally exotic. The incident portrayed by Figure 7-13, "Emperor Ming-Huang's Journey to Shu," is known to have occurred in the karst region of southern China.

The *mogotes* (haystack hills) of Sierra de los Organos, Cuba, are residual karst towers of classic beauty (Lehmann, 1960, Plate 1). The *pepino hills* of Puerto Rico are comparable (Meyerhoff, 1938; Monroe, 1976), but many show distinct asymmetry, with a gentle slope toward the northeast from which most of the wind-driven rain strikes them. Thorp (1934) proposed that solution on the windward slopes lowered that side of the pepino hills more rapidly. Monroe (1966; 1976, p. 45) subsequently proposed that solution and redeposition of surficial secondary carbonate has "case hardened" the windward slopes, whereas the lee slopes are subject to more extensive basal solution and develop steeper cliffs.

Residual karst landforms have a curious relevance to the political and social histories of the regions in which they are found. Bandits, partisans, guerrilla troops, and fugitives who are native to a karst region are able to live safely and base their operations in the many caves that penetrate karst towers. The Viet Minh operated against French colonial troops from the karst of northern Vietnam for decades; Fidel Castro and his revolutionaries based their eventual control of Cuba on the karst of the Sierra Maestra; Yugoslav partisans prevented motorized German troops from controlling large parts of the Dinaric karst; the examples could be multiplied at length. On a less significant level, illegal distilleries are traditionally hidden in the Appalachian karst of eastern United States, and Jesse James, a notorious bandit, hid his gang in the caves of Missouri. Whatever the merit of a cause, its supporters are likely to find a karst terrain that they know intimately an overwhelming advantage against numerically superior and better equipped, but foreign, opponents.

LIMESTONE CAVERNS

A **limestone cavern** (cave)[2] can be arbitrarily defined as a solution cavity large enough for people to enter. Access implies topologic continuity with the subaerial terrain, and therefore caverns are part of the karst landscape, even when they are hundreds of meters below the generally accepted "surface." Even the largest cavern is volumetrically insignificant as a landform (Thrailkill, 1968, p. 42), but the beauty of the deposits and the mysterious thrill of cave exploration give caverns a geomorphic significance out of proportion to their size.

[2]Caverns form in a variety of ways other than by solution. Lava tubes, tubes of headward "piping" by streams, and unusually wide overhanging ledges of any strong rock may be classed as caverns, or caves. Here the terms are meant to imply limestone solution features unless specifically stated otherwise.

Theories of Cave Development

This chapter has repeatedly stressed that karst is essentially a surface process. The maximum solution takes place immediately beneath or in the soil profile. How then can caves form hundreds or thousands of meters below the other landforms of a karst landscape?

Early theories of cave formation were vague about the chemistry of the process but implied that caves developed in the zone of aeration by downward-moving vadose water. Beginning with solution enlargement along bedding planes and joints, open conduits developed that were large enough to guide underground rivers and waterfalls. Great emphasis was placed on erosion by underground rivers. A surface river was necessary to accept the drainage from underground streams, and as the *vadose theory* developed, it became closely tied to theories of multiple erosion surfaces that had been rejuvenated by tectonic uplift. As the surface streams intrenched their valleys in each new cycle of erosion, the water table was lowered and caves were excavated in the newly drained vadose zone.

The vadose theory of cave formation was accepted without serious question until 1930, when W. M. Davis wrote an extended evaluation of old theories and new observations and pointed out serious theoretical deficiencies. A fundamental problem was that many caverns contain extensive dripstone deposits, which are observed to grow. If vadose water is actively filling caverns with dripstone, how could it be the agent that formerly excavated them? Davis (1930, p. 549) revived interest in long-standing hydrogeologic theories of deep flow paths in the zone of saturation (Fig. 7-4) by proposing that, because the phreatic water circulated to great depths, caverns could be excavated below the water table, perhaps far below it. Later, when the regional surface drainage network had eroded downward into the limestone terrane, the caves would be in the zone of aeration, rivers could flow through them, and dripstone could form. This became known as the *two-cycle theory* of cavern formation.

Two years after Davis published his study, A. C. Swinnerton (1932) proposed that the maximum solution takes place at or just beneath the water table, especially in the narrow zone within which the water table fluctuates. He called his variant the *water-table hypothesis*. Swinnerton postulated that water at the water table has a variety of paths along which it can flow toward a spring, but most of the volume moves along the shortest path, which is along the water table.

Swinnerton's water-table hypothesis explained the many observations that caverns are generally horizontal, even in deformed strata (Fig. 7-14). The water-table hypothesis permitted the observed development of three-dimensional networks of cavern passages, with blind passages that cannot be explained by through-flowing rivers or vadose percolation. It permitted, but did not require, second-cycle episodes of dripstone deposition and dissection by underground rivers as the regional water table and landscape was lowered. His hydrogeologic model was faulty, however

W   E

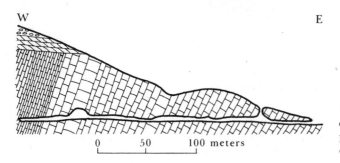

0   50   100 meters

Figure 7-14. Simplified cross section of Lehman Caves, Nevada. Horizontal passages intersect tilted strata. (Diagram courtesy of G. W. Moore.)

(Hubbert, 1940). Speleologists continued to prefer theories of cavern excavation by flowing water in the vadose zone or at the water table. Little further theoretical studies were done until after World War II, except for an excellent extension of Davis's theory by Bretz (1942).

The present opinion is that caverns form in the zone of saturation but only a short distance below the water table (Moore, 1968; Thrailkill, 1968; Moore and Sullivan, 1977). The favored cause for renewed aggressiveness of ground water is the mechanism of mixing (Fig. 7-3). Phreatic water is quite likely to have different chemical properties than vadose water, especially if the vadose water is flowing directly to the water table along a conduit from the floor of a doline above. With the addition of the temperature effect and the flow-rate effect, the mixing effect provides a theoretically sound and quantitatively reasonable solution process to excavate caverns (Thrailkill, 1968). However, by the mixing hypothesis, cavern excavation is only incidentally related to karst processes at the soil level insofar as dolines provide a mechanism for localized mixing of vadose and phreatic water at the water table. Diffuse seepage is less likely to produce contrasting water types at the water table. Those karst hydrogeologists who dislike the concept of a water table in soluble rocks, and point to evidence of multiple pipelike conduits with unlike water masses in them, flowing side by side or crossing each other, require an abrasion hypothesis for largely mechanical excavation of caverns. Their views are well summarized by Jennings (1971, pp. 90–96).

Cave Deposits

Dripstone and **flowstone** are general popular terms for cave deposits from water; the equivalent term **speleothem** is gaining popularity. A variety of delicate and beautiful shapes are assumed by speleothems (Fig. 7-15). As limestone is dissolved by vadose water, the pH of the solution rises but in equilibrium with a relatively high CO_2 partial pressure. When the water enters an aerated cave, the excess dissolved CO_2 diffuses into the cave atmosphere, and calcite is deposited. The popular idea that cave deposits are the result of evaporation should be dispelled, because wet caves have essentially 100 percent relative humidity, and evaporation is negligible. There is a possibility that biological processes may cause some speleothems. Fungal threads have been observed at the tips of growing stalactites in Lehman Caves, Nevada (Went, 1969).

Figure 7-15. Speleothems in the Postojna Caves, northern Yugoslavia. This is one of the most popular tourist caves in the famous Dinaric karst region of Yugoslavia. (Photo: Yugoslav State Tourist Office.)

The principal speleothems are *stalactites* and *helictites*, which grow downward from cave roofs, and *stalagmites*, which build upward from the floors. Helictites assume twisted or spiral forms that are related to strained calcite crystal structures. Curtains or ribbons of flowstone may coat cave walls or form a partition beneath a prominent joint-controlled seepage zone. In cave pools, a variety of delicate skeletal mineral "flowers" may form. Tufa rims and terraces line slow-moving, sediment-free underground streams.

In addition to speleothems, a variety of clastic deposits may accumulate in caves. Caves that were occupied by animals and prehistoric people may have artifact-bearing organic litter on their floors. Rock falls or weathering spalls from cave roofs and walls add sporadically to the thickness of cave-floor deposits. Caves that have flowing streams contain a variety of alluvial deposits. Many limestone caves have thick layers of red clay that was formerly thought to be the insoluble residue of subsurface solution. However, the volume of dissolved limestone has been judged insufficient to account for the amount of red clay, so it must also include washed-in surface sediment (Jennings, 1971, p. 177). Windblown silt (loess), volcanic ash, and organic detritus may accumulate in caves to produce stratigraphic records of great archeologic significance.

REFERENCES

ADAMS, C. S., and SWINNERTON, A. C., 1937, Solubility of limestone: *Am. Geophys. Un. Trans.*, pt. 2, pp. 504–8.

BRETZ, J. H., 1942, Vadose and phreatic features of limestone caverns: *Jour. Geology*, v. 50, pp. 675–811.

CORBEL, J., 1959a, Vitesse de l'erosion: *Zeitschr. für Geomorph.*, v. 3, pp. 1–28.

——, 1959b, Erosion en terrain calcaire: *Annales de Géog.*, v. 68, pp. 97–120.

CVIJIĆ, J., 1893, Das karstphänomen: *Geographische Abhandlungen herausgegeben von A. Penck*, v. 5, no. 3, pp. 218–329.

——, 1960, La géographie des terrains calcaires (French transl. by E. de Martonne): *Serbe Acad. Sci. Arts Mon.*, v. 341, 212 pp.

DAVIES, W. E., and LEGRAND, H. E., 1972, Karst of the United States, *in* Herak, M., and Stringfield, V. T., eds., *Karst: Important karst regions of the northern hemisphere:* Elsevier Publishing Co., Amsterdam, pp. 467–505.

DAVIS, W. M., 1901, An excursion in Bosnia, Hercegovina, and Dalmatia: *Geog. Soc. Philadelphia Bull.*, v. 3, pp. 21–50.

——, 1930, Origin of limestone caverns: *Geol. Soc. America Bull.*, v. 41, pp. 475–628.

DEIKE, R. G., 1969, Relations of jointing to orientation of solution cavities in limestones of central Pennsylvania: *Am. Jour. Sci.*, v. 267, pp. 1230–48.

DILL, R. F., 1971, The Blue Hole: A structurally significant sink hole in an atoll off British Honduras [Abs.]: *Geol. Soc. America Abs. with Programs*, v. 3, no. 7, pp. 544–45.

DOMENICO, P. A., and PALCIAUSKAS, V. V., 1973, Theoretical analysis of forced convective heat transfer in regional ground-water flow: *Geol. Soc. America Bull.*, v. 84, pp. 3803–13.

FOLK, R. L., ROBERTS, H. H., and MOORE, C. H., 1973, Black phytokarst from Hell, Cayman Islands (B.W.I.): *Geol. Soc. America Bull.*, v. 84, pp. 2351–60.

FOOSE, R. M., 1953, Ground-water behavior in the Hershey Valley, Pennsylvania: *Geol. Soc. America Bull.*, v. 64, pp. 623–46.

FORD, D. C., 1971, Characteristics of limestone solution in the southern Rocky Mountains and Selkirk Mountains, Alberta and British Columbia: *Canadian Jour. Earth Sci.*, v. 8, pp. 585–609.

GARRELLS, R. M., and MACKENZIE, F. T., 1971, *Evolution of sedimentary rocks:* W. W. Norton & Company, Inc., New York, 397 pp.

HACK, J. T., and DURLOO, L. H., 1962, Geology of Luray Caverns, Virginia: *Va. Div. Min. Resources Rept. Invest. 3*, 43 pp.

HERAK, M., 1972, Karst of Yugoslavia, *in* Herak, M., and Stringfield, V. T., eds., *Karst: Important karst regions of the northern hemisphere:* Elsevier Publishing Co., Amsterdam, pp. 25–83.

HUBBERT, M. K., 1940, Theory of ground-water motion: *Jour. Geology*, v. 48, pp. 785–944.

JENNINGS, J. N., 1971, *Karst: An introduction to systematic geomorphology:* Australian National Univ. Press, Canberra, 252 pp.

JOHNSON, C. H., and BIGGS, D. L., 1955, Differential surface weathering of Bighorn Dolomite: *Jour. Sed. Petrol.*, v. 25, pp. 222–25.

KOHOUT, F. A., DILL R. F., ROYAL, W. R., BENJAMIN, G. J., and HILL, R. E., 1972, Ocean-groundwater interface: Sinkholes, blue holes, and submarine springs—geologic-hydrologic windows in the Florida–Bahama platform [Abs.]: *Geol Soc. America Abs. with Programs*, v. 4, pp. 566–67.

LEHMANN, H., 1936, *Morphologische studien auf Java:* J. Engelhorn, Stuttgart, 114 pp.

———, 1960, International atlas of karst phenomena, Sheets 1, 1a, and 1b, "Sierra de los Organos, Cuba": *Zeitschr. für Geomorph.*, supp. v. 2, plates in pocket.

MEYERHOFF, H. A., 1938, Texture of karst topography in Cuba and Puerto Rico: *Jour. Geomorph.*, v. 1, pp. 279–95.

MONROE, W. H., 1966, Formation of tropical karst topography by limestone solution and reprecipitation: *Caribbean Jour. Sci.*, v. 6, pp. 1–7.

———, 1970, Glossary of karst terminology: *U.S. Geol. Survey Water-Supply Paper 1899-K*, 26 pp.

———, 1976, Karst landforms of Puerto Rico: *U.S. Geol. Survey Prof. Paper 899*, 69 pp.

MOORE, G. W., 1968, Limestone caves, *in* Fairbridge, R. W., ed., *Encyclopedia of geomorphology:* Reinhold Book Corporation, New York, pp. 652–53.

MOORE, G. W., and SULLIVAN, G. N., 1977, *Speleology:* Zephyrus Press, Teaneck, N.J., 151 pp.

PENCK, A., 1894, *Morphologie der erdoberfläche:* J. Engelhorn, Stuttgart, v. 1, 471 pp.; v. 2, 696 pp.

PLUHAR, A., and FORD, D. C., 1970, Dolomite karren of the Niagara escarpment, Ontario, Canada: *Zeitschr. für Geomorph.*, v. 14, pp. 392–410.

POWER, F. D., 1925, Phosphate deposits of the Pacific: *Econ. Geology*, v. 20, pp. 266–81.

ROGLIĆ, J., 1972, Historical review of morphologic concepts, *in* Herak, M., and Stringfield, V. T., eds., *Karst: Important karst regions of the northern hemisphere:* Elsevier Publishing Co., Amsterdam, pp. 1–18.

SANDERS, E. M., 1921, Cycle of erosion in a karst region (after Cvijić): *Geog. Rev.*, v. 11, pp. 593–604.

SAWKINS, J. G., 1869, *Reports on the geology of Jamaica:* Great Britain Geol. Survey Mem., Longmans, Green, and Co., London, 339 pp.

SMITH, J. F., and ALBRITTON, C. C., 1941, Solution effects on limestone as a function of slope: *Geol. Soc. America Bull.*, v. 52, pp. 61–78.

SWEETING, M.M., 1972, *Karst landforms:* The Macmillan Press, Ltd., London, 362 pp.

SWINNERTON, A. C., 1932, Origin of limestone caverns: *Geol. Soc. America Bull.*, v. 43, pp. 663–93.

THORNBURY, W. D., 1969, *Principles of geomorphology:* John Wiley & Sons, Inc., New York, 594 pp.

THORP, J., 1934, Asymmetry of the "Pepino Hills" of Puerto Rico in relation to the trade winds: *Jour. Geology*, v. 42, pp. 537–45.

THRAILKILL, J., 1968, Chemical and hydrologic factors in the excavation of limestone caves: *Geol. Soc. America Bull.*, v. 79, pp. 19–45.

——, 1972, Digital computer modeling of limestone groundwater systems: *Univ. Kentucky Water Resources Inst. Res. Rept. 50*, 70 pp.

VERSEY, H. R., 1972, Karst of Jamaica, *in* Herak, M., and Stringfield, V. T., eds., *Karst: Important karst regions of the northern hemisphere:* Elsevier Publishing Co., Amsterdam, pp. 445–66.

WENT, F. W., 1969, Fungi associated with stalactite growth: *Science*, v. 166, pp. 385–86.

WILFORD, G. E., and WALL, J. R. D., 1965, Karst topography in Sarawak: *Jour. Tropical Geog.*, v. 21, pp. 44–70.

WILLIAMS, P. W., 1972, Morphometric analysis of polygonal karst in New Guinea: *Geol. Soc. America Bull.*, v. 83, pp. 761–96.

CHAPTER 8

Mass-Wasting
and Hillslope Evolution

On the scale of individual landforms, where rheid flow is not a significant factor of deformation, unweathered rock resists the surface forces that act upon it. Only when rock has been broken by mechanical stresses or has reacted with water and the atmosphere can the fragments be mobilized. Weathering, then, is a necessary precondition for the movement of rock fragments downslope and for the development of the slopes. A variety of surface forces can move loose rock particles. Whenever they move, ubiquitous gravity adds a downward component to the motions produced by other forces so that they move preferentially downhill.

MASS-WASTING, GRAVITY, AND FRICTION

The collective term for all gravitational or downslope movements of weathered rock debris is **mass-wasting**. The term implies that gravity is the sole important force and that no transporting medium such as wind, flowing water, ice, or molten lava is involved. Although flowing water is excluded from the process by definition, water nevertheless plays an important role in mass-wasting by oversteepening slopes through surface erosion at their bases and by generating seepage forces through ground-water flow. This second role of water is often referred to as "lubrication" of rock or weathered rock debris on a slope. In fact, water has very little influence on the frictional properties of most rock-forming minerals, and if it were not for the seepage pressures generated by ground-water flow, subsurface water would have little influence on slope instability and mass-wasting.

163

Internal Friction

A series of simple experiments illustrates the relative importance of gravity and pore-fluid pressure in determining downslope movement. A cone of noncohesive dry sand can be built by pouring sand onto a horizontal surface through a funnel. The cone will have a surface *angle of repose* of approximately 35°, depending on the size, shape, surface roughness, and other properties of the sand grains. The tangent of the angle of repose of dry granular materials is slightly greater than, but approximately equal to, the coefficient of sliding friction of the material, or its *internal friction* (Van Burkalow, 1945). Both are determined by the balance of forces between gravity and friction. The coefficient of sliding friction of a particle is equal to the ratio between the downslope component of weight and the component of weight acting perpendicular to the slope when the particle is moving.

If in a second experiment the same sand is deposited in a saturated condition under water, the angle of repose of the cone will be virtually identical to the angle of repose of the dry sand cone. In neither example will there be any fluid flow through the sand, and consequently neither cone will be subjected to seepage forces tending to decrease its stability.

A third experiment demonstrates the influence of water flow. If water is poured gently onto the top of a cone of dry sand and allowed to seep downward and outward (the water must infiltrate and not erode the surface), failure occurs as soon as the water reaches the slope face. It can be demonstrated by experiment, or by adding a vector representing seepage pressure to a simple diagram (Fig. 8-1), that even the

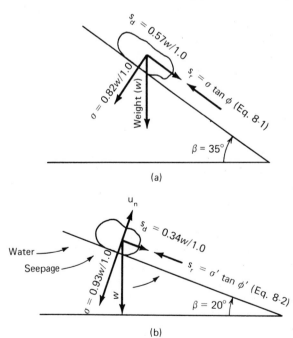

Figure 8-1. Resolution of forces on a cohesionless block of unit basal area resting on a slope. (*a*) σ (stress component perpendicular to surface) $= w \cos \beta$; s_d (stress component parallel to surface) $= w \sin \beta$. Shear strength (s_r) resists motion. When $\beta = \phi$ (angle of repose), $s_d = s_r$, and the block is ready to move. (*b*) As above, but with u_n (component of seepage pressure normal to the surface) counteracting $\sigma(\sigma' = \sigma - u)$ and reducing s_r. Particle will move when $\beta \ll \phi$.

flow of rain water runoff in the surface soil on a slope reduces the stable angle of repose to less than half of the dry value.

Cohesion

Fine-grained sediments, especially clay-size[1] particles, have strong *cohesive* surface forces between them. These forces may be strong enough to draw grains together and expel capillary water between them. That is, the cohesive forces between two grains may exceed the capillary force of water. The cohesive forces act in addition to internal friction to determine the angle at which slope failure occurs. Referring to the first experiment previously described, if the noncohesive dry sand is moistened but not oversaturated, it can be molded and shaped to a vertical slope, as any child in a sandbox will verify. Strength in that sense depends on the tension in the capillary water between sand grains.

The previous discussion was first summarized as an equation by C. A. Coulomb in 1776, in which:

$$s = c + \sigma \tan \phi \tag{8-1}$$

where s is the shear strength (in stress, or force per unit area, units), c is cohesion (also in stress units), σ is the stress component perpendicular to the surface, and ϕ is the angle of internal friction. Engineers use an equivalent form of Coulomb's equation to take into account the excess pressures in the pore fluids, including the seepage pressures. This concept is called the *principle of effective stress*. It is common engineering practice to add a superscript to all symbols in Coulomb's equation when working with effective stresses:

$$s' = c' + \sigma' \tan \phi' \tag{8-2}$$

where $\sigma' = (\sigma - u)$, the stress component normal to the surface as reduced by fluid pressure u. In noncohesive sand, gravel, or large talus blocks, $c = c' = 0$, and shear stress is proportional to the tangent of the angle of internal friction, although the relationship is complicated by the tendency of large angular blocks to stack and support steeper-than-predicted angles. Fine-grained silt and clay make cohesive soils for which c or c' are significant. As cohesion increases, so does shear strength. Many moist clays hold vertical faces of modest height without failure, which noncohesive materials can never do.

Slope angles on noncohesive weathered debris such as grus or loose rock rubble are usually between 25° and 40°. Natural slopes steeper than 40° are usually barren of rock debris, and in British terminology are classed as **cliffs**.

---

[1]"Clay" is the name used for a group of residual minerals (Chap. 6) as well as for the fine-grained fraction of sediment regardless of its mineralogy. The distinction between "clay minerals" and "clay-size sediment" always should be made clear. The geologists' classification defines clay-size particles as less than 4 μm mean diameter; the engineers' classification defines clay particles as all those less than 2 μm in diameter.

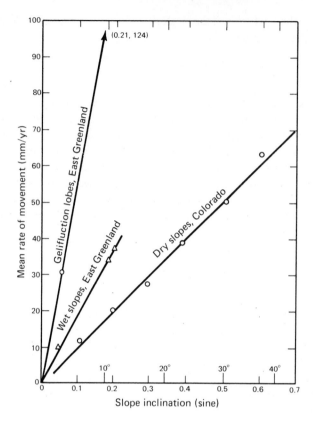

Figure 8-2.   Rate of downslope movement is proportional to the sine of the slope inclination under a wide range of conditions. (Data from Schumm, 1967; Washburn, 1967, Fig. 43.)

Rates of Downslope Movement

Coulomb's equation and the geometry of Figure 8-1 demonstrate that the shear stress acting parallel to a slope, which causes a particle to move down slope, is approximately proportional to the sine of the slope angle. Intuitively, we would expect that the *rate* of downslope movement of small, loose rock fragments would also be proportional to the sine of the slope angle (Strahler, 1956, p. 577). Most data support this inference (Fig. 8-2). Schumm (1967) measured the rate of migration of small plates of sandstone down a shale hillslope in semiarid western Colorado by marking 110 pieces with paint and measuring their distance from reference stakes driven into bedrock. Over a period of 7 years, the rate of downslope movement ranged from 2–3 mm/yr to almost 70 mm/yr in proportion to the sine of the slope angle, which ranged from 3°–40°. Rates of periglacial solifluction (p. 172) over perennially frozen ground in eastern Greenland are much higher than the sliding rate of dry rock chips on a semiarid talus (Fig. 8-2), and the modes of movement are also very different, but nevertheless, the rates of surface movement on both rapidly flowing solifluction lobes and "undifferentiated wet slopes" in Greenland are proportional to the sine of the slope angles.

It should be noted that for angles of less than $10°$, the sine and tangent functions differ by only a few percent, and Figure 8-2 would not look much different if the movement rates were plotted against the tangent of slope angle as an approximation of the internal friction (tan ϕ). Whichever function is chosen, the conclusion is obvious that downslope movement is slowest on gentle slopes and increases rapidly with increase in slope. On slopes of $40°-45°$, few detached fragments can remain at rest. All roll or slide to the base of the slope. On vertical slopes, a loosened rock particle drops under the free acceleration of gravity, its rate of fall limited only by air resistance.

DESCRIPTIVE CLASSIFICATION OF MASS-WASTING

The explanation of mass-wasting is so simple and obvious that a genetic, or explanatory-descriptive, classification of the processes and forms is not necessary. Instead, a simplified descriptive classification is generally adopted, which emphasizes certain features as diagnostic and recognizes other features as parts of continuous series. No combination of categories has been devised to classify totally the many landforms and processes of mass-wasting, nor is such a classification likely or desirable. The essence of mass-wasting is that it bridges the gap between *weathering*, which is defined as occurring in place, and *erosion*, which requires as one element transport by some agent or medium. Mass-wasting combines elements of weathering and erosion yet separates those two large categories of processes. In every sense of cause and effect, mass-wasting is a category of transitional phenomena.

Two classifications of mass-wasting are in common use in the United States and are comparable to schemes adopted in England, France, and Germany. There is broad international agreement that the definitive criteria of mass-wasting should be: (1) the type of material in motion, including its coherence and dimensions, and (2) the type and rate of movement, whether sliding, flowing, or falling.

The classification scheme most commonly seen in American geomorphology textbooks is from a valuable little book by Sharpe (1938) [Fig. 8-3(a)]. The factors by which Sharpe subdivided mass-wasting were: (1) the amount of included ice or water; (2) the nature of the movement, whether sliding or falling as a coherent mass, or flowing by internal deformation; and (3) the speed of the movement, ranging from the imperceptible to speeds of tens of meters per second under the full acceleration of gravity. The term **landslide** (or *landslip* in British usage) was not given any specific definition by Sharpe (1938, p. 64) but remains a popular nontechnical term for all perceptible forms of relatively dry mass-wasting.

A more detailed classification [Fig. 8-3(b)] of slope movements (the favored engineering term for mass-wasting) was published in the United States by a committee of engineering specialists (Transportation Research Board, 1978). The committee's responsibility was to consider the relationship of slope movements to engineering practice, so they defined slope movement as a downward and outward movement of slope-forming bedrock, rock debris, and "earth" (fine-grained fragmental debris).

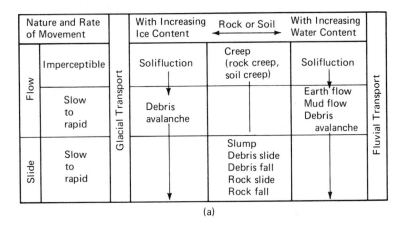

(a)

TYPE OF MOVEMENT			TYPE OF MATERIAL		
			BEDROCK	ENGINEERING SOILS	
				Predominantly coarse	Predominantly fine
FALLS			Rock fall	Debris fall	Earth fall
TOPPLES			Rock topple	Debris topple	Earth topple
SLIDES	ROTATIONAL	FEW UNITS	Rock slump	Debris slump	Earth slump
	TRANSLATIONAL	MANY UNITS	Rock block slide Rock slide	Debris block slide Debris slide	Earth block slide Earth slide
LATERAL SPREADS			Rock spread	Debris spread	Earth spread
FLOWS			Rock flow (deep creep)	Debris flow (soil creep)	Earth flow
COMPLEX			Combination of two or more principal types of movement		

(b)

Figure 8-3. Classifications of mass-wasting. (*a*) Landslides and related phenomena. (Simplified from Sharpe, 1938, p. 96.) (*b*) Types of slope movements, abbreviated version. (Varnes, 1978, Fig. 2.1.)

In the description that follows, Sharpe's classification [Fig. 8-3(*a*)] is generally followed. Some of the excellent diagrams and examples from the Transportation Research Board report are used to illustrate certain terms.

Creep

Creep is imperceptible and nonaccelerating downslope movement. It is most common, in fact ever-present, in weathered rock debris on slopes, but the term is also applied to the slow movement of otherwise unweathered joint blocks. **Soil creep** or **rock creep** are the respective subordinate terms. The material moving downslope is called **colluvium**.

Although creep is too slow to be observed, the cumulative results become obvious over a period of years (Fig. 8-4). Most stone walls and pavements on hillsides show downslope motion by tension cracks, downslope tilt, or visible displacement. Tree trunks are concave uphill, although the significance of their curvature to creep has been questioned (Phipps, 1974). All these features are due to the essentially laminar nature of creep. Each layer of soil is carried downhill by the motion of the layer beneath it, and the effect is cumulative, with the maximum rate at the surface exponentially decreasing to zero with depth (Kirkby, 1967). As a result, soil creep does not shear across immobile rock or soil at depth and is not capable of abrading a buried surface. Strata in unweathered bedrock thin abruptly as they enter the surface layer that is creeping, and instead of breaking the surface as outcrops, they are drawn out in a downhill direction (Fig. 8-5). Creep is recognized as both a problem and an aid in prospecting. If fragments of the desired coal or ore mineral are found in the "float," or colluvium, the buried layer must be in the subsoil at some point farther upslope.

Grass may be able to retain a continuous sod cover over an area of soil creep, because movement is typically only a few millimeters per year. The highest rates of creep are measured on steep slopes under tropical rainforests, where the entire sur-

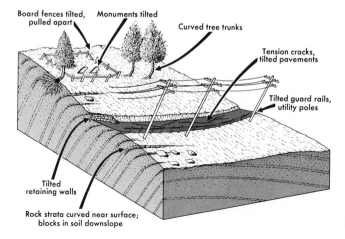

Board fences tilted, pulled apart
Monuments tilted
Curved tree trunks
Tension cracks, tilted pavements
Tilted guard rails, utility poles
Tilted retaining walls
Rock strata curved near surface; blocks in soil downslope

Figure 8-4. Common effects of creep. Not all will be present in one place.

Figure 8-5. Rock creep in roadcut along Yarra Boulevard, Melbourne, Australia. (Photo: Renate R. Hodgson.)

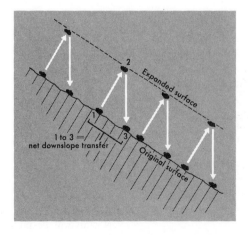

Figure 8-6. Path of a surface particle during expansion and contraction of soil on a slope. See Figure 8-8 for more complex motions.

face layer of interlocking roots moves downhill at rates of up to 4 mm/yr. An anomalous deeper soil zone of high creep velocity was measured in a Puerto Rico rainforest, just below the root zone (Lewis, 1974).

Soil creep is aided by expansion and contraction of soil by heating and cooling, freezing and thawing, or wetting and drying. Volumetric expansion by any cause displaces particles toward the free face of the expanding mass, or perpendicular to the ground surface. On contraction, however, the particle is not pulled back into its former position but settles with a gravitational component (Fig. 8-6). Only rarely are the cohesive forces of soil and water strong enough to draw particles back into the ground during contraction without some net downslope motion. Figure 8-6 illustrates that on most soil-covered slopes, which rarely have angles greater than 20°, the net downslope movement is only a small fraction of the displacement perpendicular to the surface. On steeper slopes, fluvial erosion is likely to displace creep as the dominant process.

To geophysicists and engineers, rock creep connotes slow, permanent internal rock deformation (strain) under low stress. As a process of mass-wasting, rock creep is less precisely used to describe slow movement of large coherent rock masses over a discrete basal surface. Most often, rock creep is observed on a cliff face, where a massive rock such as sandstone overlies shale. Pressure release normal to the cliff face (p. 106) produces large joint-bounded sheets or blocks, which slowly creep or lean outward until they become unstable and topple over or slide downhill (de Frietas and Watters, 1973).

The best-documented history of toppling failure that resulted from rock creep (Schumm and Chorley, 1964) is "the fall of Threatening Rock" in 1941 in Chaco Canyon National Monument, New Mexico. For 5 years prior to the fall, the gap between the cliff face and a free-standing monolith of sandstone 46 m long, 30 m high, and 9 m thick had been measured at monthly intervals. For 3 years the gap widened at approximately 25 mm/yr. During the next 2 years, the rate doubled, culminating in a violent rock fall that partly destroyed an ancient pueblo at the base of the cliff. The annual movement was concentrated during the wet winter season and was greater during years of greater precipitation. Movement had been in progress or at least recognized as possible for at least 9 centuries, proved because the prehistoric occupants of Pueblo Bonito had built an earth and masonry terrace at the base of Threatening Rock and wedged logs beneath its base about 1000 A.D. By extrapolation, Schumm and Chorley (1964, p. 1052) estimated the time required for each foot (30.5 cm) of movement away from the cliff as follows:

Movement	Years Required
1st foot	1600
2nd foot	650
3rd foot	200
4th foot	60
5th foot	8.0
6th foot	0.2

Rock creep illustrates the importance of pressure-release joints, or *sheeting* (p. 106), parallel to cliff faces or gorge walls in massive rocks. Rapid stream erosion can unbalance the stress distribution in rocks so that elastic expansion is outward toward the valley axis (Ferguson, 1967). When open joints have formed, it is only a matter of time until weathering destroys the support and a large mass of rock collapses from the cliff. Even though these events seem random and catastrophic, and may occur only a few times in a thousand years, they are part of the "normal" process of valley-side evolution. Robinson (1970, p. 2802) demonstrated that the crushing strength of massive sandstone is sufficient to support free-standing joint columns several hundred meters high on cliff faces and that failure occurs primarily by loss of basal support.

Flow

Incoherent rock debris may be sufficiently mobilized so that it flows like a viscous fluid. The criteria for defining flow are evidence of internal turbulence and either discrete boundaries or narrow marginal zones of shear. In both criteria, flow is a distinct from creep. Subdivision of flow can be based either on the kind of material, the degree of saturation, or the speed of advance (Fig. 8-3). Dry flows in sand or silt are known, but most flows are saturated with water. Rates of movement are greater than for creep but range from imperceptibly slow to tragically rapid mud flows and avalanches. Flows typically move as lobes or tongues. Some are known to erode channels. The transitional nature of mass-wasting terminology is nowhere better illustrated than in attempts to distinguish between a water-saturated mud flow and a mud-laden stream.

Solifluction. If soil or regolith is saturated with water, the soggy mass may move downhill a few millimeters or a few centimeters per day or per year. This type of movement is called **solifluction** (literally, "soil flow"). The original definition of solifluction excluded any connotation of climate but was included in a discussion of cold-climate landforms (Andersson, 1906; Washburn, 1967, pp. 10–14) and has frequently been misdefined as a cold-climate process. To clarify the confused terminology, Washburn (1967, p. 14) urged adoption of **gelifluction** for solifluction associated with frozen ground. That usage is adopted here (see also Chap. 15).

During the brief high-latitude summer thaw, an "active layer" a meter or so in thickness, composed of tundra peat, rock rubble, and other weathered debris, may flow down slopes of almost negligible gradient, because meltwater saturates the active layer but cannot penetrate the frozen ground beneath. The mass of gelifluction debris may flow with a kind of rolling motion like the endless tread of a tracked vehicle. Arcuate ridges and troughs mark the toe, or lower part of the mass. The rate of downhill motion in gelifluction lobes in East Greenland is several times more rapid than the nonlobate creep on wet slopes of similar gradient (Fig. 8-2) (Washburn, 1967, pp. 94–95). Although the annual rate of solifluction movement may be little more than 100 mm, the entire movement takes place during the brief summer season when net movement may be several mm/day on slopes of less than 10° (Fig. 8-7).

Figure 8-7. Large gelifluction lobe, Mesters Vig district, northeast Greenland. (Photo: A. L. Washburn.)

Washburn accurately surveyed lines of conical targets driven into the soil across Greenland hillsides, and by repeated resurveys that extended over a period of 9 years, he was able to separate the components of gelifluction and **frost creep** (soil creep due to the ratchetlike motion of particles during alternating freeze-thaw cycles) (Fig. 8-8). His periodic resurveys proved that frost creep was quantitatively slightly more important on most slopes than gelifluction. An unexpectedly large component of *retrograde motion* due to cohesion caused his targets to move uphill during the summer thaw season, with respect to the vertical plane through the targets at the time of maximum frost heave. The retrograde motion partly or entirely cancels downslope movement by gelifluction. Points P_1 to P_4 in Figure 8-8 trace a typical path of a soil particle on a Greenland hillside during an annual cycle. The diagram should be examined step by step, from the "jump" (P_1–P_2) during frost heave, through the downhill movement (P_2–P_3) during thaw and gelifluction, to the interaction between retrograde motion and gelifluction (P_3–P_4) that determine the net annual downslope motion. In no instance was the net motion uphill. Although retrograde motion in some years largely negated the role of gelifluction and frost creep, it cannot exceed them. By freezing insulated trays of saturated silt tilted at various low angles, Higashi and Corte (1971) were able to reproduce in a laboratory the various combinations of frost creep, gelifluction, and retrograde motion reported by Washburn.

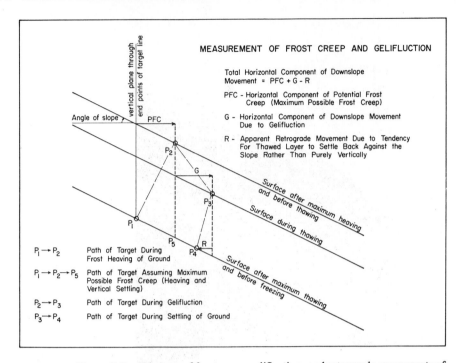

Figure 8-8.   Diagram of frost creep, gelifluction, and retrograde components of movement by a survey target on a perennially frozen hillside in northeast Greenland. (Washburn, 1967, Fig. 5.)

Solifluction is not restricted to frozen ground. It is a form of mass-wasting common wherever water cannot escape from a saturated surface layer of regolith. · A clay hardpan in a soil or an impermeable bedrock layer can promote solifluction as effectively as a frozen substratum. Under tropical rainforests, steep slopes in deeply weathered rock are scarred by large debris slides, and lower slopes are masses of solifluction lobes.

Debris Flow, Earth Flow, Mud Flow.   These three terms are applied to types of mass-wasting very similar to solifluction. They are somewhat more rapid, and they commonly flow along valleys, whereas solifluction sheets or lobes cover an entire hillside with moving debris. The three terms form a series of progressively higher water content but are often used interchangeably. Debris flows have 20–80 percent particles coarser than sand sizes, whereas earth flows and mud flows are 80 percent or more mud and sand (Varnes, 1978). Mud flow is the most liquid "end member" of the series.

Earth flows (the most general term) almost always result from excessive rainfall, and oversaturation seems to be the major factor in their formation. They may form on planar hillsides or at the toe of large slumps, but most often they develop at the heads of small gullies, where surface runoff concentrates to initiate streams. Unusually deep weathering along "seepage lines" at the head of incipient drainage networks may be the reason for the high frequency of flow initiation there (Bunting, 1961).

Case histories of earth flows from Ireland (Prior and Stephens, 1972), New Zealand (Crozier, 1969), Japan (Machida, 1966), and the United States (Williams and Guy, 1971, 1973) emphasize the combination of steep slopes (to 25°), weak or easily weathered rock or surficial cover, and abnormal precipitation in triggering earth flows. Artificial alteration, especially excavation at the foot of a slope, and vegetation clearing are very often additional factors. In seismic regions, earthquakes may trigger earth flows and avalanches. Quite commonly, aerial photographs of earth flows show many scars of earlier flows of the same general form.

Earth flows are common along the valley of the St. Lawrence River and in British Columbia, Canada, and in Scandinavia, where thick deposits of late-glacial marine clay have been uplifted by postglacial isostatic movements and now form most of the lowlands. The clay-sized particles were deposited rapidly in saline water and were flocculated into large agglomerates that have chaotic internal structure. Instead of being dense and laminated, the late-glacial marine clays have unusually high porosity and are initially held together by electrostatic forces between clay-size particles and the saline pore water. After subaerial leaching for 10,000 years or more, the soluble salts are removed, and the clays may become "quick," "sensitive," or *thixotropic* (Fig. 8-9). They retain their shear strength if they are not excessively or

Figure 8-9. Demonstration of "quick" clay. Undisturbed sample (left) supports 11 kg. Another sample of the same clay is poured from the beaker after being stirred. No water was added. (Photo: Div. Building Research, National Research Council of Canada.)

abruptly stressed, but with excessive vibration or shock, the pore water mobilizes, the weak framework of poorly arranged mineral grains collapses, and an apparently stiff plastic clay suddenly becomes a milky fluid. Along river valleys in Canada and Scandinavia, these clays in their cohesive unleached state form vertical river banks tens of meters high. Their strength degenerates, however, and sudden shocks such as earthquakes, thunder, highway traffic, blasting in excavations or log jams, or the spring breakup of ice on a river may cause an abrupt slope failure.

The form of an earth flow in quick clay is a characteristic pear or bottleneck shape, with a relatively narrow neck at the point of initial failure. As each slump mass collapses, liquefies, and flows toward the adjacent lowland, new concentric slices collapse. The flow margin retrogresses to a bowl shape from the initial narrow flow, which becomes the gorge through which a slurry of liquefied mud flows. The pattern on aerial photographs (Fig. 8-10) has a distinctive ribbed pattern of slump blocks that retained their cohesive strength briefly and then collapsed. The earth-flow bowls may continue to grow for hours or weeks to a diameter of several kilometers and are as much as 100 m deep. Their floors are hummocky, including tilted slices and pinnacles of the former surface layer that failed to liquefy (Karrow, 1972). In 1971, a disastrous flow at Saint-Jean-Vianney, Quebec, claimed 30 lives in a series of expanding slumps that progressively engulfed houses in a small village (Tavenas, et al., 1971). Personal narratives recount incidents of people outrunning the retrogressive expansion of earth-flow scarps in the Quebec lowland. It appears that the sod and soil profile stay coherent even after liquefaction has begun at depth, and people who have felt the road collapse beneath their automobile have yet had time to abandon the car and run uphill to safe ground.

A remarkable feature of Norwegian "quick" clay is the abruptness with which it loses plasticity if salt is restored and mixed into the flowing mass. Some earth flows across Norwegian roads can be stopped by throwing bags of sodium or magnesium chloride into the fluid mass and churning through it with a tracked vehicle. Almost instantly, the water-clay bonds are restored, the mass stabilizes, and it can be cleared with standard earthmoving equipment.

Three types of flows were set apart as special forms of mass-wasting by Sharpe (1938, p. 57). In semiarid and arid regions, torrential rains or *cloudbursts* may send flash floods roaring through mountain canyons and out beyond the mountain fronts across alluvial fans or pediments. Under those unique conditions, the water mass may be so charged with rock debris that when it leaves the confining canyon walls the larger fragments are dropped, and the flood spreads and becomes a mud flow. The role of flash floods in shaping arid landscapes is reviewed at length in Chapter 13.

A second unique form is the **lahar**, an Indonesian word that describes a debris flow on the flank of a volcano (Crandell, 1971). Torrential rains, sometimes triggered by pyroclastic eruptions, saturate the unstable tephra on the slopes of composite cones. Other sources of mobilizing water are breached crater lakes, melted snow, and slide-dammed rivers. Lahars may be hot or cold. As can be imagined, heavy rainfall, the steep constructional slopes of volcanic cones, and the dense population of Indonesia make lahar a word to be feared there.

Figure 8-10.   Earth flow of May, 1971, on the South Nation River, Ontario, Canada. View upstream. Head of the flow is 500 m from the river channel (Eden, et al., 1971). (Photo: Div. Building Research, National Research Council of Canada.)

The third special type of mud flow noted by Sharpe is the *alpine* type. This seems an unnecessary category, because steep slopes characteristic of glaciated mountain valleys (Chap. 17) are the obvious sites for all forms of mass-wasting. A *periglacial* category would be of greater regional significance (Chap. 15) if a climatic morphogenetic subdivision of mass-wasting were to be attempted.

In response to Sharpe's special designation of semiarid and alpine mud flows, later authors have referred to *temperate mud flows* (Hutchinson, 1968, p. 691; Prior and Stephens, 1972). Especially in the cool moist climate of the British Isles, where clay-rich bedrock or surficial clay deposits are periodically mobilized by heavy rains, the combination of present climate and Pleistocene geologic history has produced an optimal condition for mud flows, especially along seacoasts. The philosophy of defining climatic variants to geomorphic processes is discussed at length in Part III (Chaps. 13–18).

Slide

Mass-wasting wherein a mass of rock or weathered debris moves downhill along discrete shear surfaces is defined as a **slide**. The general word *landslide* is frequently extended to include all rapid forms of flow, slide, or fall, because most of the movement in all three categories is along surfaces of separation. Subcategories of slide include *slump, rock slide* or *block glide*, and *debris slide* or *debris avalanche*.

Slump.   Figure 8-11 is a composite diagram of the nomenclature of the parts of a typical "landslide." The diagram portrays a **slump** (rotational slip along a concave-up surface of rupture) passing downhill into an earthflow. This is an extremely common form of mass-wasting (the *slump-earthflow* of Varnes, 1978), although some of the named components might be absent in any single example.

Slump is the form of slide most common in thick, homogeneous, cohesive materials such as clay. The surface of failure beneath a slump block is spoon-shaped, concave upward or outward (Fig. 8-11). The upper surface of a slump block commonly is tilted backwards, because the entire mass rotates as the lower part moves outward and downhill. Vegetation or even houses may be carried intact on the surface of a large slump block. Slumps are tens or hundreds of meters wide and may be single blocks or consist of multiple slices. Ponds often form at the angle between the base of a slump scar and the head or top of the rotated slump block. This water, percolating down along the surface of rupture, may cause renewed instability on old slumps.

Slumps may be caused by currents or waves in water undercutting the foot of a slope. They are also a common result of faulty engineering design of cut embankments. They are sometimes controlled by loading the base of an unstable slope with a heavy layer of coarse rock rubble, which permits water to drain off the hill but offsets the weight of unstable earth higher on the slope. An elaborate engineering technology has been developed to predict the surface of rupture beneath a slump in order to drill into it and drain the water from the vicinity.

The most diagnostic feature of an ancient slide area is the hummocky or chaotic landforms on it (Fig. 8-12). Unless the material moved as a single slump block or is

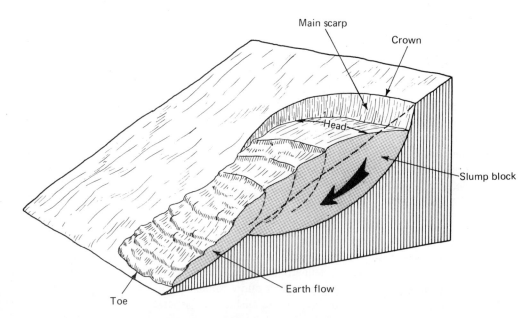

Figure 8-11.   A slump-earth flow with principal parts labeled. Components can be recognized on Figures 7-9 and 8-10.

Figure 8-12. Ancient landslide in Bailey Basin, Columbia River valley, northeastern Washington (Jones, et al., 1961). (Photo: F. O. Jones, U.S. Geological Survey.)

thoroughly fluidized as a mud flow, the region below a major slide is covered with a mass of broken rock rubble in mounds and pits, sometimes in systematic transverse ridges or lobes, but often totally unsystematic. The debris is poorly sorted and can easily be mistaken for glacial till in a mountain valley (Porter, 1970, p. 1421). Ancient slump blocks along valley walls in the Wasatch Plateau, Utah, have been mistaken for lateral moraines (White, 1973).

Rock Slide, Block Glide.   The most simple form of slide is a **rock slide** or **block glide**. The movement is relatively rapid and most commonly occurs where steeply dipping bedded strata or sheeting nearly parallels the surface slope (Fig. 8-13). Rock slides are generally shallow, because they occur along cohesionless planes, and because by Coulomb's equation, the load perpendicular to the surface (σ') increases rapidly with depth. A heavy rain or freezing and thawing provides fluid pressure, or vibration breaks off obstructions and reduces the coefficient of friction on the glide plane, and a detached slab or block slides down. It may shatter at the base of the slope, or it may remain intact. Rock slides or block glides have no specified size, but their thickness is normally only about 10 percent of their downslope length.

Rock slides can be dramatic forms of sliding mass-wasting if large masses of unweathered rock slide downhill along a sloping joint or bedding surface. Such a plane of weakness was probably involved in the Vaiont Reservoir disaster of October, 1963, in northern Italy (Kiersch, 1964). On the night of October 9, a rock slide 2 km long, 1.6 km wide, and over 150 m thick moved suddenly down the south wall of the Vaiont Canyon and completely filled the 270 m-deep reservoir for 2 km upstream from the dam to heights of 180 m above the former water level. The movement took less than a minute, so rapid that the water in the reservoir was ejected 260 m up the north canyon wall and propelled in great waves both upstream in the reservoir and downstream over the dam. The resulting floods killed 3000 people, mostly around the town of Longarone, over 2.5 km downstream from the dam and across a broad valley from the mouth of the Vaiont Canyon.

The geologic cross section of the reservoir is sketched in Figure 8-14. The most obvious feature of the geology is the bowl-shaped structure of the rocks, which dip inward toward the valley axis from both sides. The rocks are mainly limestone, with thin clay layers at intervals. The limestone is full of caves and smaller solution channels so that large amounts of rainwater can penetrate the rock and spread along the clay layers. A whole series of natural and man-made factors contributed to the disaster. The principal factors were: (1) The steeply dipping limestone beds and clay layers offered little frictional resistance to sliding. (2) All the rock types are inherently weak. (3) The river had eroded the steep inner canyon across the rock structure and removed lateral support long before the dam was built. (4) Two weeks of heavy rainfall had raised the water level in the cavernous rocks and increased both fluid pressure and weight in the potential slide mass. (5) The high water level in the reservoir had saturated the lower part of the slide, decreased frictional resistance, and increased buoyancy.

Figure 8-13. Rock slide on dipping sandstone strata near Glenwood Springs, Colorado. Sheeting in massive rocks has the same effect as dipping strata (Fig. 6-3). (Photo: D. J. Varnes, U.S. Geological Survey.)

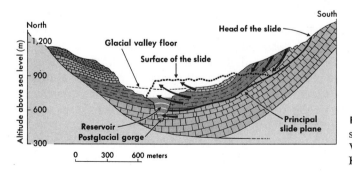

Figure 8-14. Simplified geologic cross section of the Vaiont Reservoir, Italy. View is upstream. (Redrawn from Kiersch, 1964.)

181

The Vaiont Reservoir disaster was a surprise only in its severity. The canyon is marked by the scars of ancient slides. In 1960, a smaller slide had occurred, and a pattern of cracks and slumps developed on the south valley wall that ultimately outlined the great slide of 1963. For 6 months prior to the slide, precise records of survey stations on the slide area showed that rock creep of 1 cm/week was in progress. By 3 weeks prior to the disaster, the rate of creep had increased to 1 cm/day; during the last week of heavy rains prior to the slide, the rate of creep had increased from 20–40 cm/ day. Wild animals that had grazed on the south wall of the valley sensed the danger and moved away about October 1. On the night of the disaster, 20 technicians were on duty in the control building on the south abutment of the dam, and 40 more were in the hotel and office building on the north abutment, but no one survived who witnessed the actual slide. Desperate measures were underway to lower the reservoir level, but the rock creep was apparently pinching the reservoir and actually raising the water level in spite of open outlet gates. No earthquake or other "trigger" for the Vaiont slide has been identified. The rock along the principal slide plane simply failed under excessive weight and excessive pore-water pressure.

Debris Slide.   The largest slide known on earth is the Saidmarreh slide in southwestern Iran (Fig. 8-15). It occurred more than 10,000 years ago, but because of the aridity in the region, its surface morphology can still be seen. It has been dissected by the gorge of the Saidmarreh River, which could have prepared conditions for the slide by undercutting the dip slope of a hogback. The surface of the slide is a karst landscape formed by solution of gypsum from the bedrock beneath the slide debris (Watson and Wright, 1969).

The slide came from the northern flank of the Kabir Kuh, a northwest-trending anticlinal ridge capped by massive Middle Cretaceous limestone, with an average elevation of about 2000 m above sea level. On the northern flank of the anticline is a high hogback of Oligocene Asmari Limestone. This massive unit dips toward the valley at about 20° and rests on thin-bedded Eocene limestone and marl. A segment of the hogback 15 km long, 5 km wide, and at least 300 m thick slid off the mountain into the valley below. Watson and Wright proposed (1969, p. 129) that an earthquake triggered the slide in late Pleistocene time, after the Saidmarreh River had undercut the base of the hogback ridge. The slide must have been extremely rapid. The vertical component of initial motion at the center of gravity of the slab was only about 900 m, yet the slide mass crossed a valley, rose 600 m across the nose of an adjacent plunging anticline, and continued northward for another 18 km. Shreve (1966, p. 1640) calculated that a minimum velocity of 300 km/hr was required. The slide was not turbulent, for the debris is crudely stratified with fragments from the youngest strata nearest the top of the slide debris. Pore water under dynamic stress may have counterbalanced part of the weight of the slide, or possibly the gypsum over which it moved had a very low coefficient of sliding friction.

The well-documented Sherman slide (Fig. 8-16) was triggered by the March 27, 1964, Alaskan earthquake (Shreve, 1966; Marangunic and Bull, 1968). A sheet of

Figure 8-15.   Scar of the prehistoric Saidmarreh landslide, southwestern Iran. (Photo: Aerofilms Ltd.)

debris averaging 1.3 m in thickness traveled 5 km from its origin across the gently sloping surface of the Sherman Glacier yet did not disturb the 2 m of fresh snow on the surface of the glacier. After the initial rock slide down a 40° cliff for a vertical height of 600 m, the slide crossed a bedrock ridge 150 m high and then spread out over 8.5 km² of glacier surface. The debris is thoroughly pulverized, probably by crossing the rock ridge if not in the initial slide. The ridge may have served as the "launching platform" to send the debris slide airborne over the glacier at a top speed of at least 185 km/hr (Shreve, 1966, p. 1640).

Shreve (1966, 1968) has presented convincing arguments that large landslides such as the Saidmarreh and Sherman slides must float on a layer of trapped and compressed air. His theory is also supported by evidence from other large slides. Most impressive is the simple argument that the "coefficient of friction" of large slides, defined as the maximum vertical drop divided by horizontal distance the slide travels (the tangent of a hypothetical average slope angle), is only about 0.1–0.26, much lower than the internal friction of granular materials. The slide mass must move as a buoyant flexible sheet, not as a viscous fluid.

Figure 8-16.   Sherman landslide, Alaska. Debris fell from the high peak on the right, overtopped a spur in the right foreground, and spread more than 2.5 km across the Sherman Glacier during the earthquake of March 27, 1964. (Photo: Austin Post, U.S. Geological Survey.)

Avalanche.   The most rapid flowing, sliding, or falling mass-movements are called **avalanches**, the very name being enough to frighten most mountain dwellers. In composition, an avalanche may range from entirely ice and snow to mostly rock debris. An avalanche usually begins with a free fall of a mass of rock or ice, which is pulverized on impact and flows at great speed, made fluid by the pressure-heated air and water entrapped in the mass.

Two of the worst debris avalanches in history destroyed the region around Ranrahirca and Yungay, Peru, on January 10, 1962, and again on May 31, 1970 (McDowell and Fletcher, 1962; Cluff, 1971; Plafker, et al., 1971). Observers witnessed both catastrophes from the time huge ice cornices fell from the north peak of Huascaran (6700 m) until the debris came to rest against the opposite valley wall, 14.5 km away and 4 km lower in altitude. The initial rock and ice masses, of an estimated 3 million m³ in 1962 and 50–100 million m³ in 1970, tore loose other millions of tons of rock as they roared down the valley. The shock waves produced a noise like continuous growing thunder and stripped hillsides bare of vegetation. Rocks and ice were pulverized by the turbulent flow.

The 1962 avalanche required only 7 minutes to travel 20 km. It bounced from one side of a narrow gorge to the other at least five times before it emerged onto the fertile, heavily populated valley floor at the base of the mountain. As it spread to 1 km wide over the villages and fields of the valley, the mass slowed to an estimated 100 km/hr and thinned to about 20 m. When the avalanche stopped, air and water spouted from the settling debris. Later, melting blocks of ice created pockets of soft mud in the flow that were added hazards to the nearly hopeless search operations. An estimated 3500 people were killed.

Eight years later, a major earthquake triggered another debris avalanche from the north face of Huascaran. The 1970 avalanche followed the same path as the 1962 avalanche for most of the 14 km down the gorge, but at the edge of the main valley a lobe of it jumped a bedrock ridge 200–300 m high. Within 3 minutes after the original rock fall, the debris avalanche had obliterated the town of Yungay, which had been protected from the earlier avalanche by the ridge. The average velocity must have been between 280 and 335 km/hr. Possibly 40,000 people were killed in Yungay and neighboring towns by the 1970 avalanche, but because the total casualties of the earthquake in central Peru were estimated at 70,000 killed and 50,000 injured, the exact numbers are unknown.

Like the Saidmarreh, Sherman, and other catastrophic slides, the vertical fall of both Huascaran avalanches was small relative to the horizontal distances they covered. The 1970 avalanche was airborne over the lee slopes of overtopped ridges, adding further support to the theory that these large debris masses can float on an air cushion. It is possible, however, that the extreme internal turbulence in some avalanches creates fluidized masses that do not obey the laws of sliding friction and therefore travel much greater horizontal distances at much higher velocities than might be predicted (Hsü, 1975). Evidence for avalanche deposits on the moon is also counter to an air-cushion theory but supports a theory that elastic impact between avalanche fragments makes a fluid out of solid detritus, even in a vacuum. In a sense, the avalanche debris behaves like the molecules in a perfect gas, supporting itself without an interstitial fluid.

Debris avalanches in the temperate forested region of the southeastern United States claimed a number of lives and produced major geomorphic changes during the torrential rainfall of August 19, 1969 (Williams and Guy, 1971, 1973). The storm and its results were unique in intensity but not in kind. Intense summer storms, usually of local extent, are characteristic of the Appalachian region, and their results must be regarded as part of the normal land-forming processes of the region.

Debris avalanches of the 1969 torrential summer storm originated at the heads of first-order tributary streams on slopes of 16°–39° (Fig. 8-17). The bedrock in the area is massive crystalline gneiss and granite, with sheeting joints as the major structures. The weathered layer is as much as 6 m thick. The debris avalanches stripped off the weathered layer but moved relatively little fresh rock. Bedrock, often a single joint sheet, is often exposed the entire length of an avalanche scar, which may be 250 m downslope and 10–25 m wide. In all examples of the 1969 damage that were studied by Williams and Guy, the avalanche debris was carried down the initial slope and onto the adjacent valley floor, where it became a debris flow or was carried away later as alluvium (Fig. 8-17). The obvious general cause for temperate-climate debris avalanches is saturation by heavy rain, but the precise trigger is unknown. Summer storms are known to fluctuate manyfold in intensity, even over a time span of hours

Figure 8-17.   Debris-avalanche scars in the headwaters of Davis Creek, Nelson County, Virginia. Most of the slides occurred during a few hours near the end of a torrential rainstorm on August 19, 1969 (Williams and Guy, 1973, Fig. 11). (Photo: Edwin Roseberry.)

or less, and during the brief intervals of heaviest precipitation, water alone may be the ultimate cause of failure. Lightning striking trees, the vibration of thunder, and windthrow are also suggested trigger mechanisms.

Fall

The examples just cited prove that many avalanches in alpine mountains are initiated by a mass of rock or ice breaking off and free-falling from a glacier or cliff. Considering the rarity of vertical or overhanging cliffs in natural landscapes, we can accept a rock fall or ice fall as a rare, if awesome, event. Three years before the 1970 Yungay disaster, a French mountain-climbing team reported that Huascaran peak had well-developed vertical joints behind the nearly vertical north face. The earthquake may have done nothing more than topple or shake down some huge joint blocks, which in turn shattered the lower cliffs and started the avalanche. **Fall** is a distinct landslide process, but it is rarely independent of subsequent events.

Epilogue on Processes of Mass-Wasting

It is unfortunate for our perspective that an analysis of mass-wasting becomes a description of major human disasters. Because avalanches and rock slides kill thousands of people, they deserve intensive study to save lives through better prediction, control, and warning procedures. But these spectacles are restricted to areas of high relief, steep slopes, and local special conditions. In much less conspicuous fashion, weathered rock and soil slowly creep down all slopes, whether under grass, forest, desert shrubs, or tundra. Creep is surely the dominant process of mass-wasting, but the evidence for it is subtle and often overlooked. Every tilted sidewalk slab, cracked pavement, and slumped embankment announces that creep is in progress, literally under our feet.

HILLSLOPE DEVELOPMENT AND EVOLUTION

Most of a landscape consists of curved, sloping surfaces, largely shaped by mass-wasting. How these slopes form, how they are maintained, and how they change with time are major topics of geomorphic research. Slopes are exceptionally difficult to study, because they are transitional both in processes and in form. Recall the arbitrary definitions of mass-wasting: Add a little water, and soil creep becomes earth flow; add a little more water, and earth flow becomes a muddy stream.

It is surprisingly difficult even to describe the geometry of a natural slope. One is commonly provided with only a profile surveyed down the steepest part of a hillside as a description of the slope, but obviously a profile measured along the crest of a descending ridge will have a significance very different from that of a similarly shaped profile measured along the bed of an adjacent gully. Slopes are irregular surfaces that cannot be described by simple mathematical equations. The best topographic maps are only approximations of the infinite irregularities of hillsides. We do not yet know what degree of irregularity is significant in the stability of slopes, so we are never sure that we are measuring the correct angles and distances.

Hillslope Analysis: Techniques and Theories

There have been two philosophies about studying slopes. An older school deduced the systematic changes of slope form that would accompany long-continued subaerial weathering and erosion. Because landscape evolution is too slow to be witnessed, deductions concerning the changes of slope form with time were based on assumptions that could not be tested until we had isotopic techniques for dating old land surfaces. It is no wonder that the deductive approach to slope analysis has enriched geologic literature with some remarkably opinionated and authoritarian writings. A majority of deductive geomorphologists have held that slopes, especially in humid regions, become lower and more broadly rounded with time. A vocal minority have insisted that slopes are stable forms with angles controlled by rock type and weathering processes, and when a stable slope has evolved, it persists through time, migrating backward parallel to itself unless it is eliminated by the intersection of other slopes. For an excellent review of the classic debate, see von Engeln (1942, pp. 256–67). The issues are evaluated further in Chapter 12.

Another group of geomorphologists have concerned themselves with the empirical description of slopes. With less regard for theoretical projection into the future, they have studied the processes of slope formation and the geometry of slopes. Innumerable slope profiles and descriptive texts have been published, but empirical study has suffered from the lack of a guiding theory. Only dedicated and persistent workers continue to scramble up and down hills with measuring tapes and levels.

Some progress has been made, however, by both deductive and empirical methods of study (Schumm, 1966). For some time, it has been recognized that terrestrial slope profiles generally have an upper segment convex to the sky and a lower concave segment, that some slope profiles show a straight segment between the upper and lower curves, and that if a cliff interrupts the slope, an additional segment marked by free fall of weathered debris is introduced into the profile above the straight segment (Wood, 1942). A total of nine slope segments are recognized in one recent classification (Fig. 8-18).

The straight segment of a slope below a free face (Fig. 8-18, unit 5) is usually a **talus**. A talus is a slope landform, not a kind of material. It is redundant to speak of a "talus slope," and the coarse rock debris on a talus should be called **sliderock** or *scree* (the common British usage). At fist glance, a talus may appear to be a constructional landform, built of a great wedge of sliderock at the base of a steep cliff. Rarely, except along the walls of oversteepened glacial troughs, is such the case. More often, a talus reflects a bedrock surface veneered with a layer of sliderock that is creeping. Debris that falls onto a talus from the cliff "armors" the talus and replaces fragments lost by weathering and creep. The angle of the talus is a function of fragment size, angularity of the sliderock, climate, vegetation, and rate of sliderock supply and removal.

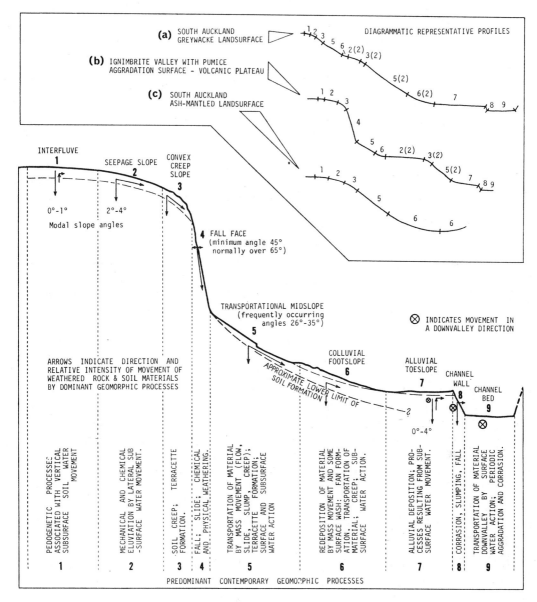

Figure 8-18. Diagrammatic representation of a hypothetical nine-unit land-surface model. (Dalrymple, et al., 1968, Fig. 1.)

Gentler slopes with straight intermediate segments (not taluses at the bases of cliffs as in Figure 8-18) seem to form where erosion is unusually rapid. In the extreme examples of gullies cut on artificial embankments by single intense storms, most of the slopes are straight. Natural landscapes scored by closely spaced V-shaped gullies with straight sides that intersect as knife-edged ridges are called **badlands**. Strahler (1950, 1956) demonstrated that straight valley-side slopes are common in areas of recent or continuing tectonic uplift where downcutting by rivers is active.

The upper convex and straight segments of slope profiles (Fig. 8-18, units 1–3) are largely controlled by mass-wasting, especially creep. Gilbert (1909) offered an explanation of summit convexity that is largely deductive but is still the best available explanation. He observed a reasonably uniform thickness of soil or regolith over the convex surface and assumed that in a given interval of time a uniform thickness of the weathered material is removed from the entire summit area. Under these conditions, larger quantities must move through cross sections progressively farther downhill. In other words, with the stated assumption, the amount of material that creeps past any point is proportional to the distance of the point from the summit. As creep is primarily a gravitational phenomenon, the slope angle must increase radially from the summit in order to move the progressively greater amount of debris. The summit curvature becomes convex to the sky. Carson and Kirkby (1972, pp. 108–9, 306–7) strongly supported Gilbert's conclusion that the hilltop convexity of humid temperate regions is primarily the result of soil creep, but they presented additional theoretical arguments that Gilbert's deduction is unnecessarily restrictive.

On lower slopes (Fig. 8-18, units 6 and 7 and part of unit 5), transportation by flowing water assumes dominance over creep. Two little rills flowing down a bare hillside during a rainstorm require a certain slope to keep flowing with their suspended-sediment load. When the two rills join, the resulting rivulet has a greater proportional increase of mass than its increase in wetted surface area. Friction is reduced in proportion to the discharge, and the larger trickle of water can transport the joint loads of the two lesser streams with no loss of velocity but on a more gentle slope. Thus slopes controlled by rain wash, sheet wash, or rill wash are generally concave skyward. At some position on a slope, rain wash becomes dominant over soil creep, and the slope profile inflects from convex near the top to concave near the base. (For an extended discussion of slope concavity, see Carson and Kirkby, 1972, pp. 310–15). Lunar landscapes seem to lack the concavity of terrestrial footslopes, perhaps in response to the total lack of water on the moon.

Thus far we have considered only slope profiles. Even from profiles alone, one can see how soil creep near the top of a slope increases the gradient downhill until rain water begins to flow over the surface instead of penetrating and saturating the creeping soil. At that level on the hillside, sheet wash and slope concavity begin. To date, most of the research on hill slope development has been confined to analyses of profiles (Schumm, 1966; Brunsden, 1971; Carson and Kirkby, 1972; Young, 1972).

Slopes also curve in directions other than downhill, and these other curvatures also affect water movement. Where contour lines bulge convexly outward on a hillside around sloping spurs or **noses**, water is spread laterally as it flows downhill. Noses and ridge crests tend to be drier than adjacent **hollows** where the contour lines swing concavely into the hill. Concave contours tend to gather water from a large area higher on the slope, and the heads of streams are localized downhill from hollows (Hack and Goodlett, 1960, p. 6).

Profile curvature and contour curvature can be combined into a single diagrammatic classification of slopes (Fig. 8-19). The horizontal axis of the diagram divides "water-gathering" slopes with concave contours (quadrants I and II) from "water-spreading" slopes with convex contours (quadrants III and IV). The vertical axis of the diagram separates slopes with convex profiles dominated by creep (quadrants II and III) from those with concave profiles dominated by rain wash (quadrants I and IV). Almost any segment of a land surface can be placed into one of the four quadrants of the diagrams (Troeh, 1965). The only exceptions are saddle-shaped surfaces, which require a higher order of mathematical analysis. Horizontal surfaces or planar slopes plot on the axes of the diagram.

Each of the slope elements of Fig. 8-19 can be represented mathematically by a quadratic equation. Each surface is generated by rotating a segment of a parabola around a vertical axis. In perfecting his classification, Troeh surveyed agricultural lands around Cornell University and calculated the best-fitting paraboloid of rotation for each small area of surveyed land. In areas where the actual land surface could be represented by a single quadratic equation with vertical deviations no greater than

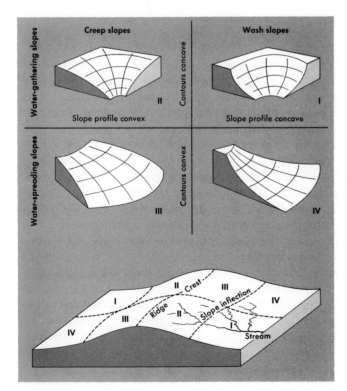

Figure 8-19. Classification of slope elements by profile curvature and contour curvature. (From Troeh, 1965.)

10–15 cm, a single soil type was represented. Where the land surface departed from the calculated surface by more than 15 cm, a new soil type appeared. Usually the difference in soil was controlled by slightly better or worse drainage (Troeh, 1964).

Mode and Rate of Slope Retreat

Most hillslopes are landforms that result when rivers cut valleys. Only tectonic scarps and sloping depositional surfaces (sand dunes, moraines, volcanoes, etc.) are not the secondary result of erosion. Ordinarily, a landscape is composed of small slope elements, each reacting in a particular way to the local effectiveness of weathering, mass-wasting, and erosion. All elements are related, however, because an accidental disequilibrium in any part of a slope affects adjacent segments above and below the site of the accident. Suppose that an animal burrowing on a hillside raises a mound of loose soil on the downhill slope. The slope is locally oversteepened, and the debris is rapidly spread downhill. Simultaneously, rain wash from uphill is trapped by the burrow, and creep is accelerated by the undermining. A dynamically stable, or *graded*, slope is an example of an open physical system through which both energy and matter move, a system that tends by self-regulating processes to maintain itself in the most efficient possible configuration. Slopes constantly change but always tend toward some central graded state appropriate to the environment of the moment.

Questions that have stimulated generations of geomorphologists concern the retreat of slopes as river valleys deepen and widen. Are slope profiles stable, migrating backward parallel to themselves, or do slope angles decrease through time as summits are lowered? Are there climatic variations in the way slopes evolve? Do slopes persist through epochs of geologic time, or are present slopes the result of present-day processes? Can slope retreat be measured, or must it be deduced from indirect evidence? These questions are so deeply embedded in the fabric of geomorphic thought that they must be repeatedly reexamined in subsequent chapters. Only a brief preview is appropriate here, while the processes that shape slopes are fresh in the reader's mind.

Some generalizations about slopes seem reasonably established. Straight valley-side slopes are associated with rapidly deepening valleys in a variety of tectonic and climatic settings and on a wide scale of sizes. In humid regions, slope profiles are commonly sigmoid, convex skyward near summits and concave on lower slopes. Creep is the dominant process on the upper convexity of the profile, and sheet wash or rill wash dominate on the concave elements. In plan view, well drained noses alternate with water-gathering hollows in which tributary streams originate. Weathering is intensified along the axes of hollows. Hollows enlarge by erosion at the expense of intervening noses.

Creep and slope wash, which determine the relative proportion of convex and concave profile elements, are in part climate-dependent. In some climates, convexities extend across low interfluves from one valley floor to the next (Fig. 15-3). In other climates, concave slopes sweep upward from valley floors to knife-edge divides (Fig. 12-1). Unless the climate changes, the regional dominance of one slope component or another will persist, although declivities might change with time.

Cliffs and taluses are usually structurally determined (Chap. 11). Vertical cliffs can be cut by valley glaciers, waves, or rivers, but they are quickly buried under their own taluses unless some erosional agent removes the debris from their bases. If this does not happen, the talus grows in height until the entire supply of sliderock at the top of the cliff is buried. Thenceforth, the slope becomes vegetated as a fossil or relict talus and gradually becomes the straight segment in a normal slope profile. If the sliderock is removed from the base of the talus as fast as it is supplied from the cliff at the top, the cliff and talus landforms migrate backward as long as the structural factors that define the cliff persist.

Colluvium of coarse rock debris on a slope may weather and change its angle of internal friction as it creeps downslope. The persistence of slope angles as valley sides retreat is thus related to the weathering character of the rock ("structure") and climate ("process"). If debris becomes finer as it moves downslope, eventually it is entrained in slope wash and transported over a concave profile to a stream. If, as is often the case with granitic rocks on desert slopes, large joint blocks "armor" slopes until they break down to coarse sandy grus, a sharp reentrant angle marks the foot of the hillslope and the head of an alluvial fan or pediment (Fig. 13-4).

Much has been said here of the climatic and structural controls of hillslope development and retreat. Slope evolution through time is deferred until Chapter 12, and additional aspects of climatic morphogenesis are reviewed in Chapters 13–18. In concluding this chapter, a few rates of slope retreat are quoted, only to suggest that such measurements can be made or deduced. Each of the five examples is from a region of dramatic slope development.

(*1*)   Cliffs of massive red or brown sandstone formations, generally of Mesozoic age, form much of the striking scenery of the Colorado Plateau. Schumm and Chorley (1964; 1966, p. 30) estimated that one cliff had retreated 30 feet in the last 2500 years (3–4 mm/yr) by rock fall. However, the cliff was exposed to weathering for an indeterminate length of time before the measured blocks began to move. The Colorado River crosses the Colorado Plateau, and its canyon is now an average of 12 km wide at the rim. By assuming that the Grand Canyon began to erode about 20 million years ago, these authors estimate the long-term widening rate of the canyon at about 0.6 mm/yr, or a rate of cliff retreat of 0.3 mm/yr on each canyon wall.

(*2*)   LaMarche (1968) measured denudation rates in the White Mountains of California by the root exposure of bristlecone pines as much as 4000 years old (Fig. 12-2). Steep hillslopes there are being lowered (parallel to the surface) at about 0.2 mm/yr; if the slope angle were 30°, horizontal recession of 0.1 mm would result. These slopes are on heavily jointed dolomite in a montane climate of intense frost action, cloudbursts, and mud flows.

(*3*)   From a comprehensive study of slope development in a subarctic mountain valley in northern Sweden, Rapp (1960, p. 186) concluded that rock fall alone causes steep slopes there to retreat at about 0.04–0.15 mm/yr, or about 1 m in the 10,000 years of postglacial time. In the cold maritime climate, hydrofracturing during freezing is the most active process. The bedrock is mica schist and limestone.

(*4*)   Mountain slopes under rainforest in New Guinea are intensely scarred by active landslides, which on some slopes cover as much as 40 percent of the total area (Simonett, 1967, p. 73). Judging by the rates of postwar vegetation recovery, slide areas are not revegetated for 40–60 years. Therefore, an estimate of regional denudation rates can be made from the area and volume of landslide scars measured on aerial photographs. A combination of intense weathering, frequent and intense earthquakes, and steep slopes results in an estimated regional denudation rate of 1.0–1.4 mm/yr by landslides alone, on granitic terrane (Simonett, 1967, p. 83). Assuming average forested slopes of clifflike 45° declivity, slope retreat here would approach 1 mm/yr. Similar rapid slope retreat and valley enlargement have been reported from Japan (Machida, 1966).

(5) As a final example, the extreme deductions of King (1962, pp. 154, 158) concerning the South African landscape can be cited. King asserted that tectonic movements initiate cycles of erosion by sending scarps retreating wavelike across the continent, each leaving a lower-level pediment behind while an older, higher land surface is consumed by the scarp. He believed that the Drakensberg scarp in Natal, 1200 m high, is one such scarp that was initiated at the coast in late Cretaceous time and has since migrated 250 km inland at a rate of about 3 mm/yr. King's hypothesis represents an extreme view of the importance of parallel scarp retreat and regional pedimentation, but the rates he cited are not excessive as compared to the previously cited four examples.

From these five examples, which are among the few that can be found in the published literature, one can conclude that steep valley walls or cliffs retreat at rates of 0.1–3.0 mm/yr in a variety of subarctic, desert, mountainous, rainforest, and savanna environments on diverse lithologies. The techniques and guiding theories of the various examples vary greatly, as do the dominant processes of slope retreat. More such measurements should be made in a wider range of climates. It is notable that no examples of the retreat of "normal" sigmoid slope profiles can be quoted, but if the examples are at all representative, most valley slopes and cliffs that now give detail to the landscape have been shaped and reshaped by the alternating climatic events of the Quaternary Period. Some may have been initiated in the early Cenozoic Era or ever earlier, far from their present positions.

REFERENCES

ANDERSSON, J. G., 1906, Solifluction, a component of subaerial denudation: *Jour. Geology,* v. 14, pp. 91–112.

BRUNSDEN, D., 1971, Slopes—form and process: *Inst. Brit. Geographers Spec. Pub. no. 3,* 178 pp.

BUNTING, B. T., 1961, The role of seepage moisture in soil formation, slope development, and stream initiation: *Am. Jour. Sci.,* v. 259, pp. 503–18.

CARSON, M. A., and KIRKBY, M. J., 1972, *Hillslope form and process:* Cambridge Univ. Press, Cambridge, England, 475 pp.

CLUFF, L. S., 1971, Peru earthquake of May 31, 1970; engineering geology observations: *Seismol. Soc. America Bull.,* v. 61, pp. 511–21.

CRANDELL, D. R., 1971, Postglacial lahars from Mount Rainier volcano, Washington: *U.S. Geol. Survey Prof. Paper 677,* 75 pp.

CROZIER, M. J., 1969, Earthflow occurrence during high intensity rainfall in eastern Otago (New Zealand): *Eng. Geology*, v. 3, pp. 325–34.

DALRYMPLE, J. B., BLONG, R. J., and CONACHER, A. J., 1968, An hypothetical nine-unit landsurface model: *Zeitschr. für Geomorph.*, v. 12, pp. 60–76.

EDEN, W. J., FLETCHER E. B., and MITCHELL, R. J., 1971, South Nation River landslide, 16 May 1971: *Canadian Geotech. Jour.*, v. 8, pp. 446–51.

von ENGELN, O. D., 1942, *Geomorphology:* The Macmillan Company, New York, 655 pp.

FERGUSON, H. F., 1967, Valley stress release in the Allegheny Plateau: *Eng. Geology*, v. 4, no. 1, pp. 63–71.

de FRIETAS, M. H., and WATTERS, R. J., 1973, Some field examples of toppling failure: *Géotechnique*, v. 23, no. 4, pp. 495–513.

GILBERT, G. K., 1909, The convexity of hilltops: *Jour. Geology*, v. 17, pp. 344–50.

HACK, J. T., and GOODLETT, J. C., 1960, Geomorphology and forest ecology of a mountain region in the central Appalachians: *U.S. Geol. Survey Prof. Paper 347*, 66 pp.

HIGASHI, A., and CORTE, A. E., 1971, Solifluction: A model experiment: *Science*, v. 171, pp. 480–82.

HSÜ, K. J., 1975, Catastrophic debris streams (sturzstroms) generated by rock falls: *Geol. Soc. America Bull.*, v. 86, pp. 129–40.

HUTCHINSON, J. N., 1968, Mass movement, *in* Fairbridge, R. W., ed., *Encyclopedia of geomorphology:* Reinhold Book Corporation, New York, pp. 688–96.

JONES, F. O., EMBODY, D. R., and PETERSON, W. L., 1961, Landslides along the Columbia River valley, northeastern Washington: *U.S. Geol. Survey Prof. Paper 367*, 98 pp.

KARROW, P. F., 1972, Earthflows in the Grondines and Trois Rivières areas, Quebec: *Canadian Jour. Earth Sci.*, v. 9, pp. 561–73.

KIERSCH, G. A., 1964, Vaiont reservoir disaster: *Civil Engineering*, v. 34, pp. 32–39.

KING, L. C., 1962, *Morphology of the earth:* Hafner Publishing Company, New York, 699 pp.

KIRKBY, M. J., 1967, Measurement and theory of soil creep: *Jour. Geology*, v. 75, pp. 360–74.

LaMARCHE, V. C., 1968, Rates of slope degradation as determined from botanic evidence, White Mountains, California: *U.S. Geol. Survey Prof. Paper 352-I*, pp. 341–77.

LEWIS, L. A., 1974, Slow movement of earth under tropical rain forest conditions: *Geology*, v. 2, pp. 9–10.

MACHIDA, H., 1966, Rapid erosional development of mountain slopes and valleys caused by large landslides in Japan: *Tokyo Metropolitan Univ. Geog. Repts.*, no. 1, pp. 55–78.

MARANGUNIC, C., and BULL, C., 1968, The landslide on the Sherman Glacier, *in* The great Alaskan earthquake of 1964, v. 3, "Hydrology": *Natl. Acad. Sci. Pub. 1603*, Washington, D.C., pp. 383–94.

McDOWELL, B., and FLETCHER, J. E., 1962, Avalanche! 3,500 Peruvians perish in seven minutes: *Natl. Geographic*, v. 121, pp. 855–80.

PHIPPS, R. L., 1974, The soil creep-curved tree fallacy: *U.S. Geol. Survey Jour. Res.*, v. 2, pp. 371–77.

PLAFKER, G., ERICKSEN, G. E., and CONCHA, J. F., 1971, Geological aspects of the May 31, 1970, Peru earthquake: *Seismol. Soc. America Bull.*, v. 61, pp. 543–78.

PORTER, S. C., 1970, Quaternary glacial record in Swat Kohistan, West Pakistan: *Geol. Soc. America Bull.*, v. 81, pp. 1421–46.

PRIOR, D. B., and STEPHENS, N., 1972, Some movement patterns of temperate mudflows: Examples from northeastern Ireland: *Geol. Soc. America Bull.*, v. 83, pp. 2533–43.

RAPP, A., 1960, Recent development of mountain slopes in Kärkevagge and surroundings, northern Scandinavia: *Geograf. Annaler*, v. 42, pp. 65–200.

ROBINSON, E. S., 1970, Mechanical disintegration of the Navajo Sandstone in Zion Canyon, Utah: *Geol. Soc. America Bull.*, v. 81, pp. 2799–805.

SCHUMM, S. A., 1966, The development and evolution of hillslopes: *Jour. Geol. Education*, v. 14, pp. 98–104.

———, 1967, Rates of surficial rock creep on hillslopes in western Colorado: *Science*, v. 155, pp. 560–61.

SCHUMM, S. A., and CHORLEY, R. J., 1964, The fall of Threatening Rock: *Am. Jour. Sci.*, v. 262, pp. 1041–54.

———, 1966, Talus weathering and scarp recession in the Colorado Plateaus: *Zeitschr. für Geomorph.*, v. 10, pp. 11–36.

SHARPE, C. F. S., 1938, *Landslides and related phenomena:* Pageant Books, Inc., Paterson, N.J., 137 pp. [Reprint, 1960].

SHREVE, R. L., 1966, Sherman landslide, Alaska: *Science*, v. 154, pp. 1639–43.

———, 1968, The Blackhawk landslide: *Geol. Soc. America Spec. Paper 108*, 47 pp.

SIMONETT, D. S., 1967, Landslide distribution and earthquakes in the Bewani and Torricelli Mountains, New Guinea, *in* Jennings, J. N., and Mabbutt, J. A., eds., *Landform studies from Australia and New Guinea:* Australian Natl. Univ. Press, Canberra, pp. 64–84.

STRAHLER, A. N., 1950, Equilibrium theory of erosional slopes approached by frequency distribution analysis: *Am. Jour. Sci.*, v. 248, pp. 673–96, 800–14.

———, 1956, Quantitative slope analysis: *Geol. Soc. America Bull.*, v. 67, pp. 571–96.

TAVENAS, F., CHAGNON, J.-Y., and LAROCHELLE, P., 1971, The Saint-Jean-Vianney landslide: Observations and eye-witness accounts: *Canadian Geotech. Jour.*, v. 8, pp. 463–78.

Transportation Research Board, 1978, *Landslides: Analysis and control:* Natl. Acad. Sci., Natl. Res. Council Spec. Rept. 176 (in press).

TROEH, F. R., 1964, Landform parameters correlated to soil drainage: *Soil Sci. Soc. America Proc.*, v. 28, pp. 808–12.

———, 1965, Landform equations fitted to contour maps: *Am. Jour. Sci.*, v. 263, pp. 616–27.

VAN BURKALOW, A., 1945, Angle of repose and angle of sliding friction: An experimental study: *Geol. Soc. America Bull.*, v. 56, pp. 669–707.

VARNES, D. J., 1978, Slope movement types and processes, *in Landslides: Analysis and control:* Transportation Research Board, Natl. Acad. Sci., Natl. Res. Council Spec. Rept. 176 (in press).

WASHBURN, A. L., 1967, Instrumental observations of mass-wasting in the Mesters Vig District, northeast Greenland: *Meddelelser om Grønland*, v. 166, no. 4, 296 pp.

WATSON, R. A., and WRIGHT, H. E., Jr., 1969, The Saidmarreh landslide, Iran, *in* Schumm, S. A., and Bradley, W. C., eds., U.S. contributions to Quaternary research: *Geol. Soc. America Spec. Paper 123*, pp. 115–39.

WILLIAMS, G. P., and GUY, H. P., 1971, Debris avalanches—a geomorphic hazard, *in* Coates, D. R., ed., *Environmental geomorphology:* State Univ. of New York Publications in Geomorphology, Binghamton, N.Y., pp. 25–46.

———, 1973, Erosional and depositional aspects of hurricane Camille in Virginia, 1969: *U.S. Geol. Survey Prof. Paper 804*, 80 pp.

WHITE, S. E., July 14, 1973, Personal communication.

WOOD, A., 1942, Development of hillside slopes: *Geologists' Assoc. Proc.*, v. 53, pp. 128–40.

YOUNG, A., 1972, *Slopes:* Oliver and Boyd, Ltd., Edinburgh, 288 pp.

CHAPTER 9

Fluvial Erosion and Transport: River Channels

THE FLUVIAL GEOMORPHIC SYSTEM

Water, flowing down to the sea over the face of the land, is the dominant agent of landscape alteration. Surface weathering and ground-water solution provide a load for flowing streams, and mass-wasting may dump great quantities of rock debris at the foot of slopes, but eventually rivers must carry all but a small fraction of the total rock waste from the lands to the sea. Estimates of the relative importance of various agents of continental denudation can hardly be better than orders of magnitude, but one careful evaluation (Garrels and Mackenzie, 1971, p. 114) credits rivers with 85–90 percent of the total present sediment transport to the sea, glaciers with about 7 percent, ground water and waves with about 1–2 percent, and wind and volcanoes with less than 1 percent each. Therefore, to understand how landscapes evolve, one must understand how rivers do their work. The useful adjective **fluvial** (from L. *fluvius*: "river") pertains to the work of rivers but, in the context of landscape development, also includes the water-dominated preconditioning of rock debris by weathering and mass-wasting before reaching a river channel (Leopold, et al., 1964).

Rivers are the return to the ocean of the excess water that falls as rain or snow on the landscape and does not evaporate during the hydrologic cycle (Fig. 5-5). Hence, the fluvial system is powered by the conversion of the potential energy of solar distillation and gravity to the kinetic energy of motion and heat. Most of the energy is lost to friction of internal turbulence in flowing water, but perhaps 2–4 percent (Rubey, 1938, p. 138) of the total potential energy of water flowing downhill is converted to the mechanical work of erosion.

Fluvial processes obviously vary in intensity among climatic regions and along gradients of temperature, precipitation, altitude, and seasonality (Chap. 5). The annual flow of the Amazon River alone is estimated to be almost 15 percent of the total annual runoff from all the land (Oltman, et al., 1964; Gibbs, 1967, p. 1206; Oltman, 1968). Oddly, on this hydrous planet, perhaps one-third of the land surface has no runoff to the ocean (Chap. 13). However, even arid regions with drainage into closed intermontane basins have landscapes of branching stream valleys. Infrequent or brief seasonal stream flow can shape otherwise dry landscapes. Furthermore, on the time scale in which landscapes evolve, climatic regions have shifted and changed in intensity, so that many regions now intensely arid show evidence of previous fluvial erosion (Chap. 18).

The hydrologic cycle (Fig. 5-5) offers abundant evidence that more water annually falls on the land than evaporates from it and that most landscapes evolve under conditions of excess water runoff. The fluvial system is so widespread and effective that it has been commonly regarded as the "normal" process of landscape evolution on earth (Cotton, 1948). W. M. Davis (1905a, reprinted 1954, p. 288; 1905b, reprinted 1954, p. 296) based his classic geomorphic cycle on the certainty that "the greater part of the land surface has been carved by . . . the familiar processes of rain and rivers, of weather and water." He defined a "normal" climatic region as "not so dry but that all parts of the surface have continuous drainage to the sea, nor so cold but that the snow of winter all disappears in summer."

It is regrettable that Davis's definition betrays a certain provinciality that was common in eastern North America and western Europe at the time he wrote. Nevertheless, the subsequent recognition that vast areas of terrestrial landscape are shaped by processes not dominant in the cool, humid climates of Cambridge (Massachusetts), London, and Paris has not diminished the importance to geomorphology of fluvial, or "normal," processes. Use of the word "normal" to describe fluvial processes should cause no more inconvenience to the geomorphologist than it does when it is used to describe a "normal" fault.

INFILTRATION, OVERLAND FLOW, AND SOIL EROSION

The proportion of surface and subsurface water that feeds a stream varies greatly with climate, soil type, bedrock, slope, vegetation, and many other factors. One estimate is that one-eighth of the annual runoff of the hydrologic cycle goes directly overland to the sea, and seven-eighths of the water goes underground at least briefly. The rapidity with which infiltrating water reacts with minerals (Chap. 6) explains why "pure" spring water closely reflects the chemistry of the local rocks. Even as river water begins its downhill flow, its chemical energy has been largely expended.

During a gentle rain on bare soil, all the water may infiltrate into the ground. The *infiltration capacity* of soil is controlled by duration and intensity of precipitation, prior wetted condition of the soil, vegetation, soil mineralogy and texture, slope, and other factors (Horton, 1945, p. 307). When the surface layer becomes saturated, overland flow begins, and soil particles loosened by raindrop impact or turbulence are entrained in the flow (Smith and Wischmeier, 1962). Direct runoff from well vegetated slopes rarely exceeds a few percent of the precipitation, but on sun-dried, clay-rich surface crusts, runoff may begin as soon as a thin surface layer is wetted by capillary absorption.

Overland flow, or **sheet flow**, can be observed during a rainstorm, but it occurs at the level of grass roots, in a layer a few millimeters thick, and is easily overlooked. In this film, flow may be laminar or turbulent. The velocity and mass of sheet flow might not be sufficient to erode and transport soil particles were it not for the violent impact of raindrops and the tendency of sheet flow to develop "surges" as it forms and breaks minor dams of vegetation and soil. In an excellent evaluation of the relation between sheet flow and soil erosion, Horton concluded (1945, pp. 320–23) that a critical length of overland flow is required before sheet flow acquires sufficient velocity to erode. This critical length, measured in hundreds of meters downhill from the crest line, he called the "belt of no erosion." The relation of this belt to the summit convexity of slope profiles, where creep dominates, is obvious. However, Horton seems to have ignored the ability of raindrop impact to erode soil particles, and few subsequent authors have accepted his concept of a "belt of no erosion" of such width.

Surface runoff will not flow for long as a sheet, even on a smooth surface. Internal friction and surface tension in flowing water soon generate ripples and kinematic waves where turbulent energy is concentrated. Dominant threads of current develop and erode ephemeral channels called **rills**. At the end of a rain, the water soaks into the channel floors and locally intensifies weathering. The little rill channels quickly fade through creep, but beneath each one a little more rock or soil has been weathered and prepared for removal (p. 175).

When cut to a size large enough to survive the interval between rains (Fig. 9-1), a rill collects water at its steep headward end, and the resulting waterfall migrates

Figure 9-1. Rill erosion from less than 4000 m² of drainage on late-seeded winter wheat after summer fallow, near Mica, Washington. Rill in foreground is 45 cm deep. (Photo: U.S. Dept. Agriculture-Soil Conservation Service.)

Figure 9-2.   Gully formed in 26 years, Madison County, Georgia. (Photo: U.S. Dept. Agriculture-Soil Conservation Service.)

upstream, extending the rill into previously undissected upland and capturing other rills. On barren slopes, rills can be observed growing headward at a rate of centimeters per minute or faster. They tend to parallel each other, rather than to form dendritic networks, because each is growing headward on the steepest local gradient.

Only some vague notion of size and permanency distinguishes a **gully** from a rill (Fig. 9-2). A gully is a steep- or vertical-sided ephemeral stream valley with a steep head that is actively eroded headward, usually into a water-gathering wash slope (Fig. 8-19). Gullies are known to develop rapidly as a result of poor agricultural practices. One in Georgia, the size of which impressed the great British geologist Sir Charles Lyell in 1846, subsequently grew to a length of 230 m, a width of 150 m, and a maximum depth of 20 m, in little more than 100 years. It is eroded in the deeply weathered crystalline rocks of the Georgia piedmont and began to form after the area was deforested (Ireland, 1939). Gullies grow headward rapidly enough to capture adjacent rills and commonly develop a branching network of tributaries. They are the first step in the fluvial dissection of landscapes.

[handwritten margin notes: "rills grow headward very quickly" and "Gullies are larger than rills and behave like them"]

DRAINAGE NETWORKS

Quantitative Analysis and Probability

> Every river appears to consists of a main trunk, fed from a variety of branches, each running in a valley proportioned to its size, and all of them together forming a system of valleys, communicating with one another, and having such a nice adjustment of their declivities, that none of them join the principal valley, either on too high or too low a level; a circumstance which would be infinitely improbable, if each of these valleys were not the work of the stream that flows in it [Playfair, 1802, p. 102].

Playfair's law, in addition to being written in a style only rarely achieved in scientific prose, illustrates the mathematician's appreciation of probability theory. In discussing the quotation, Playfair noted (1802, pp. 353, 355):

> The truth of the proposition . . . is demonstrated on a principle which has a close affinity to that on which chances are usually calculated. . . . we must conclude, that the probability of such a constitution having arisen from another cause, is, to the probability of its having arisen from the running of water, in such a proportion as unity bears to a number infinitely great.

For nearly 150 years, Playfair's law was quoted as a defensible and acceptable logical proposition, without any attempt to quantitatively test it. In 1945, Horton (1945, p. 280) demonstrated that enough hydrologic measurements were available to quantify the description and theories of developing river networks and drainage basins. "Horton's laws," a series of exponential or geometric equations, were expanded by Strahler (1952a, 1952b, 1954) and others, and a new subject of *quantitative fluvial geomorphology* was established (Salisbury, 1971; Gregory and Walling, 1973). The goal was to establish quantitative, rather than qualitative, relationships between geomorphic processes and landforms. The subject matter continues to be a productive area of research, especially now that computers can portray calculated results in a display comparable to the familiar map or terrain model.

One of the early successes of quantitative geomorphology was the analysis of branching drainage networks (Fig. 9-3). An ingenious numbering technique was devised whereby the fingertip tributaries of streams, those that originate from overland or ground-water flow and flow to a junction without receiving any tributaries themselves, are called **first-order** streams. A **second-order** stream begins at the junction of two first-order streams; it may receive additional first-order tributaries, but if it joins another second-order tributary, a **third-order** stream is formed, and so forth. In subsequent jargon (Shreve, 1966, 1967; James and Krumbein, 1969), first-order streams were called **external links** in topological networks, and higher-order segments were called **internal links**.

The statistical analysis of parameters such as link lengths, bifurcation (branching) frequency, and bifurcation angles can become a highly technical and practical subject. For example, suppose that a cluster of houses is to be provided with water by a network

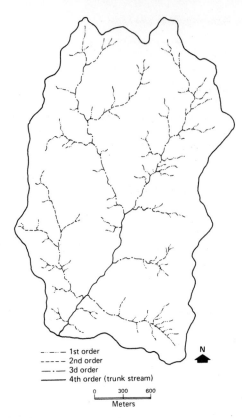

--·--·— 1st order
-------- 2nd order
--·--·— 3d order
———— 4th order (trunk stream)

N

0   300   600

Meters

Figure 9-3. Drainage network of Mill Creek, Ohio, from topographic maps and field surveys (Morisawa, 1959). Inset shows system of stream ordering. Statistical data are listed in Table 9-1 and graphed in Figure 9-4.

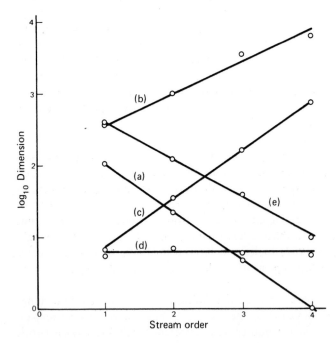

Figure 9-4. Graph of stream order in the Mill Creek drainage network compared to the channel and basin dimensions in Table 9-1.

[handwritten margin notes: "tain parameters", "the", "tern of", "channels", "efficient", "nature", "in nature"]

of underground pipes. Each size of pipe has a stated cost per unit length, the cost of laying each size of pipe is specified, and each coupling costs a certain additional amount. What is the distribution array from a trunk line to individual houses that can be laid for the least cost? If we rephrase such a question in terms of natural drainage networks and make the important initial *assumption* that nature is inherently conservative in expending energy, we ask: What is the most efficient array of branching channels that can carry water and sediment from an area downhill to the sea?

Early in the studies of drainage networks it was discovered that on homogeneous terrains they are highly ordered systems in which parameters such as mean link length, bifurcation ratio, down-link slope, drainage density (total length of links per unit area of drainage basin), stream azimuth, and many other dimensions and dimensionless ratios (Table 9-1; Fig. 9-4) could be predicted from information about some segment of the system (Strahler, 1958). Hydraulics engineers and geomorphologists have a common interest in being able to predict such practical facts as the annual and peak discharges of a river, flood frequency, and duration of flood crest from a minimum number of observation stations within a river system.

TABLE 9-1. ANALYSIS OF SOME CHARACTERISTICS OF THE
MILL CREEK DRAINAGE NETWORK*

Stream Order	*(a)* No. Streams	*(b)* Ave. Length *(ft)*	*(c)* Ave. Basin Area *($10^5 ft^2$)*	*(d)* Stream Density *(mi/mi^2)*	*(e)* Ave. Channel Slope *($tan \angle \times 10^3$)*
1	104	364	6.97	5.45	396
2	22	993	33.73	7.02	123
3	5	3432	161:97	6.06	39
4	1	6283	747.14	5.66	10

SOURCE: Morisawa, 1959, Table 12.
*See Figures 9-3 and 9-4.

Random Walks and Infinite Topologically Random Networks

Quantitative analysis of fluvial systems took a turn from the empirical (and tedious) analysis of real drainage basins to theoretical and statistical considerations of two-dimensional networks or systems in general. Leopold and Langbein (1962, p. 14) used a "random walk" technique to generate drainage networks that had dimensions and ratios similar to the natural networks analyzed by Horton, Strahler, and others. It is an instructive exercise to generate a river system by a "random walk" game. Start a number of markers at equal spaces along one edge of a piece of graph paper. Move the markers either ahead, left, or right, one space at a time, by the cast of a die. One restraint to random movement is that a marker cannot move backward. This is the game equivalent of gravity. If a marker intersects the path of another marker, it must follow the previously defined path from that point. This rule of the game represents the surface tension and viscosity of water. The paths of the markers trace drainage nets in which first-order tributaries join to form second-order streams and so on until either all paths have merged into a single master stream or have diverged beyond any probable junction. The frequency with which both game markers and streams join, the average length of path between successive junctions, and other parameters of the networks depend only on the original rules. If one part of the net is known, other parts can be predicted.

The significance of "random walk" models is that they demonstrate the highly probable organization of drainage networks. High probability, in the concept of general systems theory (Chap. 5), represents the minimum energy expenditure to the system (maximum entropy) and the optimal distribution of energy expenditure within the system. The highly probable nature of drainage networks was repeatedly and emphatically demonstrated by Shreve (1966, 1967, 1969, 1975). He showed that all of the empirical relationships of links and junctions that had been determined by earlier workers analyzing real drainage networks could also be demonstrated in "infinite topologically random channel networks." In other words, real drainage networks (Fig. 9-5), like many other branching systems such as mammalian lungs or oak trees, are essentially random although highly probably in their structure.

Orderliness and randomness are not mutually exclusive (Mann, 1970). Although it is not possible to predict the exact position or origin of a component in a random system, it is quite possible to state precisely the general properties and subdivisions of the system and the number of components of a certain rank that are present, especially if some information is available about a small part of the system. The serious deficiency of quantitative fluvial geomorphology for explanatory description is that even though real drainage networks on homogeneous or structureless terranes can be demonstrated to be topologically random and are highly probabalistic systems, there is no way to prove or disprove that real networks form in the same way that random network models are generated (Leopold, et al., 1964, p. 421). The techniques of quantitative fluvial geomorphology give an excellent *description* of drainage networks but no *explanation*.

The deficiency was illustrated by a quantitative study (Howard, 1971a) in which stream networks were simulated by headward growth and branching. The technique is essentially the reverse of a random walk in which paths merge when they touch; at intervals specified by the rules of headward growth, each headward-expanding channel branches randomly. Howard's models, which are the equivalent of gully systems expanding into undissected uplands, satisfactorily simulate natural systems. His results were even better with models in which growth took place on a matrix that corresponded to an area of previously dissected upland. Howard (1971a, pp. 48–49) concluded that it is dangerous to generalize about genetic processes from computer simulations. Many different simulation procedures generate similar topologically random networks. It is not possible to determine whether a given network was formed by (1) headward branching growth of a gully system, or (2) progressive intersection

Figure 9-5. Ephemeral branching networks of intertidal creeks draining the mud flats at the mouth of the Colorado River near San Felipe, Mexico. Tide range is 8–10 m. (Photo: U.S. Navy.)

and consolidation of small rills and channels as water flowed downhill, with or without preexisting irregularities (noses and hollows) that guided the evolution of the network. Leopold, et al. (1964, p. 421) agreed that "a river or a drainage basin might best be considered to have a heritage, rather than an origin."

The Principle of Least Work

Some important generalizations have emerged from the evaluation of drainage networks as random, highly ordered systems. For example, Howard (1971b) provided a new explanation for the old observation (Playfair, 1802, pp. 113–14) that tributary streams normally enter the trunk stream at an acute angle that points downstream. The only previous explanation was that the angle of junction was determined at the time of origin of the network by the ratio between the hillslope gradient and the gradient of the trunk stream (Horton, 1945, pp. 349–50). However, junction angles have been observed to change during dissection of rapidly eroding terrains and thus cannot be totally inherited, and the predicted angles were not supported by field measurements on finely dissected shale badlands in Utah (Howard, 1971b, p. 865). Instead, a relationship was found between the junction angles and discharge of the tributary and other tributaries adjacent to it, so that the angles minimize the rate of work (power) expended by gravity at the junction. Erosion and sedimentation change the junction angle with time, so that the tributary neither enters at an angle so steep that excessive velocity and bank erosion occur nor so gentle that loss of velocity and deposition at the junction occur. This *principle of least work* is reconsidered frequently in the remainder of this chapter.

Another important generalization to emerge from computer models of drainage networks concerns the balance of energy expenditure between overland flow and channel flow (Woldenberg, 1969). Suppose that two watersheds of equal area are each crossed by a single stream channel of equal size. If the watershed is long and narrow parallel to the channel, overland flow is minimized, but length of channel flow is maximized. In an equant or circular watershed, the channel length is decreased, but overland flow is greater. Because the energy consumption per unit area of overland flow is not in the same ratio as the consumption per unit length of channel flow, the "least work" shape of watersheds can be deduced. Woldenberg (1969, p. 100) argued that circular watersheds, compressed by the requirement of continuity into honeycomblike hexagons, offer the advantages of least work. He further deduced that if first-order streams and their watersheds can be approximated by hexagons, higher-order drainage networks are also least-work assemblages of close-packed, or nested, hexagons.

To Woldenberg and others, drainage networks are not random but are determined by physical laws of conservation of energy and energy distribution in open systems. The issue of *randomness* versus *determinism* in natural systems is a fundamental topic in the philosophy of science, but many interesting discussions have used drainage networks of fluvial systems as examples (Mackin, 1963; Leopold, et al., 1964, pp. 420–28; Mann, 1970).

RIVERS AND CHANNEL GEOMETRY

Whether formed by chance or necessity, by headward erosion or downslope convergence, whether inherited or newly formed, systems of branching river channels dissect most of the subaerial landscape, each "in a valley proportioned to its size." Some channels are occupied by *permanent streams* that flow throughout the year, others have *intermittent streams*, and still others are used only briefly during and immediately after precipitation by *ephemeral streams.* Channel size is obviously related to river discharge, but it is common for rivers to overflow their channels at *flood stage.*

Fluvial Classification Systems

Rivers and their channels can be descriptively categorized in several ways. Some river segments have *bedrock channels*, eroded into the regional rock with little or no sedimentary veneer. Others have *alluvial channels*, cut in river-transported rock debris, or **alluvium**. Bedrock channels are likely to be much more irregular than alluvial channels.

Whether in bedrock or alluvium, the **thalweg**, or line connecting points of maximum water depth in a general downstream direction along the channel, is rarely straight. Structural control usually determines the details of bedrock channels. Plucking action by flowing water can break away exposed layers of bedrock, or mechanical abrasion from the solid load can grind the channel into bedrock. An abrasive bedload may be caught in eddies and abrade circular pits in the channel floor. This process is called *pothole drilling.* If the bedrock is soluble, it may be chemically dissolved by flowing water as well as mechanically abraded. The net effect is a series of pools alternating with ledges or depositional bars on alternating sides of the channel.

In humid regions, the mean annual discharge of rivers increases downstream unless structural controls such as karst divert some of the water. Not all of the annual runoff from the land is poured into rivers at the heads of first-order tributaries, equally spaced from river mouths like the markers at the start of a game. Water can be added anywhere along a river from surface runoff and from subsurface seepage. Rivers in humid regions are called **effluent**, because they receive contributions of ground water. Rivers in arid regions generally lose water to the ground in addition to losing it by evaporation, and often they dry up entirely without reaching the sea. These are called **influent** streams; their distinctive channel characteristics are described in Chapter 13.

The shape or pattern of river and channel segments, or *reaches*, offers another set of descriptive terms. Some segments may be *straight* in plan view. *Crooked* or *zigzag* channels are known, but where found they are structurally controlled, and the shape is externally determined rather than being a description of the channel character. *Sinuous* channels develop naturally in bedrock or alluvium, either by rounding the corners of crooked channels or alternately eroding and depositing sediment as *pools* and *riffles* along former straight reaches. In alluvial reaches, smooth free-swinging

*Meanders
and braids
tell something
about fluvial
activity*

sinuosities called **meanders** (Fig. 10-4) may form. Rivers that carry large amounts of coarse sediment (sand and gravel) build midstream bars at frequent intervals and divide into numerous separate but intersecting and shifting minor channels. Such a channel is anastomosing or **braided** (Figs. 15-8, 17-10). The meandering and braiding habits are important aspects of channel geometry, interrelated with channel shape, river discharge, and other hydraulic properties. These two basic habits may alternate in space or time but do not intergrade. They seem to be separate and distinct modes of fluvial activity.

Longitudinal Slope

*the amount
of discharge
determines
the long
profile.*

One of the interesting and significant aspects of stream flow is that as the quantity of water in a stream increases, the downstream slope of the water surface decreases. As an empirical rule, *slope is an inverse function of discharge* (Fig. 9-6). Apparently, water flows more efficiently in larger channels and therefore requires less slope to maintain its velocity. Langbein and Leopold (1964, p. 786) expressed the relationship between longitudinal slope (s) and mean annual discharge (Q) as lying within the range of two power functions: $s \propto Q^{-0.5}$, and $s \propto Q^{-1.0}$ (Fig. 9-11). That is, slope varies in a range between the reciprocal of discharge and the reciprocal of the square root of discharge. For effluent rivers, in which discharge increases downstream, the long profile should be concave to the sky, as a parabolic curve.

Bedrock channels commonly have irregular long profiles with nearly horizontal segments in lakes or pools separated by high-gradient segments of waterfalls or rapids. Alluvial channels are more likely to have regular, concave-skyward, long profiles. Carlston (1969b) measured the long profiles of numerous rivers in the Mississippi and Atlantic-coast watershed in the eastern United States and confirmed that their longitu-

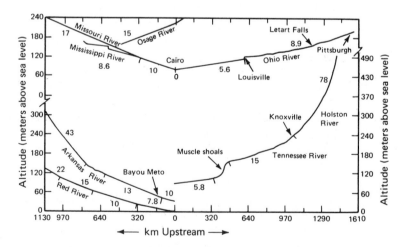

Figure 9-6.   Longitudinal profiles of seven rivers in the Mississippi drainage basin. Gradient units are slope tangents $\times$ 10^5 (meters of drop per 100 km stream length). (From Carlston, 1969b, Fig. 1.)

dinal profiles are overall concave upward unless dominated by structural or tectonic controls. However, he noted that many rivers have straight segments ($s \propto Q^{0.0}$) or even convex-skyward segments as part of the overall concavity. For instance, downstream from the confluence of the Missouri and the Mississippi Rivers, the relatively coarse sediment load of the Missouri steepens the slope of the Mississippi (Fig. 9-6).

The inverse relationship between slope and discharge can be illustrated by a practical application, if not explained by a general theory. In irrigation systems, each ditch must slope steeply enough to keep water moving and keep mud from settling in the channel but not so steeply as to cause the water to erode the banks of the ditch. A delicate energy balance must be maintained. Through centuries of trial and error and experiment, farmers learned that successively smaller distributary irrigation ditches must be given successively steeper gradients to keep the water moving at the proper speed. The mathematical concept of a *regime canal*, which neither scours nor fills its channel, was first worked out for canals in India (Leopold and Maddock, 1953, pp. 43–48), but the principles must date from the beginning of human history on the irrigated alluvial plains of the Tigris-Euphrates, Nile, Indus, and Ganges Rivers. In an irrigation system, the discharge decreases downstream, as the water from a single large feeder canal is divided and subdivided, but the inverse rule between slope and discharge holds, even in this reversal of the pattern of a normal river system.

Base-level Control

Where rivers enter the sea, the gravity potential of the falling water reaches zero. No further conversion of potential energy to stream work is possible, so sea level, and its projection under the land, is called the **ultimate base level** of stream erosion. Actually, most streams enter the sea with a considerable velocity and therefore have kinetic energy available to erode their channels well below sea level, but this observation does not invalidate the use of sea level as the reference level for the limit of potential energy conversion. Within river systems there are also **local or temporary base levels** that delay stream erosion but do not halt it.

Because rivers have their greatest discharge and therefore their most gentle gradients near their mouths, they enter the sea at smooth, almost undetectable, tangents. The Mississippi River at New Orleans is at sea level 170 km upstream from its multiple mouths in the Gulf of Mexico. The mighty Amazon River is only about 65 m above tidewater at the Peruvian border, 3000 km from its mouths (Oltman, et al., 1964, p. 4).

Repeated continental glaciations during the Quaternary Period introduced serious complications into the concept of base-level control. All large rivers now enter the sea either in *estuaries*, which are obviously the drowned segments of their former lower valleys, or across *deltas*, where borings demonstrate that a hundred meters or more of aggradation to a rising base level has filled former valleys (Fig. 10-1). Sea level has fluctuated through a range of at least 100 m in harmony with glaciation, introducing cyclic perturbations in the hydrologic cycle on a time scale shorter than the age of most river valleys. Thus, when glaciers expanded, sea level fell, and rivers eroded their valleys deeper. During the retreat of glaciers, sea level rose, and rivers with minor solid loads were drowned. Those with large loads, especially if glacier meltwater contributed to the discharge, built deltas. It is not possible to demonstrate ultimate base-level control by modern sea level in any river system, yet the theoretical significance of base-level control is an important concept of geomorphology.

Channel Width, Depth, and Current Velocity

For many years, agencies of governments have maintained *gaging stations* along rivers all over the world. At these stations, the water-surface level, channel shape, stream velocity, amount of dissolved and suspended mineral matter, and other variables are periodically or continuously recorded. The discharge of a stream is measured indirectly by multiplying the cross-section area of the channel at the gaging station by the average velocity of the current. In the United States, discharge is expressed in cubic feet/second (cross-section area in square feet × velocity in feet per second). The equivalent unit in the metric system is m³/s.[1]

The records of stream gaging stations are essential for the prediction of potential flood damage, stream pollution, and other disasters (Fig. 10-2). These records also provide a voluminous history of stream flow. In 1953, L. B. Leopold and Thomas Maddock published an analysis of thousands of measurements from stream gaging stations all over the United States. They called their analysis of the relationships among stream discharge, channel shape, sediment load, and slope the *hydraulic geometry of stream channels*.

The first step in analyzing the hydraulic geometry of stream channels was to study the changes in channel width and depth, stream velocity, and suspended load at selected gaging stations during conditions ranging from low flow to bankfull discharge and flood (Leopold and Maddock, 1953, pp. 4–9). Over a wide range of conditions, it was found that width, depth, velocity, and suspended load increase as simple power functions of discharge. That is, all increase as some small, positive, exponential function of discharge. Some of the gratifyingly simple equations are:

$$w = aQ^b, \qquad d = cQ^f, \qquad v = kQ^m$$

where Q = water discharge, w = width, d = mean depth, and v = mean velocity. The changes of load with discharge are described later in this chapter.

The numerical values of the arithmetic constants a, c, and k are not very significant for the hydraulic geometry of streams, but the numerical values of the exponents b, f, and m are very important. Leopold and Maddock (1953, p. 9) found that the average of 20 representative gaging stations in central and southwestern United States gave values of the exponents b, f, and m as follows:

$$b = 0.26, \qquad f = 0.40, \qquad m = 0.34$$

[1]In this section, all equations are expressed as in the original publications, which used the nonmetric U.S. units of measurement.

These values signify that as the discharge of water past a gaging station increases, as after a heavy rain, the width of the channel (in feet) increases approximately as the fourth root of discharge (in ft^3/sec) ($w = aQ^{0.26}$), the mean depth increases approximately as the square root of discharge ($d = cQ^{0.40}$), and the velocity increases approximately as the cube root of discharge ($v = kQ^{0.34}$). Channel width, depth, and current velocity all increase at gaging stations during rising water. This conclusion is no surprise to anyone who has seen a river in flood, but the regularity of the changes is significant.

More surprising are the results of comparing the changes in channel shape and stream velocity in a downstream direction (Leopold and Maddock, 1953, pp. 9–16). River discharge in humid areas increases downstream. When the mean annual discharge of many rivers past successive gaging stations was compared to the width, depth, and velocity at each station, the same equations were found to apply that had been derived for changes in flow past a single point. *As mean discharge of a river increases downstream, channel width, channel depth, and mean current velocity all increase.*

Everyone knows that effluent rivers get both wider and deeper as they grow larger downstream, but until Leopold and Maddock published their work, no one had guessed that average current velocity also increases downstream. The conclusion violates our poetic impressions about wild, rapidly flowing mountain streams and deep, wide, placid rivers like the Mississippi. We do not immediately realize that much of the current in a mountain torrent flows in circular eddies, with almost as much backward as forward motion. Figure 9-7 reproduces an example of the evidence that mean current velocity increases downstream with mean discharge (Leopold, 1953).

The amount that velocity changes is exponential to the discharge

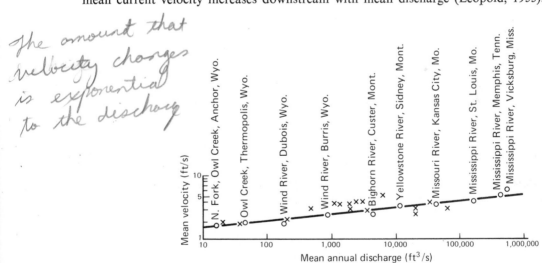

Figure 9-7. Velocity and discharge along the Yellowstone-Missouri-Mississippi River system, demonstrating that mean velocity increases downstream with increasing mean discharge. Additional gaging stations, unnamed, are shown by X. (Data from Leopold and Maddock, 1953, App. A.)

Both velocity and discharge are plotted on logarithmic scales to show as a straight line the exponential relationship between the variables.

The numerical values of the three exponents b, f, and m are not the same for changes downstream and for changes with discharge past a point. In the downstream direction, the average values for the exponents were found by Leopold and Maddock (1953, p. 16) to be:

$$b = 0.5, \quad f = 0.4, \quad m = 0.1$$

In the downstream direction, channel width increases most rapidly with discharge, depth next most rapidly, and mean velocity increases only slightly, although it definitely increases. Leopold and Maddock (1953, p. 14) proposed that the increasing depth downstream permits more efficient flow in a river and overcompensates for the decreasing slope, thus providing a slight net increase in velocity at mean annual discharge.

A mathematical test of the hydraulic geometry equations suggests useful applications of the principles. Discharge is defined as area times velocity, or $Q = wdv$. If

$$w = aQ^b, \quad d = cQ^f, \quad v = kQ^m$$

then by substitution,

$$Q = (aQ^b)(cQ^f)(kQ^m)$$

or

$$Q = ackQ^{b+f+m}$$

It follows that

$$a \times c \times k = 1.0$$

and

$$b + f + m = 1.0$$

The arithmetic constants a, c, and k, satisfy this test but are not of concern here. It is more interesting to verify that in the examples given for both single gaging stations and a downstream succession of gages, $b + f + m = 1.0$.

Carlston (1969a, p. 500) reviewed the statistical problems of evaluating these hydraulic geometry equations and confirmed their validity. He recalculated least-square solutions for the same data from 10 river basins used to determine downstream changes by Leopold and Maddock. His more precise values of the exponents are:

$$b = 0.461, \quad f = 0.383, \quad m = 0.155$$

Leopold (1953) noted that at flood discharge, the exponent $m = 0$ in the downstream direction. That is, flood velocity is essentially constant downstream along the length of a river. The downstream increase in flood discharge is accommodated by increasing depth and especially width, so that when $m = 0$, $b + f = 1$. Mackin (1963, pp. 143–45) and Carlston (1969a) argued that because the significant work of rivers is accomplished during flood, when downstream velocity is constant, the increase in mean velocity downstream at mean annual discharge is of statistical, but not practical, significance. Nevertheless, to the average person, the fact that rivers at least *do not decrease* in velocity downstream is a striking and sometimes disturbing revelation. Mackin observed (1963, pp. 146–47) that in some segments of the rivers studied by Leopold and Maddock, even the mean velocity actually decreased with increased discharge. Careful selection of data points on Figure 9-7 can be used to support that observation, but the trend of the line nevertheless confirms the overall increase in mean velocity downstream.

SEDIMENT TRANSPORT AND EROSION

Capacity and Competence; Solid and Dissolved Loads

The theoretical maximum amount or mass of sediment load that a stream can transport has been called its **capacity**. The grain size of the detritus may partly determine how much can be carried, but capacity is primarily a measure of the maximum amount, not grain size, of the load. **Competence**, on the other hand, is the measure of a stream's ability to transport a certain maximum grain size of sediment. Competence depends primarily on velocity, although channel shape, the shape and degree of sorting of the sediment particles, amount of suspended load, and water temperature can also affect competence. For coarse sediment moving on the bed of a channel, the boundary shear stress, which is approximately proportional to water depth and surface gradient, may be a better measure of competence than velocity (Baker and Ritter, 1975).

Weathered rock is carried by rivers in three forms. The compounds in solution or colloidal mixtures are the **dissolved load**. The solid matter is either fine-grained particles in suspension (**suspended load**) or coarse-grained particles that slide, roll, or bounce along the stream bed (**bed load**, or traction load). The division of the load varies greatly, controlled by climatic and structural factors. Tables 12-1 to 12-3 give some estimates for major rivers of the world. Rivers that are subject to large fluctuations in flow, or that drain poorly vegetated regions, or that have generally large loads, tend to have high percentages of solid load. Rivers that drain heavily forested regions, or overflow from lakes that act as settling basins, or drain karst regions, tend to have mostly dissolved loads. Their total loads are generally small.

Both the dissolved and suspended loads in river water are routinely measured at gaging stations, and excellent data are available. Bed load defies attempts to measure it accurately. Any device lowered to the bed of a stream to measure or collect the sediment in motion also deflects the boundary-flow conditions and distorts the measurement. Bed loads as great as 50–55 percent of the total solid load have been estimated for alluvial rivers on the Great Plains (Leopold and Maddock, 1953, pp. 29–30), but for most purposes bed load is simply assumed to be 10 percent of the total solid load, or about an additional 11 percent of the measured suspended load.

Dissolved load has no detectable effect on stream flow. The concentrations average only 130 ppm (Livingstone, 1963), so the solutions are too dilute to affect viscosity, turbulence, or the density of river water. Therefore, dissolved load, which can represent over half of the total work of fluvial denudation (Table 12-1), gets a "free ride" to the sea by rivers. No kinetic energy is required to move it.

Competence of Rivers

Figure 9-8 shows graphs of river *competence*, the relationship between mean velocity in a river and the size of particle that can be eroded or transported by it (Hjulström, 1935, 1939; Nevin, 1946; Sundborg, 1956, pp. 177–80). The assumptions should be noted, because the average velocity in a stream is not the velocity at the bank or bed of the stream, where much of the erosion and transport occur. Furthermore, the curves were experimentally derived from flume experiments on well sorted sediments. If coarse and fine particles are mixed, the fine particles fill spaces between the coarse ones and are thereby protected, but they also help prevent the coarse grains from moving (Blatt, et al., 1972, chap. 4).

Figure 9-8 demonstrates that the grain size of sediment that will stay in suspended transport is an exponential function of mean velocity (approximately a straight line

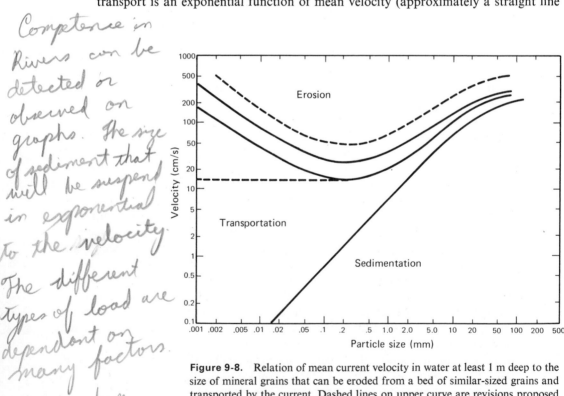

Figure 9-8.   Relation of mean current velocity in water at least 1 m deep to the size of mineral grains that can be eroded from a bed of similar-sized grains and transported by the current. Dashed lines on upper curve are revisions proposed by Sundborg (1956).

Competence in Rivers can be detected or observed on graphs. The size of sediment that will be suspend in exponential to the velocity. The different types of load are dependant on many factors.

1) bed shape
2) actual load
3) Discharge
4) velocity

on the graph). A point on the graphed line between the transport and sedimentation fields represents the mean velocity in a stream at least 1 m deep that will provide sufficient turbulence to keep a particle of a given size from settling out of suspension. The upper curve ("Hjulström's curve") is actually a field or zone of values for the velocity necessary to initially dislodge or erode particles. For pebbles, cobbles, and boulders, the erosion threshold is only slightly greater than the transportation velocity, because they barely rise from the bottom as they bounce or roll as bed load. Actually, large stones on the bed of a stream are overturned or rolled by mean velocities somewhat less than those indicated by Hjulström's curve. The exact relationship between eroding velocity and large particle sizes has yet to be worked out (Novak, 1973). The minimum eroding velocity of 15 cm/sec is for fine sand, but sand can be kept in transport by a mean velocity an order of magnitude less if some other process stirs it from the bottom. Probably 95 percent of the total work done by stream transport is attributable to the few percent of the load that is in the coarsest grain sizes (Bagnold, 1968, p. 53). Silt and clay sizes (collectively called **mud** when wet and **dust** when dry) are kept in suspension by very slight currents, but the velocities necessary to erode a consolidated clay bank are comparable to those that move pebbles and cobbles.

A muddy river bed may resist stream flow and induce friction as much as a gravel bed although in a different way. *Channel roughness* is an important but complex parameter of fluvial process that is related not only to velocity and grain size but also to a depth parameter, bed form (rippled, smooth, etc.), and channel slope (Leopold and Maddock, 1953, pp. 135–43; Leopold, et al., 1964, p. 158; Blatt, et al., 1972, pp. 85–90; Rhodes, 1977).

In alluvial rivers, the downstream changes in bed roughness and the increased width and depth of the channel provide less turbulence and therefore less competence to move coarse sediment. In bedrock channels, no such interaction among the several hydraulic geometry variables is possible.

Capacity of Rivers; Variations of Suspended Load

Capacity of rivers is the theoretical maximum sediment load that can be carried. It is not a useful concept, but a discussion is included because of frequent past confusion of the idea with that of competence (Nevin, 1946). The maximum possible suspended load cannot be specified. At some arbitrary mixture of water and mud, a muddy river is simply called a mud flow instead. One of the greatest reported suspended loads is that of the Hwang Ho, or Yellow River, of northern China. It dissects a great region of windblown silt and is reported to carry a suspended load of 40 percent sand, silt, and clay by weight during high discharge (Cressey, 1963, p. 125). Capacity might have significance for bed loads, but it can also be argued that if sufficient sediment of the proper size is available to be moved as bed load, it will be moved. Unless its bed is armored, any alluvial stream is likely to be moving its capacity bed load at all times.

Because dissolved load has no effect on the hydraulic geometry of rivers, and bed load defies accurate measurement, the load of a river that is usually measured is the amount of suspended load. The unit of measurement is the dry weight of sediment per volume of water or more commonly units such as tons per day of sediment for a specified discharge, Q.

The hydraulic geometry of streams involves the amount of solid load as well as discharge, width, depth, velocity, slope, bed roughness, and grain size of the load. The equation relating suspended-sediment load to discharge is similar in form to the equations for width, depth, and velocity. The equation given by Leopold and Maddock (1953, p. 21) is $L = pQ^j$, where L is suspended-sediment load, Q is discharge, and p and j are numerical constants.

In general, as the discharge increases at a gaging station, the suspended-sediment load increases. Values for the exponent j range from 2.0–3.0. These large exponential values mean that as discharge at a station increases tenfold, the suspended load may increase a hundredfold to a thousandfold! Figure 9-9 is a typical graph of suspended-sediment load compared with discharge. The suspended-sediment load at a station

suspend-sed. load increase exponentially to velocity

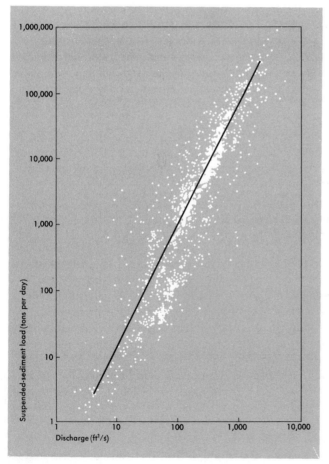

Figure 9-9. Relation of suspended-sediment load to discharge, Powder River at Arvada, Wyoming. (Leopold and Maddock, 1953, Fig. 13.)

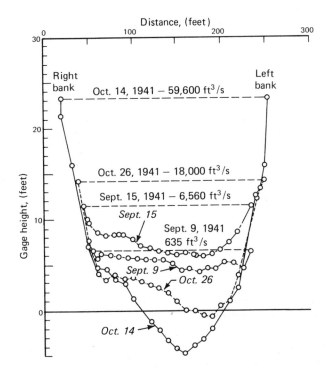

Figure 9-10. Channel cross sections during progress of a flood, September–December 1941, San Juan River near Bluff, Utah. Horizontal line is the water surface on the indicated date. (Leopold and Maddock, 1953, Fig. 22.)

increases much more rapidly with discharge than either channel width or depth; therefore, the enlargement of the channel by erosion cannot account for all of the increased load. Most of the suspended sediment comes from the watershed upstream from the gaging station. The sediment is newly delivered to the stream by mass-wasting and rill wash during the same rains or snowmelts that swell the discharge of the river.

Measurements of channel shape and suspended-sediment load confirm that streams move most of their loads during times of higher than average discharge. As the water rises during a flood, alluvial channels initially aggrade their bed (Fig. 9-10). Presumably, the aggradation results from the sudden influx of new sediment from the uplands and the decreased roughness effect of the more dense water-sediment mixture. Aggradation increases the bed slope in order to move that sediment downstream. During the peak flood discharge, the channel may be scoured even deeper than the preflood level and then aggrade again during the declining discharge late in the flood. Thus competence is not to be simply correlated with velocity. At a given velocity during increasing discharge, bed aggradation occurs, but when that same velocity is reached during decreasing discharge after the flood crest has passed, the bed may be eroded (Leopold and Maddock, 1953, pp. 30–35).

In some rivers, the net channel erosion of the spring runoff is gradually restored by deposition during the low-water summer season. If channel shape is measured at the same location at the same season in successive years, one might believe that no permanent change has taken place in the river channel during a period of high discharge, unless one realizes that the mud, sand, and gravel that previously formed the stream bed at the location have been moved downstream toward the sea and have been replaced by new sediment from upstream. The geomorphology machine has surged ahead slightly.

The change of suspended-sediment load downstream with increasing discharge has not been measured directly, but it can be estimated indirectly from other parameters of hydraulic geometry (Leopold and Maddock, 1953, pp. 21–26). The value for exponent j in a downstream direction is 0.8, which implies that the total suspended-sediment load increases downstream slightly less rapidly than discharge, and the concentration of suspended sediment becomes more dilute toward the river mouth. Two factors are probably involved. First, few first-order tributaries and negligible overland flow enters a trunk stream directly, and these are the original source of most of the sediment load. Second, effluent ground water enters the trunk stream to increase discharge but carries no suspended load, thus diluting the concentration.

Channel Shape and Solid Load

The bed of an alluvial channel is similar in grain size to the solid load of the river at that point, for if the stream is competent to move a fragment on the stream bed, it will do so, and that fragment becomes part of the bed load or suspended load. If the stream lacks competence to move fragments of a given size, they settle and become part of the alluvial bed. A constant exchange between bed and load takes place. Even though the channel is primarily shaped during high discharge (Fig. 9-10), for every lesser discharge there is a velocity that has a certain competence to selectively move grains up to a certain size limit. To the extent that channel roughness is determined by the grain size or bed forms of the alluvial bed, and roughness affects turbulence and competence, a complex readjustment between discharge, grain size of the load (and the bed), velocity, and the amount of load is constantly in progress.

Rivers that carry primarily sand and gravel as bed load develop wide, shallow alluvial channels, often braided (Figs. 15-8, 17-10). It could be visualized that the river optimizes bed-surface area as a device to efficiently move the load by shear. However, a more accurate statement is that because bed load is carried near the bottom of the stream, the velocity gradient increases rapidly upward in such a way that the relatively clear, faster-moving upper water can erode banks made of grain sizes similar to those already in transport as bed load. A related factor is that sand and gravel are noncohesive, and banks of coarse alluvium collapse readily, to be spread out on the flat, shallow bed.

Rivers that carry mostly mud (silt and clay) as suspended load develop deeper, narrower channels with a catenary or trapezoid cross section. The *wetted perimeter* approaches the minimum length, that of a semicircle. A channel shape of this sort minimizes surface area and friction and therefore provides for the maximum transport of suspended load, which is carried by internal turbulence, not bed shear. A deep, narrow channel has steep banks, which can be undercut, but the cohesive strength of consolidated mud (Fig. 9-8) resists bank erosion.

Schumm (1960a, 1960b) found that the width-to-depth ratio of rivers on the Great Plains is inversely proportional to the percentage of fine-grained sediment in the banks, and therefore also in the load. The relationship between channel shape and sediment grain size is given by the equation,

$$F = 255 \ M^{-1.08}$$

where F is the width–depth ratio, and M is the weighted mean percent of silt and clay in the sediment. In many Great Plains rivers, the bed load may exceed half of the total load, so the value for M is low and, inversely, F is high. The braided habit of these rivers is partly an internal adjustment among semidependent variables of width, depth, and sediment grain size and partly a response to the independent variables of discharge and load.

The meandering habit of many rivers, especially those that flow on fine-grained alluvium in humid regions, is also related to the width–depth ratio of the channel and sediment grain size. As the suspended-sediment load (mud) increases in proportion to bed load, F decreases, and the channel narrows and deepens. By these interrelated adjustments, more of the energy of the stream is expended against the banks and less against the deep bottom. The sinuosity of the channel increases, and meanders form.

In flume experiments, Schumm and Khan (1972) attempted to produce meandering channels by varying slope, discharge, and sediment loads. They were able to produce channels with alternating bars and pools so that the thalweg meandered, but the channel banks remained essentially straight until the stream abruptly developed mid-channel bars and became braided. The sediment in the flume was poorly sorted sand. A truly meandering channel could not be formed until 3 percent by weight of kaolinite clay was mixed into the water. The clay coated and stabilized the banks and bars and allowed the thalweg to deepen and expose the stabilized sandbars on alternating sides of the channel. True meanders formed. The experiment illustrates well the action of alluvial grain size as a factor in the hydraulic geometry of rivers.

It has been noted previously that transitional forms are rare between wide, shallow channels with a braided habit and narrower, deeper channels that meander. Some reaches of alluvial rivers have changed historically from braiding to meandering or back (Schumm, 1969), but the changes are abrupt. Perhaps the lack of intermediate forms is related to the fact that the minimum threshold of sediment erosion on Hjulström's curve (Fig. 9-8) is in the fine sand size. If the alluvium is noncohesive sand or coarser sediment, rising velocity drags it as bed load, and a wide, shallow braided channel develops. If the alluvium is cohesive, an equivalent velocity increase may erode it, but it is transported as suspended load in a deep, meandering channel.

The exponents b, f, and m at 315 gaging stations were plotted on a ternary diagram (Rhodes, 1977) to demonstrate that certain patterns of channel behavior can be recognized and classified. Ten classes of channel and flow responses to varying discharge were hypothesized by Rhodes on the basis of various ratios of b, f, and m and their combinations. A large amount of scatter around mean values is apparent, but many of the responses described in the previous paragraphs can be recognized in the diagram.

THE CONCEPT OF GRADE IN FLUVIAL SYSTEMS

The concept of an open system that maintains itself in a steady state of most efficient configuration by internal self-regulation among variables was introduced in Chapter 5, by analogy to a rotary cement kiln. A river is an excellent example of an open system through which matter and energy flow, but within which are inherent tendencies toward self-regulation. Numerous examples of the interaction between discharge, load, channel shape, and other variables of hydraulic geometry have been cited. It is appropriate now to review the interaction of all the known variables of fluvial systems in the tendency to achieve **grade**, or long-term self-regulation, in a river channel.

Variables of the Graded River

At least 10 variables are involved in the tendency for a river to maintain a graded state. Not all are of equal significance, and some are not within the self-regulatory ability of the river. Leopold and Maddock (1953) divided the variables of hydraulic geometry into three classes: *independent, semidependent, and dependent*.

Discharge, sediment load, and ultimate base level are the three independent variables in the graded state. The stream has little control over these factors; rather, it must adjust to them. Discharge is determined by precipitation and evaporation in the watershed, the permeability of the soil, the amount and type of vegetation, and the area of the watershed. Only the area of the watershed is affected by other changes in the river system. Headward erosion of first-order tributaries can enlarge the watershed area and thereby increase discharge, but even this process is limited, because adjacent drainage nets are most probably enlarging, too. Capture of part of a network by a competing system can cause an abrupt decrease in discharge. Quite early in the erosional development of a landscape, the boundaries of each river's watershed become defined.

Sediment load is also nearly independent of the other variables of stream flow. Many of the same climatic, soil, and biologic factors that determine discharge also determine the amount of sediment that slopes deliver to streams. The type of bedrock is an additional powerful control for sediment load. Some rocks weather quickly to sand-size particles; others produce only silt and clay. Limestone weathering produces mostly a dissolved load, with little solid detritus. Streams erode their channels and thereby have some self-regulation of their load, but we have seen that most of the load reaches the river "ready-made" by weathering and mass-wasting.

The ultimate base level of erosion is the third independent variable of stream flow. When a stream reaches the sea, it loses its identity. The potential energy of the stream is set by the altitude above sea level at which precipitation falls. Regardless of the discharge, load, or any other variable, a river that rises on a coastal plain only a few hundred feet above sea level will never be a mountain torrent.

The semidependent variables that interact to achieve the graded state include channel width, channel depth, bed roughness, grain size of the sediment load, velocity, and channel "habit" (the tendency for a stream to either meander or braid). These are semidependent inasmuch as they are partly determined by the three independent variables, but they are also partly capable of mutual self-regulation in a river. Width, depth, and mean velocity have been shown to be power functions of discharge. Competence, or grain size of the load, is a function of velocity, but the mixture of grain sizes, especially the "mud ratio" M, has been shown to determine the width-depth ratio. Bed roughness is determined by the alluvial grain size or bed forms. The alluvial bed is built of grain sizes determined by the stream competence, yet that competence is in turn determined by such factors as velocity, channel shape, and the amount of load and discharge. In turn, bed roughness creates turbulence that affects the competence of the stream to move suspended load.

Alluvial meanders are remarkably regular, and their dimensions are proportional to channel width. The radius of curvature of meanders is usually between 2–3 times the channel width. The wave length of most meanders varies between 7–10 times the channel width. Meandering involves inherent properties of flowing water, as well as size and shape of the channel, erodibility of the stream banks, proportion of suspended and bed load, and probably other factors. In turn, meandering increases the channel length between two points and thus decreases the slope of the stream. Slope influences velocity and sediment-transporting capacity, so meanders not only are affected by other variables of stream flow but in turn affect those same variables. An analogous argument can be made for the role of the braided habit as a semidependent variable of hydraulic geometry.

Only one variable of the fluvial system, the downstream slope of the water surface, is regarded as being dependent on all other variables. Slope can be changed only by building up one part of the channel, cutting down another, or changing channel length as by meandering or delta-building. All these changes take time, so slope is usually the final adjustment the stream makes in becoming graded. If slope could change suddenly, it would be mutually semidependent with the variables previously listed. Because it ordinarily cannot do so, it is subject to the influences of all the other known variables.

Progressive Development of Grade

We can visualize the achievement of grade, or the graded state, in a river by first imagining an ungraded river that flows over a tectonic landscape, newly raised from the floor of the sea. Rainfall, and therefore discharge, varies over the new landscape. Water flows downhill along chance hollows and other water-collecting slopes. Drainage networks are eroded, and valleys are widened by mass-wasting on slopes along the entrenching channels. Variables of the stream system are wildly out of equilibrium. Half of the potential energy of such a river might be expended in one great vertical waterfall. The load of the stream is determined by chance rock slides anywhere along the banks. Channel width and depth are restricted by the erodibility of the rocks that are crossed by the channel. The slope of the river approximates the initial slopes of the landscape.

Rather quickly, the semidependent variables of stream flow interact to form a channel system appropriate to the work at hand. Many of the hydraulic geometry equations for stream flow apply to both graded and ungraded streams. The independent variables of discharge, load, and base level may be the same for a graded and an ungraded river. The critical factor in the achievement of the graded condition is that the stream must flow on "adjustable" materials, so that changes of one variable can produce the appropriate changes in others. Alluvium is the ideal bed for a river, and the establishment of an alluvium-lined channel on a continuous flood plain signals the achievement of grade in that part of a river.

Alluvium not only permits stream channels to adjust toward equilibrium; it is also a powerful absorbent for peak energy inputs into the river system. Discharge may increase by many orders of magnitude during a flood (Fig. 9-9). If the excess energy of the flood is not fully absorbed by the increased load carried into the river from headwater slopes, then alluvium is eroded until the ability of the river to do work is balanced by the work it is doing (Fig. 9-10). Alluvium is analogous to a chemical buffering compound that is added to a solution in great excess to ensure that some property of the solution remains constant during a reaction. Alluvium is an excellent buffer for variations in stream energy, because every fragment on the flood plain has some critical energy threshold of transportation. Having been once transported by the river, it will move again when the energy level is high enough.

Grade, as defined by an alluvial bed on an eroding valley flat (suballuvial rock floor), is established first in the downstream segments of rivers and is gradually extended upstream (Davis, 1902). Larger rivers attain grade earlier than smaller streams. A trunk stream may be at grade when its first-order tributaries are still gnawing headward into undissected slopes. A *graded reach*, or graded segment of a river, may form on easily eroded material upstream from a gorge across a resistant rock barrier. The resistant rock then forms the local or temporary base level for the graded reach. As the barrier is slowly eroded, the graded reach remains at grade, because it can adjust easily during the slow lowering of the temporary base level.

Grade as a Thermodynamic Equilibrium

Another approach to the concept of grade is to regard the graded condition of a river from the viewpoint of theoretical thermodynamics. In any steady-state physical system through which matter and energy move, we find a tendency for the least possible work to be done and also a tendency toward a uniform distribution of work. Nature is inherently conservative in these matters. In a river system, which derives its energy from water flowing downhill, the tendency for minimum work opposes the tendency for a uniform distribution of work in several different aspects of hydraulic geometry.

The Long Profile as a Graded Form.   If all the water of a river were added at the head of a single tributary, the "least-work" profile would be a waterfall straight down to sea level. In humid regions, where rivers gain water downstream, the "least-work" profile is a curve in which the greatest loss of altitude takes place where the discharge is least, near the head of the river. The profile is very steep near the head and almost horizontal near the mouth (Fig. 9-11).

[handwritten marginal note: "Effluent" type rivers tend to lose their altitude at the head and almost horizontal at the mouth]

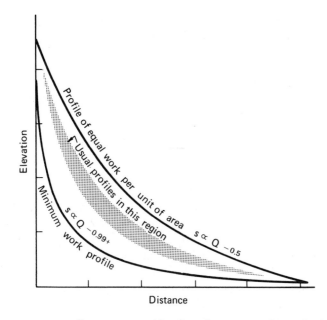

Figure 9-11. Schematic longitudinal profiles of rivers showing tendency to balance minimum total work with maximum distribution of work. (From Langbein and Leopold, 1964, Fig. 1.)

In contrast to the sharply concave theoretical profile of "least work," the theoretical profile for uniformly distributed work has a more nearly constant slope (Fig. 9-11). The profile is only slightly concave skyward, because as the rate of doing work increases downstream with discharge, the surface area of the stream bed also increases. The rate of doing work per unit area of stream bed is constant on a river that widens downstream but decreases its slope only gradually.

If two tendencies toward equilibrium oppose each other, the resulting equilibrium is most likely to be some predictable intermediate condition. Langbein and Leopold (1964) concluded from their studies of river dynamics that both the downstream profiles (Fig. 9-11) and channel cross sections of rivers approach the equilibrium form predicted from the principles of "least work" and uniform distribution of work, *provided that the channels are in material that is adjustable.* The presence of an alluvial flood plain is again implied as a condition of grade. The long profile of a river flowing in an alluvial channel is a smooth concave-skyward profile that allows water to flow downhill to the sea with slight or negligible acceleration yet to expend only a few percent of the potential energy of the system on channel erosion or sediment transport (Bagnold, 1968).

Meanders in Graded Rivers. The free, regular meanders of alluvial rivers have aroused curiosity and artistic senses for a long time. It seems a waste of potential energy for a river to avoid the most direct path downhill, which in plan view would be a straight channel. However, meanders, like the long profile of alluvial rivers, have been demonstrated to be the result of balance between least work and maximum distribution of work (Langbein and Leopold, 1966; Leopold and Langbein, 1966). Meanders are *sine-generated curves* that have the statistical property of minimum variance. The work done on each segment of stream bank is uniform, and a minimum amount of total work is done. Furthermore, the sine-generated curve is the shape of the average, or most probable, random walk between two points. As in the random walk game described earlier in this chapter, randomness is not most likely to take the straightest path. Thus meanders represent highly probable shapes, again provided that the channel bed is adjustable. Meltwater streams on glaciers meander, although they may be sediment-free between ice walls. The Gulf Stream also meanders up the east coast of the United States and eastward into the Atlantic.

Accordant Junctions and Least-work Drainage Networks.   Two other examples of the balance between least work and uniform distribution of work were described in an earlier section of this chapter on drainage networks. The acute angles at stream junctions were shown to be most probable, least-work configurations (Howard, 1971b). Overland flow and the ratio of basin area to first-order stream length was also shown to be a highly ordered, most-probable condition (Woldenberg, 1969). In these examples, alluvium is not required to achieve the graded condition, but as the drainage systems evolve, the junctions and basin areas change as appropriate to maintain the self-regulation.

The application of thermodynamic principles to the description of graded streams is not simple, and rivers are sufficiently complex so that rigorous "laws of stream flow" may never be written, but it is satisfying to discover that graded rivers with alluvial flood plains apparently meet certain broadly defined conditions of equilibrium that are typical of the steady-state open systems maintained during carefully controlled laboratory experiments.

Summary of Grade

The most concise definition of a graded stream was carefully phrased by Mackin (1948, pp. 471, 484). The hydraulic geometry of Leopold and Maddock required only a slight transposition of phrases in Mackin's definition. It is quoted here, as modified by Leopold and Maddock (1953, p. 51), as a summary of the concept of the **graded river**:

> A graded river is one in which, over a period of years, slope and channel characteristics are delicately adjusted to provide, with available discharge, just the velocity required for the transportation of the load supplied from the drainage basin. The graded stream is a system in equilibrium; its diagnostic characteristic is that any change in any of the controlling factors will cause a displacement of the equilibrium in a direction that will tend to absorb the effect of the change.

A few additional comments must be made about the concept of grade. Grade is a condition, not an altitude or a certain slope angle. It develops first near the mouths of rivers and gradually extends headward. Erosional lowering of the land continues for a long time after grade is achieved, because as long as rivers carry sediment to the sea, they continue to lower the landscape over which they flow. A graded river is in a steady state only with regard to short-term changes. Over a time scale of millions of years, typical of the time intervals in which landscapes evolve, the potential energy of an undisturbed river system gradually approaches zero, and the rate of change of the system also decreases. The river remains at grade, but the characteristics of the graded condition change with time. The significance of the concept of grade to the life history of regional landscapes is emphasized in Chapter 12.

REFERENCES

BAGNOLD, R. A., 1968, Deposition in the process of hydraulic transport: *Sedimentology*, v. 10, pp. 45–56.

BAKER, V. R., and RITTER, D. F., 1975, Competence of rivers to transport coarse bedload material: *Geol. Soc. America Bull.*, v. 86, pp. 975–78.

BLATT, H., MIDDLETON, G. V., and MURRAY, R. C., 1972, *Origin of sedimentary rocks:* Prentice-Hall, Inc., Englewood Cliffs, N.J., 634 pp.

CARLSTON, C. W., 1969a, Downstream variations in the hydraulic geometry of streams: Special emphasis on mean velocity: *Am. Jour. Sci.*, v. 267, pp. 499–509.

———, 1969b, Longitudinal slope characteristics of rivers of the midcontinent and the Atlantic East Gulf slopes: *Internat. Assoc. Sci. Hydrology Bull.*, v. 14, no. 4, pp. 21–31.

COTTON, C. A., 1948, *Landscape, as developed by the processes of normal erosion*, 2d ed.: Whitcombe and Tombs, Ltd., Wellington, N.Z., 509 pp.

CRESSEY, G. B., 1963, *Asia's lands and peoples*, 3d ed.: McGraw-Hill Book Co., Inc., New York, 663 pp.

DAVIS, W. M., 1902, Base-level, grade, and peneplain: *Jour. Geology*, v. 10, pp. 77–111. (Reprinted 1954 in *Geographical essays:* Dover Publications, Inc., New York, pp. 249–78).

———, 1905a, Complications of the geographical cycle: *Internat. Geog. Cong., 8th, Washington 1904, Repts.*, pp. 150–63. (Reprinted 1954 in *Geographical essays:* Dover Publications, Inc., New York, pp. 279–95.

———, 1905b, Geographical cycle in an arid climate: *Jour. Geology*, v. 13, pp. 381–407. (Reprinted 1954 in *Geographical essays:* Dover Publications, Inc., New York, pp. 296–322).

GARRELS, R. M., and MACKENZIE, F. T., 1971, *Evolution of sedimentary rocks:* W. W. Norton & Co., Inc., New York, 397 pp.

GIBBS, R. J., 1967, Geochemistry of the Amazon River system: *Geol. Soc. America Bull.*, v. 78, pp. 1203–32.

GREGORY, K. J., and WALLING, D. E., 1973, *Drainage basin form and process: A geomorphological approach:* John Wiley & Sons, Inc., New York, 456 pp.

HJULSTRÖM, F., 1935, Studies of the morphological activity of rivers as illustrated by the River Fyris: *Geol. Inst. Univ. Uppsala Bull.*, v. 25, pp. 221–527.

———, 1939, Transportation of detritus by moving water, *in* Trask, P. D., ed., *Recent marine sediments:* Am. Assoc. Petroleum Geologists, Tulsa, pp. 5–31.

HORTON, R. E., 1945, Erosional development of streams and their drainage basins: *Geol. Soc. America Bull.*, v. 56, pp. 275–370.

HOWARD, ALAN D., 1971a, Simulation of stream networks by headward growth and branching: *Geog. Analysis*, v. 3, pp. 29–50.

————, 1971b, Optimal angles of stream junction: Geometric, stability to capture, and minimum power criteria: *Water Resources Res.*, v. 7, pp. 863–73.

IRELAND, H. A., 1939, "Lyell" gully, a record of a century of erosion: *Jour. Geology*, v. 47, pp. 47–63.

JAMES, W. R., and KRUMBEIN, W. C., 1969, Frequency distributions of stream link lengths: *Jour. Geology*, v. 77, pp. 544–65.

LANGBEIN, W. B., and LEOPOLD, L. B., 1964, Quasi-equilibrium states in channel morphology: *Am. Jour. Sci.*, v. 262, pp. 782–94.

————, 1966, River meanders—theory of minimum variance: *U.S. Geol. Survey Prof. Paper 422-H*, 15 pp.

LEOPOLD, L. B., 1953, Downstream change of velocity in rivers: *Am. Jour. Sci.*, v. 251, pp. 606–24.

LEOPOLD, L. B., and LANGBEIN, W. B., 1962, Concept of entropy in landscape evolution: *U.S. Geol. Survey Prof. Paper 500-A*, 20 pp.

————, 1966, River meanders: *Sci. American*, v. 214, no. 6, pp. 60–70.

LEOPOLD, L. B., and MADDOCK, T., 1953, Hydraulic geometry of stream channels and some physiographic implications: *U.S. Geol. Survey Prof. Paper 252*, 57 pp.

LEOPOLD, L. B., WOLMAN, M. G., and MILLER, J. P., 1964, *Fluvial processes in geomorphology:* W. H. Freeman and Company, San Francisco, 522 pp.

LIVINGSTONE, D. A., 1963, Chemical composition of rivers and lakes: *U.S. Geol. Survey Prof. Paper 440-G*, 64 pp.

MACKIN, J. H., 1948, Concept of the graded river: *Geol. Soc. America Bull.*, v. 59, pp. 463–512.

————, 1963, Rational and empirical methods of investigation in geology, *in* Albritton, C. C., ed., *The fabric of geology:* Freeman, Cooper and Company, Stanford, Cal., pp. 135–63.

MANN, C. J., 1970, Randomness in nature: *Geol. Soc. America Bull.*, v. 81, pp. 95–104 (discussions, pp. 3185–96).

MORISAWA, M. E., 1959, Relation of quantitative geomorphology to stream flow in representative watersheds of the Appalachian Plateau province: *Office of Naval Research Project NR 389-042, Tech. Rept. 20*, 94 pp.

NEVIN, C., 1946, Competency of moving water to transport debris: *Geol. Soc. America Bull.*, v. 57, pp. 651–74.

NOVAK, I. D., 1973, Predicting coarse sediment transport: The Hjulström curve revisited, *in* Morisawa, M., ed., *Fluvial geomorphology:* State Univ. of New York Publications in Geomorphology, Binghamton, N.Y., pp. 13–25.

OLTMAN, R. E., 1968, Reconnaissance investigations of the discharge and water quality of the Amazon River: *U.S. Geol. Survey Circ. 552*, 16 pp.

OLTMAN, R. E., STERNBERG, H. O'R., AMES, F. C., and DAVIS, L. C., 1964, Amazon River investigations reconnaissance measurements of July, 1963: *U.S. Geol. Survey Circ. 486*, 15 pp.

PLAYFAIR, J., 1802, *Illustrations of the Huttonian theory of the earth:* Dover Publications, Inc., New York, 528 pp. [Facsimile reprint, 1964.]

RHODES, D. D., 1977, The b-f-m diagram: Graphical representation and interpretation of at-a-station hydraulic geometry: *Am. Jour. Sci.*, v. 277, pp. 73–96.

RUBEY, W. W., 1938, The force required to move particles on a stream bed: *U.S. Geol. Survey Prof. Paper 189-E*, pp. 120–41.

SALISBURY, N. E., 1971, Threads of inquiry in quantitative geomorphology, *in* Morisawa, M., ed., *Quantitative geomorphology: Some aspects and applications:* State Univ. of New York Publications in Geomorphology, Binghamton, N.Y., pp. 9–60.

SCHUMM, S. A., 1960a, The effect of sediment type on the shape and stratification of some modern fluvial deposits: *Am. Jour. Sci.*, v. 258, pp. 117–84.

———, 1960b, Shape of alluvial channels in relation to sediment type: *U.S. Geol. Survey Prof. Paper 352-B*, pp. 17–30.

———, 1969, River metamorphosis: *Hydraulics Div. Jour., Am. Soc. Civil Engineers Proc., Paper 6352*, v. 95, no. HY 1, pp. 255–73.

SCHUMM, S. A., and KHAN, H. R., 1972, Experimental study of channel patterns: *Geol. Soc. America Bull.*, v. 83, pp. 1755–70.

SHREVE, R. L., 1966, Statistical law of stream numbers: *Jour. Geology*, v. 74, pp. 17–37.

———, 1967, Infinite topologically random channel networks: *Jour. Geology*, v. 75, pp. 178–86.

———, 1969, Stream lengths and basin areas in topologically random channel networks: *Jour. Geology*, v. 77, pp. 397–414.

———, 1975, Probabilistic-topologic approach to drainage-basin geomorphology: *Geology*, v. 3, pp. 527–29.

SMITH, D. D., and WISCHMEIER, W. H., 1962, Rainfall erosion: *Advances in Agronomy*, v. 14, pp. 109–48.

STRAHLER, A. N., 1952a, Dynamic basis of geomorphology: *Geol. Soc. America Bull.*, v. 63, pp. 923–38.

———, 1952b, Hypsometric (area-altitude) analysis of erosional topography: *Geol. Soc. America Bull.*, v. 63, pp. 1117–42.

———, 1954, Quantitative geomorphology of erosional landscapes: *Internat. Geol. Cong., 19th, Algiers 1952, Comptes Rendus,* sect. 13, fasc. 15, pp. 341–54.

———, 1958, Dimensional analysis applied to fluvially eroded landforms: *Geol. Soc. America Bull.*, v. 69, pp. 279–300.

SUNDBORG, Å., 1956, The River Klarälven, a study of fluvial processes: *Geograf. Annaler*, v. 38, pp. 125–316.

WOLDENBERG, M. J., 1969, Spatial order in fluvial systems—Horton's laws derived from mixed hexagonal hierarchies of drainage basin areas: *Geol. Soc. America Bull.*, v. 80, pp. 97–112.

CHAPTER 10

Fluvial Deposition:
Processes and Landforms

An important group of fluvial landforms are created not by erosion but by alluvial deposition. For many reasons, we consider alluvium to be in transit toward the sea, and therefore the landforms on alluvium must be ephemeral as compared to the erosional forms. Yet in terms of human lifetimes, or even the length of human history, landforms built of alluvium appear permanent. Written human history began on the alluvial plains and deltas of the Middle East. An overwhelming majority of the human population today lives on alluvium. Although constructional rather than destructional in origin (Chap. 2), alluvial landforms are so closely, and almost incidentally, related to fluvial erosion that this chapter is logically part of the larger review of fluvial processes in general (Price, 1947; Russell, 1958).

ALLUVIUM AND FLOOD PLAINS

Alluvium is unconsolidated sediment of relatively recent geologic age that was deposited by flowing water. The term does not include sediments permanently submerged (i.e., submarine sediments or lacustrine bottom sediments) but may be applied to at least the upper part of deltaic sediments in either a lake or the ocean.

Alluvium covers the bedrock floors of river valleys and builds fans or deltas where stream competence abruptly decreases. It may be mapped as a ribbonlike deposit on an upland, marking the ancient course of a now-diverted stream. Rarely, alluvial gravel forms ridges where gravel-floored former channels protected easily eroded bedrock while the adjacent unprotected substrate was windblown or otherwise eroded away. *Topographic inversion* of this type is sufficiently rare to demonstrate that most alluvium is in transit to the sea and is not likely to find a permanent resting

alluvium is unconsolidated sediment

an alluvium covering is rarely known to form ridges etc. This alluvium is mostly in transit

228

place on land. In the geologic record, alluvial conglomerates and sandstones are well known, but they do not comprise a large fraction of the total volume of sedimentary rocks. Alluvial deposits, ancient and modern, are studied in greater proportion than their abundance would indicate, because they may contain valuable placers of heavy ores, precious metals, or gems. Further, most terrestrial vertebrate fossils are collected from alluvium, and the artifacts of early people are most often found in Quaternary alluvial deposits.

Active and Relict Alluvium

Quaternary tectonism and climatic fluctuations have made alluvium a complex material to interpret. The distinction between alluvium actively "in transit" and "relict" is especially critical in areas of former glaciation, where **glacial outwash** (a type of alluvium characteristic of heavily loaded meltwater streams with highly variable discharge) underlies much finer recent alluvium. If the modern stream is not competent to move all grain sizes in the outwash, then properly speaking, the outwash is not the alluvium of the present stream but rather the substrate on which the modern alluvium is deposited. The Mississippi River illustrates that interpretive problem well (Fig. 10-1 and p. 420). In full-glacial time, the lower river entrenched a deep valley to the Gulf of Mexico. As the ice margin retreated from the upper midwestern states, sand and gravel alluvium of a braided Mississippi River filled the lower trench. Later, when the ice margin was fringed by proglacial lakes in the present Great Lakes region, the coarse alluvium dropped out far upstream, and only mud and fine sand went down the Mississippi. By that stage, rising sea level was causing swampy deltaic deposition in Louisiana. In the last few thousand years, with sea level rising only slowly and with the modern delta prograding across the continental shelf, mud and organic deposition predominate on the flood plain. How much of the cross section in Figure 10-1 should be called alluvium, or what qualifications to the term are required to explain the entire cross section?

Many modern streams in formerly glaciated regions are grossly "*underfit*," both to the size of their valleys and channels, and in their competence to transport the sediment over which they flow (Dury, 1964a, 1964b, 1965). As a working definition, we can restrict the definition of "active" alluvium to that sediment now capable of being transported by the stream that flows on it. Such a definition relates to both the depth of the alluvium and its grain size. The thickness of the modern alluvium is then defined by the flood-scour depth of the modern channel and its maximum grain size by the flood-stage competence of the modern stream. Anything deeper or coarser is "relict" alluvium. The depth of modern channel scour should not be underestimated, however. Todd (1900) cited soundings of the bed of the Missouri River during high discharge that proved the channel to be 15–30 m deep in flood where it is only 3 to 15 m deep at low water (compare Fig. 9-10). The flood-eroded channel was floored by bedrock, thus proving that the entire alluvial thickness is potentially in transit. The greatly exaggerated vertical scale of Figure 10-1 shows that the modern Mississippi channel is as deep as the uppermost stratigraphic unit of braided-stream and backswamp deposits, which thereby become classed as the modern alluvium.

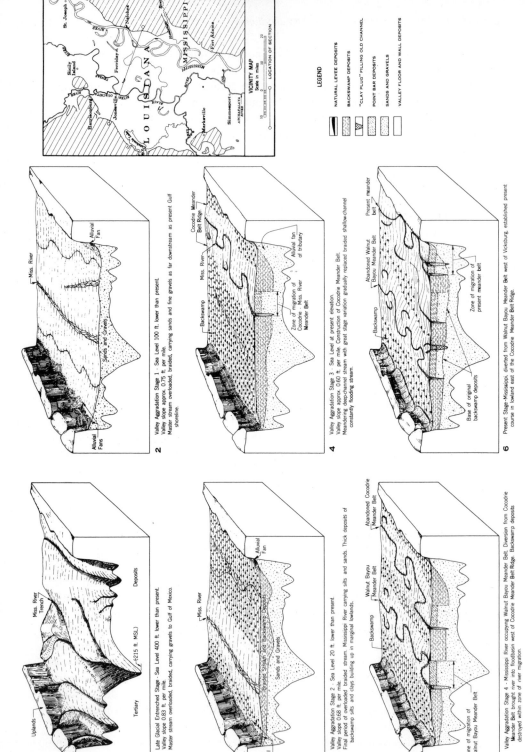

Figure 10-1. Stages in the late-glacial and postglacial history of the Mississippi River valley near Natchez, Mississippi. (Fisk, 1947, Plate 5.)

Downstream Changes in Alluvium Grain Size

Coarse alluvium decreases in mean grain size downstream by sorting and abrasion. The complex interplay between bed roughness, velocity, slope, and grain size was discussed in Chapter 9. Size-sorting by progressive loss of competence downstream is the major cause of decreasing mean grain size in alluvium. A given grain size simply cannot be transported downstream beyond the point where the stream is competent to move it.

Abrasion also reduces the grain size of bed-load alluvium in a downstream direction, although abrasion is not a factor in reducing the grain size of the suspended load. In 260 km of transport down the Colorado River of Texas below Austin, granitic pebbles and cobbles from a known source area diminish in mean diameter by about 50 percent, quartz by about 30 percent, and chert by about 20 percent (Bradley, 1970, p. 68). The glacial outwash of the Knik River in Alaska shows a more dramatic decrease in grain size downstream (Bradley, et al., 1972). Cobbles and boulders that are readily moved by periodic floods from an ice-dammed lake are reduced in size by 87 percent (to 13 percent of initial mean diameter) in only 26 km of river transport. This rapid reduction is consistent with many other observations of the rapid downstream change in grain size of glacial outwash.

An attempt was made to verify both sets of field measurements just cited by abrading gravel in laboratory conditions. Selected stones were dragged over the roughened bed of a toroidal abrasion tank by a current of water until they had traveled a distance equivalent to the downstream movement in the Colorado and Knik Rivers. Fresh granitic pebbles from the Colorado River were reduced by only 10 percent of mean grain size in the tank in contrast to their 50 percent reduction in an equivalent field transport distance. Knik River gravel was reduced by only 8 percent in 26 km of laboratory transport in contrast to the 87 percent reduction in the field.

The authors of these studies concluded that sorting rather than abrasion is the major cause of the rapid downstream decrease in grain size in an aggrading glacial outwash stream bed. Sorting was also judged to be important to the downstream change of grain size in the Colorado River alluvium, but even more important in that river is the abrasion of gravel that weathers slightly during temporary storage in the flood plain. Although fresh pebbles and cobbles were not notably abraded in the tank experiments, stones with a surficial weathered zone were abraded much more rapidly. A third factor in the downstream reduction of bed-load alluvium is the chipping and abrasion of cobbles as they shift and vibrate under current velocities that are not strong enough to transport them. Conceivably, cobbles could be abraded and rounded by vibration "in place" until they become small enough to be transported. The greater size reduction observed in actual rivers as opposed to flume experiments may be due in part to this kind of abrasion without net transport (Schumm and Stevens, 1972).

It can be concluded that sorting, abrasion, and weathering during temporary storage all contribute to the observed decrease in mean grain size of bed-load alluvium in a downstream direction. The abraded and weathered detritus is also transported but as dissolved and suspended load. As a very crude guess, the average alluvial fragment might take 1000 years to move downstream to the sea, most of that time being at rest (and weathering) in some part of the flood plain (Leopold, et al., 1964, p. 328).

Flood Frequency

Hydraulic-geometry equations and direct observation prove that the water surface of a river rises during times of high discharge. At some excess discharge, depending on the channel geometry at the point of observation, a river will exceed the height of the channel banks, and overbank flow, or flooding, will occur. A **flood** is synonymous with discharge greater than bankfull capacity, although for general descriptive purposes it is common to refer to any high discharge interval as a flood. Flood frequency is defined by a **recurrence interval**, a statistical parameter that describes the probable interval, usually in years, between floods of a specified magnitude (Fig. 10-2).

To establish the magnitude and frequency of floods, stream-gage records for several decades are required. The most common analytical procedure is to compile and sequentially rank by magnitude the highest discharge for each year of record. Such a list establishes a ranking order in an **annual series**. The recurrence interval of a flood equal to or greater than one of a certain magnitude in the list is given by

$$ri = \frac{n+1}{r}$$

where *ri* is the recurrence interval in years for a flood at least equal to the annual event that has ranking order *r* from a list of *n* years. For example, the recurrence interval of a flood at least equal to. the fifth greatest annual discharge from a 25-year record is $(25 + 1)/5 = 5.2$ years.

The recurrence interval of a flood at least equal in magnitude to the average flood of an annual series is 2.33 years (Leopold, et al., 1964, p. 64). This event is called the **mean annual flood**. Floods at least equal to the median discharge in an annual series have a recurrence interval of slightly over 2 years $[2(n + 1)/n]$. The statistically significant **most probable annual flood** in an annual series has a recurrence interval of 1.58 years.

In practice, the magnitude and recurrence intervals of floods in an annual series are plotted on a special kind of probability paper (Gumbel, 1958). From such a graph (Fig. 10-2), the probability of a flood of any intermediate magnitude can be interpolated. The graphic solution is considered reliable for events with recurrence intervals up to one-half the length of the annual series.

As an alternate procedure, all discharges above a selected magnitude are recorded and ranked, even though there may be more than one such event per year. A minimum duration for each event must be specified, however. Such a list is a *base-stage series* or **partial-duration series**. The recurrence interval calculated from a partial-duration series is the average time interval between two successive discharges at least as great as the magnitude of the selected base stage. The magnitude of the flood with a recur-

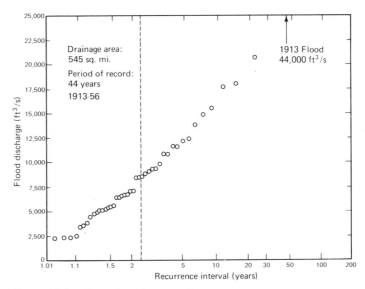

Figure 10-2.   Recurrence interval of annual floods of a specified magnitude, Miami River, Sydney, Ohio. [From Cross and Webber, 1959, Fig. 337. (Later supplemental data do not significantly change the intervals.)]

rence interval of 1.58 years in an annual series (the most probable annual flood) is identical to the magnitude of the flood with a recurrence interval of 1.0 year in a partial-duration series (Langbein, 1949; Dury, 1967, 1973). Dury hypothesized, and successfully verified by some examples, that in graded rivers the most probable annual flood is similar in magnitude to bankfull discharge. We thus have the interesting hypothesis that bankfull discharge is likely to be reached or exceeded once each year on the average, and that bankfull flow, dominant channel-shaping flow, and the most probable annual flood are equivalent (Dury, et al., 1963). In rivers where the most probable annual flood does not go overbank, the underfit condition is implied.

Hydrologic Significance of Flood Plains

Well established rivers usually have their valley floors covered with alluvium, into which the normal-flow channel is carved. The surface of low relief on the alluvium from the banks of the low-water channel to the base of the valley walls is called the **flood plain** of a river. The name is appropriate, because in flood part or all of the flood plain becomes the bed of the river.

Several of the semidependent variables of hydraulic geometry change abruptly when flood stage is reached. Most obvious are the changes of width and depth with discharge. Prior to overbank flooding, with increasing discharge at a station, depth increases more rapidly than width (Chap. 9). However, when overbank flooding begins, width abruptly increases manyfold, so that depth and velocity need not increase at all in order that $Q = w \cdot d \cdot v$. The power functions for width, depth, and velocity during flood become approximately:

$$w = aQ^{1.0}, \qquad d = cQ^{0.0}, \qquad v = kQ^{0.0}$$

The width of the flooded plain increases in linear proportion to discharge, and if the flood plain is very wide, depth and velocity need not increase at all. Actually, depth continues to increase slightly, but the mean velocity of a river in flood decreases to less than the mean velocity at bankfull stage, because the broad, shallow sheet of water over the flood plain has a very low velocity that reduces the average of the entire cross section.

The slope along a meandering channel is always less than the down-valley slope of the flood plain because of the greater length of the meandering channel over the straight-line distance along the flood plain. Rivers in flood have a water-surface gradient that is several times as steep as when the river is confined in a meandering channel. This increased slope enables the flood waters to move down the valley efficiently, even though the mean flood velocity need not be as high as in channeled flow—thus the peculiarity that rivers in flood flow on an increased average down-valley slope but at decreased mean velocity. The greater width of the flood plain more than compensates for the loss of velocity. Mean current velocity in the flood stage is approximately constant along the entire length of rivers (p. 212).

The sediment-transporting competence and capacity of a river also change abruptly in flood. Although enormously greater sediment load is in transit (Fig. 9-9) along the axis of the permanent channel, the shallow, low-velocity sheet of flood water over the adjacent flood plain has little competence. As velocity is checked, sand, silt, and clay may settle out of the flood waters, and the shallowness over the flood plain ensures that the fine sediment will reach the bottom of the water column before turbulence sweeps it downstream. People who live on flood plains know that a single flood may deposit a meter or more of muddy alluvium on their fields or in their houses. It may seem anomalous at first reading that a river in flood is flowing down a steeper gradient but may not be competent to transport sand and mud. Yet these interactions of the several semidependent variables of hydraulic geometry of rivers in flood illustrate well the fundamental role of channels and flood plains: to provide a low-gradient, efficient, sinuous or braided channel for average discharges and loads, and a steep, shallow, broad, constant-velocity flood channel for peak annual discharges. Truly, flood plains are for rivers. People occupy flood plains only at great risk, but judging from historical patterns, it is a risk that many find tolerable or inevitable.

Flood Plain Morphology

Flood plains and their minor morphologic subdivisions are primarily depositional landforms. As such, their forms are genetically related to specific depositional processes (Fig. 10-3; Table 10-1). Not all deposits create landforms, however. The lag deposits shown beneath the channel in Figure 10-3 might well be relict alluvium, for instance.

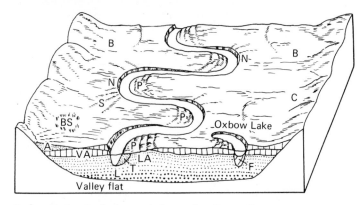

Figure 10-3. Typical associations of valley sediments. (Vanoni, 1971, Fig. 2-Q-1.) A—Alluvial fan; B—Backland; BS—Backswamp; C—Colluvium; F—Channel fill; L—Lag deposit; LA—Lateral accretion; N—Natural levee; P—Point bar; S—Splay; T—Transitory bar; VA—Vertical accretion.

TABLE 10-1. GENETIC CLASSIFICATIONS OF VALLEY SEDIMENTS*

Place of Deposition	Name	Characteristics
Channel	Transitory channel deposits	Primarily bed load temporarily at rest; part may be preserved in more durable channel fills or lateral accretions.
	Lag deposits	Segregations of larger or heavier particles, more persistent than transitory channel deposits, and including heavy mineral placers.
	Channel fills	Accumulations in abandoned or aggrading channel segments; ranging from relatively coarse bed load to fine-grained oxbow lake deposits.
Channel margin	Lateral accretion deposits	Point and marginal bars that may be preserved by channel shifting and added to overbank flood plain by vertical accretion deposits at top.
Overbank flood plain	Vertical accretion deposits	Fine-grained sediment deposited from suspended load of overbank flood water; including natural levee and backland (backswamp) deposits.
	Splays	Local accumulations of bed-load materials, spread from channels onto adjacent flood plains.
Valley margin	Colluvium	Deposits derived chiefly from unconcentrated slope wash and soil creep on adjacent valley sides.
	Mass movement deposits	Earth flow, debris avalanche, and landslide deposits commonly intermix with marginal colluvium; mud flows usually follow channels but also spill overbank.

SOURCE: Vanoni, 1971.
*Compare with Figure 10-3.

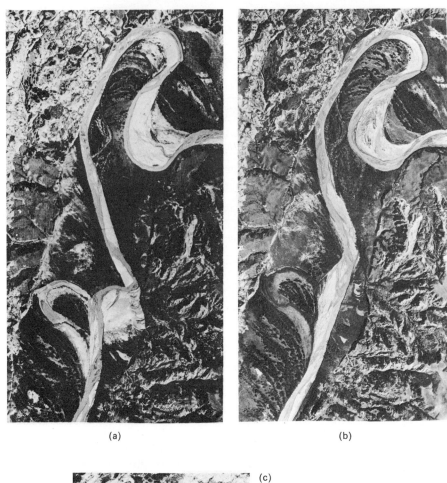

(a)   (b)

(c)

Figure 10-4. Flood plain and channel of the Little Missouri River near Watford City, North Dakota. For orientation and scale, see Figure 10-5. (*a*) 1939 vertical aerial photograph; (*b*) 1949 vertical aerial photograph; (*c*) 1958 vertical aerial photograph of the northernmost meander. (Photos: U.S. Dept. Agriculture, from Everitt, 1968.)

Lateral Accretion.   Flood plains are built in two fundamental ways, by *lateral accretion* and by *vertical accretion*. **Point bars** (Fig. 10.3) are the most important component of lateral accretion. As a meandering channel migrates across the flood plain, the **cut bank** on the convex side is undercut and eroded. As it collapses the derived bed load is carried a short distance downstream and deposited as a submerged bar, usually on the same side of the stream. The result is a cross-stratified deposit, with a subdued relief of low ridges and intervening swales, that may record many episodes of meandering channel migration (Figs. 10-4, 10-5). The crests of point bars approach the level of the former flood plain on the cut-bank side unless the stream is actively

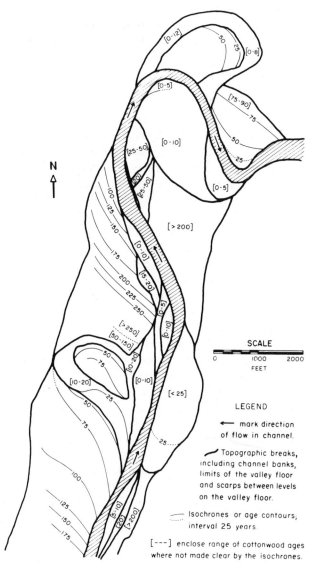

SCALE

0   1000   2000
FEET

LEGEND

← mark direction of flow in channel.

⌐ Topographic breaks, including channel banks, limits of the valley floor and scarps between levels on the valley floor.

— Isochrones or age contours; interval 25 years.

[- - -] enclose range of cottonwood ages where not made clear by the isochrones.

Figure 10-5.   Age map of the flood plain of the Little Missouri River near Watford City, North Dakota, based on ages of cottonwood trees. (Everitt, 1968.)

entrenching or aggrading, so the level of the flood plain need not change significantly by lateral accretion. In many streams that carry relatively coarse alluvium and lack natural levees, lateral accretion may account for 80–90 percent of the flood-plain deposition (Wolman and Leopold, 1957, p. 96).

The age of point-bar deposits along the Little Missouri River was determined by Everitt (1968) by tree-ring chronology of cottonwood trees growing on various parts of the flood plain. An age map of the valley floor (Fig. 10-5) shows that only relatively small areas of flood plain are older than 200 years. Everitt concluded that the reach of the river he studied redistributes every century a volume of sediment equal to the total present volume of alluvium along that reach, yet the flood plain has probably been within 1 m of its present elevation for at least 150 years. Even though a flood plain has retained a stable form for centuries, the alluvium that composes it is constantly moving through the system, weathering while it is in storage and being abraded as it moves in the river. Sigafoos (1964) gave an excellent summary of the technique of using flood-plain trees to document flooding and flood-plain deposition on a time scale of about a century.

Vertical Accretion.   Large rivers with gentle gradients and a predominantly suspended load typically have meandering channels on broad flood plains. When they flood, the velocity in the overbank water may be very low, and the flood waters may be weeks in receding back into the channel. Under these conditions, flood-plain deposition is largely by *vertical accretion*. The deposits are usually sand or mud, although splays from breached channel banks may bring coarse sediment. From April to June, 1973, the Mississippi River flooded large areas of its lower flood plain to depths of up to 4 m. During the 2-month interval, vertical accretion averaged 53 cm along natural levees but decreased rapidly to less than 2 cm beyond 400 m from the natural levees and averaged only slightly more than 1 cm in the backswamps (Kesel, et al., 1974). The abrupt loss of velocity at the edge of the flooded channel and the abundant sediment supply there commonly cause more and coarser deposition on the natural levees, which grade laterally into finer backswamp deposits. A result of vertically accreting natural levees may be that the river surface within the levees is many meters above the level of adjacent backswamps. Under these conditions, a flood level high enough to overtop a segment of natural levee has a pronounced gradient into the backswamp, and floods are violent and severe. The current at a point of overflow erodes a gap or **crevasse** in the levee, and a **splay** spreads far across the backswamp (Fig. 10-3). Sometimes the erosion is sufficient to begin an entirely new master channel, and a segment of the former channel is abandoned, to slowly fill with channel deposits and peat along a curved **oxbow lake**. If a meander recurves until it intersects an upstream portion of the channel, a **neck cutoff** occurs, wherein the abrupt shortening of channel length locally increases the gradient and velocity in the cutoff, and eventually seals the truncated ends of the abandoned meander with bank deposits or **plugs**.

If a master stream flows between high natural levees, few tributaries can join it but must flow parallel to the trunk stream along the lowest axis of the backswamp.

In older American terminology, such streams were called **Yazoo tributaries,** from the Yazoo River in Mississippi. A Yazoo tributary may be forced to extend downstream by the migrating meander belt of the trunk stream, but eventually it will be intersected by a meander and will be shortened again.

Vertical accretion cannot continue indefinitely unless base level is rising. The later history of the lower Mississippi River (Fig. 10-1) has accentuated vertical accretion and the formation of natural levees and backswamps. In the absence of the postglacial rise of sea level, we would expect much less vertical accretion in major rivers. The striking regular frequency of the annual flood on most upland rivers suggests that vertical accretion is not a major contributor to flood-plain development; otherwise, rivers with larger flood plains should flood progressively less frequently. We may suspect that thick alluvium deposited by vertical accretion, like the relict alluvium found beneath many large streams in their lower valleys, is the result of postglacial rise of sea level and is not typical of hypothetical rivers that enter a stable sea level. Thin vertical-accretion alluvium may accumulate slowly on a flood plain if the river is constrained from migrating laterally (Ritter, èt al., 1973).

ALLUVIAL FANS AND DELTAS

Two of the most common depositional landforms of alluvium result from the same cause: an abrupt loss of competence in a stream. *Alluvial fans* are subaerial forms, deposited at an abrupt decrease in velocity where a stream channel loses water by infiltration or widens abruptly. *Deltas* are mostly subaqueous forms, although their upper surfaces are usually slightly above water level. They result when a river enters a lake or the ocean and its velocity is abruptly checked. The two forms may merge; many deltas are capped by low-gradient alluvial fans. The "delta" of the Colorado River into the Gulf of California has separated the Salton Basin from the upper end of the Gulf, and on its north side is a sea-level alluvial fan 70 m above the floor of the closed, below-sea-level Salton Basin.

Alluvial Fans

In its simplest classic form, an **alluvial fan** is a segment of a low cone with its apex at the mouth of a gorge or canyon (Fig. 10-6). Contours are arcs of circles concentric on the head of the fan. In longitudinal or radial profile, fans are slightly concave skyward. They are excellent examples of water-spreading wash slopes (Fig. 8-19).

An alluvial fan usually forms where a high-gradient tributary stream enters directly onto the flood plain of a trunk stream or the flat floor of an intermontain basin. The causes for abrupt changes of gradient are numerous. A stream dissecting a fault scarp typically builds a fan at the foot of the scarp. Streams flowing down glacially oversteepened valley walls build fans on valley trains of outwash.

Figure 10-6. A new fan forming on a braided flood plain in front of an erosionally truncated earlier fan at the mouth of Hutt Stream, Rakaia valley, Canterbury, New Zealand (Carryer, 1966). (Photo: W. D. Sevon and S. J. Carryer.)

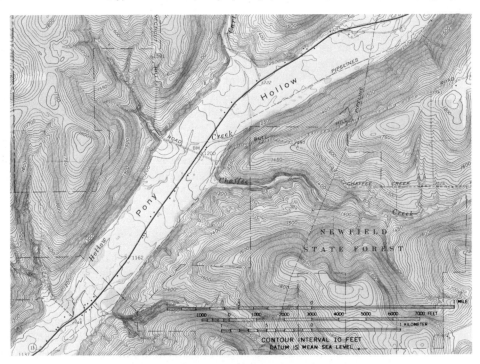

Figure 10-7. Portion of Alpine, New York, 1:24,000 topographic map showing Pony Hollow Creek displaced from one side of a valley train to the other by postglacial alluvial fans.

ns are mostly
bedload alluvium Fans are normally built of bed-load alluvium, especially gravel. Stratification may be sheetlike parallel to the fan surface or may contain many cut-and-fill structures of buried channels. Their surface morphology is characterized by radiating, branching *distributary channels*, most often braided rather than meandering. The permeable character of fan alluvium allows considerable infiltration, so that desert fans may contain mudflow deposits recording the progressive loss of water until the suspended load becomes too thick to flow as a muddy stream.

Fans are *aggradational*, or built-up, forms. One radial channel may carry most of the discharge for a time until its gradient becomes oversteep by deposition, when the stream abruptly shifts to a lower-gradient radius. The process ensures the uniform slope development along all radii and a resulting smooth conical form. Fans may block a trunk stream and cause a lake. Many underfit trunk streams of glaciated plateaus have low gradients that cannot transport the coarse alluvium delivered by valley-side tributaries, and the master stream is pushed laterally from one valley wall to the other by fans (Fig. 10-7). Very steep alluvial fans are called **alluvial cones** and are comparable in form, if not in internal structure, to talus cones. Alluvial fans are especially abundant and significant land forms in arid intermontane regions (Fig. 13-2).

Deltas

Deltas
are where
stream competency
starts to drop.
Multiple Dist.
channels are
common A level surface of water, whether a lake or the sea, has a powerful trapping effect on the suspended and bed loads of a river. The incoming fresh water spreads out as a sheet on more dense salt water but may sink beneath the warmer fresh water of a reservoir. In either event, its velocity is sharply reduced, and competence drops. "Trap efficiencies" of 50–98 percent have been reported for artificial reservoirs on small drainage basins (Colby, 1963, p. 35). All of the bed load, and most of the suspended load, is dropped close to the point where a stream enters a lake or the sea. As the sediment accumulates, the stream must extend its channel, or multiple *distributary channels*, across the accumulating alluvium. The resulting landform is a **delta**.

Deltas may be the subject of the oldest geomorphic description. Herodotus (fifth century B.C.) observed the growth of the Nile Delta by the annual floods and correctly inferred that all of lower Egypt was the delta of the Nile. The name *delta* refers to the Greek letter Δ, which is the shape of the Nile Delta (Fig. 10-8).

delta shape
is influenced Not all deltas have the same shape (Fig. 10-9). The shape is influenced by river discharge and load, grain size, flood frequency and intensity, climate, vegetation, wave energy, tidal range, salinity, water temperature, and tectonics (Morgan, 1970a, 1970b). Some deltas have shifting sublobes that mark separate periods of growth (Fig. 10-10). Others have been fairly stable in postglacial time, gradually building forward along their entire seaward perimeter through a maze of distributaries.

The cross section of a simple or small delta shows three basic depositional units: *bottomset beds, foreset beds,* and *topset beds* (Gilbert, 1890). Only the third is subaerial. A surface gradient must be maintained to carry alluvium to the perimeter, and

Deltas are composed of three depositional units
bottomset beds, foreset beds, and topset beds

Damietta Mouth

Rosetta Mouth

Figure 10-8. The Nile Delta, Suez Canal, and the Sinai Peninsula, viewed toward the southeast. Rosetta and Damietta distributary mouths are prominent. (Photo: NASA Technology Application Center, Gemini IV S-65-32752.)

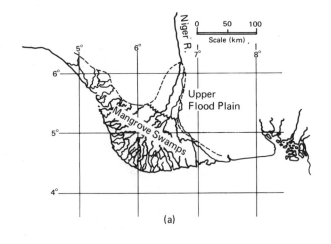

(a)

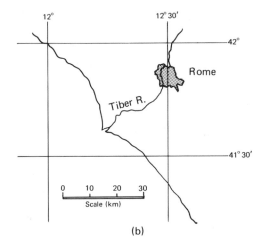

(b)

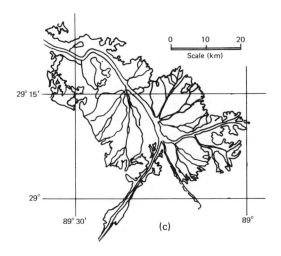

(c)

Figure 10-9. Some common delta forms. (*a*) Lobate (convex seaward) Niger Delta. (*b*) Cuspate (concave seaward) Tiber Delta. See also the Rosetta and Damietta mouths of the Nile River (Fig. 10-8). (*c*) Digitate ("birdfoot") delta of the Mississippi River. Another common form, the bay-head delta, is shown in Figure 3-15.

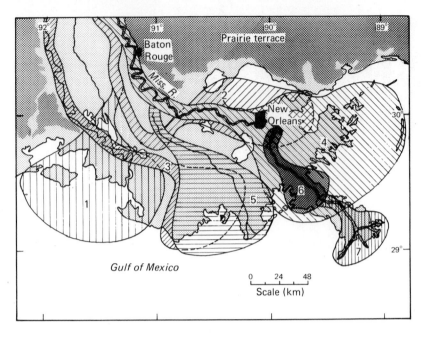

	1	Sale Cypremort	>4600 yrs. B.P.
	2	Cocodrie	*ca.* 4600-3500 yrs. B.P.
	3	Teche	*ca.* 3500-2800 yrs. B.P.
	4	St. Bernard	*ca.* 2800-1000 yrs B.P.
	5	Lafourche	*ca.* 1000-300 yrs. B.P.
	6	Plaquemine	*ca.* 750-500 yrs. B.P.
	7	Balize	<550 yrs.

Figure 10-10. Chronology of the deltas that compose the Mississippi delta plain (Morgan, 1970a). The modern digitate delta is shown in Figure 10-9(*c*).

aggradation may extend many miles upstream as a delta grows forward and reduces the river gradient.

The Mississippi River has one of the largest deltas and certainly the most studied one (Morgan, 1970a, 1970b). It has built forward to the edge of the continental shelf after having filled a deeply entrenched glacial-age valley (Fig. 10-1). The internal structure consists of linear, branching channel sands flanked by paired natural levees of poorly sorted sand and mud, which in turn grade laterally into highly organic backswamp mud. The entire delta complex, called a **delta plain** (Fig. 10-10), resembles a pile of maple leaves, with the veins represented by channel and levee deposits and the intervein areas represented by backswamps. When a set of channels is abandoned, compaction and subsidence lower the area until only the drowned levee crests are exposed. Simultaneously, waves erode the delta front and concentrate the sand as barrier spits and islands. Some later generations of distributary channels reoccupy segments of earlier drowned levee systems, in a few instances flowing in the opposite direction. Gradients are negligible throughout the modern delta.

Lakes in Fluvial Systems

A lake must be regarded as a temporary feature in the fluvial system. It is a highly efficient sediment trap and thus tends to fill with deltaic alluvium. At the same time, the local and temporary base level that causes the lake is certain to be eroded, so the long profile of the river channel tends to become smoothly concave and eliminate the lake basin by erosion.

Lakes may be caused by tectonic movements, volcanism, glacial erosion and deposition, landslides, aggradation by a trunk stream, or a host of other causes. Rivers that are temporarily dammed create large lakes that eventually break out with catastrophic severity. The greatest floods known to have occurred on earth are the Pleistocene Lake Missoula floods that repeatedly broke through a glacial lobe in eastern Washington and gouged thousands of square kilometers of anastomosing river channels across the Columbia Plateau to the Columbia River (Bretz, 1969; Baker, 1973). Another great flood resulted from the overflow of pluvial Lake Bonneville into the Snake River (Malde, 1968). These great floods occurred more than 12,000 years ago, but their eroded channels and enormous alluvial deposits are the dominant features of the present landscape on the Columbia Plateau and the Snake River Plain. Accounts of these great natural catastrophes should be read by all geomorphologists as a basis for speculation and contemplation on the role of infrequent but spectacular events in the shaping of "normal" landscapes.

The scale of the phenomena associated with these floods exceeds any other known fluvial events. The maximum discharge of the Missoula floods was 21.3×10^6 m³/s ($= 752 \times 10^6$ ft³/s; compare with Fig. 9-7). An alluvial fan built by the Lake Bonneville flood includes well-rounded basalt boulders 3 m in diameter (Malde, 1968, p. 13). A considerable extrapolation of Hjulström's curve (Fig. 9-8) is required to estimate the flood velocity that could move such boulders. Midchannel alluvial bars as much as 100 m thick, 2 km long, and 1 km wide were built of boulder gravel. Postflood weathering and erosion have been insignificant in modifying the shape of the flood deposits.

Piedmont Alluvial Plains

Large regional landforms develop from the lateral merging of adjacent low-gradient alluvial fans at the base of dissected mountains. The High Plains of the central United States are veneered by an apron of late Cenozoic sand and gravel named the Ogallala Formation (Thornbury, 1965, pp. 300–301; Frye, 1971), a region 1300 km long southward from South Dakota to Texas and 500 km wide (Fig. 10-11). Most of the alluvium is sandy, but gravel channel deposits and fine-grained vertical-accretion deposits are also known. The Ogallala Formation is as much as 150 m thick along the axes of buried valleys. Some combination of tectonic upwarp of the Rocky Mountains, and perhaps climatic change as well, caused the rivers of the Great Plains to aggrade until they entirely filled their valleys and coalesced across former interfluves. Pleistocene climatic changes and renewed tectonism probably

[handwritten margin note: alluvial plains are capable in altering a rivers shape or path, line or extend as are required]

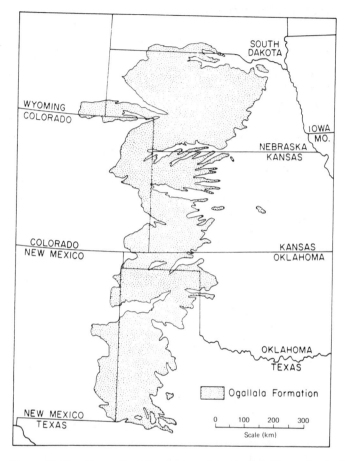

Figure 10-11.   Generalized distribution of the Ogallala Formation in the Great Plains province of the United States. (Frye, 1971, Fig. 1.)

ended the aggradation phase (Stanley and Wayne, 1972). The present relief of the unglaciated Great Plains is essentially the depositional surface of the Ogallala Formation and similar, but slightly younger, Pleistocene alluvium. The rivers that now flow east to the Mississippi are at grade on this ancient alluvial surface. The regional slope is only about 0.1 percent.

Similar piedmont alluvial plains form the Argentine Pampas on the eastern side of the Andes Mountains, the Indo-Gangetic Plain on the southern piedmont of the Himalaya Mountains, and the Canterbury Plains of New Zealand. Late Cenozoic tectonism has played an important role in initiating these and other great alluvial plains. Climatic changes during the Quaternary Period repeatedly caused the rivers to change their hydraulic geometry, however, and much of the alluvium is probably relict.

AGGRADATION IN THE FLUVIAL CYCLE

Aggradation as a Normal Phase of Valley Evolution

The primary function of river systems is to carry away weathered debris fed into them by mass-wasting and overland flow. If any potential energy is available beyond that required to move the load, the river will erode its channel. The deduced normal cross section of a river valley thus should have an erosional rock surface, the **valley flat,** beneath alluvium in transit or temporarily stored in the flood plain. Examples of relict alluvium have previously been described in this chapter, where aggradation has been caused by rising base level, climatic change, or tectonism. All three of these factors have been so important in shaping the present landscape that we are hard pressed to find a river that has a "normal" flood plain underlain by alluvium in transit to the sea and that is slowly eroding a valley flat beneath the alluvium. For example, compare the discussion of alluvium and grade (p. 222) with Figure 10-1.

In view of the abundant accidental events that can cause aggradation, it is appropriate to consider whether an aggradational phase can occur during the normal development of fluvial landscape evolution. A trunk stream might become graded early in the dissection history, while headward erosion is still in progress by first-order streams. Then, as the drainage network expands, greater discharge and load might require the longitudinal profile to be steepened by aggradation in the middle portion of the river system. When the uplands are totally dissected by competing drainage networks, no additional expansion can occur except by capture, and presumably the trunk stream will resume its slow downcutting (Davis, 1899, reprinted 1954, p. 260). Alluvial terraces might be left along the valley walls to record an aggradational phase that is a normal part of the continuous regional erosional evolution.

Cotton (1948, pp. 80–82) considered other possible causes of aggradation as a temporary phase of normal fluvial landscape development. Channel lengthening by both meandering and delta progradation causes a decrease of gradient, but the principles of hydraulic geometry suggest that aggradation need not result. Neither does it seem likely that the increased channel size of a graded river or subsurface flow through alluvium could cause a loss of competence and result in aggradation (Cotton, 1948, p. 81). The interaction of the many semidependent variables of the fluvial system is much too well balanced to permit such a disequilibrium to develop. Thus the only likely cause for an aggradational phase during fluvial dissection of a landscape, barring climatic and tectonic "accidents," seems to be the development of a graded trunk stream while drainage networks are still expanding.

Climatic Change and Aggradation

Rivers in alluvial channels on the Great Plains have been observed to change from a braiding to a meandering habit in response to decade-length climatic changes (p. 219). A generalization about the response of fluvial systems in semiarid or subhumid regions to slight shifts in climate was set forth by Huntington (1907, p. 358) and is now called *Huntington's principle* (Fairbridge, 1968, p. 1125). It suggests that increased aridity in areas of marginal rainfall leads to a loss of vegetative cover and increased mass-wasting (Fig. 12-6). Stream channels become choked with the excessive loads supplied to them concurrent with a reduction in discharge. Aggradation results as rivers build up their alluvial beds until gradients are steep enough to move the greater load of often larger grain size. Huntington's enunciation of this principle of fluvial response to climatic change was conditioned by his observations in the interior of Asia, where decreased total rainfall often coincides with increased seasonality of precipitation. This is often true but need not be incorporated into the principle as a requirement. Clearly, climates with an intense dry season followed by concentrated precipitation are conducive to a maximum sediment load in streams (Chap. 13).

Huntington deduced that, during times of greater precipitation, vegetation would stabilize landscapes, inhibit mass-wasting, promote infiltration, and result in more sustained flow of rivers. Peak flood discharges would be minimized, and rivers would erode channels of more gentle longitudinal gradients. Most geomorphologists would mentally associate a braided habit with the aggradational phase and meandering with the more humid phase of valley deepening. Huntington suggested that the great gravel terraces that line the valleys of central Asia represent multiple cycles of arid or semiarid aggradation alternating with entrenchment during more humid intervals. The association with Quaternary glacial and interglacial episodes is obvious, but the current evidence suggests that in some climatic zones the wetter intervals were associated with high sea levels of interglacial ages, and the arid aggradational phases are recorded by coarse alluvium graded to low sea levels that are correlative with glacials (pp. 425–430). In other regions, especially the presently arid southwestern United States, glacial ages were the more humid, or **pluvial,** ages. Huntington's principle is an important key to relating Quaternary climatic change to landscape evolution (Chap. 18).

Fluvial Terraces

Most flood plains have a microrelief of a few meters or less, including low *flood-plain terraces* eroded during floods of varying severity. The conditions of fluvial deposition during variable discharge, by either meandering or braided rivers, ensure such minor flood-plain terraces. However, minor low terraces are not likely to be preserved for more than a few centuries before channel migration removes them (Fig. 10-5).

If a river is slowly downcutting as its channel meanders laterally from one side of the valley to the other, portions of older flood plain or valley flat may be preserved as **unpaired terraces** (Fig. 10-12), especially if they are *rock-defended* by a buttress of the valley wall. Unpaired terraces need not represent a change in the rate or nature of fluvial processes. They are little more than accidentally preserved fragments of former flood plains. Long ago, Playfair (1802, p. 103) cited such terraces as strong proof that valleys are indeed the work of the streams that flow in them.

Figure 10-12. Terraces of the lower Rakaia River valley, Canterbury, New Zealand. Two high terraces in the distance are aggradational glacial outwash surfaces. Terraces in the foreground are unpaired erosional terraces, cut in glacial deposits. (Photo: National Publicity Studios, New Zealand.)

Paired terraces, which occupy the same level on opposite valley walls, are significant evidence that a river has cut downward in an intermittent fashion. Paired terraces, like unpaired, may be remnants of alluvium or of the former valley flat cut into rock. The distinction in the nature of terrace material is useful for descriptive purposes but does not have much genetic significance, because most rock-cut terraces retain at least a veneer of alluvium. Alluvial terraces may represent aggradation alternating with downcutting, as might be caused by climatic change, or simply pulses of entrenchment alternating with times when the valley flat was veneered with alluvium in transit. Even though a former flood plain was built by aggradation, it does not become a terrace except by subsequent erosion. It is important to remember that *fluvial terraces are cut, not built, by rivers* (Cotton, 1948, p. 187).

Flights of terraces on a weak-rock lowland may end downstream at a gorge through resistant rock that is the local and temporary base level for the upstream graded reach (p. 222). Because of slow downcutting at the resistant structure, the river upstream has time to open out a broad terraced valley. Such terraces are a special class of rock-defended terrace, where a single resistant obstruction controls and preserves multiple terraces (Cotton, 1948, p. 197). They are normally unpaired.

Thus far, only the cross-valley profile of fluvial terraces has been considered. The longitudinal profiles of terraces should also be studied for clues to river history. For instance, terraces may converge or diverge downstream. Convergence suggests progressive rejuvenation of the headwaters, perhaps by continuing tectonic uplift. Divergence suggests a progressive lowering of base level, either temporary or ultimate, more rapidly than the regional erosion rate. Changes in the amount and grain size of the alluvium carried by a river when it flowed at each successive terrace level may also explain differences in longitudinal slope of terraces.

Fluvial terraces on a single river system may be misinterpreted if the regional setting is ignored. Ritter (1967, 1972) described examples of rivers at the foot of the Beartooth Mountains in Montana. Rivers that rise in the mountain range have relatively steep gradients to carry coarse alluvium, whereas rivers that rise on the plains carry finer alluvium and have lower gradients. Therefore, even though two adjacent rivers drain to the same base level, near the mountain front the "mountain bred" stream is at a higher level than, and is likely to be captured by, the adjacent stream that heads on the plain. However, the low-gradient plains river cannot carry the coarse alluvium, and when capture occurs, an active aggradational phase begins, while the beheaded stream adjacent to it begins downcutting and forming terraces.

Valleys that now head in the mountains and also those that now originate on the plains have prominent terraces of coarse alluvium derived from the Beartooth Range, proving that capture has caused alternating episodes of aggradation and entrenchment, with terrace formation in each valley. Although a single valley would seem to record a complex history that might be given a tectonic or climatic explanation, the larger view demonstrates normal or "single-cycle" regional geomorphic development. The regional history of the Beartooth Range terraces is complicated further by outwash terraces from former valley glaciation.

REFERENCES

BAKER, V. R., 1973, Paleohydrology and sedimentology of Lake Missoula flooding in eastern Washington: *Geol. Soc. American Spec. Paper 144*, 79pp.

BRADLEY, W. C., 1970, Effect of weathering on abrasion of granitic gravel, Colorado River (Texas): *Geol. Soc. America Bull.*, v. 81, pp. 61–80.

BRADLEY, W. C., FAHNESTOCK, R. K., and ROWEKAMP, E. T., 1972, Coarse sediment transport by flood flows on Knik River, Alaska: *Geol. Soc. America Bull.*, v. 83, pp. 1261–84.

BRETZ, J. H., 1969, The Lake Missoula floods and the channeled scablands: *Jour. Geology*, v. 77, pp. 503–43.

CARRYER, S. J., 1966, Note on the formation of alluvial fans: *New Zealand Jour. Geol. Geophys.*, v. 9, pp. 91–94.

COLBY, B. R., 1963, Fluvial sediments—a summary of source, transportation, deposition, and measurement of sediment discharge: *U.S. Geol. Survey Bull. 1181-A*, 47 pp.

COTTON, C. A., 1948, *Landscape as developed by the processes of normal erosion:* Whitcombe and Tombs, Ltd., Wellington, N. Z., 509 pp.

CROSS, W. P., and WEBBER, E. E., 1959, Floods in Ohio, magnitude and frequency: *Ohio Department of Natural Resources, Division of Water, Bull. 32*, 325 pp.

DAVIS, W. M., 1899, The geographical cycle: *Geog. Jour.*, v. 14, pp. 481–504. (Reprinted 1954 in *Geographical essays:* Dover Publications, Inc., New York, pp. 249–78.)

DURY, G. H., 1964a, Principles of underfit streams: *U.S. Geol. Survey Prof. Paper 452-A*, 67 pp.

———, 1964b, Subsurface exploration and chronology of underfit streams: *U.S. Geol. Survey Prof. Paper 452-B*, 56 pp.

———, 1965, Theoretical implications of underfit streams: *U.S. Geol. Survey Prof. Paper 452-C*, 43 pp.

———, 1967, Bankful discharge and the magnitude-frequency series: *Australian Jour. Sci.*, v. 30, p. 371.

———, 1973, Magnitude-frequency analysis and channel morphometry, *in* Morisawa, M., ed., *Fluvial geomorphology:* State Univ. of New York Publications in Geomorphology, Binghamton, N.Y., pp. 91–121.

DURY, G. H., HAILS, J. R., and ROBBIE, M. B., 1963, Bankfull discharge and the magnitude-frequency series: *Australian Jour. Sci.*, v. 26, pp. 123–24.

EVERITT, B. L., 1968, Use of the Cottonwood in an investigation of the recent history of a flood plain: *Am. Jour. Sci.*, v. 266, pp. 417–39.

FAIRBRIDGE, R. W., 1968, Terraces, fluvial-environmental controls, *in* Fairbridge, R. W., ed., *Encyclopedia of geomorphology:* Reinhold Book Corporation, New York, pp. 1124–38.

FISK, H. N., 1947, *Fine-grained alluvial deposits and their effects on Mississippi River activity:* U.S. Army Corps of Engineers, Mississippi River Commission, Vicksburg, Miss., v. 1, 82 pp., v. 2, 74 plates.

FRYE, J. C., 1971, The Ogallala Formation—a review, *in* Ogallala Aquifer Symposium: *Texas Tech. Univ. Internat. Center for Arid and Semi-arid Land Studies Spec. Rept. no. 39*, pp. 5–14a.

GILBERT, G. K., 1890, Lake Bonneville: *U.S. Geol. Survey Mon. 1*, 438 pp.

GUMBEL, E. J., 1958, Statistical theory of floods and droughts: *Inst. Water Engineers Jour.*, v. 12, pp. 157–84.

HUNTINGTON, E., 1907, Some characteristics of the glacial period in non-glaciated regions: *Geol. Soc. America Bull.*, v. 18, pp. 351–88.

KESEL, R. H., DUNNE, K. C., McDONALD, R. C., ALLISON, K. R., and SPICER, B. E., 1974, Lateral erosion and overbank deposition on the Mississippi River in Louisiana caused by 1973 flooding: *Geology*, v. 2, pp. 461–64.

LANGBEIN, W. B., 1949, Annual floods and the partial-duration series: *Am. Geophys. Un., Trans.*, v. 30, pp. 879–81.

LEOPOLD, L. B., WOLMAN, M. G., and MILLER, J. P., 1964, *Fluvial processes in geomorphology:* W. H. Freeman and Company, San Francisco, 522 pp.

MALDE, H. E., 1968, The catastrophic late Pleistocene Bonneville flood in the Snake River Plain, Idaho: *U.S. Geol. Survey Prof. Paper 596*, 52 pp.

MORGAN, J. P., 1970a, Deltas—a résumé: *Jour. Geol. Education*, v. 18, pp. 107–17.

———, 1970b, Depositional processes and products in the deltaic environment, *in* Morgan, J. P., ed., Deltaic sedimentation, modern and ancient: *Soc. Econ. Paleontologists and Mineralogists Spec. Pub. 15*, pp. 31–47.

PLAYFAIR, J., 1802, *Illustrations of the Huttonian theory of the earth:* Dover Publications, Inc., New York, 528 pp. [Facsimile reprint, 1964.]

PRICE, W. A., 1947, Geomorphology of depositional surfaces: *Am. Assoc. Petroleum Geologists Bull.*, v. 31, pp. 1784–1800.

RITTER, D. F., 1967, Terrace development along the front of the Beartooth Mountains, southern Montana: *Geol. Soc. America Bull.*, v. 78, pp. 467–83.

——, 1972, Significance of stream capture in the evolution of a piedmont region, southern Montana: *Zeitschr. für Geomorph.*, v. 16, pp. 83–92.

RITTER, D. F., KINSEY, W. F., and KAUFFMAN, M. E., 1973, Overbank sedimentation in the Delaware River Valley during the last 6000 years: *Science*, v. 179, pp. 374–75.

RUSSELL, R. J., 1958, Geological geomorphology: *Geol. Soc. America Bull.*, v. 69, pp. 1–21.

SCHUMM, S. A., and STEVENS, M. A., 1972, Abrasion in place: A mechanism for rounding and size reduction of coarse sediment in rivers: *Geology*, v. 1, pp. 37–40.

SIGAFOOS, R. S., 1964, Botanical evidence of floods and floodplain deposition: *U.S. Geol. Survey Prof. Paper 485-A*, 35 pp.

STANLEY, K. O., and WAYNE, W. J., 1972, Epeirogenic and climatic controls of early Pleistocene fluvial sediment dispersal in Nebraska: *Geol. Soc. America Bull.*, v. 83, pp. 3675–90.

THORNBURY, W. D., 1965, *Regional geomorphology of the United States:* John Wiley & Sons, Inc., New York, 609 pp.

TODD, J. E., 1900, River action phenomena: *Geol. Soc. America Bull.*, v. 12, pp. 486–90.

VANONI, V. A., 1971, Sediment transportation mechanics: Q. Genetic classification of valley sediment deposits: *Hydraulics Div. Jour., Am. Soc. Civil Engineers Proc., Paper 7815*, no. HYl, pp. 43–53.

WOLMAN, M. G., and LEOPOLD, L. B., 1957, River flood plains: Some observations on their formation: *U.S. Geol. Surv. Prof. Paper 282-C*, pp. 86–109.

CHAPTER 11

Structural Control of Fluvial Erosion

INTRODUCTION

No geomorphologist doubts the importance of rock structure to the shape of erosional landforms. *Structure*, along with *process* and *time*, is a cornerstone of explanatory description (Chap. 1). Structural factors of rock weathering were discussed in Chapters 6 and 7. This chapter is concerned with the effects on fluvial erosion of macroscopic or large-scale *rock structure*, including stratification and other primary layering, foliation and similar metamorphic fabrics, joints, faults, and folds.

The previous two chapters have stressed processes and depositional forms, so it has not been necessary to consider structural control of fluvial processes except incidentally. Drainage networks were assumed to develop over homogeneous terranes in order to simplify the study of fluvial processes. However, on the real earth, rivers cross belts or zones of contrasting structures, and their hydraulic geometries change in response to the erodibility of each structure and the type of load it delivers to the river.

The present subaerial landscape is nothing more than an instantaneous and arbitrary interface between atmosphere and lithosphere, marking a transitory phase of the erosion process. Erosion has probably removed more than 10 km from many continental landscapes and as much as 30 km from some mountain ranges (p. 106). During such tremendous amounts of erosion, whole structural terranes can be removed. For instance, coastal plain sediments can be stripped off as streams entrench into the older rocks beneath. Folded and faulted sedimentary structures can be stripped from basement complexes of igneous and metamorphic rocks. Buried relief under unconformities and thrust sheets, perhaps part of ancient landscapes, can be exhumed.

During the progress of erosion, it is inevitable that some structures will be more resistant to destruction than others. Relative erodibility, or rock resistance, is a subjective property, not simply defined. For instance, gravel alluvium may survive on very old river terraces, not because it became strongly cemented or lithified but because it is so permeable that no surface streams could form to erode it (Rich, 1911). Emerged coral reefs on tropical coasts (Fig. 2-6) may have their relief preserved or even enhanced, because solution lowers the adjacent lagoon-floor carbonate sand and mud, whereas the permeable reef rock is lowered only slightly. In both examples, passive resistance is more effective than strength.

FRACTURES AS STRUCTURAL CONTROLS

Joints

Joints provide for greater weathering and mass wasting

Systematic or nonsystematic joint sets cut most rocks. Some are the result of contraction on cooling (Fig. 11-1). Others are sheeting joints, the result of pressure release (Fig. 6-3). Still others are due to regional tectonic stresses (Fig. 11-2). Whatever their origin, joints provide openings for ground-water movement, and weathering and mass-wasting are intensified. Joints that open parallel to gorges and steep-sided valleys probably aid in enlarging and perpetuating the valleys.

Figure 11-1. Bruneau Canyon, Idaho. Cliff-and-talus canyon walls are controlled primarily by vertical columnar joints in multiple lava flows. (Photo: Idaho Dept. Commerce and Development.)

Figure 11-2. Jointed sandstone and shale near Ithaca, New York. The joints belong to two vertical sets, one striking northeast and the other northwest. The conjugate joint sets are systematically related to fold axes in the Appalachian Mountains, which in this region strike approximately east-west.

During weathering, joint-bounded blocks become progressively more rounded. **Corestones** of fresh rock, initially surrounded by weathered spheroidal exfoliation shells (Fig. 11-3), may actually weather free to become loose boulders on the surface or in regolith. It has been proposed that the large boulders of crystalline rocks that are found in glacial drift are not necessarily quarried free by glacial erosion but might be corestones from a preglacial or interglacial weathering profile (Feininger, 1971).

Masses of jointed bare rock, stacked like well-worn childrens' blocks, are called **tors** (Fig. 11-4). They are not transported but result from differential weathering and erosion along joints, at least one set of which is subhorizontal. Tors may form at depth by joint-controlled weathering within the regolith, or they may form subaerially by differential weathering and erosion of structures (Wahrhaftig, 1965; Eggler, et al., 1969). Both the tropical savanna (p. 325) and periglacial (p. 352) climates have been claimed to be especially favorable for tors (Thomas, 1968). A regional landscape of schist tors in New Zealand has been described and illustrated by Ward (1952), Turner (1952), and McCraw (1965). Tors are of interest to geomorphologists because they indicate that structurally controlled subsurface weathering may precondition the shape of a landscape long before the subaerial surface is lowered by erosion to expose the differentially weathered terrane (Fig. 13-8).

Figure 11-3. Corestones of fresh quartz diorite surrounded by concentric shells of partly decomposed rock. Sheeting joints exert some control on weathering patterns. Roadcut 35 km northeast of Medellín, Colombia. (Photo: Tomas Feininger.)

Figure 11-4. Tors in a tropical savanna landscape, Townsville, northern Queensland, Australia. See also Figures 13-7 and 13-8.

Fault-Line Scarps

Fault line scarps require regional geology [margin handwritten note]

The distinction between *fault scarps*, of tectonic origin, and **fault-line scarps**, formed by structurally controlled erosion at a fault contact, has been called "the indispensable prerequisite to a correct interpretation of geologic history" (p. 33). The distinction is not based on the observable morphology of a scarp but on full understanding of the regional geology. Any scarp, whether formed by recent tectonic movement or differential erosion of unlike rock types, is likely to be a facet or surface at an appropriate angle of repose.

Two genetic classes of fault-line scarps are distinguished (Fig. 11-5). **Obsequent** ("opposite to the consequent") scarps are unambiguous, because the geologic relationships prove that movement on the fault would have produced a scarp facing in the opposite direction. **Resequent** ("renewed consequent") fault-line scarps are difficult to distinguish from tectonic fault scarps, because in both, the upthrown fault block is also the high side of the scarp. Geomorphic criteria alone cannot distinguish the two.

Differential erosion along faults does not necessarily produce scarps. In a faulted terrane of massive igneous or metamorphic rock, only the *crushed zone* along the faults may be subject to more rapid weathering and erosion. On aerial photographs, crystalline terranes show a characteristic network of differentially eroded zones on fault lines and joints (Fig. 11-6).

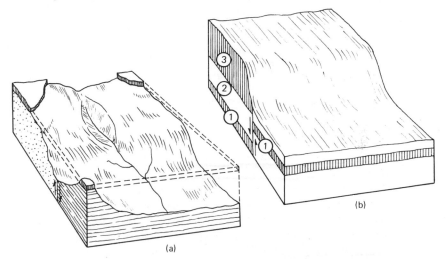

(a)

(b)

Figure 11-5. Two classes of fault-line scarp. In both examples, the stipulation is made that no tectonic movement has occurred in the present erosion cycle. (*a*) Resequent fault-line scarp. The former extent of a post-faulting, but dissected, lava flow has been added to suggest that evidence can sometimes by found to prove that such a landform is a fault-line scarp rather than a fault scarp. (*b*) Obsequent fault-line scarp. At a still later stage of erosion, the scarp might become resequent if bed 1 resists erosion on the front half of the block and beds 2 and 3 are eroded from the rear half.

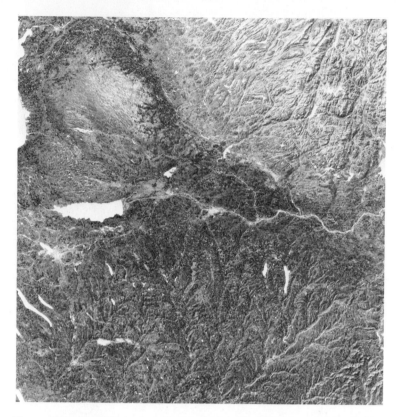

Figure 11-6. Satellite image of central New York. Water is white in this image. Lake Oneida, left center. Mohawk River drains east (right) across the center. Metamorphic terrane of the southwest Adirondacks, upper right, shows strong fault-line and joint control of weathering and erosion. (NASA ERTS E-1080-15180.)

mélanges are hard to understand just by their own character

A confused or chaotic landscape is likely to result from structurally controlled erosion of tectonic *mélanges*. In many regions of the Mediterranean and circum-Pacific tectonic belts, great masses or blocks of rigid igneous, metamorphic, and sedimentary rocks, some several kilometers in length, are embedded in a more plastic matrix of shale or low-grade, fine-grained metamorphic rock (Hsü, 1968). The tectonic interpretation of mélanges is a matter of debate, but they probably are formed by thrust faulting and submarine slumping in an orogenic marine sediment-ational environment. Hsu (1971) described a mélange landscape in Cyprus and compared it with parts of eastern Taiwan. In both regions, resistant blocks of graywacke, chert, ultrabasic igneous rock, or limestone form hills or boulder-strewn ridges surrounded by claystone lowlands. Maxwell (1974, p. 1201) described the mélange topography of northern California as "accidental, characterized by scattered blocks jutting out of a rounded or land-slid surface." The mélange structure may extend to a depth of 10 km. A unique geomorphic aspect of mélange terrane is that the hills of resistant rock have no "roots," and no prediction of sequential landscape evolution can be made. Although not strictly an example of fault-line erosional scarps, the abrupt structural boundaries of mélange hills do not seem to fit into any other available descriptive category.

DIFFERENTIAL FLUVIAL EROSION ON LAYERED ROCKS

Most sedimentary rocks, and many igneous and metamorphic rocks, have layers of contrasting erodibility. The layering may be the result of sedimentary stratification, igneous intrusion, metamorphic differentiation, or a host of other causes. Weathering emphasizes even slight contrasts in lithology, but on the larger scale, belts of distinctive erosional topography form on the exposed edges of rock layers. Where layers are horizontal, their exposed edges, or *outcrops*, are parallel to contours and form altitudinally zoned structural landforms. Where layers are tilted, much more complex geometry develops. Much of the geomorphic diversity of landscapes is due to the structural control of layered rocks, and an extensive descriptive and explanatory terminology has evolved. Much of the terminology is shared with the field of structural geology, because geomorphic form is one important way of recognizing structures in the field.

Layered Igneous and Metamorphic Rocks

Volcanic plateaus are built of a pile of lava and tephra sheets and intrusive sills crossed by numerous dikes. Columnar jointing provides high-angle sets of closely spaced joints that promote rock falls, cliffs, and taluses (Fig. 11-1). The margins of flows may cool so rapidly that the columnar jointing does not extend to the base and surface of a flow. Also, one or more flows that were erupted in close temporal proximity may share a central core zone of columnar joints and glassy or vesicular margins. The glassy or shattered marginal zones commonly form taluses separating joint-controlled cliffs. The spectacular effects of the Missoula and Bonneville floods (p. 245) were in large part structurally controlled by the plateau basalts of the region (Baker, 1973).

Metamorphic terranes commonly have zones or layers of more resistant rock alternating with more erodible belts. A typical example would be quartz-rich gneiss alternating with biotite schist. Weathering and erosion more rapidly lower the schistose belts, and the resistant layers are etched out. The scale may be measured in meters, or, as in the coastal region of Maine, in strike ridges of resistant gneiss 100 m or more in height and as much as 50 km in length (Fig. 11-7). Metamorphic terranes have a dominant regional structural grain. Hills have their long axes on strike. Streams follow erodible belts or cross ridges at fracture zones. Entire regional landscapes such as the New England geomorphic province of the northeastern United States become a series of structurally controlled subparallel ridges and valleys (Flint, 1963). The Precambrian Grenville metamorphic terrane of eastern North America

Figure 11-7.   Satellite image of Casco Bay, Maine. North is to the top; water is black on this image. Gneiss ridges form long peninsulas and islands projecting into the Gulf of Maine. (NASA ERTS E-1472-14530.)

provides some especially impressive examples of differential erosion. Marble belts between schists and gneisses have formed weak-rock lowlands that curl with the metamorphic grain (Fig. 11-8).

Folded Sedimentary Rocks

Most of the present-day landscape is eroded on stratified sedimentary rocks. In the United States, for instance, only 8.4 percent of the area is of Precambrian terrane, but strata of the Carboniferous, Cretaceous, and Quaternary Systems each cover about 14 percent of the area (Gilluly, 1949, p. 576). The sedimentary *cover strata* range from easily eroded, unconsolidated Cenozoic glacial drift or coastal-plain sedimentary rocks to stronger, better lithified, and possibly folded and faulted Mesozoic and Paleozoic rocks. Thickness ranges from a few meters to more than 10 km. Usually the thickest sections of cover strata are now contorted in orogenic belts. Cenozoic epeirogeny has, however, uplifted large continental areas with only slight warping of the cover strata, which are only 1–2 km thick. Thus Cenozoic epeirogeny and orogeny have created regionally extensive subaerial landscapes that are largely eroded from sedimentary structures. The differential erodibility of shale, limestone, and sandstone is the dominant structural control of landscapes over very large portions of all continents.

[handwritten margin notes: Sedimentary rocks cover the largest area. They also differ a lot. Thus their erodability is dominant as to structural control]

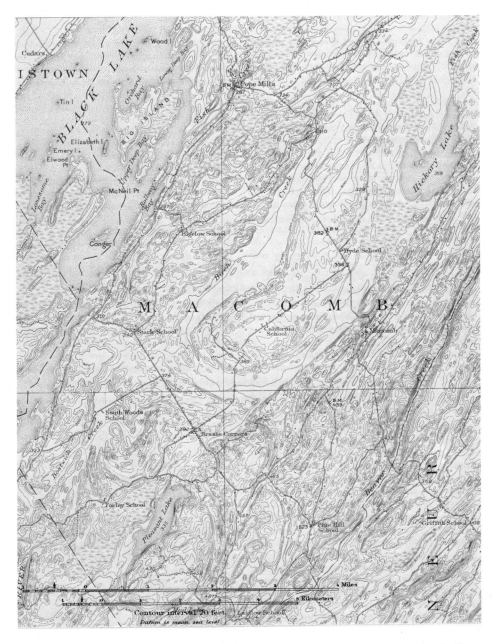

Figure 11-8. Contorted drainage pattern following marble and calc-silicate lowlands among gneiss hills, northwestern Adirondack Mountains, New York. (Part of Hammond, N.Y. 15′ topographic map quadrangle, 1935 edition.)

Flat-lying sedimentary rocks of contrasting erodibility form *structural benches* on valley walls. Many plateau surfaces are structurally controlled by resistant *caprock* that persists at a high elevation, while the terrain around it is being lowered. The caprock is usually a sedimentary rock, but it can also be a lava flow or a concretionary weathered layer such as caliche or laterite (p. 131).

Tilted or folded strata provide more complex structural controls to erosion. Generally the outcropping edge of a resistant stratum will form an escarpment or cliff, often joint controlled. The *dip slope*, away from the crest of the escarpment, is parallel to the stratification and tends to resist dissection as weaker overlying strata are eroded. A single thin layer of sandstone may cap an extensive shale dip slope, for instance.

A loose terminology has evolved to differentiate structurally controlled ridges on tilted resistant layers (Fig. 11-9). If the more resistant unit is horizontal and caps a broad flat-topped hill, the hill is called a **mesa**. If the diameter of the cap rock is less than the height of the hill above the surrounding terrain, the term **butte** is more often applied. If the resistant bed dips gently, an asymmetric **cuesta** results, with a steep escarpment and a gentle dip slope. When the dip of the layer is comparable to the angle of repose of the sliderock on the escarpment face, the ridge becomes approximately symmetric and is known as a **hogback**.

The structural control of tilted strata imposes a powerful asymmetry on drainage networks. Escarpment streams are steep, short, and have high gradients. Dip-slope streams are likely to have more gentle gradients, larger watersheds, more tributaries, and more sustained flow (Fig. 11-10). As erosion progresses, a trunk stream flowing along *weak-rock lowland* between two cuestas will migrate laterally in the down-dip direction, because the escarpment face can deliver waste to the trunk river faster than the lower gradient dip-slope streams can. The entire ridge and valley system migrates laterally as well as downward with time in a process termed **homoclinal shifting**. The significance of the process was first recognized a century ago by Gilbert (1877, pp.

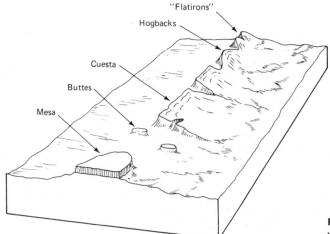

Figure 11-9. Terminology of structurally controlled scarps and ridges.

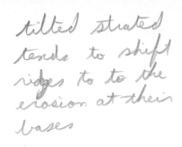

*tilted strata
tends to shift
ridges to to the
erosion at their
bases*

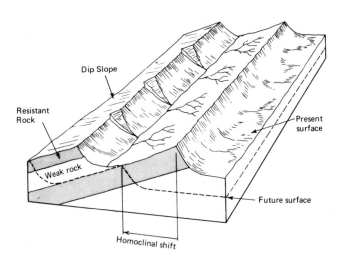

Dip Slope

Resistant Rock

Weak rock

Present surface

Future surface

Homoclinal shift

Figure 11-10. Asymmetric drainage networks on homoclinal ridges. Ridge crests and valleys will migrate laterally downdip as the landscape is lowered by erosion.

135–40) in his report on the Henry Mountains (p. 50), although his original term was *monoclinal shifting*. The terminology is not important, except that the lateral migration will stop when a monoclinal flexure flattens with depth, whereas it should continue indefinitely where strata have a uniform (homoclinal) dip. By this process, structurally controlled ridges and valleys must migrate laterally many times their own width as a region of dipping strata is eroded.

Beyond homoclinal dips, the next degree of structural complexity in stratified rocks is a series of parallel anticlines and synclines with horizontal fold axes (Fig. 11-11). Tectonic hills (Chap. 3) are normally anticlines or domes, but in structurally controlled eroded landscapes, the hills are commonly on synclinal structures. An early observation of this fact led to the concept of **topographic inversion**, whereby structural depressions such as synclinal axes become topographic elevations (Fig.

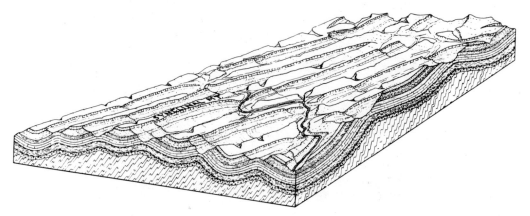

Figure 11-11. Block diagram of dominantly synclinal ridges in a terrane of folded sedimentary rocks. (From von Engeln, 1942, Fig. 183.)

11-11). An impressive amount of erosion is implied, but, as demonstrated earlier (p. 106), the amount of rock eroded from the continental masses can be measured in kilometers or even tens of kilometers, and multiple topographic inversions must have resulted during the dissection of fold mountains such as the central Appalachians. Therefore, it is doubtful whether eroded anticlines are any more indicative of topographic inversion than eroded synclines, or synclinal mountains more a measure of the depth of erosion than anticlinal mountains.

One may safely infer that erosion is in progress while an orogeny is in progress. The rising anticlines of tectonic landscapes (Fig. 3-21) are cut by gullies as soon as runoff begins to drain from them. When a resistant stratum becomes exposed in a rising anticline, it is still deeply buried in the axes of adjacent synclines. By the time the regional landscape has been eroded sufficiently to expose the resistant stratum at the level of the synclinal axes, the anticlines have been breached, and the synclinal floors become flat or gently basined summits, flanked by escarpments that face outward. If only a single resistant unit is embedded in a thick sequence of erodible rocks such as shale, the resistant stratum is very likely to cap synclinal ridges as erosion progresses. If resistant and erodible strata are interlayered at close intervals, the landscape will have numerous synclinal and anticlinal ridges formed on the various resistant units, as is shown by the Zagros Mountains of Iran (Fig. 8-15, 11-14). Where folding becomes extremely complex and metamorphism has reduced the contrast in erodibility, dissection favors neither synclines nor anticlines (Fig. 11-12). The abun-

Figure 11-12. Tight folding in Cambrian limestone and shale, Canadian Rockies, Kootenay District, British Columbia. (Photo: Geological Survey of Canada.)

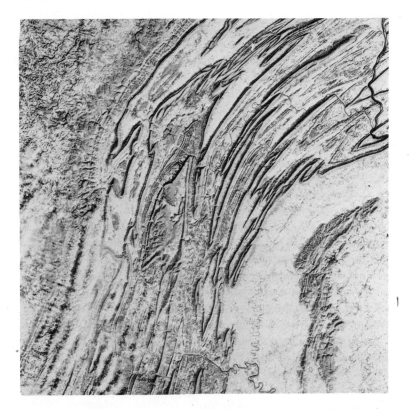

Figure 11-13. Zigzag ridges in the Valley and Ridge province of the Appalachian Mountains in Pennsylvania. North is toward the top of the image; Harrisburg, Pennsylvania, is on the Susquehanna River just south of the major fold belt. Low-angle lighting from the southeast may cause the ridges to appear inverted. If so, invert the page. Most of the ridges in the north-central part of the image are breached anticlines. (NASA ERTS E-1513-15220.)

dance of synclinal mountains has led some authors to suggest that unusually intense strain during folding encourages the subsequent erosional breaching of anticlines. Breached anticlines are not the result of some peculiar fracture pattern at the crests of folds but are simply a consequence of Powell's dictum (p. 84) that early-formed topographic prominences, most commonly anticlines, are the first structures to be attacked by erosional agents.

Plunging structures introduce a still higher level of complexity to structurally controlled landforms. Cuesta- and hogback-forming resistant strata not only dip laterally from anticlinal toward adjacent synclinal axes, but the structures plunge along strike. The result is subparallel belts of *zigzag ridges* such as characterize the folded central Appalachians (Fig. 11-13). The same resistant rock, such as the Tuscarora quartzite of the Appalachians, may form ridge after ridge as it is repeatedly intersected by the land surface. Faults that cause strata to be offset laterally also displace ridge crests controlled by the resistant layers.

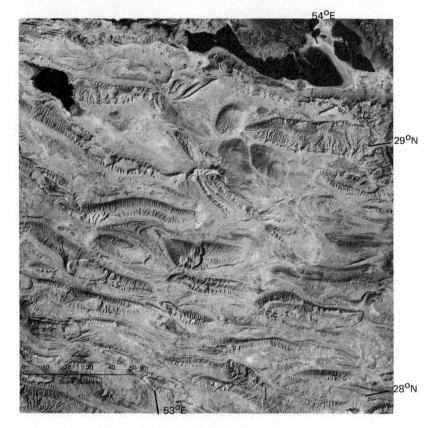

Figure 11-14.   Doubly plunging anticlines and domes of limestone east of Shiraz, Iran, in the Zagros Range. Some domes are breached, others show only consequent valleys down their flanks. (NASA ERTS E-1221-06293.)

Structurally controlled *domes* and *basins* are the result of erosion on double-plunging folds. Although common in fold belts of orogenic regions (Fig. 11-14), broad domes and basins are important features of epeirogeny as well. In the Central Lowland of the United States, the Michigan basin has a central structural depression of more than 4000 m. To the south, the Nashville dome is 80 km by 200 km across but has only a gentle central uplift. Resistant strata form cuestas encircling the crests of many epeirogenic structures, facing inward onto an eroded lowland on the domes and facing outward from the basins. The Paris basin of northern France is ringed by outward-facing escarpments, especially in the northeast quadrant. The low wooded ridges and intercuesta lowlands between Paris and the Rhine graben were the scene of bitter fighting during World Wars I and II. Control of the cuesta ridges meant tactical advantage to one of the opposing armies. The English Channel nearly transects a broad structural dome that is ringed by inward-facing cuestas of chalk, which are called **downs** in England. The core of the dome is deeply eroded in Jurassic clay shales with Cenozoic sediments deposited on the basin floor.

In most of the examples just cited, topographic inversion has produced a topographic basin on erodible shale in the center of an epeirogenic dome. In the Black Hills of South Dakota, the Henry Mountains of Utah, and elsewhere, however, sharp domal uplift over a core of strong crystalline rock has created a central mountain range with ringing hogbacks and cuestas. Similar rings of upturned strata are common along the fronts of many mountain ranges (Fig. 11-15).

The most extreme orogenic deformation of layered rocks consists of great overturned, recumbent, often thrust-faulted folds known as **nappes** from their type locality in the Alps. Alpine mountain sides may show strata more than 1000 m in thickness, all overturned. Detached pieces of nappes or thrust plates may cap ridges underlain by much younger rocks. Nappes, like mélanges, produce complex structurally controlled landforms, usually well exposed because they are within great mountain ranges. As the structures are eroded to progressively deeper levels, the increasing metamorphic alteration, joints, and igneous intrusions assume the dominant structural control, and stratification is of reduced importance to the land forms.

sharp domal uplift in an area of topographic inversion results in ⟵

Figure 11-15. Sheep Mountain, Wyoming, looking southeast. A plunging anticlinal ridge with surrounding hogbacks. (Photo: J. S. Shelton.)

STRUCTURALLY CONTROLLED DRAINAGE PATTERNS

In regions of folded sedimentary rocks, structurally controlled strike ridges form most of the higher terrain. As with all landforms, these are best regarded as mere residuals of fluvial erosion, with the intervening valleys the actively developing landforms. Systems of valleys eroded on contrasting structures conform to those structures, giving to both valley profiles and drainage networks distinctive, structurally controlled geometry. Both empirical and genetic classifications of structurally controlled drainage networks have evolved and are among the more useful and important tools of geomorphic analysis.

Empirical Classification by Form

Arthur D. Howard (1967) compiled and classified drainage patterns into categories of basic patterns, modifications to the basic patterns, and varieties of the modified patterns. His very thorough classification is summarized in Table 11-1 and Figs. 11-16, 11-17, and 11-18. The classification refers to the regional patterns of an aggregate of valleys and gullies, not to the patterns of single streams, which are described by such terms as crooked, straight, meandering, and braided.

TABLE 11-1. SIGNIFICANCE OF BASIC AND MODIFIED BASIC DRAINAGE PATTERNS

Basic	Significance	Modified Basic	Added Significance or Locale
Dendritic	Horizontal sediments or beveled, uniformly resistant, crystalline rocks. Gentle regional slope at present or at time of drainage inception. Type pattern resembles spreading oak or chestnut tree.	Subdendritic	Minor secondary control, generally structural.
		Pinnate	Fine-textured, easily erodible materials.
		Anastomotic	Flood plains, deltas, and tidal marshes.
		Distributary (Dichotomic)	Alluvial fans and deltas.
Parallel	Generally indicates moderate to steep slopes but also found in areas of parallel, elongate landforms. All transitions possible between this pattern and type dendritic and trellis.	Subparallel	Intermediate slopes or control by subparallel landforms.
		Colinear	Between linear loess and sand ridges.
Trellis	Dipping or folded sedimentary, volcanic, or low-grade metasedimentary rocks; areas of parallel fractures; exposed lake or sea floors ribbed by beach ridges. All transitions to parallel pattern. Type	Subtrellis	Parallel elongate landforms.
		Directional Trellis	Gentle homoclines. Gentle slopes with beach ridges.
		Recurved Trellis	Plunging folds.
		Fault Trellis	Branching, converging, diverging, roughly parallel faults.

TABLE 11-1. (Continued)

Basic	Significance	Modified Basic	Added Significance or Locale
Trellis (*cont.*)	pattern is regarded here as one in which small tributaries are essentially same size on opposite sides of long parallel subsequent streams.	Joint Trellis	Straight parallel faults and/or joints.
Rectangular	Joints and/or faults at right angles. Lacks orderly repetitive quality of trellis pattern; streams and divides lack regional continuity.	Angulate	Joints and/or faults at other than right angles. A compound rectangular-angulate pattern is common.
Radial	Volcanoes, domes, and erosion residuals. A complex of radial patterns in a volcanic field might be called multiradial.	Centripetal	Craters, calderas, and other depressions. A complex of centripetal patterns in area of multiple depressions might be called multi-centripetal.
Annular	Structural domes and basins, diatremes, and possibly stocks.		Longer tributaries to annular subsequent streams generally indicate direction of dip and permit distinction between dome and basin.
Multibasinal	Hummocky surficial deposits; differentially scoured or deflated bedrock; areas of recent volcanism, limestone solution, and permafrost. This descriptive term is suggested for all multiple-depression patterns whose exact origins are unknown.	Glacially Disturbed Karst Thermokarst Elongate Bay	Glacial erosion and/or deposition. Limestone. Permafrost. Coastal plains and deltas.
Contorted	Contorted, coarsely layered metamorphic rocks. Dikes, veins, and migmatized bands provide the resistant layers in some areas. Pattern differs from recurved trellis (Fig. 11-17, H) in lack of regional orderliness, discontinuity of ridges and valleys, and generally smaller scale.		The longer tributaries to curved subsequent streams generally indicate dip of metamorphic layers and permit distinction between plunging anticlines and synclines.

SOURCE: Howard, 1967.

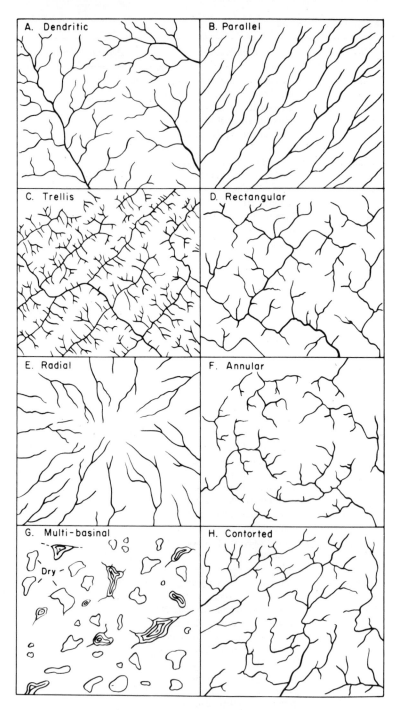

Figure 11-16. Basic drainage patterns. See Table 11-1. (Howard, 1967, Fig. 1.)

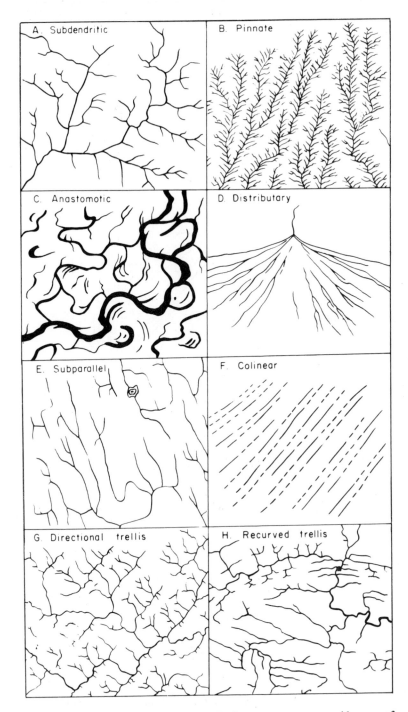

Figure 11-17. Modified basic patterns. Each pattern occurs on a wide range of scales. See Table 11-1. (Howard, 1967, Fig. 2.)

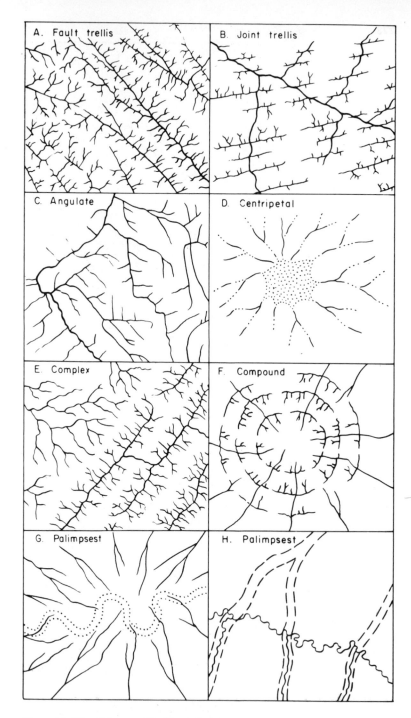

Figure 11-18. Other modified basic patterns (A–D). See Table 11-1. Complex, compound, and palimpsest patterns (E–H) are combinations or superpositions of basic or modified basic patterns. (Howard, 1967, Fig. 3.)

Basic drainage patterns (Fig. 11-16) follow regional structural trends and cover a wide range of scales. They are subjectively defined and transitional from one to another. Nevertheless, the basic and modified terms in Table 11-1 are part of the working vocabulary of geomorphologists and aerial photographic interpreters. They can be best studied by comparing Figures 11-16 through 11-18 with a selection of topographic maps of various scales.

Genetic Classification of Drainage Patterns

Consequent and Subsequent Rivers and Their Valleys.   River systems that develop as a direct result of, and in harmony with, a pre-existing land surface are said to be **consequent**. One may consider that the flow of water was a direct consequence of whatever slopes were initially present. (A less-common term for small valleys or gullies is **insequent**, shortened from "initial consequent.") Even newly emerged sea floors or fresh basalt plateaus are not likely to be devoid of initial relief (Fig. 12-7), and runoff will fill initial hollows to the lowest point on the rim and then drain down to the next hollow. Only in arid regions will the depressions on an undissected landscape remain free of lakes. Each basin joins the next lower one until an entire drainage network is established.

Initially, all runoff is consequent. As valleys are eroded, however, structural control may be exerted. Rivers obliquely crossing weak-rock belts tend to follow the zone of easy erosion before crossing the next resistant structure. Tributaries that expand into initial uplands underlain by regions of erodible rocks expand their drainage basins more rapidly than those that erode resistant rocks. Thus structurally controlled **subsequent** valleys evolve, which develop independent of and subsequent to the primary, or consequent, drainage (Fig. 11-19). Of the eight basic drainage patterns in Table 11-1, only dendritic, parallel, and radial are likely to be consequent. The others are all linked closely to control by structures. The contorted pattern of Figure 11-8 is a good example of subsequent drainage on a complex terrane, although disrupted somewhat by glacial erosion and deposition.

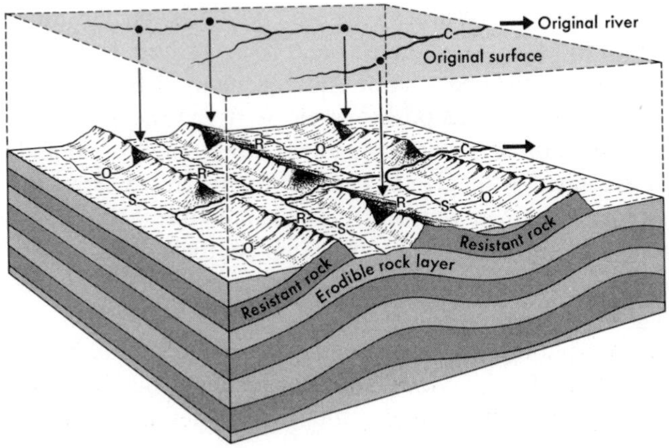

Figure 11-19. Expansion of a consequent valley system over an initial landscape, with a later stage of erosion when subsequent tributaries have become dominant. Letters C, S, R, and O signify consequent, subsequent, resequent, and obsequent rivers and valleys. Arrows project downward from valley segments of the consequent system that were defeated by resistant layers and stream capture to become wind gaps.

Two special categories of consequent streams are often confused. **Resequent** (a shortened form of the phrase "renewed consequent") and **obsequent** (a shortened form of the phrase "opposite to the consequent") streams are low-order tributaries that develop on new structurally controlled hillslopes as the result of expanding and deepening fluvial erosion. If, as a result of subsequent erosion along a weak-rock belt, new escarpments and dip slopes are formed, the gullies on these slopes are called either resequent or obsequent by their orientation with respect to the original consequent trunk stream. A resequent stream flows approximately parallel to and in the same direction as the trunk stream, and an obsequent stream flows in the opposite direction. The terms should not be applied with reference to the dip of the controlling strata, as is frequently and erroneously done. Figure 11-19 illustrates that resequent or obsequent streams can be either on an escarpment or on a dip slope.

The terms resequent and obsequent are troublesome, because their use requires prior knowledge of the direction of the original consequent drainage. It is possible that capture or some other diversion has reversed the direction of the trunk stream. The entire geomorphic history of a region must be studied carefully before the relatively trivial designations of resequent and obsequent can be applied. The words are also used with reference to fault-line scarps (Fig. 11-5) with similar connotation.

Antecedent, Superposed, and Captured Drainage.   In orogenic regions, the valleys that are first eroded over a landscape may later have new tectonic ridges rise across their trends. The rivers may be *defeated* (forced to abandon their valleys), or they may be able to maintain gorges across the land mass rising in their paths. Streams that have maintained their valleys across tectonic ridges are said to be **antecedent**, because the river is older than or antecedent to the deformation. Antecedent rivers and their valleys are common only in actively orogenic regions.

Yet another class of rivers and valleys are called **superposed**. The word is a contraction of "superimposed" and implies that a river system has been let down or laid over a landscape. Suppose that a land newly emerged from the sea has a thin cover of sediments that bury an older terrane of complexly folded rocks. On emergence, a consequent drainage system, perhaps with a dendritic pattern, might form. As the valleys deepen, they are eventually eroded into the old land under the covering strata. The dendritic pattern becomes superposed on the older rocks without regard to structural control. When the covering beds have been entirely removed by erosion, the only clue to superposition is the anomalous positions of the river valleys.

Superposition is extremely common in glaciated regions, where interglacial and preglacial drainage networks are filled with glacial drift, and the present streams almost randomly occupy alternating segments of old re-excavated valleys and new, postglacial valleys. Short gorgelike segments superposed on bedrock form scenic attractions in the Finger Lakes region of central New York and in the "Dells" region of southern Wisconsin, among other places.

A common problem in geomorphology is the distinguishing between antecedent and superposed rivers. In the Middle Rocky Mountains, for instance, major rivers cross mountain ranges in narrow gorges, although easy alternate routes are available around the ends of the ranges a few miles away. Early geologists thought the rivers were antecedent, implying that the mountain blocks had risen very recently, after the rivers had become established. Later work proved that a mid-Cenozoic interval of deep lacustrine and alluvial filling buried large parts of the mountains. Rivers like the Yellowstone, Bighorn, and Laramie formerly flowed over alluvium and were superposed on the resistant rocks of the buried ranges as the area was epeirogenically uplifted and dissected. The evidence for superposition is (1) remnants of former alluvium now on the mountainsides at suitable altitudes, and (2) the similarity of the alluvium on opposite sides of the ranges, both of which suggest a former continuous cover.

Stream capture (or "piracy") has been a popular concept of geomorphologists for generations. Either by chance or aided by a structural advantage, one stream may erode into the drainage basin of another and either progressively or abruptly divert, or capture, new tributaries. The trunk stream that is thereby deprived of a former watershed is said to have been **beheaded**.

Quite commonly, subsequent drainage systems capture segments of earlier consequent drainage. Figure 11-19 suggests how the process could occur, especially if the original drainage on cover strata had been superposed on a folded belt. As subsequent valleys expand along weak-rock belts, they progressively intersect the earlier consequent drainage, which by that stage would be a series of graded reaches alternating with gorges or **water gaps** across the resistant strata. After capture, the former course across the resistant ridges might be marked by **wind gaps**, notches that mark the former drainage line. Figure 11-19 shows only a single trunk stream that continues to drain the same region during the change from consequent to subsequent pattern. It is easy to visualize that in the competition between regional trunk streams, substantial areas of watershed might be permanently gained or lost by capture, however.

The effects of a capture on the hydraulic geometry of both the capturing and the beheaded system can be dramatic. Most commonly, the capturing stream has a relatively steep longitudinal gradient in a low-order tributary at the point of capture, whereas the defeated stream is likely to have a lower gradient. The sudden diversion of a greatly increased discharge through the capturing system causes rapid entrenchment, perhaps with a waterfall or rapids that migrate rapidly up the captured segment from the point of initial capture. Any sharp change in direction of a stream, downstream from an anomalously steep section, suggests an **elbow of capture**.

Another common mechanism of capture is provided by the unequal gradients of two parallel streams emerging from a mountain front. The one carrying the greater bed load may build an alluvial fan until its channel has been aggraded above the level of the adjacent stream into which it then easily diverts. The terraces on rivers at the foot of the Beartooth Mountains in Montana, described in the previous chapter (p. 250), exemplify this process.

Cenozoic orogeny has created many mountain belts that generally parallel the margins of landmasses (Fig. 11-14). During deformation, antecedent streams may maintain their valleys across fold belts, or *cross-axial* drainage may be superposed on the fold belts from thin cover strata. In either case, drainage networks are likely to develop with their long axes roughly normal to the coast, or transverse across orogenic belts (Oberlander, 1965). Later, as subsequent valleys deepen and lengthen, regional drainage networks are likely to develop trellis patterns, parallel to the coast. Their great length makes the long subsequent tributaries of a trellis system susceptible to capture by a short, high-gradient stream valley expanding inland from the coast if it happens to find a weak or narrow zone in the resistant ridge that separates it from the subsequent valley inland behind the ridge. If a capture occurs, the length of flow to the sea of the captured tributaries is shortened, and as the steep-gradient segment migrates upstream from the elbow of capture and smooths out, stream terraces may be left behind, graded toward the beheaded trunk stream rather than toward the capturing river.

Gilbert (1877) proposed a "law of equal declivities," which stated that if two streams of unequal longitudinal gradient drain the opposite sides of a ridge, the divide between them will shift laterally, or *migrate*, until their gradients are equal. If one of the streams is deepening its valley more rapidly than the other, by virtue of a structural advantage, the divide may shift laterally until the expanding stream entirely captures the drainage of the lesser competing stream. Possibly, the migrating divide may intersect a trunk stream, suddenly diverting a larger headwater region into the capturing stream. At that instant, the divide "leaps" abruptly to a new regional position. Slowly or abruptly migrating divides are part of the regional adjustments among drainage systems that result from progressive adjustments to structure as a landscape evolves by fluvial erosion.

REFERENCES

BAKER, V. R., 1973, Erosional forms and processes for catastrophic Pleistocene Missoula floods in eastern Washington, *in* Morisawa, M., ed., *Fluvial geomorphology:* State Univ. of New York Publications in Geomorphology, Binghamton, N.Y. pp. 123–48.

EGGLER, D. H., LARSON, E. E., and BRADLEY, W. C., 1969, Granites, grusses, and the Sherman erosion surface, southern Laramie Range, Wyoming: *Am. Jour. Sci.*, v. 267, pp. 510–22.

VON ENGELN, O. D., 1942, *Geomorphology:* The Macmillan Company, New York, 655 pp.

FEININGER, T., 1971, Chemical weathering and glacial erosion of crystalline rocks and the origin of till: *U.S. Geol. Survey Prof. Paper 750-C*, pp. 65–81.

FLINT, R. F., 1963, Altitude, lithology, and the Fall Zone in Connecticut: *Jour. Geology*, v. 71, pp. 683–97.

GILBERT, G. K., 1877, *Report on the geology of the Henry Mountains [Utah]:* U.S. Geog. Geol. Survey Rocky Mtn. Region (Powell), 160 pp.

GILLULY, J., 1949, Distribution of mountain building in geologic time: *Geol. Soc. America Bull.*, v. 60, pp. 561–90.

HOWARD, A. D., 1967, Drainage analysis in geologic interpretation: A summation: *Am. Assoc. Petroleum Geologists Bull.*, v. 51, pp. 2246–59.

HSU, T. L., 1971, Mélange occurrence in Cyprus: *Geol. Soc. China Proc. for 1970*, no. 14, pp. 155–64.

Hsü, K. J., 1968, Principles of mélanges and their bearing on the Franciscan-Knoxville paradox: *Geol. Soc. America Bull.*, v. 79, pp. 1063–74.

Maxwell, J. C., 1974, Anatomy of an orogen: *Geol. Soc. America Bull.*, v. 85, pp. 1195–1204.

McCraw, J. D., 1965, Landscapes of central Otago, *in* Lister, R. G., and Hargreaves, R. P., eds., Central Otago: *New Zealand Geog. Soc. Misc. Ser.*, no. 5, pp. 30–45.

Oberlander, T., 1965, The Zagros streams: A new interpretation of transverse drainage in an orogenic zone: *Syracuse Geog. Ser.*, no. 1, Syracuse Univ. Press., Syracuse, N.Y., 168 pp.

Rich, J. L., 1911, Gravel as a resistant rock: *Jour. Geology*, v. 19, pp. 492–506.

Thomas, M. F., 1968, Tor, *in* Fairbridge, R. W., ed., *Encyclopedia of geomorphology:* Reinhold Book Corporation, New York, pp. 1157–59.

Turner, F. J., 1952, "Gefugerelief" illustrated by schist tor topography, central Otago, New Zealand: *Am. Jour. Sci.*, v. 250, p. 802–7.

Ward, W. T., 1952, The tors of central Otago: *New Zealand Jour. Sci. and Technology*, ser. B., v. 33, pp. 191–200.

Wahrhaftig, C., 1965, Stepped topography of the southern Sierra Nevada, California: *Geol. Soc. America Bull.*, v. 76, pp. 1165–89.

CHAPTER 12

Landscape Evolution
by Fluvial Processes

DEDUCTIVE GEOMORPHOLOGY

Chapters 5 through 11 have been concerned with the destructive processes that change landscapes, the rates at which the processes work, and the observational evidence that landscapes differ from place to place because of the nature, intensity, and duration of the processes of change. The reasoning has been from specific examples toward statements of general principles. This is the *inductive* approach, and it is a vital part of the system of logical thinking we call science.

The other side of the logical coin of science is the *deductive* approach. Here, we reason from general principles toward analysis of specific problems. The deductive approach to geomorphology primarily concerns the changes of landscapes through time. We cannot watch a landscape evolve, even though we have abundant reasons to think that it does. In our deduction, we apply the principles we have learned from studying many places, each at a brief interval of time, to the prediction of events at a single place, during many successive time intervals. We wish to make a motion picture of a landscape's life history, but our source material is a series of still photographs of many different landscapes.

Biologists or philosophers might argue that a landscape is nonliving and therefore has no "life history." But in the same sense that one commonly speaks of the life of an automobile, an electric light bulb, or a book, life history is used here for the events that occur during the recognizable duration of a landscape. The initial form and time of origin of a landscape, as well as the conditions of its ultimate destruction, must be specified. Between these limits, it has a life history to be deduced.

The deductive study of landscapes has been severely criticized for having exceeded the restraints of observational evidence. It is true that in the first decades of this century the deductive system of "explanatory description" was pushed well beyond what had been proved experimentally. The great danger of deductive reasoning is that if the general principles are wrong or incorrectly applied, even the most careful logical procedures will inevitably lead to erroneous final deductions. Suppose, for instance, that a general principle had been established through repeated observations in humid regions (it has not) to the effect that hillside slopes decrease in steepness through time. If this principle were incorrectly used to predict the future shape of a desert slope, the deduced shape would have little relation to real landforms, because different processes of mass-wasting dominate in arid and in humid climates.

In recent decades, much new evidence has been accumulated; the repeated glaciations of lands in the middle and high latitudes were but one aspect of repeated general changes in climate during the late Cenozoic Era (Chap. 18). We do not know the causes of these climatic changes, but we recognize their effects on landscapes. Other evidence supports the theory that some continents have been carried across substantial arcs of latitude and into different climatic zones by the motions of lithospheric plates, even while the subaerial portions of the continents were being reduced by erosion.

Some geomorphologists feel that we should not attempt to deduce the advanced stage of landscape evolution under a certain climatic condition, because long before that advanced stage is reached, new climatic conditions will be imposed that will alter the course of landscape development. They forget that we recognize the effects of climatic change of landscapes only because the landscapes are not what we predict should have formed under the conditions we now observe. Figure 5-5 gives adequate evidence that more water annually falls on the land than evaporates from it and that "normal" or "typical" landscapes evolve under conditions of excess water runoff. In this chapter, the sequential evolution of landscapes under conditions of humid climate and abundant flowing rivers is given first importance. Subsequently, Chapters 13 through 18 evaluate the climatic controls of landscape evolution in detail, especially the impressive and widespread landscapes dominated by processes that are emphasized by aridity, perennially frozen ground, and glaciers. As in a musical composition, the complexities of the variations cannot be appreciated until the theme is known.

PROOFS THAT LANDSCAPES EVOLVE SEQUENTIALLY

No one who has seen a muddy river or a gullied hillside doubts that erosional processes consume the subaerial landscape. Nevertheless, the burden of proof rests on geomorphologists to demonstrate that landforms are produced, or that landscapes evolve, in a systematic fashion through time. As was noted in Chapter 1 (p. 8), one of the fundamental justifications for geomorphology is that changes can be predicted. There are at least six experimental or observational proofs that landscapes evolve in predictable sequences of forms. Some are more rigorous than others, but all are useful.

Small-Scale Experiments

Small-scale experiments on landscape evolution can be conducted on sand tables or other models. A groove scratched in a sloping sand surface conducts water and, as it does so, changes its shape to accommodate the discharge. A mound of fine sand under a spray mist develops drainage networks down its slopes. By careful choice of materials and experimental conditions, many landforms can be reproduced in miniature in a laboratory.

The difficulty with all scale models is that changes of dimensions in length, mass, and time do not alter the intrinsic properties of matter. Also, physical constants such as gravitation cannot be scaled down. For example, water in a model channel a few inches deep has the same density and viscosity as water in a real river; therefore, turbulence in the water due to these properties is grossly out of scale in the model. Furthermore, as the linear dimension of an object is decreased, the volume decreases as the cube of the length, but the surface area decreases only as the square of the length. Thus small particles have much larger surface areas in proportion to their masses than large particles of the same material. Surface effects can cause very fine-grained wetted particles in a model landscape to cling strongly together in a fashion totally inappropriate to the sand or gravel of the real landscape that they are meant to represent in scale. An experimental river channel on a sand table proves how river channels form on sand tables but proves very little more.

Real Landscapes Evolving Under Accelerated Conditions

When a former equilibrium is disturbed by natural or man-made disaster, new landforms may develop quickly. By observing the changes, we can infer the sequences to be expected when changes are much slower. Complex drainage networks may evolve on tidal mud flats during the few hours of each low tide (Fig. 9-5). After volcanic tephra falls, new drainage systems develop within a period of months or years. Everyone has seen gullies erode the artificial fill at construction sites. Larger gullies, up to 10 m deep and several hundred meters in length, have been known to develop in less than a century after land was cleared for agriculture (Fig. 9-2).

Accelerated erosion due to natural or man-made catastrophe is good proof that landscapes evolve, because the changes are full-scale except in time. Unfortunately, the "accident" that induces accelerated change commonly affects only one or a few processes of change. Accelerated erosion due to forest clearing and farming is caused by more rapid and concentrated runoff, but soil formation is not speeded up correspondingly, and mass-wasting may change from soil creep to slumping and earth flow. Thus the resulting landforms are not those that would have formed if all processes of change had been accelerated proportionally. Accelerated erosion cannot be directly extrapolated to interpret landscape evolution on the scale of geologic time.

Playfair's Law of Accordant Junctions, and Other Probabilistic Deductions

A third defense for the proposition that fluvial landscapes evolve sequentially is embodied in Playfair's law (p. 201) and more modern statistical studies of drainage networks (Chap. 9). Playfair's statement is not a rigorous proof of a natural law but only an observation about a highly probable condition. If valleys did not evolve by the work of the streams that flow in them, then the "nice adjustment of their declivities" would be a highly improbable state. Having considered the hydraulic geometry of streams and the concept of graded rivers, the average reader of this book is better equipped to appreciate the significance of Playfair's words than the greatest scholars of 1802.

However, the inherent problems of interpreting topologically random networks should be recalled (p. 204). An integrated system of river valleys can be explained by either headward growth and branching of a stream network or by progressive integration of small gullies and rills down an initial hillslope. Many computer simulations generate branching systems that look like drainage networks, yet the programs do not tell us how real networks have evolved or how they might change in the future.

Progressive Loss of Potential Energy in Fluvial Systems

As long as a river is carrying sediment to the sea, it is lowering the landscape that provides the gravity potential for flow. There can be no long-term steady state in a physical system of declining energy supply. We observe sediment in transport to the sea by fluvial systems and can prove that the sediment is derived from erosion of the landscape. These facts are powerful support for the assertion that landscapes evolve with time and, furthermore, that the rate of evolutionary change is likely to decrease with time as the energy of the system declines.

Just as the sediment in transport by rivers is a powerful proof that landscapes evolve, so is the alluvium that is temporarily left behind as river terraces during valley deepening (Chap. 10). Alluvial terraces have surface features and internal structure that prove they were once part of a river's flood plain. If they are not now reached by floods, either the valley has been deepened since they formed or there has been some diversion of water from the river. This proof of valley erosion is another of Playfair's contributions (1802, p. 103).

Climatic morphogenesis introduces a serious complication into deductions of sequential landscape evolution. Climatic change can cause a river to aggrade when it formerly was eroding (p. 247). River terraces that mark a climate-controlled aggradational phase might be misinterpreted as part of a continuous valley evolution. It has even been proposed (Cotton, 1963) that the "normal" system of landscape evolution, deduced for midlatitude cool–humid climates, is largely relict from a time of much colder climate with perennially frozen ground and intense frost action. We must learn to recognize the influence of climatic change on landforms, without losing sight of the fundamental observation that rivers carry by far the bulk of the eroded debris from the lands to the sea except under the most extreme climatic disruptions.

Landscapes Can Be Arranged in Series

The fifth method of proving that landscapes evolve is not based on rigorous logic but on pragmatic observation. It is still the best practical demonstration of landscape evolution. At the beginning of this chapter, it was noted that the conceptual goal is a motion picture of a single landscape evolving, but the source material is a series of still photographs of many landscapes. The fact that such a motion picture can be made is the most convincing evidence of its validity. We can assemble any group of photographs of eroding landscapes into progressive or transitional sequences, making due allowance for regional differences of climates and rock types. However, the ability to arrange landscapes in some sequential order does not tell us in which direction the order proceeds. Even if we make our motion picture, we do not know which way to run the film. For the establishment of a unidirectional series, we return to the preceding proof of landscape evolution, that rivers carry sediment. If landscapes form a developmental series, it must be in the direction of larger valleys and smaller residual hills.

The practical fact that landscapes can be sequentially arranged has a theoretical basis in thermodynamic principles. As noted in Chapter 9, the graded state results from a tendency toward equilibrium between opposing principles of least work and uniform distribution of work. As a landscape erodes, progressively better adjustment between opposing tendencies must result. In their discussion of river-channel morphology, Langbein and Leopold (1964, p. 793) made the significant inference that as rivers tend toward equilibrium between the two opposing tendencies, "deviations in time will approximate those which can be observed in an ensemble of places." Some conditions of landscapes are more probable than others, and nature favors tendencies toward the probable. The two most probable changes in landscapes are, first, that the rivers will become graded, and second, that their potential energy will decline through time. Both changes permit predictions to be made about the changes in appearance of individual landscapes with the passage of time.

It may seem a waste of time and intelligence to outline these first five proofs of landscape evolution, because they are all quite apparent. Yet only 200 years ago, the concept of slow, orderly development of landscapes under the same conditions that operate today was a challenge to the established religious and philosophical order. The acceptance of Hutton's concept of uniformity and continuity of process, and the corollary concept of the enormity of geologic time, set the intellectual stage for the theory of organic evolution in the mid-nineteenth century.

The Sixth Proof: Radiometrically Dated Landscapes

A few landscapes, such as those eroded from volcanic rocks or emerged coral reefs, can have their structure dated by radiometric means. Knowing the length of time that has elapsed since inception, a rate of denudation can be determined and used to estimate the survival time of the subaerial landscape. Knowing that a landscape has a finite life does not aid in deducing its appearance at various stages of the aging process, but by dating the rocks in landscapes of similar structure but of different ages, a precise chronology of the erosional stages can be made.

For example, some of the oldest volcanic rocks on the island of Hawaii have been dated by the potassium-argon method at about 0.7 million years old (Dalrymple, 1971). Most of the island has been built since then. The island is a type example of a basalt shield volcano, and the initial constructional shape is obvious. By contrast, the original shield volcano of Oahu, from which the Koolau Range is carved (Fig. 12-1), is now about one-half gone. The basalts of the Koolau Range are 2.5 million years old, but the younger Castle Volcanics, with an age of 0.8 million years, rest on the eroded surface near present sea level (Gramlich, et al., 1971). Thus the windward side of Oahu was eroded to low relief, and a mass equal to approximately one-half of the original subaerial shield was removed, in little over 1.5 million years. One could confidently predict that when the eruptive phase on the island of Hawaii has ended, the view after a million years or so will be like that of Figure 12-1.

Other datable erosional landscapes, such as dissected till plains of the last glaciation, surfaces covered by tephra of a known age, or coral reefs elevated above sea level (Chappell, 1974), are available to measure the actual progress of erosional dissection. They give a proof of landscape evolution that was not available to the great geomorphologists of earlier generations. It should be employed extensively in the future.

Figure 12-1. View northwest from the Nuuanu Pali, Honolulu, Hawaii. Erosion on the northeastern side of Oahu is rapid because of high rainfall brought by the trade winds.

RATES OF LANDSCAPE EVOLUTION

The progress of sequential evolution of landscapes traditionally has been described in relative, not absolute, terms. The words "youth," "maturity," "old age," and "senility" were applied to landforms by analogy to organisms, noting that the absolute interval of time required to attain maturity may vary by orders of magnitude among the various forms of life. The stages were defined only by observed criteria, not by time in years. There are a number of ways, however, by which denudation rates, or the volume of rock removed in a specified time interval, can be measured. By applying the appropriate rate to the total volume of a landscape, quantified estimates of the durability of subaerial landforms can be made.

Methods of Measuring Denudation Rates

Degradation and Exposure of Historical Monuments.   The stone monuments and walls of civilizations more than a few centuries old commonly show loss of surface detail and mass through weathering and erosion (Fig. 6-4). Marble tombstones in the humid northeastern United States become illegible after 150–175 years of exposure (Judson, 1968b, p. 357). Red sandstone monuments in cemeteries near Middletown, Connecticut, are weathering at a rate of about 6 mm per century (Matthias, 1967; Rahn, 1971), almost the same rate as the present average rate of landscape lowering for the entire United States. Carved and polished tombstones may not weather in the same way as naturally exposed rock, but they are exposed to equivalent weathering conditions and provide a good index of relative weathering by various lithologies.

In addition to losses of their own surface detail, ancient monuments and walls may mark the former ground level, so that regional denudation can be measured. Neolithic earthworks on the chalk hills of southern England have preserved the chalk from solution, while around the structure the land surface has been lowered at a rate of 10–12.5 cm/1000 yr (Pitty, 1971, p. 191). Soil erosion has exposed the foundations of Roman ruins and lowered the land surface adjacent to paved Roman roads by 20–50 cm/1000 yr in about the last 2000 years (Judson, 1968a). The post-Roman rates are high, probably as the direct consequence of human interference.

Sediment Loads of Rivers.   The most obvious way to measure denudation rates by fluvial processes is to monitor the amount of load a river carries from its watershed. Although simple in principle, the actual measurements are difficult. Only the suspended load of rivers is routinely measured. Dissolved load is less frequently recorded, and bed load has thus far defied direct measurement (Chap. 9). Corrections for volumetric changes during weathering must always be kept in mind, because much of the dissolved load is combined with atmospheric carbon dioxide or oxygen when in transport.

Careful estimates of river loads and regional denudation for the entire United States were first attempted by Dole and Stabler (1909). They concluded that the United States was being lowered at a rate of 1 inch in 760 years (3.3 cm/1000 yr, the most useful units for these data). Their analysis was very thorough, and their study is still widely quoted. However, a later study (Judson and Ritter, 1964) had much more data available, especially for dissolved loads. The solid load of Table 12-1 includes a bed load equal to an assumed 10 percent of the suspended load. The regional denudation rates in Table 12-1 vary from 17 cm/1000 yr lowering of the Colorado River Basin to 4 or 5 cm/1000 yr for Atlantic and Gulf Coast drainage basins. Judson (1968b) estimated that owing to erosion induced by agriculture the present average denudation rate for the United States of 6 cm/1000 yr is approximately double the precolonial rate.

Total sediment loads for most of the world's rivers are poorly known, but some major rivers at least have had their average annual load of suspended sediment estimated (Holeman, 1968). A few are listed in Table 12-2, with denudation rates calcu-

TABLE 12-1. RATES OF REGIONAL DENUDATION IN THE UNITED STATES*

Drainage Region	Drainage Area $(10^3 \ km^2)$	Runoff (m^3/sec)	Annual Load						Denudation Rate $(cm/1000 \ yr)$
			Dissolved		Solid†		Total		
			(tonnes/ km^2)	(%)	(tonnes/ km^2)	(%)	(tonnes/ km^2)		
Colorado	629	0.6	23	5.2	417	94.8	440		17
Pacific Slopes, California	303	2.3	36	14.7	209	85.3	245		9
Western Gulf	829	1.6	41	28.9	101	71.1	142		5
Mississippi	3238	17.5	39	29.2	94	70.8	133		5
S. Atlantic and E. Gulf	736	9.2	61	55.5	48	44.5	109		4
N. Atlantic	383	5.9	57	45.2	69	54.8	126		5
Columbia	679	9.8	57	56.4	44	43.6	101		4
Totals	6797	46.9	43	26.5	119	73.5	162		6

SOURCES: Judson and Ritter, 1964; Judson, 1968b.
*Great Basin, St. Lawrence, and Hudson Bay drainage are not included.
†Solid load includes assumed bed load equal to 10 percent of suspended load.

TABLE 12-2. SELECTED RIVERS OF THE WORLD RANKED BY DENUDATION RATES DUE TO SUSPENDED LOAD ONLY

River Name	Location	Drainage Area $(10^3 \ km^2)$	Ave. Annual Suspended Load $(10^3 \ tonnes)$	Denudation Rate $(cm/1000 \ yr)$
Ching (Yellow trib.)	China	57	408,000	271
Kosi (Ganges trib.)	India	62	172,000	106
Yellow	China	673	1,890,000	100
Ganges	India	957	1,452,000	53
Yangtze	China	1940	501,000	19
Colorado	USA	637	135,000	14
Mississippi	USA	3222	312,000	4
Amazon	Brazil	5780	363,000	2

SOURCES: Judson and Ritter, 1964; Holeman, 1968; and other sources.

lated from the suspended loads only. The tonnages of suspended load were converted to cm/1000 yr of denudation over the drainage area on the assumption that the density of the denuded rock is 2.65 g/cm³, a typical value for silicate and carbonate rocks. The actual denudation rates must be higher owing to the amount of combined bed load and dissolved load.

Among many other attempts to document regional denudation rates by fluvial erosion, one by Corbel (1959) is notable. Table 12-3 is his summary of fluvial denudation, classifying regions by climate and relief. Corbel attributed more denudation to chemical solution than most other authorities accept, but even so, his rates are comparable to those estimated by many others. The climatic factor in denudation rates is reviewed further in a later section of this chapter.

Abundant evidence has been presented in previous chapters that most of the sediment transported by rivers is derived from the slopes adjacent to first-order tributaries. The smaller and more headward the drainage basin, the larger the proportion of total stream load it supplies. In fact, the sediment load of a trunk stream never is as great as the sum of the loads from its tributaries. This anomaly may be a phenomenon of either climatic change or the explosive impact of humans on landscapes during the last few centuries and millenniums (Trimble, 1975, 1977). Alternatively, flood plains may naturally store alluvium much longer than has been suspected.

The rates of slope retreat by mass-wasting (p. 193), when converted to approximate denudation rates, imply regional lowering at a rate an order of magnitude more

TABLE 12-3. DENUDATION RATES FOR RIVERS IN VARIOUS CLIMATES

Regional Climate	Denudation Rate (cm/1000 yr)	Denudation By Solution (%)
Lowlands:		
Climate with cold winter	2.9	93
Intermediate maritime climate (Lower Rhine, Seine)	2.7	83
Hot–dry climate (Mediterranean; New Mexico)	1.2	10
Hot–moist climate with dry season	3.2	34
Equatorial climate (dense rainforest)	2.2	70
Mountains:		
Semihumid periglacial climate	60.4	34
Extreme nival climate (SE Alaska)	80.0	24
Climate of Mediterranean high mountain chains	44.9	18
Hot–dry climate (SW United States; Tunesia)	17.7	4
Hot–moist climate (Usumacinta, Mexico)	9.2	33

Source: Corbel, 1959.

rapid than the estimates based on stream sediment loads. The contradiction could be explained by hypothesizing that only about 10 percent of the area of a fluvial eroding landscape actually furnishes the stream load, and the other 90 percent is nearly immobile under colluvial or alluvial cover. Some evidence supports that hypothesis. In the Amazon Basin, for instance, about 85 percent of the total stream load is furnished by the 12 percent of the drainage area that is high in the Andean headwaters (Gibbs, 1967). River-water chemistry and submarine cores off the mouths of the Amazon provide no hint that the largest river on earth traverses an enormous rainforest lowland for most of its length.

The contradictions between present-day upland erosion rates and river sediment loads remain to be resolved. New hypotheses could lead to drastic revisions of deductions about the sequential evolution of fluvial landscapes.

Submarine Sedimentation Rates. Sediment in fluvial transport from a populated agricultural region may give an exaggerated measure of fluvial denudation. Because nearly all fluvial sediment is deposited on the continental margins, geophysical studies of the sedimentary thickness on the continental margins can be used to derive longer term rates of denudation. More assumptions are involved, including major uncertainties about the exact age of the sediments in the wedge under the continental margins, the area of continent that furnished the sediment, the contribution of marine biogenic sediments, volume and density changes during diagenesis, loss by uplift and recycling of previously deposited sediments, and many other factors. Representative papers on the subject are those by Menard (1961), Gilluly (1964), Curray and Moore (1971), and Mathews (1975). This approach is promising, because as seismic reflection profiles and offshore exploratory drilling records become more available, the record of continental denudation should become better known. Unconformities and abrupt changes in grain size in offshore sediments should be capable of correlation with significant changes in the geomorphic evolution of the adjacent landmass. Although the record of previous erosional episodes may be lost when a landscape is uplifted and vigorously eroded, the sediments deposited around the landmass should preserve that record.

Depth of Erosion Over a Known Interval of Time. In the arid to semiarid White Mountains of California, basalt lava flows with an age of 10.8 million years were erupted onto a dissected landscape of low relief. Later, the area was block faulted and eroded. Most of the present landscape lies 50 m or more below the base of the lava remnants. By drawing topographic profiles of the present landscape and projecting the base of the former lava flows above the profiles, cross-section areas can be obtained that, when divided by profile lengths, give the mean lowering of the landscape since the flow was removed. The average thickness of the basalt, 37 m, must be added to the average amount of lowering to give the total denudation since the flows stopped and dissection began. By this technique, a denudation rate of 1–3 cm/1000 yr was derived for the region (Marchand, 1971). The rates were probably greater on granitic igneous rocks than on basalt and dolomite in this cold, dry upland climate. Another study of the same region (LaMarche, 1968) concluded that the denudation rate during

Figure 12-2.   Bristlecone pines, White Mountains, Inyo County, California. Centers of roots extend about 60 cm above ground level, a measure of degradation while the tree has been growing. (Photo: V. C. LaMarche Jr., U.S. Geological Survey.)

the last few thousand years was an order of magnitude more rapid, because the roots of bristlecone pine trees as old as 4000 years have been progressively exposed by mass-wasting and erosion at rates of 15–120 cm/1000 yr (Fig. 12-2). Holocene climatic change may have intensified the denudation rates, however.

The foregoing studies, from the same region but varying greatly in the scales of time that were studied and the evidence that was used, have a common basis in that a reference horizon, either the base of a lava flow or the former ground level on roots of long-lived trees, was used to measure regional denudation over an extended interval of time. In Chapter 6, several other studies of reconstructed fold geometry, metamorphic facies, and geothermal heat flow were cited as proof that many landscapes are now developed on rocks that were formerly buried 20 km or more. Denudation rates derived from those studies are: 2 cm/1000 yr for 1 billion years of erosion in the Adirondack Mountains; 4 cm/1000 yr for 200 million years of erosion in the folded Appalachians, 10 cm/1000 yr for post-Cretaceous erosion of the Sierra Nevada, and 100 cm/1000 yr for 30 million years of Alpine erosion. The rates, although they are gross approximations that do not take into account any changes in rate of erosion over great intervals of geologic time, are rather similar to those derived by the methods previously described.

Papuan Erosion: A Mathematical Exercise

Only one study has been made of an eroding mountain mass where denudation rates were established for various altitudinal zones (Ruxton and McDougall, 1967). The study and some extended conclusions from it are reviewed here as an example of the usefulness of known denudation rates to geomorphic research.

The Hydrographers Range is a dissected andesitic composite cone in northeastern Papua New Guinea. Annual rainfall in the region probably ranges from 2250 mm at sea level to more than 3000 mm above an altitude of 1000 m. Most of the mountains are covered by tropical rainforest. Potassium-argon dates establish that the volcano ceased eruption about 650,000 years ago, when it began its erosional history. Most of the central summit portion is now dissected, and its initial shape is not known, but it probably was at an altitude of about 2000 m and had a considerable amount of constructional relief around multiple eruptive centers. Remnants of the youngest lava flows form the interfluves below an altitude of 975 m, and the predissection shape of the volcano can be reconstructed with confidence below that altitude. By measuring the areas of numerous valley cross sections at selected altitudes, the volume of rock eroded in 650,000 years was calculated. Using assumptions similar to those used in the previous example of the White Mountains in California, denudation rates were tabulated for each of a succession of reconstructed contour lines between 975 m and 60 m above sea level (Fig. 12-3). Above 1000 m, in the summit area, only rough estimates of the amount of erosion could be made. The denudation rates range from 52 cm/1000 yr above 900 m altitude to only 8 cm/1000 yr at 60 m above sea level. The

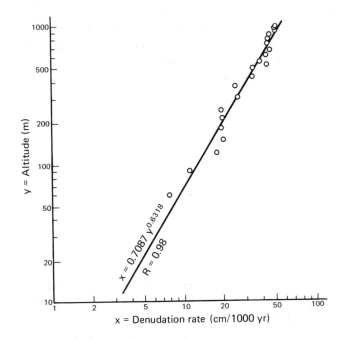

Figure 12-3. Denudation rate as a function of altitude, eastern Papua New Guinea. (From Ruxton and McDougall, 1967.)

best-fitting equation for the relationship between denudation rate and initial altitude is shown in Figure 12-3.

Because denudation rate is so strongly correlated with altitude in the range of 60–1000 m, it is possible to extrapolate the equation given in Figure 12-3 to 2000 m, the probable initial constructional height of the Hydrographers Range. By such extrapolation, the denudation rate near the original summit must have been about 86 cm/1000 yr, sufficient to lower the central region about 560 m since it was built. This extrapolation is close to the 490 m of summit-area lowering that can be roughly estimated from reconstructing the original cone-shaped profile. Both estimates suggest that the central part of the range has been lowered by fully one-quarter of its initial altitude in only 650,000 years.

The exponential equation (Fig. 12-3) that links denudation rate to altitude can be integrated between any successive altitudes to give the amount of time necessary to lower the range the specified amount. Figure 12-4 graphs the denudation history of the region but assumes an initial altitude of only 1000 m to avoid the uncertainties of the unknown configuration of the actual summit region. At the denudation rates reported by Ruxton and McDougall (1967), a mountain range could be lowered from 1000 m to 500 m in 1.1 million years, to 100 m in 2.8 million years, to 10 m in 4.0 million years, and to sea level in 4.9 million years. Although the denudation rate approaches zero as altitude approaches zero, the rate remains finite, and the time for total erosion of the volcanic terrane to sea level is also finite.

The denudation rates for the Hydrographers Range are not excessive as compared to other regions (compare the rates shown in Figure 12-3 with Tables 12-1 through 12-3), but the integrated time for denuding the entire region is unusually short. Judson and Ritter (1964) estimated that the entire United States could be removed to sea level in 11–12 million years if the present rates (Table 12-1) are maintained and if no isostatic or other tectonic uplift occurs. Schumm (1963) estimated that with corrections for isostatic recovery and decreased rates as denudation proceeds, regional denudation

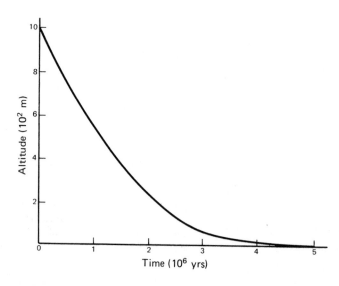

Figure 12-4. Time required to lower the summit of the Hydrographers Range, eastern Papua New Guinea, from an initial assumed altitude of 1000 m. Rates are shown on Figure 12-3.

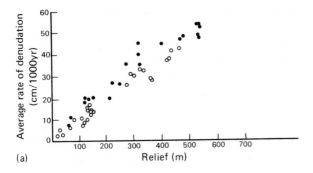

(a)

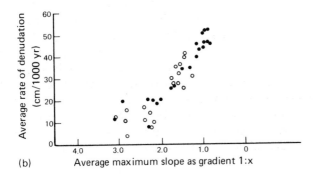

(b)

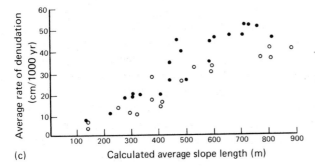

(c)

Figure 12-5. Relation between denudation rate and (*a*) local relief, (*b*) average maximum slope, and (*c*) average slope length in the Hydrographers Range. Data from more dissected sectors are shown as closed circles; data from less dissected sectors are shown as open circles. (From Ruxton and Mc-Dougall, 1967, Fig. 5.)

to near sea level might require a time interval of 15–110 million years. The apparent rapid time for total removal of the Hydrographers Range might be caused by the initial assumption that denudation rate is only a function of altitude and not of some other variable such as area of watershed above a certain altitude. As a mountainous area is reduced, the total orographic rainfall might decrease, even at low elevations.

Figure 12-5 gives additional evidence for inferring the sequential evolution of a mountainous terrain during denudation. In the Hydrographers Range, local relief (vertical difference between ridge crests and valley floors), slope gradients, and average slope length all decrease with decreasing denudation rates, and therefore also with altitude. Extrapolating in time, as the Hydrographers Range is lowered, its valleys will become broader and shallower and will develop decreased side-slope gradients. Highest areas are eroding most rapidly, so the original conic form should become lower and more dissected, and the interstream ridges should become lower and less steep-sided. After only a few million years, this constructional volcanic cone will become a low range of radiating ridges, separated by broad, shallow valleys. No mention is made in the original description of structural controls being exerted on the denudation rate by massive lava flows, dikes, or sills, but such controls could convert the initial radial pattern to an annular or centripetal pattern (Table 11-1).

Climatic Factors in Denudation Rates.   Table 12-3 gives some estimates of regional denudation rates characteristic of various climates in mountains and lowlands. As expected, mountainous regions erode an order of magnitude more rapidly, regardless of climate type. The figures offer impressive support for Powell's dictum (p. 84) that mountains are ephemeral landforms. Mountains with heavy snowfall seem to erode most rapidly of all, probably because freezing water and glaciers add to the erosional work of rivers. Cold, dry mountains with intense frost action, and mountains in seasonally wet and dry climates where rains cause seasonal floods, also show high denudation rates. Corbel (see Table 12-3) was especially impressed with the ability of cold water to carry a large dissolved load, but other authorities have been skeptical of the large percentages of total denudation he attributed to solution (p. 142). Studies of denudation rates in cold climates are complicated by the fact that many of the regions are only recently deglaciated, and landforms shaped primarily by glaciers are now out of equilibrium with fluvial processes. Under such conditions, rapid geomorphic change is inevitable (Mills, 1976). With these several limitations or qualifications in mind, Table 12-3 is a useful summary of opinion that mountains, especially snow-covered ones, may be denuded at rates approaching 1 m/1000 yr, whereas lowlands are eroded an order of magnitude less rapidly in any comparative climatic region.

Another way of comparing the climatic control of fluvial denudation rates is to compare the sediment yield per unit area of drainage basins of the same size but located in a variety of climates (Langbein and Schumm, 1958; Wilson, 1973). Figure 12-6 shows the sediment yields of 94 small watersheds, grouped into six categories of similar effective precipitation. (For this study, effective precipitation was defined as the annual precipitation required to produce a runoff equal to the known runoff of regions with a mean annual temperature of 10°C.) From the graph, it can be concluded that within the climatic range found in the conterminous United States, an effective precipitation of about 25–40 cm/yr, typical of the semiarid to subhumid grasslands, produces the maximum denudation rates. As annual precipitation decreases, less fluvial denudation occurs, because less water is available. As precipitation increases, vegetation becomes more effective in stabilizing slopes, and less denudation occurs in spite of increased runoff. Figure 12-6 is compatible with other observations that strongly seasonal precipitation, as in Mediterranean or savanna climates, maximizes erosion. However, the very high denudation rates on the heavily forested

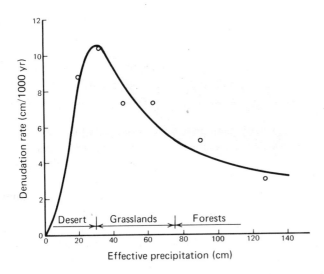

Figure 12-6. Variation of denudation rate with annual precipitation in the United States. Effective precipitation is defined as the precipitation necessary to produce a given amount of runoff. (After Langbein and Schumm, 1958.)

slopes of Papua New Guinea mountains are inconsistent with the decline of the denudation-rate curve in forested middle latitudes. Obviously, a forested mountain slope in the humid tropic erodes much more rapidly than a forested small drainage basin in the humid temperate United States. Other variables, such as seasonality and human activity, are surely involved.

Although opinions may conflict concerning denudation rates under moderate to high precipitation, all authors agree that as precipitation approaches zero, the fluvial denudation rate must also approach zero. With no net runoff from the land, no net lowering of the landscape can occur unless by wind. The lowering of mountains by denudation is neutralized or even exceeded by aggradation of desert basins with lower density sediments. Arid mountains are literally buried in their own detritus. Erosion to a rising base level of aggradation is the most important factor in the deduced erosional history of desert landscapes (Chap. 13).

RATES OF DENUDATION AND TECTONISM COMPARED

In Chapters 2 and 3, numerous examples were cited of orogenic and epeirogenic uplift increasing the height of landscapes by as much as a few millimeters per year, rarely even 10 mm/yr. The denudation rates reported in the preceding sections of this chapter should be compared with constructional rates in order to determine whether newly rising landscapes can outpace denudation and increase regional relief or whether landscapes are doomed to be kept low by erosion in spite of tectonic uplift. Most denudation rates for mountainous regions in various climates range from 10–100 cm/1000 yr, roughly 10 percent of the reported orogenic uplift rates. Denudation rates for large watersheds such as the Amazon or Mississippi are generally between 1 and 10 cm/1000 yr, again roughly an order of magnitude less than epeirogenic uplift rates.

It would appear that only extremely high mountain ranges might suffer denudation rates so great that their absolute elevation could not increase in spite of strong uplift. For example, an intriguing but unwarranted extrapolation of the best-fitted equation for Figure 12-3 gives a denudation rate of 233 cm/1000 yr for a 10,000 m mountain range in Papua New Guinea. The rate is consistent with the denudation rate of more than 100 cm/1000 yr for the Kosi River (Table 12-2), which drains the high parts of the Himalaya Mountains where the highest peaks are almost 10 km above sea level. Because of the exponential increase of denudation rate with altitude (Fig. 12-3), perhaps the Himalayas are roughly as high as any mountain range on earth can be, given the present effectiveness of destructive geomorphic processes. (The strength of rocks may also set a comparable limit to terrestrial relief. See p. 6.)

All of these estimates and extrapolations lead to the conclusion that if tectonic movements raise new land above the sea, the land is likely to increase in area and altitude much faster than erosion can destroy it. However, when tectonic uplift ceases, the landscape must be reduced toward sea level at a declining rate, perhaps approaching total reduction after an interval of 5–100 million years. The time scale seems vast only until it is compared with a scale of geologic time. Most of our present landscape has taken shape in only about 25 million years of late Cenozoic time. The few mountain ranges that are older than that are deeply dissected. A few landscapes in arid regions such as Australia and South Africa may preserve Paleogene or even Cretaceous landscape remnants in the form of laterite-capped mesas or plateaus, but these landscapes are judged to be the oldest on earth that are not exhumed from beneath a former cover.

SEQUENTIAL DEVELOPMENT OF FLUVIAL LANDSCAPES

The necessary foundation has now been laid for deducing the sequential development of a regional landscape under a climate that ensures year-round runoff by rivers that increase their discharge as they flow to the sea. In such a fluvial climate, continuous vegetational cover by grass and trees is assumed. It is necessary to deduce the development of individual river valleys and drainage basins prior to considering the regional landscape, because in fluvial landscapes networks of valleys are the basic landform units.

Sequential Development of River Valleys

As drainage networks expand over a subaerial landscape, most of the erosion is accomplished by the first-order gullies and streams (Fig. 9-1). Gradients are high, and as these streams grow headward and organize their subsidiary drainage basins, the landscape changes rapidly. New slopes shed waste directly into the first-order channels, which can deepen only if their streams can first move out the detritus supplied by mass-wasting from their own valley walls. A wavelike front of intense erosion sweeps inland across the old landscape as new drainage networks expand.

If we could watch a particular area as an outward expanding drainage network sweeps across it, we would see the initial slopes steepen toward the encroaching tributaries. The area would become a complex of small interfluves, side slopes, and gully floors. Then, as the network expands headward, the minor dissection would be consolidated into larger and less complex landforms, perhaps a single valley wall as a former first-order stream sends out its own headwater tributaries and becomes a higher-order trunk stream. At this stage, regional lowering of relief would become less rapid, although the local relief between interfluve and valley floor should continue to increase.

As a stream network expands, trunk streams and major tributaries change their functions from initial dissection to transportation. Downcutting may be deferred by an aggradation phase, as a former V-shaped gully is choked by sediment from its own expanding watershed. Quite early in the deduced sequence of events, we can expect the beginning of flood plains, because alluvium is temporarily or seasonally stored and weathered before it is removed downstream. Dams of alluvium may divert the streams against their banks to begin the process of spur trimming and valley-floor widening. By the time a stream has developed its own tributary system and is draining a watershed of 1 km², it is likely to be in a well-formed valley approximately 1 to 10 m deep, its low-water channel meandering or braiding around masses of poorly sorted colluvium and alluvium. Valley slopes will be at the angle of repose of the local bedrock or weathered debris and will extend down either to the channel banks or to fragments of flood plains.

Eventually, valley-side slopes will be trimmed back by stream erosion, and alluvium will be deposited on the valley floor so that the river flows between alluvial banks except at peak discharge. Midchannel bars may divert the main current to one or the other valley wall, so that for a time the stream has one bank of alluvium and one of bedrock. This process aids in valley widening and the development of a flood plain.

A significant event in the deduced sequential evolution of a stream valley is the development of a continuous flood plain. Depending on the grain size of the load and the complex interplay of hydraulic geometry variables, the stream channel will meander or braid in a layer of its own alluvium. At this stage, the mutual adjustment of channel shape, discharge, load, and slope becomes possible, and the stream is graded. Erosion is not completed, but from this time onward, the river must mutually or sequentially readjust the entire complex of variables in response to changing conditions. Local perturbations such as landslides are less likely to affect a graded river, because any debris from the valley wall will accumulate as a cone or sheet on the edges of the flood plain and may weather there for years before the stream begins to rework it. As a consequence, valley profiles will change from initially straight slopes directly to the stream banks to complex curved forms, blending above into convex summits of divides and below into concave wash slopes that in turn blend into flood plains (Fig. 8-18).

When streams carry increased discharge, their longitudinal gradients decrease (Chap. 9). Therefore, in an expanding fluvial network such as that being used as the model for this deduction, an initial high-gradient, first-order tributary or gully will be replaced in time by successively lower-gradient valley segments. A net regional lowering of altitude must result. If the example of the Hydrographers Range is valid, with lower altitude and decreasing denudation rate, local slope gradient, slope length, and local relief will also decrease (Fig. 12-5). Slower mass-wasting on more gentle slopes should permit more intense chemical weathering and thicker soils, with a corresponding reduction in the amount and grain size of sediment delivered to the river.

In general then, the changes we observe in a downstream direction along present river valleys are those we can deduce for the temporal changes at any point along the river valley: a reduction in down-valley gradient, progressive growth of a continuous flood plain, loss of waterfalls and rapids, progressive widening of the cross-valley profile by spur trimming and adjacent valley-side retreat, and progressive loss of local first-order tributaries draining adjacent slopes.

Sequential Evolution of Regional Fluvial Landscapes

A regional fluvial landscape is simply an assemblage of individual drainage systems. Whereas the deduced development of a single river valley can assume a continued headward growth of the drainage network, on the regional scale there must come a time when independent drainage systems intersect, interfere, or compete. Deductions about the sequential evolution of regional fluvial landscapes must be strongly conditioned by the restraint of a finite limit for basin expansion.

There is no generalization possible about the initial appearance of a regional landscape just beginning to be dissected by rivers. The favorite initial stage postulated by deductive geomorphologists is the former floor of a shallow sea, newly emerged above sea level. The assumptions are that the sea floor is smooth and featureless and that it emerges quickly relative to erosional processes. Figure 12-7 suggests such a landscape, with scale omitted. The figure could represent 1 or 100 km². Any other constructional landscape—volcanic, tectonic, or depositional—could as well be taken as the initial form. Equally, or perhaps more probably, a landscape already undergoing fluvial dissection becomes the initial form for a new sequence of changes. It is well to recall the quotation (p. 206) that a landscape is best considered to have a heritage rather than an origin.

Regional landscapes require additional deductions about the sequential development of divide areas between adjacent or opposed drainage networks. Initially, the

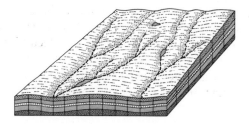

Figure 12-7. A hypothetical initial landscape, newly emerged from the sea. The initial surface is conformable to the sedimentary beds that underlie it. Drainage is consequent on the surface irregularities. (Based on drawings by W. M. Davis and C. A. Cotton.)

divide area is large relative to the expanding fluvial networks. As they meet, however, the old land is converted to a series of fragments isolated between vigorous headward streams. Drainage divides become more sharply defined at this stage, but whether the divides become knife-edged ridges between adjacent steep valleys or smooth, convex hilltops depends on complex interrelations between structure, climate, and stream gradients.

As river valleys dissect a landscape, they are influenced by many subtleties of structural control. Weak-rock belts develop subsequent stream valleys. Divides become localized on the most resistant rocks, although superposition from cover strata onto a resistant structure may introduce complexities into the deduced evolutionary patterns.

Eventually, an evolving region will consist of a series of divides, valley sides, and valley floors, all mutually adjusted and graded with respect to each other. The drainage density is determined by structure and climate; drainage patterns reflect structural control or lack of it; divides are on the most resistant rock type between adjacent trunk streams; and the highest terrain is the region farthest inland, at the headwaters of the largest streams. A critical stage in regional landscape evolution is the final elimination of whatever initial forms were present. In significance, this stage of complete upland dissection is analogous to the development of a continuous flood plain along a segment of stream valley. In no way can the two events be considered simultaneous, however, because flood plains are constantly under development along a valley segment somewhere along each river.

Local, Available, and Critical Relief

Deductions concerning the development of regional fluvial landscapes subsequent to complete upland dissection depend on the interaction between mean regional elevation above sea level and the drainage density. With high initial elevation or closely spaced streams, valley-side slopes are likely to intersect as narrow divides while the valleys are still being deepened. With low regional elevation or a coarse drainage pattern, individual rivers will have completed their initial phase of rapid deepening and will have become the trunk streams of expanding networks while large interfluve areas are as yet unaffected by the new dissection. Glock (1932) defined **available relief** as the vertical distance between the initial upland flats and graded valley floors of dominant rivers and argued that if the available relief exceeded the local relief (the vertical distance from ridge crests to adjacent stream channels), then total regional dissection of the oldland would be complete before flood plains began to accumulate on the local valley floors [Fig. 12-8(a)]. If the available relief were less than the local relief, then downcutting by trunk streams would be inhibited, because they become alluvial-lined and graded, and upland denudation would be very slow [Fig. 12-8(b)]. Glock envisioned a special case of **critical relief**, such that regional disappearance of relict terrain on the interfluves would exactly coincide with the development of graded segments in the adjacent valleys. The concepts of available and critical relief need redefinition to incorporate the factor of drainage density (Chap. 9), but the basic

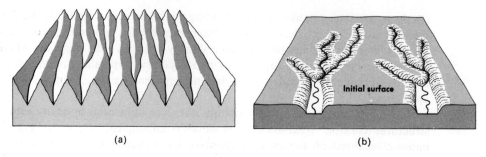

(a)   (b)

Figure 12-8. Complexities in comparing the deduced development of individual valleys with the development of the regional landscape. (*a*) With high available relief, upland is totally dissected but streams are vigorously downcutting. (*b*) With low available relief, trunk streams are not downcutting, and upland dissection is very slow.

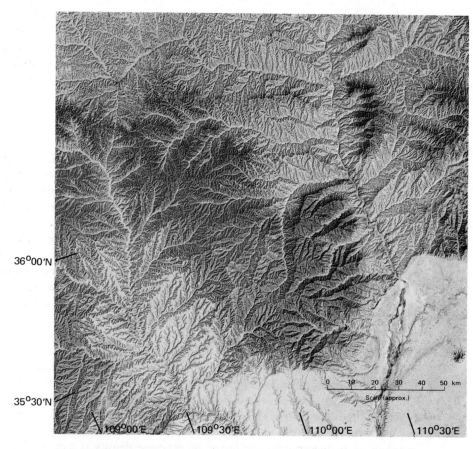

Figure 12-9. Complete regional fluvial dissection of loess, Shenhsi Province, China. (NASA ERTS E-1525-02455.)

concepts are fruitful nonetheless. With high available relief (or a high drainage density), regional landscapes will quickly be completely dissected to the stage called *ridge-and-ravine topography* by Hack (1960, p. 89). Subsequently, until alluvial flood plains extend upstream into the region, a long interval of time might pass, and a great amount of regional denudation might occur, without any significant sequential changes in the appearance of the landscape (Fig. 12-9). Competing streams might capture or be captured, resistant structures might develop as regional divides, and initially buried structures might be exposed and exert new controls on sequential development. But unless climate or structure change radically, the drainage pattern and local relief will seem "timeless." Not until regional denudation has brought the terrain low enough so that trunk streams into the region can become graded on alluvial plains can valley cross profiles soften, summits become rounded, and lower valley walls become concave wash slopes.

The Later Stages of Sequential Development

Accepting the premises that the initial constructional surface was at a finite altitude and that tectonic uplift eventually ceases, we can deduce the later stages of sequential landscape development. When graded, a river flood plain widens only slowly, because the meandering channel only rarely encounters the bedrock of the valley wall. The loss of local relief in the headwater regions will decrease the sediment load and probably the grain size of the load, so all the hydraulic variables of the stream system must continue to change, although in mutually compatible ways. The net effect must be a gradual lowering of the longitudinal valley profile as the river slowly swings laterally, trimming the valley flat beneath the flood plain. As relief is lowered, the slopes in interstream areas become more gentle. As flood plains become significant parts of the total landscape, the proportion of stream-dissected upland terrain must necessarily decrease.

When summit convexities are graded to lower slope concavities, which in turn are graded to flood plains, which are graded down-valley to the sea, what else can happen? Surely, as long as solar energy is supplied and sediment is exported, the landscape must continue to evolve. Local relief will decrease, slope angles will decrease, chemical weathering will increasingly dominate mechanical weathering, mass-wasting will become slower and slower, and each increment of change will take a longer time. The extrapolations provided by the denudation rates in the Hydrographers Range (Fig. 12-4) offer some clues to the later stages of landscape development. The exponential loss of relief with time is such that of the nearly 5 million years of subaerial denudation that is predicted for an initial elevation of 1000 m, nearly one-half of the time is required to remove the final 10 m of relief. Unless tectonic uplift or volcanism renews the available relief, fluvial landscapes seem doomed to a condition of low elevation above sea level and low local relief for most of their life histories. A corollary to Powell's dictum is that if mountains cannot long remain mountains, they are very likely to become low rounded hills, deeply mantled with weathered residuum and drained by low-gradient river networks that are thoroughly adjusted to the slight but still declining available energy.

PENEPLAINS AND OTHER DEDUCED ENDFORMS OF EROSION

The History of an Idea

During the nineteenth century, geomorphologists gradually accepted the ideas of Hutton and Playfair that vast amounts of subaerial denudation had preceded the shaping of the modern landscape. The enormity of geologic time was demonstrated by stratigraphy and faunal succession. The erosion of valleys by rivers and glaciers became an accepted uniformitarian concept, and the progressive changes of landscapes through extended time became the subject for geomorphic speculation and debate. By 1874, G. K. Gilbert could propose that the basin-range fault blocks were much younger than the folding and metamorphism of the rocks within the ranges, (p. 39), and in 1876, J. W. Powell used the phrase "the great denudation" for this erosion interval. Powell's dictum (p. 84), which has been noted repeatedly in this chapter, illustrates the attitudes of the time.

By 1883, W. M. Davis had begun development of his *geographical cycle*, or *cycle of erosion*, inductively generalizing from his own field work and that of his contemporaries that fluvial landscapes evolve in a predictable genetic sequence, which he characterized by analogy with the biological terms *youth*, *maturity*, and *old age* (Davis, 1885, 1899, 1902). Present readers should note that Davis wrote during the peak of intellectual enthusiasm for Darwinian evolution. Determinism was a popular philosophical scheme, and Davis's use of the term "cycle" implied not so much a steady-state condition but a directed sequence of events from an assumed initial state to a deduced end form. It was cyclic only by the assumption that the initial state could be recreated by tectonism or volcanism.

Davis assumed that rapid uplift initiated a "cycle" of erosion. The prevailing tectonic theory of his time called for worldwide pulses of orogeny followed by stillstand, so his assumption was readily accepted. After uplift, he deduced a stage of youth, when rivers rapidly deepened and extended their valley systems and regional landscapes were dissected. He chose as the criterion of maturity in valleys the development of free-swinging meanders in trunk streams, presumably on alluvial flood plains. For the regional landscape, total removal of the former land surface from interfluves was the primary criterion of maturity. Other criteria, such as total elimination of waterfalls and lakes from river valleys, sharp-crested divides, and high drainage density, were also deduced by Davis but are probably better omitted because they are usually determined by structure or climate.

Davis emphasized that old age is not separated from maturity by any specific criteria. As in organisms, where maturity is defined by the specific condition of reproductive capacity, in geomorphic analysis the twin criteria of meandering rivers and complete dissection are specific features that can be observed. Old age is less easily defined. In organisms it is only an extension of maturity, with a decrease in metabolic activity, loss of facilities, and an intangible decline of ability. As a criterion of old age in landscapes, Davis suggested that flood plains would become several times wider than the meander belt. No recent studies of fluvial mechanics give any evidence that meander belts develop higher-order "supermeanders" or that flood plains increase their width through time beyond the amplitude of the meanders. Also, Davis's old landscapes were deduced to be covered with broad sheets of alluvium over which the rivers sluggishly meandered. A more likely deduction is that low, degraded hills would form more of the regional terrain than Davis implied. The hills would be the drainage divides and would either be localized by structural controls or be at the maximum distance from the sea. As noted earlier (p. 287), the apparent contradictions between upland denudation rates and river sediment loads may lead, when resolved, to important new deductions about the later stages of erosional landscape evolution.

In an attempt to describe the form of landscapes that have undergone long-continued weathering and erosion in humid climates, Davis (1889) introduced the elegant word **peneplain**. He used as a root the word "plain," in the geographical context of a regional surface of very low relief near sea level. Realizing that the ultimate base level is the limit of subaerial erosion, which like a mathematical limit may be approached but might not be achieved, he prefixed to the word "plain" the Latin derivation "pene," meaning almost. Thus peneplain was introduced to the scientific literature as *a surface of regional extent, low local relief, and low absolute altitude, produced by long-continued fluvial erosion.*

How can we define a landscape that is "almost" gone? Apparently, the peneplain that W. M. Davis had in mind was far from a mathematical plane, or even a "peneplane," as some have proposed to spell the word. He envisioned an erosionally dissected landscape, one that includes the drainage networks of several master streams that enter the sea, in which the total relief is no more than about 100 m at the regional divides. One story is that he described a peneplain as a surface over which a horse could draw a carriage at a trot in any direction. In our modern era of excessive horsepower, it is hard to envision such a gently sloping surface of erosion, but it would not be planar.

Other Concepts of Sequential Development and Endforms

Davis's cycle of erosion made only passing reference to tectonic movements other than initial uplift followed by stillstand and erosion. He did allow for interruptions to the cycle by renewed intermittent uplift, but these were treated only as temporary disturbances that would initiate new downcutting and develop river terraces and compound valley forms. The most notable attempt at formulating a systematic geomorphology that related tectonic movement to erosion can be found in the writings of Walther Penck (Penck, 1924; English transl., 1953). His major works, written during a terminal illness and published posthumously in fragmentary form, are rightfully known for their complex and obscure German prose. Nevertheless, they attracted much attention among geomorphologists as an alternative or even a challenge to Davis's cyclic concept. Davis himself wrote a critical review of Penck's proposals (Davis, 1932) that became the primary source for most English-speaking geomorphologists. Unfortunately, certain errors of translation and interpretation in Davis's review have been misattributed to Penck ever since (Simons, 1962). The impact of Walther Penck's contributions on American geomorphology can be best judged by the published papers and discussions of a 1939 symposium on the subject (von Engeln, 1940). One of the better reviews of Penck's theories was by O. D. von Engeln (1942, pp. 256–68).

Penck accepted the idea that landscapes could be reduced to endforms of low relief but maintained that these endforms never became the initial forms for new episodes of dissection. By his theory, uplift was not rapid and then zero, as Davis supposed, but always began slowly, waxed to a maximum, and then waned gradually to a stop. During the long stage of initial slow uplift, all prior forms were destroyed, and a new surface of low relief, adjusted to a balance between uplift and degradation, developed. This initial surface of low relief was called **Primärrümpf** by Penck and was the basis for his deductions. Instead of hypothesizing sequential forms, he maintained that slopes, once developed, retreat parallel to themselves and persist until they intersect. During downcutting by rivers, such as might be caused by accelerating uplift, the initial slopes would be broad and gentle, but as the rate of uplift accelerated, inner valley slopes would be steeper. Because each increment of slope profile retreated from the valley axis and was followed by a lower slope element of steeper gradient, convex skyward slopes would be diagnostic of accelerated uplift. Similarly, straight slopes would indicate a constant rate of stream downcutting, and concave slopes would indicate a waning rate of entrenchment. Thus, as uplift died away, landscapes were theorized to be an assemblage of concave forms, to be converted to a *Primärrümpf* during the next episode of initial slow uplift.

No erosional sequence of forms was allowed by Penck's scheme, because each morphologic assemblage was related to a certain tectonic condition. His commendable primary goal was to use geomorphology to determine regional tectonic history. Penck denied any climatic control of geomorphic processes other than glaciation, believing that tectonics alone determined landform assemblages. Yet it is a curious fact that Penck's ideas of slope retreat have been found most analogous to processes that shape mountain fronts and pediments in semiarid climates (Chap. 13).

A later generation of geomorphologists have used the pediment landform as the basis for deductions about sequential evolution of landscapes. Maintaining that hillslopes retreat parallel to themselves and leave a residual low slope of transportation (the pediment) at their base, these authors propose that huge areas can be reduced to a series of coalescing pediments capped by residual rock knobs and that successive generations of coalescing pediments (or "*pediplains*," by analogy with peneplains) can migrate across a landscape, each one consuming higher and older surfaces by parallel scarp retreat, while in turn being consumed by the scarp retreat of a lower and younger generation of pediments (p. 194). No sequence of landforms is admitted by this theory either, because when formed, the pediments are essentially immutable, stable forms (King, 1953). The origin and significance of pediments and related landforms is considered in the next chapter, but it is worth noting that very large areas of the earth's surface are pediment-prone, and in the geologic past, before the evolution of vascular plants, the entire subaerial landscape may have been shaped by pediment-forming processes that we now associate with semiarid climates.

The emphasis on internal self-adjustment among complex variables in open physical systems, such as characterizes modern stream hydrology (Chap. 9), has led to a hypothesis of "steady-state" landscapes (Hack, 1960). Critics of Davis's cycle of erosion argue that when alluvial flood plains develop, even in small streams, the landscape is essentially graded and therefore deny the progressive development of graded master streams as a criterion of the stage of regional evolution. The only sequential change they envision is an early one from initial disequilibrium to subsequent equilibrium. Thereafter, as long as the energy in the open system remains constant, the landforms also remain constant, even though mass is removed from the eroding terrane. Hack (1960, p. 85) applied the name "dynamic equilibrium" to this concept of a time-independent, open-system landscape in a steady state. He regarded most landscapes that had previously been called "maturely dissected peneplains" as nothing more than erosionally graded steady-state surfaces that carry no connotation of sequential development. Only when the energy of the whole system ultimately declines, as when the available relief is consumed, will the landscape evolve to one of reduced local relief, decreased slope gradients, and blunted divides.

Hack's landscape in dynamic equilibrium could be accommodated in Davis's cyclic theory by assuming extreme available relief, during the reduction of which great volumes of rock must be removed by rivers whose spacing and gradient are climate- and structure-dependent and, therefore, time-independent. However, his avowed purpose was to explain landscapes in an essentially noncyclic conceptual framework, and Hack rightfully challenged some of the hypotheses of multiple residual peneplains that have been inferred from accordance of summits in the Appalachian region. Summit accordance and many other landscape features, such as gravel-covered interfluves and summit regions, can well be explained by long-continued erosion and progressive adjustment of structure. Multiple cycles of erosion, each one terminating in a peneplain that was uplifted and dissected, are not required.

One can very well agree with Hack that most landscapes are in an uneasy dynamic equilibrium between the energy available for work and the work being done. However, if regional elevation, and therefore the energy available for geomorphic change, continues to decline at a measurable rate, then no landscape can be said to represent an open system in a steady state. As in the analogy of a rotary cement kiln (Chap. 5), the landscape must be constantly evolving toward equilibrium with the declining energy. The major contrast between Hack's "steady-state" landscape and Davis's cycle is that Hack deduced total regional dissection and graded rivers as a nearly permanent condition rather than as the mature stage of an evolving series. The choice between these models depends on the relative rates of regional denudation and late Cenozoic tectonism. If, as in the example of the Hydrographers Range, total denudation in 5 million years can be inferred, evolving landforms can be deduced. If, however, tectonics and climatic change invalidate the assumption of initial uplift or other constructional process followed by stillstand and landscape evolution, then the dynamic equilibrium model, changing only from disequilibrium to equilibrium, is most suitable as a basis for interpreting the present landscape. Until late Cenozoic tectonic history and climatic change are better known, there seems to be room for multiple hypotheses in the interpretation of fluvially eroded landscapes.

REFERENCES

CHAPPELL, J., 1974, Geomorphology and evolution of small valleys in dated coral reef terraces, New Guinea: *Jour. Geology*, v. 82, pp. 795–812.

CORBEL, J., 1959, Vitesse de l'erosion: *Zeitschr. für Geomorph.*, v. 3, pp. 1–28.

COTTON, C. A., 1963, A new theory of the sculpture of middle-latitude landscapes: *New Zealand Jour. Geol. Geophys.*, v. 6, pp. 769–74.

CURRAY, J. R., and MOORE, D. G., 1971, Growth of the Bengal deep-sea fan and denudation in the Himalayas: *Geol. Soc. America Bull.*, v. 82, pp. 563–72.

DALRYMPLE, G. B., 1971, Potassium-argon ages from the Pololu Volcanic Series, Kohala Volcano, Hawaii: *Geol. Soc. America Bull.*, v. 82, pp. 1997–2000.

DAVIS, W. M., 1885, Geographic classification, illustrated by a study of plains, plateaus, and their derivatives (Abs.): *Am. Assoc. Adv. Sci. Proc.*, v. 33, pp. 428–32.

———, 1889, Topographical development of the Triassic formation of the Connecticut valley: *Am. Jour. Sci.*, v. 37, pp. 423–34.

———, 1899, The geographical cycle: *Geog. Jour.*, v. 14, pp. 481–504. (Reprinted 1954 in *Geographical essays:* Dover Publications, Inc., New York, pp. 249–78.)

———, 1902, Base-level, grade, and peneplain: *Jour. Geology*, v. 10, pp. 77–111. (Reprinted 1954 in *Geographical essays:* Dover Publications, Inc., New York, pp. 249–78.)

———, 1932, Piedmont benchlands and Primärrümpfe: *Geol. Soc. America Bull.*, v. 43, pp. 399–440.

DOLE, R. B., and STABLER, H., 1909, Denudation: *U.S. Geol. Survey Water-Supply Paper 234*, pp. 78–93.

VON ENGELN, O. D., 1940, Symposium: Walther Penck's contribution to geomorphology: *Assoc. Am. Geographers Annals*, v. 30, no. 4, pp. 219–84.

———, 1942, *Geomorphology:* The Macmillan Company, New York, 655 pp.

GIBBS, R. J., 1967, Geochemistry of the Amazon River system: Part I: *Geol. Soc. America Bull.*, v. 78, pp. 1203–32.

GILLULY, J., 1964, Atlantic sediments, erosion rates, and the evolution of the continental shelf: Some speculations: *Geol. Soc. America Bull.*, v. 75, pp. 483–92.

GLOCK, W. S., 1932, Available relief as a factor of control in the profile of a landform: *Jour. Geology*, v. 40, pp. 74–83.

GRAMLICH, J. W., LEWIS, V. A., and NAUGHTON, J. J., 1971, Potassium-argon dating of Holocene basalts of the Honolulu volcanic series: *Geol. Soc. America Bull.*, v. 82, pp. 1399–404.

HACK, J. T., 1960, Interpretation of erosional topography in humid temperate regions: *Am. Jour. Sci.*, v. 258-A, pp. 80–97.

HOLEMAN, J. N., 1968, Sediment yield of major rivers of the world: *Water Resources Res.*, v. 4, pp. 787–97.

JUDSON, S., 1968a, Erosion rates near Rome, Italy: *Science*, v. 160, pp. 1444–46.

———, 1968b, Erosion of the land, or what's happening to our continents?: *Am. Scientist*, v. 56, pp. 356–74.

JUDSON, S., and RITTER, D. F., 1964, Rates of regional denudation in the United States: *Jour. Geophys. Res.*, v. 69, pp. 3395–401.

KING, L. C., 1953, Canons of landscape evolution: *Geol. Soc. America Bull.*, v. 64, pp. 721–52 (discussion and replies, v. 66, pp. 1205–14).

LaMARCHE, V. C., 1968, Rates of slope degradation as determined from botanic evidence, White Mountains, California: *U.S. Geol. Survey Prof. Paper 352-I*, pp. 341–77.

LANGBEIN, W. B., and LEOPOLD, L. B., 1964, Quasi-equilibrium states in channel morphology: *Am. Jour. Sci.*, v. 262, pp. 782–94.

LANGBEIN, W. B., and SCHUMM, S. A., 1958, Yield of sediment in relation to mean annual precipitation: *Am. Geophys. Un. Trans.*, v. 39, pp. 1076–84.

MARCHAND, D. E., 1971, Rates and modes of denudation, White Mountains, eastern California: *Am. Jour. Sci.*, v. 270, pp. 109–35.

MATHEWS, W. H., 1975, Cenozoic erosion and erosion surfaces of eastern North America: *Am. Jour. Sci.*, v. 275, pp. 818–24.

MATTHIAS, G. F., 1967, Weathering rates of Portland arkose tombstones: *Jour. Geol. Education*, v. 15, pp. 140–44.

MENARD, H. W., 1961, Some rates of regional erosion: *Jour. Geology*, v. 69, pp. 154–61.

MILLS, H. H., 1976, Estimated erosion rates on Mount Rainier, Washington: *Geology*, v. 4, pp. 401–6.

PENCK, W., 1924, *Die Morphologische Analyse:* J. Engelhorn's Nachfolger, Stuttgart, 283 pp.

———, 1953, *Morphological analysis of landforms* (transl. by H. Czech and K. C. Boswell): St. Martin's Press, New York, 429 pp.

PITTY, A. F., 1971, *Introduction to geomorphology:* Methuen & Co., Ltd., London, 526 pp.

PLAYFAIR, J., 1802, *Illustrations of the Huttonian theory of the earth:* Dover Publications, Inc., New York, 528 pp. [Facsimile reprint, 1964.]

RAHN, P. H., 1971, Weathering of tombstones and its relationship to the topography of New England: *Jour. Geol. Education*, v. 19, pp. 112–18.

RUXTON, B. P., and McDOUGALL, I., 1967, Denudation rates in NE Papua from K-Ar dating of lavas: *Am. Jour. Sci.*, v. 265, pp. 545–61.

SCHUMM, S. A., 1963, The disparity between present rates of denudation and orogeny: *U.S. Geol. Survey Prof. Paper 454-H*, 13 pp.

SIMONS, M., 1962, The morphological analysis of landforms: A new review of the work of Walter Penck: *Inst. Brit. Geographers Trans.*, no. 31, pp. 1–14.

TRIMBLE, S. W., 1975, Denudation studies: Can we assume stream steady state?: *Science*, v. 188, pp. 1207–8.

———, 1977, The fallacy of stream equilibrium in contemporary denudation studies: *Am. Jour. Sci.*, v. 277, pp. 876–87.

WILSON, L., 1973, Variations in mean annual sediment yield as a function of mean annual precipitation: *Am. Jour. Sci.*, v. 273, pp. 335–49.

PART III

Climatic Morphogenesis

A **morphogenetic system** is "the assemblage of processes or agencies locally or regionally operating under a prevailing climatic condition and combining to produce a particular type of landscape [Cotton, 1958, p. 125; see also Tricart and Cailleux, 1972]."[1] A **morphogenetic region** is then characterized by certain climatic factors that produce a distinct combination of geomorphic processes, which in turn may produce a landscape distinct in appearance from other regions (Birot, 1968). The concept of morphogenetic regions dominated by processes subordinates the factors of structure and time in geomorphology. Followed to the extreme conclusion, it suggests that each climatic region produces a unique assemblage of landforms independent of time and structure. No geomorphologist would accept such an extreme view, but within the science there is abundant room for debate about the relative importance of climatically controlled landforms and, for instance, structurally controlled landforms.

The dominance of fluvial processes on the subaerial part of this planet is the justification for giving those processes such emphasis in Part II. In some regions, however, water has never been known to flow. Sandy deserts may show only landforms shaped by wind; many glaciated terrains show evidence of ice erosion alone. Even within regions dominated by fluvial processes, the range of climatic variables, such as temperature and the amount and seasonality of precipitation, validates the study of climatic morphogenesis.

W. M. Davis (1905a, 1905b) considered the cool, humid, midlatitude climate as "normal" and wrote at length about two "climatic accidents" that could affect landscapes: glaciation and aridity. We now realize that subtropical arid belts and

[1]For citations, see References, Chapter 13.

307

polar icecaps are very much a part of the planetary climatic zonation and are not accidents. Even more significantly, the theory of plate tectonics provides a mechanism to move continents long distances over the earth's surface during the Cenozoic Era. In the last 55 million years, for instance, Antarctica and Australia are believed to have separated by the present 30° of latitude. Probably the Antarctic ice cap and the dominantly arid landscape of Australia are both Cenozoic phenomena that are best explained by plate tectonics rather than by climatic "accidents" (Kennett, et. al., 1974) (p. 407).

Even without considering the perplexing problems of past climatic change and plate motions, the present diversity of terrestrial climates is impressive and significant to geomorphology. The following table shows that a significant percentage of land area falls in each of five major climatic regions. According to other compilations, about 10 percent of the land area is now ice-covered, 20 percent is underlain by perennially frozen ground, and 33 percent is characterized by drainage that does not reach the sea. Unique combinations of processes are imposed on each landscape.

APPROXIMATE PROPORTIONS OF THE LAND SURFACE
FALLING WITHIN KÖPPEN'S VARIOUS TYPES AT THE
PRESENT TIME

Climatic Group	Percentage of Land Surface
A. Tropical rain climates	20
B. Arid and semiarid climates	26
C. (Warm) temperate rain climates	16
D. Boreal forest and snow climates	21
E. Cold snow climates (treeless)	17

SOURCE: Modified from Lamb, 1972, p. 513.

Not so obvious is the length of time available for the present morphogenetic systems to have shaped regional landscapes. In many regions, such as the 20 percent of the land area that was glacial as recently as 15,000 to 20,000 years ago but is now ice free, most of the landforms are the work of former glaciers and periglacial processes, not of the rivers that now flow there. Many, perhaps most, landscapes are *palimpsests* (p. 9) of forms shaped during different climatic conditions. In areas of former glaciation, relict and contemporary forms are relatively easy to distinguish, because glaciation imposes totally new processes on fluvial landscapes. Quaternary climatic changes in the lower latitudes apparently changed only the relative intensity of processes, not their kind. Areas now perennially humid became seasonally dry and *vice versa*. One of the next great challenges to geomorphologists is to recognize the late Cenozoic sequence and succession of processes and landforms in the tropics and subtropics (Tricart, 1972). Because most of the developing nations of the world are in these regions, the geomorphic information can be applied to problems of water supply, land use, and mineral resources.

The six chapters of Part III build on the processes and deductions of Part II but reorganize the information with systematic emphasis on processes that are unique to or uniquely intense in certain climatic regions. For each region, an inquiry is made as to whether the assemblage of landforms is also unique, that is, whether the concept of morphogenetic regions applies only to processes or to landforms as well.

CHAPTER 13

Arid, Semiarid,
and Savanna Landscapes

DRY CLIMATES

The largest single identifiable morphogenetic region on earth is the dry region, where either seasonal or annual precipitation is insufficient to maintain vegetational cover and permit perennial streams to flow. Climatologists have devised various empirical formulas to define dry climates in terms of the ratio of precipitation to evapotranspiration. The temperature of a region must be incorporated into such formulas, because it is a major factor in evapotranspiration, but otherwise temperature is not a diagnostic parameter of arid climates. For instance, the Köppen system (Köppen and Geiger, 1936) defined the boundary between semiarid and humid climates with winter rains by the relationship,

$$P \leq 20T$$

where P is the annual precipitation in millimeters and T is the mean annual temperature in Celsius degrees. The truly arid (desert) climate was defined as one in which precipitation is only one-half as great as the amount that locally separates the semiarid from the humid climate, or,

$$P \leq 10T$$

Köppen and his assistants calculated other similar empirical formulas for climates in which precipitation is nonseasonal or is concentrated in the summer. Although apparently related only to mean annual precipitation and evaporation, Köppen's

Figure 13-1. Dry regions of the world. (After Meigs, 1953.) Areas with mean annual temperature < 10°C are excluded. Definitions of degrees of aridity are discussed in the text.

Extremely arid regions

Arid regions

Semiarid regions

Sandy deserts

choice of boundaries was primarily based on vegetation. The parameters that he chose to define aridity also define areas in which insufficient moisture is available to support continuous vegetational cover of the ground. By Köppen's definitions, 26 percent of the land area is dry (See table, p. 308).

In another climatic classification, Thornthwaite (1948) attempted to correlate potential evapotranspiration to other climatic variables, such as temperature and length of day, and thereby to more closely define the limits of dry climates as expressed in plant growth. His formulas apply primarily to the United States, where the minimum annual precipitation required for a humid climate increases southward from about 500 mm in North Dakota to about 750 mm in Texas. Less than 250 mm of annual precipitation produces a desert in almost any temperature range. By Thornthwaite's definitions, fully 36 percent of the land area is dry, even excluding the high-latitude frozen deserts.

The most recent attempt at classifying and delimiting arid regions was by Meigs (1952, 1953; Cooke and Warren, 1973). He used Thornthwaite's definitions but remapped areas based on better climatic data. Meigs' map (Fig. 13-1) shows 34 percent of the land area to be dry, including 4 percent defined as extremely arid, where no precipitation has fallen for at least 12 consecutive months at least once in the period of record and where rainfall lacks a seasonal rhythm.

By these various definitions, from one-quarter to more than one-third of our land is dry. Recent estimates by the United Nations expand the area of aridity to 43 percent of the land, owing in part to overgrazing and other unwise agricultural practices. Although not yet scientifically documented, these estimates suggest that aridity is expanding rapidly in our generation and emphasize the significance of the arid and semiarid morphogenetic regions.

In low latitudes, two belts of aridity coincide with the subtropical anticyclonic belts of high atmospheric pressure about 15° to 30° north and south of the equator. They are fringed by relatively narrow belts of semiarid transitional climates. In middle latitudes, dry climates are localized in the interior of the large continents. Temperature and evaporation are not as high in these regions as in the subtropics. Some precipitation may fall as snow. Midlatitude dry climates are characterized by relatively large areas of semiaridity around smaller cores of true desert. Small areas of dry climate also are on the west sides of the continents, where cold ocean currents flow offshore, and on the downwind, or rain shadow, side of mountain ranges. The polar deserts are excluded from consideration in this chapter, because they are largely ice- or snow-covered.

It should be noted that when climatologists attempt to define dry climates, they use the criterion of discontinuous vegetative cover. However defined, in arid regions erosion is not inhibited by a continuous cover of plants. Furthermore, rivers flow only seasonally at best, and even when flowing, they decrease in discharge downstream, in contrast to the streams in humid regions. Moisture deficiency causes ground water to move very slowly and upward rather than downward in the soil, with the result that secondary minerals accumulate in soil profiles, and soil structure is strongly affected. For the geomorphologist, discontinuous flowing rivers, the lack

of ground cover, and the accumulation of secondary minerals in soil profiles define
the dry morphogenetic region. Each has a significant effect on landscape evolution.

The tropical savanna climate is dry for a significant part of each year, even
though total rainfall exceeds the definitions used to define the dry regions by climatologists. In the savanna climate, which covers about 15 percent of the earth's
land area, precipitation is strongly seasonal, usually coming during the hottest season.
Toward the equator, the savanna grades into tropical rainforest, and poleward it
grades into the subtropical dry belts. The savanna is a region of sparse open wood-
lands, thorny shrubs, and grass. Most photographs of savanna landscapes convey
the impression of aridity, because most are taken during the dry season when the
grass is brown and many trees have shed their leaves. During the wet months, however, very heavy rainfalls are common, and vast areas of savanna plains are under
standing water. The year-round high temperatures and seasonally abundant water
promote intense chemical weathering, even though rivers flow only during the rainy
season and for a brief interval thereafter.

GEOMORPHIC PROCESSES IN DRY CLIMATES

The geomorphic significance of dry climates begins with the vegetation. Semiarid
climates usually support a sparse grassland or **steppe** vegetation. In the United States,
the boundary between humid and semiarid climates is approximated by the transition
westward from medium-height grasses with a continuous turf or sod in the humid
regions to short, shallow-rooted bunch grasses on otherwise bare ground in the
semiarid regions. In arid regions, even the bunch grasses disappear, and the vegetation is, at best, widely spaced shrubs and salt-tolerant bushes. Wind then assumes
importance as an agent of erosion and transportation (Chap. 14).

In savanna regions, the seasonal rejuvenation of plant growth provides a ground
cover during the rainy season, but barren ground is widespread during the dry months.
Fires are common during the dry season; some are started by natural causes, but
many are started by people in order to clear the land or to drive wild game. Some
biogeographers suspect that the entire area of savanna grassland is due to human
activity. If so, this important morphogenetic region is unique to the later Cenozoic
Era.

Soils and Weathering

With a deficiency of water, dry soils do not develop strong diagnostic horizons
except for salt crusts or concretionary layers. The Mollisols of the humid temperate
grasslands (Table 6-1) give way to the Vertisols, Aridisols, or Entisols of semiarid
and arid regions. The definitive climatic criterion—precipitation less than evaporation
—means that if soil moisture is available by lateral infiltration, as for instance in
a desert basin flanked by more humid mountain slopes, then the water is drawn

upward through the soil by capillarity and evaporates either in the soil profile or at the ground surface. Dissolved mineral matter is precipitated when the soil water evaporates. Calcium carbonate (**caliche**), gypsum, or alkali (sodium and potassium carbonates) may accumulate in desert soils in this manner. Soils develop a strong columnar structure. No humus layer develops under the sparse vegetation.

Steppe soils have better defined horizons than desert soils, but nevertheless they are poorly developed and thin as compared to soils in humid regions. Evaporation is rarely so extreme that the more soluble salts in the soil water are precipitated, but relatively insoluble calcium carbonate may form nodules or layers in the soil profile.

Savanna soils are as yet little studied. The seasonal wet-and-dry climate, continuously warm, probably causes the maximum development of Ultisols and Oxisols (Table 6-1). Lateritic weathering (p. 131; Fig. 6-9) extends to a depth of 100 m or more, even though the seasonal runoff is not adequate to remove the weathered debris. The intensity of chemical weathering sets the savanna region apart from the arid and semiarid regions as a morphogenetic unit.

The general rate of arid-climate weathering is very slow. The rare rainfalls and night dew apparently provide the moisture for hydrolysis and hydration. The puzzling fractured stones of hot deserts were described on p. 109. Mechanical weathering becomes relatively important in dry regions as compared to chemical weathering, but many researchers believe that even in the driest deserts, chemical decomposition involving water dominates over mechanical disintegration.

Mass-Wasting and Erosion

Soil creep in deserts is not the same as in humid climates, primarily because the surface layers are not laced together with a turf of interconnecting roots. Also, desert soils are likely to be broken into prisms or weakly indurated blocks by vertical dessication cracks, so that creeping masses are blocks rather than granular layers. Hillsides are likely to be a series of small scarps and benches, even in colluvium. Taluses are obvious at the bases of cliffs. Summit convexities develop on easily weathered rocks such as shale, but resistant strata, joints, dikes, and other structural units are clearly exposed.

In semiarid regions, slopes are at least partly stabilized by vegetation, usually grass. The association of plains (*steppe, prairie, pampas,* and *veld* are a few of the regional terms) with semiarid grasslands is notable and may be genetic. In hills or mountains, trees can be sheltered from sun, winds, and fire, but open plains offer no such protection. Fire, burning uninhibited across plains, may be the controlling factor of their vegetational patterns, even more important than climate (Wells, 1970).

Precipitation inadequate for trees is also typically erratic and variable from season to season or year to year. Semiarid regions are frequently cursed by droughts and floods, sometimes in the same year. The rate of landscape denudation is rapid under these conditions. Some geomorphologists believe that mountains in the semiarid climate have the most rapid rates of denudation found anywhere on earth (Fig. 12-6). Others, however, suspect that denudation in the tropical rainforest or the periglacial zone is even more rapid (Tables 12-1 through 12-3).

Stream channels in arid and semiarid regions are likely to be flat floored, wide (to carry massive bed loads), and steep walled (because of the strong vertical or columnar jointing in dry soils). Their longitudinal profiles are steep. When dry, which is most of the time, they are floored with coarse alluvium. In Spanish-speaking regions, these distinctive channels are called **arroyos**. The equivalent Arabic term **wadi** (or **ouadi** in French) is widely used in North Africa and the Middle East.

The intensely seasonal precipitation of the tropical savanna briefly causes massive bed-load transport in braided channels that are dry for most of the year. Hillslopes are often bare joint-controlled rock, washed free of weathered debris. The savanna landscape has a unique assemblage of landforms that is described later in this chapter.

LANDFORMS OF DESERT REGIONS

Desert Plains and Plateaus

By far the greatest proportion of the world's arid regions are monotonous plains and plateaus. There is no compelling reason why dry regions should be nonorogenic, but nonorogenic regions are likely to form the large interior parts of continents where moist winds cannot penetrate, and nonorogenic regions also lack the relief necessary for orographic precipitation. In addition, internal drainage characteristic or even diagnostic of deserts causes alluviation that buries minor relief. Wind has been suggested as a leveling agent of desert plains as well, but evidence presented in the next chapter suggests that wind is more likely to increase desert relief by the excavation of basins and construction of major dune landforms.

Where alluvial fans merge in a desert basin, a saline lake (**playa** or **sebkha**) may form. Here, mud and salts accumulate to form the most level topographic surfaces developed on earth. Water less than a meter in depth may briefly accumulate, but when the water has evaporated, a great flat of polygonal-cracked mud or salt is all that remains. Residual brines may ooze from beneath the encrusted surface of the playas, too concentrated to evaporate further but liquid enough to engulf motor vehicles that are driven onto the surface. On other playas, hard crusts of salt and mud provide nearly perfect surfaces for high-speed racing cars.

Dust and sand are swept from desert fans and flood plains, leaving behind a lag concentrate of gravel called a **reg**, or stony desert. A **desert pavement** may form a stable surface of closely fitted pebbles, each polished by sand blast and often faceted as well. Approximately one-fourth of the Sahara and the Arabian desert are sandy (Chap. 14). The remainder is either **hammada** (barren rock) or reg. Rock ledges are pitted and polished by abrasive windborne sand, but the dominant relief is fluvial. Even in the driest desert plains, drainage networks cover the landscape. Some are used by present runoff, even if rain falls less than once in 12 months, but an unknown portion of the desert dissection is probably due to formerly greater runoff. The issue of climatic change (Chap. 18) is of major significance to theories of landscape evolution in arid regions. We do not yet know enough about either present processes or Quaternary climatic change in the dry regions to confidently distinguish between active and relict landforms.

Mountainous Deserts

The desert Basin and Range province of the southwestern United States and Mexico, the Pacific coastal desert of Chile and Peru, and the fold mountains adjacent to the Persian Gulf (Fig. 11-14) are examples of desert mountains (Fig. 13-2). Their area is small as compared to the continent-sized deserts of Australia and the Sahara and the great desert plains of Arabia, trans-Caspian USSR, and western China.

Mountains in deserts are likely to be islands of nondesert climate rather than part of the desert. Orographic precipitation and sometimes snowmelt feed streams that dissect mountains in deserts much as they dissect mountains in other climatic zones. The contrast in morphogenesis is to be noted at the base of desert mountains, where the streams discharge from the mountains onto the surrounding dry *piedmont* ("foot-of-the-mountain") slopes.

Figure 13-2. Death Valley, California, from Dante's View. Note saline playa at the base of massive alluvial fans and the exposed regolith in the foreground. (Photo: U.S. Borax Co.)

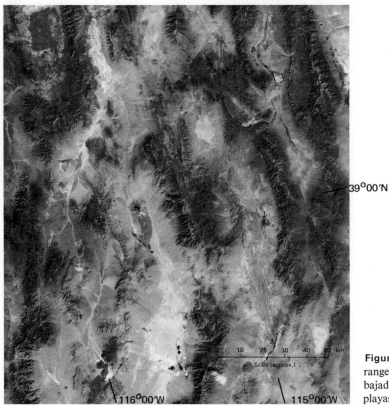

Figure 13-3. Satellite image of basin ranges in east-central Nevada. Massive bajadas converge on salt-encrusted playas. (NASA ERTS E-1071-17540.)

The dominant landform of intermontane deserts must be the alluvial fan. Ephemeral rivers decrease in discharge as they enter desert basins and deposit their entire load as they evaporate or infiltrate. It is no exaggeration to say that desert mountains are progressively buried in their own waste. The great apron of coalescing fans at a mountain front is called a **bajada** (or bahada).

The alluvial fans of Death Valley (Fig. 13-2) have been analyzed as equilibrium, or steady-state, landforms (Denny, 1965, 1967). Each fan has an area equal to one-third to one-half of the mountain watershed that feeds it. Fans grow until their surface area becomes so large that the rate of sediment supply from the mountain is balanced by the rate of erosion of the fan surface by gullies that originate on the fan itself. If the catchment area is large and the depositional basin is small, fans will comprise nearly the entire piedmont area (Fig. 13-3). If, however, small mountainous regions are shedding detritus onto a large lowland, fans may form only on the lower slopes, separated from the mountain fronts by eroded bedrock surfaces. Frequent shifts in distributary channels and capture of aggrading, sediment-transporting channels by eroding fan channels ensure the symmetric evolution of fan forms. Segmented fans, with progressively steeper straight-line radial profiles toward the fan apices, have been attributed to intermittent tectonic uplift of the mountain front (Bull, 1964).

The best-known mountainous desert is the Sonoran region of the southwestern United States and adjacent Mexico. Much of the area is graded to sea level by intermittent streams and by tributaries to the lower Colorado River and Gila River. Although the region appears to consist of mountains rising above broad alluvial plains or fans, much of the intermontane surface is of planed rock, similar to that which composes the mountains. An estimated 20 percent of the region consists of mountains, 40 percent of barren or thinly veneered bedrock plains, and 40 percent of thick alluvium (McGee, 1897, p. 91). Early explorers were surprised to find that the gently sloping piedmont surfaces 5–8 km from the mountain front were cut across hard granitic rocks, with only a thin layer of alluvium on the surface: "so thin that it may be shifted by a single great storm (McGee, 1897, p. 91)." The wide extent of barren piedmont plains in the Sonoran district is probably permitted by the external drainage, which exports alluvium to the sea rather than permitting it to accumulate in closed intermontane basins.

The significance and extent of such rock surfaces in the Sonoran desert and elsewhere has led to extensive further study. Bryan (1922, p. 88) formally adopted the elegant term **mountain pediment** for:

> A plain which lies at the foot of mountains in an arid region or in headwater basins within a mountain mass. The name is applied because the plain appears to be a pediment upon which the mountain stands. A mountain pediment is formed by erosion and deposition of streams, usually of the ephemeral type, and is covered with a veneer of gravel in transit from higher to lower levels. It simulates the form of an alluvial fan.

In classic architecture, a pediment is the triangular end of a low-pitched gable roof, on which a frieze is often carved. The term *mountain pediment* was chosen because in profile, when viewed from a distance, each mountain peak seems to stand at the crest of a low conical plain, triangular in profile (Fig. 13-4). Today, the form is referred to simply as a **pediment**.

Figure 13-4. Mountains near Maricopa, Arizona, surmounting a pediment. The boulder-controlled mountain front has an abrupt junction angle with the pediment. (Photo: R. E. Moeller.)

The slope on pediments, as on alluvial fans, increases toward the mountain front. Typical gradients range from 1°–7°. Headward portions of pediments may penetrate into a mountain mass, isolating segments of the range and possibly intersecting with pediments from the other side of the range at *pediment passes* (Howard, 1942). The pediment form is an excellent example of a water-spreading wash slope (Fig. 8-19). Functionally, it is an ideal surface to distribute and dissipate a mass of water and sediment introduced at its apex. By original definition and conventional use, the pediment is a landform of dry regions. However, some authors have unfortunately extended the term to include all water-eroded plains.

As first used by Gilbert (1882, p. 183), the term pediment was casually applied to the surfaces of the alluvial fans that encircle desert mountains. Bryan's definitive study, following McGee's observations in the same general area, emphasized that pediments are slopes of transportation cut on bedrock, usually covered with a veneer of alluvium in transit from high to lower levels. In form and function, a pediment is similar to an alluvial fan, the difference being that a pediment is an erosional landform and a fan is constructional.

The origin of pediments is closely related to morphogenetic processes of dry regions. McGee's dramatic eye-witness account of a sheetflood (McGee, 1897, pp. 100–101) was especially influential in establishing the sheetflood as an important geomorphic process.

> The shower passed in a few minutes and the sun reappeared, rapidly drying the ground to whiteness. Within half an hour a roar was heard in the foothills, rapidly increasing in volume; the teamster was startled, and set out along the road up the valley at best speed; but before he had gone 100 yards the flood was about him. The water was thick with mud, slimy with foam and loaded with twigs, dead leaflets and other flotsam; it was seen up and down the road several hundred yards in either direction or fully half a mile in all, covering the entire surface on both sides of the road, save a few islands protected by exceptionally large mesquite clumps at their upper ends. The torrent advanced at race-horse speed at first, but, slowing rapidly, died out in irregular lobes not more than a quarter of a mile below the road; yet, though so broad and tumultuous, it was nowhere more than about 18 inches and generally only 8–12 inches in depth, ... For perhaps five minutes the sheetflood maintained its vigor, and even seemed to augment in volume; the next five minutes it held its own in the interior, though the advance of the frontal wave slackened and at length ceased; then the torrent began to disappear at the margin, the flow grew feeble in the interior, the water shrank and vanished from the margin up the slope nearly as rapidly as it had advanced, and in half an hour from the advent of the flood the ground was again whitening in the sun, save in a few depressions where muddy puddles still lingered.

McGee's description led to a general assumption that sheetfloods erode weathered bedrock surfaces to the perfection of pediments. Other geomorphologists proposed that lateral corrasion by streams (or **streamfloods**, episodic high intensity discharges confined to channels) is the dominant process of pediment formation (Johnson, 1932).

The controversies between proponents of these and other hypotheses of pediment evolution were summarized by Hadley (1967).

Field observations of desert storms are sufficiently rare that processes of erosion and deposition are poorly known. Rahn (1967) described examples of flood activity on piedmont slopes in Arizona and concluded that the floods on pediments were streamfloods, confined to channels but exhibiting high velocity and capable of impressive erosion and deposition. Only when rain fell directly onto bajadas were sheetfloods produced. Rahn concluded that pediments are likely to be "born dissected" by channeled flow near mountain fronts and later smoothed by weathering and sheetfloods as the mountain front retreats.

Apparently, pediments widen by the lateral migration of rills and ephemeral channels that periodically impinge on the mountain front and undercut taluses and cliffs. They also expand by the normal mass-wasting retreat of cliffs and taluses and by abrasion of the rock-cut pediment surface by sheetfloods. The relative importance of the three processes seems to vary from region to region. Analysis of these details is difficult, because all erosion is slow in dry regions, and some pediments probably formed under a different climate and are now being slowly dissected. A strong contemporary opinion is that many, if not all, pediments were first formed by more humid morphogenetic processes, were buried, and are now being exhumed (Tuan, 1959; Oberlander, 1974). Evidence of climatic change in presently dry climates is reviewed in Chapter 18.

As pediments expand into the mountain slopes, the lower part of the pediments in turn may be progressively buried by the rising bajada alluvium (Fig. 13-5). For

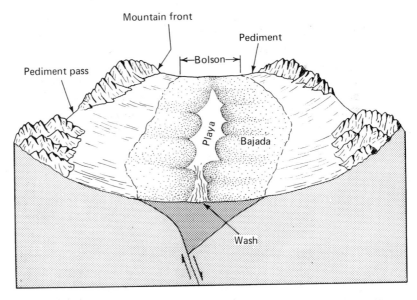

Figure 13-5. Typical assemblage of mountainous desert landforms.

this reason, pediments in truly arid regions, where the local base level is a playa, are generally only narrow rock-cut fringes between the mountain fronts and the bajada. The pediments plunge downslope beneath the thickening alluvial fill toward the center of the **bolson**, or intermontane basin. By those who insist on precise definitions, the smoothly graded wash slope at such a mountain front would be called a *pediment* in its upper part, where the entire thickness of alluvium is in transit, and a *bajada* in its lower part, where the alluvium has permanently come to rest in the bolson. The boundary between pediment and bajada is then defined by the thickness of alluvium that can be periodically reworked by fluvial processes.

The geometry of desert basins and mountains is such that as a basin is filled, its surface area increases, whereas the area of the adjacent mountain decreases by erosion. Therefore, successive increments of mass eroded from the mountain cause a decreasing rate of vertical accretion in the receiving basin and the development of a convex-skyward buried bedrock surface, even though the piedmont surface is always concave-upward (Fig. 13-6). Lawson (1915) referred to the convex buried rock surface as the **suballuvial bench** and to the concave exposed portion as the *subaerial bench*. The latter term is synonymous with pediment. Wells on the margins of many bolsons in the Great Basin demonstrate that their subsurface geometry is as Lawson inferred. The original fault scarp of the basin-range structure is now far out under the bajada, and large portions of pediments truncate older sedimentary formations or indurated fan gravels. Figure 13-6 illustrates that a slight lowering of base level by tectonism or climatic change could exhume or regrade large areas of a pediment.

The most extensive pediments are in regions that have sufficient precipitation to allow at least occasional direct surface runoff to the sea. Under these conditions, which are technically semiarid rather than arid, weathered sediment can be transported to the sea, instead of filling bolsons and progressively burying the lower edges of developing pediments. The Sonoran Desert in Arizona and in adjacent Mexico, the classic region of pediment development in North America, has pediments, alluvium-floored plains, and river channels all graded to sea level in the Gulf of California. Water rarely flows in the channels. The entire Sonoran landscape may be relict from a time of greater precipitation.

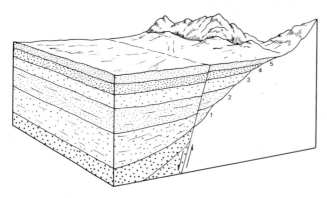

Figure 13-6. Lawson's theory for the origin of a convex-upward suballuvial rock bench. Equal volumes of alluvium cause successively smaller increments of aggradation in the widening bolson. Minor continued faulting demonstrates that the original fault-block boundary is now well away from the mountain front. Vertical scale is exaggerated.

The interior plateaus of large parts of Africa are reported to be great, gently sloping pediments, and large parts of the dry continent, Australia, are also pediment prone. The arid interior of Asia is less well known than other dry regions, but there, too, pediments can be predicted to be among the dominant landforms. Considering that 26–36 percent of the land area is arid or semiarid, the pediment may be the most common erosional landform on earth.

If there is a landscape characteristic of the arid morphogenetic region, it should consist of barren rock or gravel-armored plains scarred by braided ephemeral channels and arroyos. Eolian processes produce deflation basins and dune fields (Chap. 14). Mountains are fringed with cliffs, taluses, pediments, fans, and playas. Large areas of the earth do have such landform assemblages, and the arid zone is one of the more widely accepted morphogenetic regions. Even so, the degree to which the forms were shaped by processes that were formerly more intense has yet to be resolved. Quaternary climates formerly more humid than now can be demonstrated in many desert regions (Chap. 18).

LANDFORMS OF SEMIARID REGIONS

The semiarid *steppes*, *prairies*, *veld*, and *pampas* of the world are characterized by: (1) grass cover; (2) annual precipitation up to twice the amount that locally defines the desert boundary; and (3) rivers that are graded externally to the region, usually to the sea. Slopes and drainage networks are integrated regionally, and erosional landforms evolve toward sea level rather than the rising base level of a closed basin.

Steppe regions are generally plains or low dissected plateaus. Large areas of grassland, such as the Great Plains of North America and the Pampas of Argentina, are the constructional surfaces of huge piedmont alluvial fans (Fig. 10-11). In other regions, especially Australia and southern Africa, grassy plains are largely erosional surfaces. Pediments are presumed to be the dominant erosional landforms of semiarid regions.

No grasslands are very old, because grass is believed to have evolved only in the Miocene Epoch At that time, climatic change (p. 407) and plant evolution produced a new ecosystem dominated by grass. Herbivorous mammals, perhaps including hominids, evolved rapidly in response. One can only speculate about the appearance of the world's plains in pre-Miocene time.

SAVANNA LANDFORMS

The savanna landscape of Africa was first made known through the work of German geomorphologists early in this century (Bornhardt, 1900; Passarge, 1928). They called it an *Inselberglandschaft*, or **inselberg landscape**, with reference to the steep-sided, isolated hills and mountains of barren rock that rise like islands from

Figure 13-7.   Castle Hill, Townsville, northern Queensland, Australia. A coastal inselberg fringed by thin colluvium and alluvium.

nearby flat plains (Fig. 13-7). During the wet months, wide areas of the plains are flooded, and some of the hills are truly islands.

The term *inselberg* subsequently came to be used for the residual knobs on pediments in arid and semiarid regions, and the distinction between morphogenetic processes in truly dry regions and the savanna wet-and-dry region became blurred. In an effort to restore precision, Willis (1934; King, 1948) proposed that the barren, domelike "sugar-loaf" hills without taluses so common in savanna regions should be called *bornhardts*, after the pioneer geomorphologist. Unfortunately, this proposal substituted a distinct morphologic form, unique to massive rocks and structurally controlled by sheeting joints, for the original concept of a morphogenetic landform, controlled by processes of the savanna climate. This usage of "bornhardt" should be abandoned and the word **inselberg** restored to its original morphogenetic sense.

General agreement now has been reached that inselbergs are truly the result of savanna morphogenetic processes (Thomas, 1974). With high year-round temperatures and heavy seasonal rainfall, chemical weathering extends to great depths (Fig. 6-10). With only seasonal runoff, however, the savanna-zone rivers cannot remove the massive sediment loads furnished by weathering. Rather than entrenching their valleys, rivers braid or flood across extremely flat **wash plains**. It has also been proposed that deep tropical weathering so thoroughly reduces rocks to clays that no abrasive tools are available for river corrasion (Büdel, 1957; Cotton, 1961), and therefore savanna rivers cannot deepen their valleys even though they may have considerable available relief. The result of rapid chemical weathering and inhibited stream erosion is that wash plains are underlain by deep zones of altered rock. Although the ground surface is a monotonous plain of alluvium and weathered bedrock, at a depth of perhaps 100 m an irregular surface is in the process of formation at a "weathering front," perhaps in the zone of seasonally fluctuating water table. Masses of fresh rock that are bypassed by the deep weathering are eventually exhumed as tors or inselbergs (Figs. 11-4, 13-7, and 13-9). The huge size of some savanna inselbergs leads to the conviction that they are of considerable age, possibly having survived several episodes of rejuvenation and lowering of the adjacent wash plains. When the weathering front has bypassed a monolith, its preservation and eventual expression as an inselberg seem assured. Thus structural control and climate combine to produce one of the more striking landscapes of the earth. The process has been referred to as a "double surface of leveling" (*"doppelten Einebnungsflächen"*) by Büdel (1957). Figure 13-8 outlines Büdel's hypothesis of inselberg development.

In the savanna climate, fractures and other structural controls are the only obvious ways of creating irregularities on the weathering front and potential inselbergs. Thomas (1974, p. 196) listed 12 lines of evidence to support Büdel's hypothesis, including a description of a large excavation in the weathered layer that artificially exposed a typical inselberg. It is possible that purely chance events at the surface, such as a landslide, might strip the regolith off an emerging rock mass, allow it to be washed clean, and thereby preserve it from future deep chemical weathering. The development of a benched landscape on the Sierra Nevada without structural control has been attributed to similar randomness of exposure (Wahrhaftig, 1965).

Some inselbergs are encircled by shallow moats at their bases, probably due to concentrated weathering by runoff from the inselberg (Twidale, 1968, pp. 105–24;

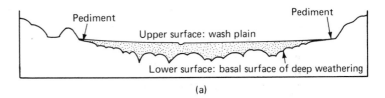

(a)

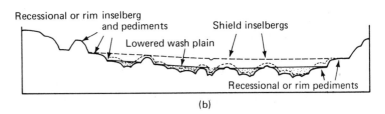

(b)

Figure 13-8. Hypothesis of a "double surface of leveling" in the tropical savanna. (*a*) Wash plain of seasonal flooding is as much as 100 m above the weathering front. Pediments fringe the wash plain. (*b*) Wash plain is lowered by rejuvenation or climatic change. Inselbergs and marginal pediments are exhumed or regraded to the lowered wash plain. (From Büdel, 1957, Figs. 5 and 7.)

Figure 13-9.   Steepened margin of a granite inselberg caused by multiple episodes of subsurface weathering and subsequent exposure of the weathering front.
Pildappa Hill, South Australia. (Photo: C. R. Twidale.)

1976, pp. 7–17; Twidale and Bourne, 1975). As wash plains are lowered, these moats
and weathered notches in the flanks of steep-sided inselbergs (Fig. 13-9) are exhumed
to record former ground levels. Other inselbergs have narrow pediments around their
bases (Fig. 13-8), which led to the confusing extension of the word *inselberg* to residual
hills that are surrounded by much larger pediments in other types of dry regions.
The savanna pediments are narrow and of low gradient, 3° being the maximum
reported in Ghana (Clayton, 1956). The minor development of pediments and the
great extent and flatness of wash plains distinguish the savanna or inselberg landscape
from the pedimented landscapes of dry regions. The contrast is due to the intensity
of chemical weathering in the savanna climate, by which exposed rock is so weakened
that it is easily eroded. The detritus is seasonally distributed across the wash plains.

There seems to be adequate evidence to suspect that the inselberg landscape
defined by Bornhardt is a valid example of a morphogenetic landscape unique to
the savanna climate. Certainly its best expression is in the seasonally wet-and-dry
tropics. On the wetter, equatorial side of the savanna regions, the wash plains become
forest covered, stabilized, and eroded by perennial streams. On the dry poleward
side of the savanna belt, desert landforms replace the inselbergs and wash plains,
because chemical weathering becomes less effective. However, both climatic boundaries
of the savanna zone have shifted during the Quaternary, and some undetermined
amount of savanna landscape is found relict in both adjacent regions (Chap. 18).

REFERENCES

BIROT, P., 1968, *Cycle of erosion in different climates* (transl. by C. I. Jackson and K. M. Clayton): Univ. California Press, Berkeley and Los Angeles, 144 pp.

BORNHARDT, W., 1900, *Zur Oberflächengestaltung und Geologie Deutsch-Ostafrikas:* D. Reimer, Berlin, 595 pp.

BRYAN, K., 1922, Erosion and sedimentation in the Papago country, Arizona: *U.S. Geol. Survey Bull. 730-B*, pp. 19–90.

BÜDEL, J., 1957, Die "doppelten Einebnungsflächen" in den feuchten Tropen: *Zeitschr. für Geomorph.*, v. 1, pp. 201–28.

BULL, W. B., 1964, Geomorphology of segmented alluvial fans in western Fresno County, California: *U.S. Geol. Survey Prof. Paper 352-E*, pp. 89–129.

CLAYTON, R. W., 1956, Linear depressions (Bergfüssniederungen) in savanna landscapes: *Geog. Studies*, v. 3, pp. 102–26.

COOKE, R. V., and WARREN, A., 1973, *Geomorphology in deserts:* Univ. of California Press, Berkeley, 394 pp.

COTTON, C. A., 1958, Alternating Pleistocene morphogenetic systems: *Geol. Mag.*, v. 45, pp. 125–36.

————, 1961, Theory of savanna planation: *Geography*, v. 46, pp. 89–101.

DAVIS, W. M., 1905a, Complications of the geographical cycle: *Internat. Geog. Cong., 8th, Washington, 1904, Repts.*, pp. 150–63. (Reprinted 1954 in *Geographical essays:* Dover Publications, Inc., New York, pp. 279–95.)

————, 1905b, Geographical cycle in an arid climate: *Jour. Geology*, v. 13, pp. 381–407. (Reprinted 1954 in *Geographical essays:* Dover Publications, Inc., New York, pp. 296–322.)

DENNY, C. S., 1965, Alluvial fans of the Death Valley region, California and Nevada: *U.S. Geol. Survey Prof. Paper 466*, 62 pp.

————, 1967, Fans and pediments: *Am. Jour. Sci.*, v. 265, pp. 81–105.

GILBERT, G. K., 1882, Contributions to the history of Lake Bonneville: *U.S. Geol. Surv. 2nd Ann. Rept.*, pp. 167–200.

HADLEY, R. F., 1967, Pediments and pediment-forming processes: *Jour. Geol. Education*, v. 15, pp. 83–89.

HOWARD, A. D., 1942, Pediment passes and the pediment problem: *Jour. Geomorph.*, v. 5, pp. 2–31, 95–136.

JOHNSON, D. W., 1932, Rock fans of arid regions: *Am. Jour. Sci.*, v. 223, pp. 389–420.

KENNETT, J. P., HOUTZ, R. E., ANDREWS, P .B., EDWARDS, A. R., GOSTIN, V. A., HAJOS, M., HAMPTON, M. A., JENKINS, D. G., MARGOLIS, S. V., OVENSHINE, A. T., and PERCH-NIELSEN, K., 1974, Development of the Circum-Antarctic Current: *Science*, v. 186, pp. 144–47.

KING, L. C., 1948, A theory of bornhardts: *Geog. Jour.*, v. 112, pp. 83–87.

KÖPPEN, W., and GEIGER, R., 1936, *Handbuch der Klimatologie*, v. 1, part C: Gebrüder Borntraeger, Berlin, 44 pp.

LAMB, H. H., 1972, *Climate: Present, past and future*, v. 1: Methuen & Co. Ltd., London, 613 pp.

LAWSON, A. C., 1915, The epigene profiles of the desert: *Calif. Univ. Pub. Geol. Sci.*, v. 9, pp. 23–48.

McGee, W J, 1897, Sheetflood erosion: *Geol. Soc. America Bull.*, v. 8, pp. 87–112.

Meigs, P., 1952, Arid and semiarid climatic types of the world: *Internat. Geog. Cong., 17th, Washington 1952, Proc.*, pp. 135–38.

———, 1953, World distribution of arid and semiarid homoclimates, *in Reviews of research in arid zone hydrology:* UNESCO, Paris, pp. 203–10 and maps.

Oberlander, T. M., 1974, Landscape inheritance and the pediment problem in the Mohave Desert of southern California: *Am. Jour. Sci.*, v. 274, pp. 849–75.

Passarge, S., 1928, *Panoramen afrikanischer; Inselberglandschaften:* Dietrich Reimer, Berlin, 15 pp., 25 plates.

Rahn, P. H., 1967, Sheetfloods, streamfloods, and the formation of pediments: *Assoc. Am. Geographers Annals*, v. 57, pp. 593–604.

Thomas, M. F., 1974, *Tropical geomorphology:* John Wiley & Sons, Inc. New York (a Halsted Press book), 332 pp. (In U. K. by The Macmillan Press, Ltd.)

Thornthwaite, W., 1948, An approach toward a rational classification of climate: *Geog. Rev.*, v. 38, pp. 55–94.

Tricart, J., 1972, *Landforms of the humid tropics, forests and savannas* (transl. by C. J. K. de Jonge): Longman Group Limited, London, 306 pp.

Tricart, J., and Cailleux, A., 1972, *Introduction to climatic geomorphology* (transl. by C. J. K. de Jonge): Longman Group Limited, London, 295 pp.

Tuan, Yi-Fu, 1959, Pediments in southeastern Arizona: *Univ. Calif. Pub. Geog.*, no. 13, 140 pp.

Twidale, C. R., 1968, *Geomorphology, with special reference to Australia:* Thomas Nelson (Australia) Ltd., Melbourne, 406 pp.

———, 1976, *Analysis of landforms:* John Wiley & Sons Australasia Pty. Ltd., Sydney, 572 pp.

Twidale, C. R., and Bourne, J. A., 1975, Episodic exposure of inselbergs: *Geol. Soc. America Bull.*, v. 86, pp. 1473–81.

Wahrhaftig, C., 1965, Stepped topography of the southern Sierra Nevada, California: *Geol. Soc. America Bull.*, v. 76, pp. 1165–89.

Wells, P. V., 1970, Postglacial vegetational history of the Great Plains: *Science*, v. 167, pp. 1574–82.

Willis, B., 1934, Inselbergs: *Assoc. Am. Geographers Annals*, v. 26, pp. 123–29.

CHAPTER 14

Eolian Processes and Landforms

ENERGETICS OF EOLIAN PROCESSES

In Chapter 5, it was noted that a substantial amount of the solar energy received at the earth's surface is converted into thermal and density contrasts in the atmosphere, which produce winds. Perhaps as much as 0.7 percent of the thermal energy exchange is accomplished by wind (Fig. 5-2), which, considering the enormity of the solar energy available for geomorphic processes, should give eolian processes and landforms a significant place in geomorphology. Yet the total effectiveness of eolian processes is rated quite low (p. 197), less than rivers, glaciers, and waves. The present general opinion of the minor role of eolian geomorphic processes is especially surprising in view of the very substantial proportion of earth's land area that is arid and semiarid, where vegetation is restricted in its ability to cover protectively the weathered detritus at the ground surface.

Wind is a comparatively minor agent of geomorphic change primarily because of the low density of air as compared to rock and water. The relatively low contrast in densities between immersed rock and water (about 1.6:1) means that most of the sediment load can be carried in suspension, buoyed by the turbulence of the moving liquid. Rock fragments are 2000 times heavier than the air they displace, however. Only very fine dust particles are moved in suspension except at extremely high wind velocities. Also because of its low density, moving air exerts a relatively low pressure against obstructions, whereas water can accomplish significant erosion by simple hydraulic force against submerged rock ledges.

Eolian processes and landforms are most obvious in arid regions, where lack of vegetation not only makes wind erosion, transportation, and deposition possible but enables the resulting landforms to be observed. Eolian landforms are safely regarded as part of the arid morphogenetic system if due provision is made for those restricted areas in other climatic regions where sediment supply and exposure to winds also permit eolian landforms to develop. For example, belts of coastal dunes are on sandy coastlines in all climatic regions, although their frequency is less in the humid tropics (p. 477). The braided channels made by glacial meltwater are notable sources for windblown dust and sand owing to their variable discharges and their extensive nonvegetated exposed bars and abandoned channels. The massive sheets of loess (Fig. 14-14) downwind from outwash channels in glaciated areas attest to the effectiveness of meltwater streams to provide a renewable supply of fine-grained alluvium for subsequent wind transport. Dry periglacial regions are especially susceptible to eolian processes. Sandy soils in any climatic region are likely to have minor eolian landforms such as blowouts and dunes, especially if they have been disturbed by inappropriate agricultural practices.

WIND EROSION

Wind erosion occurs either by **deflation**, the process of removing loose sand and dust, or by **abrasion**, using the entrained sand grains as tools against rock surfaces or other grains. The widespread industrial use of sandblasting by an air jet with entrained fine sand illustrates the potent abrasive capacity of windblown sand. Rock surfaces are fluted, grooved, pitted, or polished, probably within decades after exposure (Hickox, 1959; Whitney and Dietrich, 1973, p. 2571). Cobbles and pebbles that show wind abrasion are called **ventifacts** (Fig. 14-1). On long exposure, perhaps measured in thousands of years, rock surfaces become *faceted* (Verstappen and van Zuidam, 1970). Some facets are joint-faced initially, but even well rounded alluvial gravel may develop a surface concentration of faceted stones on which the facets clearly postdate the rounding.

If undisturbed, a ventifact develops a facet facing the effective wind direction and sloping down to windward. Most ventifacts have multiple facets that intersect along sharp keel-like ridges, testifying to more than one direction of effective abrasion. Large wind-eroded boulders and talus blocks, unlikely to have moved, may be used to infer one or more effective wind directions. Smaller cobbles and pebbles are likely to be moved by a great variety of processes, so their facets have little significance for determining effective winds (Sharp, 1949). The detailed fluting, pitting, and rill marks on some ventifacts have been attributed to abrasion by dust rather than by sand (Whitney and Dietrich, 1973).

Undercut, mushroom-shaped pedestal rocks in desert areas are commonly attributed to wind abrasion. However, experiments (p. 334) prove that the mean height of windblown sand transport is only about 1 cm over a layer of similar-sized grains and little more than 10 cm over hard-rock surfaces. Most authorities now doubt that wind abrasion could cut the notches or overhangs on desert cliffs or large boulders that are 1–2 m in height. More likely, desert weathering processes (p. 130)

Figure 14-1. Ventifacts on lag gravel between lake-shore dunes, Sleeping Bear, Leelanau County, Michigan. North is at the bottom. P = prevailing winds; H = high-velocity winds; unlabeled arrow at bottom indicates other effective winds. (Photo: M. I. Whitney.)

involving exfoliation or spalling on sheltered surfaces are the cause of the undercut surfaces. Some contain rock paintings or are encrusted with gypsum crystals, neither of which could survive if abrasion were an active process.

The only generally accepted eolian abrasional landforms are the chutes or flumes that notch low escarpments around playas and other desert plains. They are aligned with the locally effective winds and are probably scoured by windborne sand being swept through small gullies. They are usually cut in weakly indurated sediments and are separated by tonguelike or streamlined, wind-abraded ridges called **yardangs** (Blackwelder, 1934).

Deflation hollows of various sizes are known to form quickly by wind erosion on exposed sandy terranes. When the vegetation on stabilized dunes is destroyed by fire or overgrazing, *blowouts* develop within months. When excavated below root depth, a blowout can grow as deep as the effective winds can lift sand (Fig. 14-2). Dry creep keeps the adjacent slopes at the appropriate angle of repose. The limiting depth for such blowouts is usually the water table, because moist sand resists deflation, and plants may grow on the floor of the basin as well. Over the more arid sections of the American Great Plains, thousands of small basins mark sandy districts. They were termed "buffalo wallows" by early explorers because animals frequented the ponds found in them. Although primarily the result of deflation during dry years, buffalo and domestic cattle surely aid deflation by loosening the sun-dried soil in the floors of the blowouts as they seek water. After being farmed for decades, some of the Great

Figure 14-2. An active blowout in coastal dunes on Cape Cod, Massachusetts. Vegetated blowout in background is at the water table.

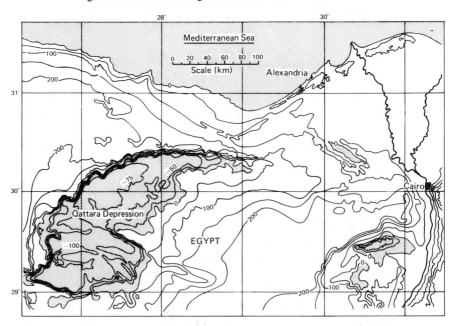

Figure 14-3. The Qattara Depression, a desert basin with an area below sea level of 19,500 km², within 56 km of the Mediterranean coast. The depression has been suggested as a site for hydroelectric power generation. (Redrawn from Ball, 1933.)

Plains blowouts were observed to actively deflate during the dry, fallow years of the mid 1930s.

Larger closed basins in arid regions, some of them many kilometers in diameter and over 100 m in depth, have variously been attributed to solution, differential compaction, faulting, and deflation. Tectonic causes can usually be affirmed or eliminated by geologic mapping or drilling, but the other possibilities are difficult to separate. Big Hollow, near Laramie, Wyoming, is about 5 km wide, 15 km long, and about 100 m deep. It is considered an inactive blowout, developed on fine-grained Mesozoic sedimentary rocks. The enormous Qattara Depression (Fig. 14-3) in the western Egyptian Sahara is estimated to have an excavated volume of 3200 km³ (Peel, 1966, p. 21) and no trace of a tectonic origin. Its floor reaches 134 m below sea level, less than 100 km inland from the Mediterranean coast. A combination of mass-wasting and slope retreat on the surrounding escarpments and wind deflation of the resulting sediment from the basin floor is involved (Ball, 1927, 1933). Many other Saharan depressions have been discovered by analyses of the wartime aerial photographs of the region. Numerous very large depressions in the Mongolian Desert, the P'ang Kiang Hollows described by Berkey and Morris (1927, pp. 336–37, 347), also are confidently attributed to a similar combination of Pleistocene deflation and arid scarp retreat.

In a general review of desert landscapes, Peel (1966) urged renewed attention to the origin of large desert basins, suggesting that wind erosion, rather than tending to reduce the relief of desert landscapes, might actually increase it. Only the water table can limit the depth of deflation if structures do not produce a gravel lag, and wind strength is adequate. The water table might actually be lowered as basins develop (Murray, 1952). It is notable that erosional desert landscapes, unlike fluvial landscapes, do not evolve toward the ultimate base level of the sea. Although tectonic desert basins seem to be sediment traps (pp. 293 and 318), it is possible that some basins could deepen almost without limit by deflation alone. The Qattara Depression and the P'ang Kiang Hollows are cut into Pliocene rocks and therefore are Quaternary landforms. If the Qattara Depression is assumed to have formed in the last 2 million years, the eolian denudation rate is 9 cm/1000 yr, impressively comparable to reported rates of fluvial denudation (Chap 12).

THE MECHANICS OF EOLIAN SEDIMENT TRANSPORT

Wind transport occurs when previously loosened grains of sand and dust are entrained in air and carried to a new site of deposition. Most eolian sediment is derived by weathering or from alluvium.

Eolian sediment transport can be compared to fluvial transport only to a limited degree. In both air and water, some load is suspended, some bounces or rolls at the bottom of the moving fluid, and some particles are dragged along in contact with the substrate. Because of the great density contrast between air and rock fragments, only very fine detritus, with an effective diameter of less than 0.2 mm, can be maintained in suspension by the upward component of atmospheric eddies. This is *dust* that grades into extremely fine particulate matter called *haze* or *smoke* and can be carried entirely around the world by planetary winds before settling out or being washed out of the air.

Loose sand particles with a diameter of 0.2 mm or larger can be stirred by a light wind of 5 m/s (11 mph) but do not remain in suspension. They roll or slide along the ground or bounce, transferring momentum to other grains by collisions. Bagnold (1941) made a thorough laboratory and field study of the physics of windblown sand, and by ingenious experiments first demonstrated that most sand grains transported by wind do so in short asymmetric trajectories, never far from the ground. The process is termed **saltation**. A falling grain strikes one or more other grains and makes a visible dimple or crater in the sand surface. Although the disturbed particles are scattered in various directions, as they rise into the air stream and fall again they are swept forward. Just prior to impact they are usually approaching the ground at an angle of only 10°–16° (Bagnold, 1941, p. 19). Above some threshold velocity, the inertia transferred from the wind to saltating grains is sufficient to maintain continuous forward transport of saltating sand. A lesser amount of sand is driven forward over the ground surface without saltating. This component Bagnold called **surface creep**. It is roughly analogous to the traction or bed load transported by rivers.

Bagnold was able to describe the geometry of the characteristic path or trajectory of a saltating sand grain in terms only of its initial upward velocity (derived from prior impact) and wind-velocity distribution. The important variable is the velocity gradient, or wind shear, near the ground. As wind velocity increases, so does the wind shear near the ground, where most saltating grains move. The effect is to impart a higher forward velocity to saltating grains, which in turn produces a rapid increase in the amount of sand in transport (Fig. 14-4). Most of the sand saltating over a surface of similar-sized grains does not rise higher than 1 or 2 m, regardless of wind speed. Because of the exponential decrease in wind velocity close to the ground, the median height of the trajectories of all saltating sand grains is only about 1 cm above the ground.

Saltating sand grains absorb the kinetic energy of moving air and convert it to kinetic energy of moving sand. Over loose sand, an increase in average wind velocity does not result in an increase in wind velocity at the sand surface. Instead, the velocity gradient, or wind shear, increases sharply, and much more sand is set in motion (Fig.14-4). This response of loose sand to wind energy is a remarkable illustration of the tendency for sediment transport to increase sharply during times of high energy inputs, absorbing much of the energy. As in the case of river alluvium during a flood, or beach sand during a coastal storm, a sheet of sand on a desert absorbs energy during windstorms by moving downwind. If an upwind supply of new sand is available, the desert looks just the same after the storm, even though the depositional landforms are composed of sand that formerly rested somewhere upwind.

The impact of a saltating sand grain can move a much larger grain forward by surface creep. Bagnold (1941, pp. 33–37) estimated that significant creep occurs in grains six times the diameter of saltating grains. However, creeping grains move only a few millimeters with each impact, whereas saltating grains have characteristic

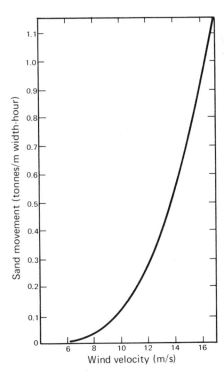

Figure 14-4. Relation between the flow of average dune sand and the wind velocity as measured at a standard height of one meter (Bagnold, 1941, Fig. 22).

trajectories a meter or more in length. Here is a potent method of size sorting, in which the size fraction moving by saltation greatly outdistances coarser grains that move only by surface creep. During strong winds over poorly sorted alluvium, the finest fraction is completely exported from the region as dust; a selected size fraction (perhaps including three-fourths of the total volume moved) is driven downwind by saltation; a comparatively small amount of coarse sand (perhaps one-fourth of the total sand volume) is driven downwind by creep; and a gravel lag is left behind. Wind passing from a sheet of moving sand onto a barren rock surface is no longer slowed by the drag of loose, saltating sand. The abrupt increase in surface wind velocity sweeps the hard surface clean, but when any swept-up grains again pass onto a sand surface, the wind velocity is checked and deposition results. Deserts thus consist of various combinations of barren, windswept, and sandblasted exposed rocks (*hammada*), gravel or stony lag deposits (*reg*), and sand sheets or dune fields (*erg*). The boundaries between these various terranes are abrupt.

DEPOSITIONAL LANDFORMS OF EOLIAN SAND

A patch of sand on a desert plain has a curiously definitive size limitation. Under gentle winds, barely strong enough to stir the sand grains, the patch is spread downwind and grows thinner or disperses. No new sand is delivered to the patch from upwind, because the available sand there is sheltered by bedrock or gravel irregularities. Under strong winds, however, a pre-existing sand patch larger than a critical size grows in thickness and in size as its upwind border extends still further upwind. The change in behavior is related to wind-transport dynamics described in the preceding paragraph. Strong winds create surface velocities over hard surfaces that can transport sand grains downwind. Over a free sand surface, however, the energy of strong wind is absorbed by saltating sand, and the new sand derived from upwind is deposited. But for net accumulation to occur on an exposed sand patch, it must have a minimum initial downwind length of 4–6 m (Bagnold, 1941, p. 183).

A smooth sheet of sand is inherently unstable under gentle wind. The saltation method of sand movement results in most grains of a given size following trajectories of similar length. At the upwind edge of a sand patch, where transport first begins, a lag of coarser grains is left behind. The saltating grains moving downwind from that lag area impact at the mean distance of saltation and produce another lag area. By this process, **sand ripples** develop, transverse to the wind direction. Their typical wavelength is 1 meter or less, very similar to the characteristic flight length of saltating grains. Sand ripples are characterized by coarse grains at their crests and finer grains in their troughs. Because of the rather low angle of impact of saltating grains, each ripple crest creates a shadow zone of fewer saltation impacts in its lee.

Under stronger winds, the surface of a sand patch loses its rippled form, because less size sorting takes place. Then the entire surface layer is sheared forward by saltation and surface creep. However, under such conditions, new sand is also trapped from upwind sources, and the sand sheet is not eliminated but increases in area and thickness as long as an upwind sand supply is available.

Large accumulation landforms of windblown sand are called **dunes**. The dynamics just described make a sharp distinction between ephemeral ripples on a sand surface, which form under gentle winds, and dunes, which are distinctly larger and stable or growing landforms (Fig. 14-5). Even larger-scale complex or multiple dunes are called **draas** (Wilson, 1972a, 1972b). Dunes, in order to survive and grow, must be at least 4–6 m in downwind length and are commonly much larger. Ripples, dunes, and draas lack intermediate forms (Fig. 14-5), although it is common for the sand on dunes to be rippled, just as large swells on the ocean surface have small wind waves on their surfaces in turn. Ripples have the coarsest sand on their crests, but dunes have coarse grains in the troughs and the finest sand on their crests.

Bagnold (1941, p. 188) classified sand accumulations under the following five headings:

Sand Shadows and Sand Drifts

Wherever wind velocity is checked by a fixed obstruction, sand may strike the obstruction and then fall at its windward base or be swept into the lee of the obstruction and accumulate as a streamlined mound. Both landforms are classed as **sand shadows**. They are fixed in form and size by the size and aerodynamic shape of the obstruction. For instance, a small tree raised above the ground on a narrow trunk will not accumulate a shadow, whereas a tuft of grass or a ground-hugging shrub may produce a substantial lee shadow.

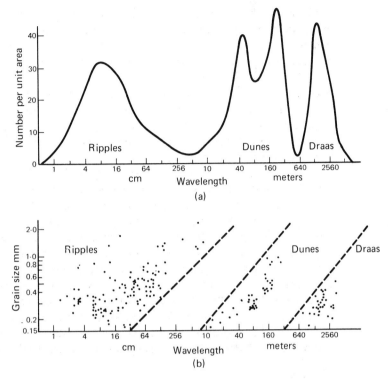

Figure 14-5.   (*a*) Areal frequency of different wavelengths of desert sand waves. (*b*) Grain size of the deposits constituting sand waves of different wavelengths. Ripples, dunes, and draas form three distinct clusters without transitional forms. (Wilson, 1972a, Fig. 2. This figure first appeared in *New Scientist*, London, the weekly review of science and technology.)

Sand drifts are the accumulations in the lee of a local high-velocity zone between obstacles. Most commonly they accumulate where wind is funneled through a gap at a higher-than-average velocity and then resumes the mean velocity of the regional wind. Another important site of sand drifts is at the foot of a downwind cliff or escarpment. Sand blown off the plateau surface sinks into the still air in the lee of the cliff. When such sand drifts grow large enough, they begin to trail off downwind from the base of the cliff.

Dunes

True dunes can best be considered as deformable obstructions to air flow. They are free to move, divide, grow, and shrink, but they are not dependent on a fixed obstruction for their maintenance. Two principles that have been previously emphasized are critical to dune formation. First is the principle that a free sand patch larger than a few meters is likely to increase in size and thickness and thus become a mound or hill. The second principle is that a dune, like a cliff, develops a lee-side pocket of low-velocity air that promotes deposition there. As soon as a mound of sand grows to moderate size, it develops the classic dune asymmetry of a gentle windward slope and a steep, angle of repose *slip face* to leeward (Figs. 14-6, 14-11). The sand on the windward slope is firm and compact, but as it blows over the crest and cascades down the slip face, it may become loose enough to make walking difficult. Note that the internal structure of a dune is fixed by the original deposition and slight compaction; whereas the dune *form* may move without changing shape, the internal *structure* is immobile. Eroded edges of slip-face bedding are commonly exposed on windward slopes (Fig. 14-7). Less rarely, gently dipping sheetlike accumulations of windward-slope deposits are preserved in dune structures.

Dunes have a wide range of forms, some of which were unknown until aerial photography became available for large areas of hitherto unexplored sand regions in the Sahara and Arabia. Bagnold recognized only two basic dune forms, the *barchan* or *crescentic dune*, and the *seif* or *longitudinal dune*. He maintained that *transverse dunes*, with their long axes at right angles to the effective wind, are inherently unstable, and that they break up into shorter segments as barchans or become elongate in the direction of the wind as longitudinal dunes (Bagnold, 1941, p. 205). Other geomorphologists recognize additional dune forms. Transverse dunes are actually very common in areas of copious sand supply and weak winds, in spite of Bagnold's deemphasis (Hack, 1941; Wilson, 1972a, 1972b). In semivegetated areas, blowouts are paired with downwind *parabolic dunes*. Very large *stellate* and *rosette dunes* (Fig. 14-8) may be several hundred meters in height and several kilometers in diameter, with radiating sinuous ridges culminating at a common crest. Their origin is as yet unknown.

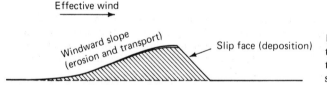

Figure 14-6. Cross section of an active dune. Much of the slip-face deposition finally results from sand flow and slumping.

Figure 14-7. Dipping strata of an earlier dune cycle truncated by cross winds. Barchan dune area, White Sands, New Mexico. (Photo: Tad Nichols.)

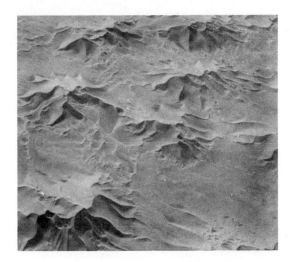

Figure 14-8. Saharan erg near 30°10′N, 8°15′E, dominated by closely spaced peaked or stellate dunes. (Photo: U.S. Air Force, from collection of H. T. U. Smith.)

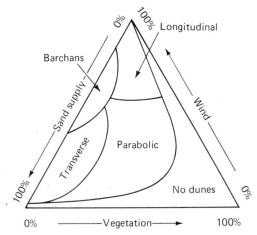

Figure 14-9. Dune forms as a response to three semiquantified variables: wind effectiveness, vegetative cover, and sand supply. Wind direction is assumed to be nearly constant. (Modified from Hack, 1941, Fig. 19.)

The basic dune forms can be classified by reference to three controlling variables: wind effectiveness, vegetative cover, and sand supply (Fig. 14-9). **Longitudinal dunes** may be huge landforms. extending hundreds of kilometers in length, a kilometer or more in width, and several hundred meters in height. They are especially well developed in the heart of the trade-wind deserts, where the wind is either from a constant direction or varies seasonally. Bagnold proposed that the longitudinal dunes of the Sahara form by an alternation of winds from the north and northeast: one season the sand trails off downwind as elongate sand patches and the next season is swept into ridges by obliquely blowing winds. Great systems of longitudinal dunes cross the arid interiors of Australia and South Africa (Fig. 14-10), but most are now covered by sparse vegetation and soil and seem to be relict from one or more earlier periods of greater aridity. Active longitudinal dunes are found in areas nearly devoid of vegetation, where winds are persistent for many months and sand supply is irregular. The dune ridges are separated by gravel-armored reg or hammada.

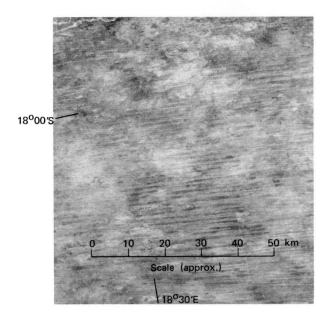

18°00'S

0   10   20   30   40   50 km

Scale (approx.)

18°30'E

Figure 14-10. Longitudinal dunes a-bout 1 km wide and at least 100 km long, south of the Okavango River, South-West Africa (Namibia). These dunes are stabilized by thin soils and sparse vegetation. (NASA ERTS E-1183-08143.)

Figure 14-11. Transverse coastal dunes breaking inland into barchans. Guerro Negro dune field, Baja California, Mexico. View northward, Laguna Manuela in background. Dunes in foreground are about 6 m high. (Photo: D. L. Inman)

Transverse dunes (Fig. 14-11) are associated with massive quantities of sand and relatively ineffective winds. They are not common in the great sandy deserts but form along coastlines (Cooper, 1958, 1967) and on zones of erodible sandy bedrock or alluvium. Although their crests are ridges roughly perpendicular to the effective wind, in detail the crests have sinuous or "fish-scale" outlines or break up into fields of isolated barchan or longitudinal dunes.

Barchan dunes (Fig. 14-11) are the classic eolian landforms. Their crescentic form consists of a gently inclined windward slope and a steep lee side around which the horns or cusps of the dune project downwind, making the slip face *concave* to the downwind direction. Barchans are isolated dunes that migrate freely across rock or gravel desert plains, usually downwind from some other dune form (Inman, et al., 1966). They are large enough to maintain themselves as they migrate and represent a remarkable balance between accumulation, transportation, and erosion.

Parabolic dunes are characteristic of partially stabilized sandy terranes that develop blowouts (Fig. 14-2). The crescentic shape of parabolic dunes has a very different orientation from that of barchan dunes, likely to confuse a novice. On the lee side of an elongate blowout a dune may accumulate, its windward slope rising from the blowout and the slip face *convex* downwind in plan view. A parabolic dune may become elongated (a "hairpin" or "upsiloidal" dune) by downwind migration of a blowout until it is split into two minor longitudinal ridges paralleling the long axis of the blowout. These dunes are usually much longer and narrower than barchan dunes and are always associated with a central blowout. They should not be confused with barchans, especially on aerial photographs, because the resulting interpretation of effective wind direction will be reversed.

Whalebacks

Where longitudinal dunes have migrated downwind, the coarser sand is left behind as a prominent ridge. These **whalebacks** attest to the stability of longitudinal dunes in form and spacing over long periods of time. Either very large longitudinal dunes migrated a long distance downwind or longitudinal dunes repeatedly followed the same tracks across the desert. Either case implies former wind effectiveness, sand supply, or both, far in excess of present conditions. Those described by Bagnold (1941, p. 230) in the Egyptian desert are 1–3 km in width and 50 m high and extend for 300 km. Modern longitudinal dunes may surmount these great ridges but are much smaller and segmented. Sometimes as many as four chains of modern parallel longitudinal dunes surmount a single whaleback (for similar Arabian forms, see Holm, 1960). These forms represent one of the more intriguing aspects of the effect of climatic change on landforms (Chap. 18).

Undulating Fixed Sand Sheets

Large areas of semiarid land, especially in the middle latitude interiors of the large continents, are under sparse grass cover and have intermittent rain or winter snow. As the result of either drought or fire, local areas of shallow deflation basins develop, alternating with subdued parabolic dune forms. Winds are not sufficiently effective to create dunes with active slip faces, but the combination of deflation and deposition makes a landscape of undulating or "billowing" hills, somewhat analogous to the top of a cloud cover as viewed from an airplane (Fig. 14-12). During wet years or seasons, ponds accumulate in hollows between the hills. Permeability and low precipitation prevent stream dissection even when the water table is near the surface. The Sand Hills region of Nebraska (Smith, 1965) is a typical example. Large areas of the Russian steppe and Iran are also of this origin. As a transitional landscape between arid and subhumid climates, these large regions can be expected to show many relict landforms of conditions both more moist and more arid than now experienced. Buried soil profiles, sand-filled drainage systems, polycyclic drainage networks, and buried sites of human occupance attest to the mobility of sandy terranes under marginal precipitation.

Sand Sheets (Ergs)

In contrast to the fixed dunes of grassy steppes, from one-third to one-quarter of the true deserts consist of vast "sand seas" or **ergs** (Figs. 14-8, 14-13). Their surfaces consist of a variety of dune forms, ranging from low mounds to huge star dunes. These largest of all the sandy landscapes are primarily fixed by lithologic factors. Those in Africa, Arabia, and Australia are on easily weathered, sandy sedimentary rocks, probably deposited in Cenozoic or Mesozoic marine embayments or freshwater lakes. They are more nearly residual than constructional or erosional landforms. Wind transport and deposition constantly shift the surface layers, but the regional landscape is one of awesome monotony.

LOESS

Thick deposits of homogeneous unstratified silt are draped over or bury other landforms in the middle latitudes. The extremely well-sorted, fine-grained sediment is generally known by the German term, **loess**, meaning simply "loose," or unconsolidated. Loess has an eolian origin, derived from glacial outwash and other alluvium (Fig. 15-8) or fixed sand sheets, although minor water transport, masswasting, and pedogenic processes are accepted as part of the overall process of loess accumulation (Pécsi, 1968; Schultz and Frye, 1968). The extremely good size sorting during long-distance transport of windblown silt accounts for most of the peculiar and significant properties of loess. It is a material, not a kind of landscape, but where thick enough to bury the underlying landscape, its surface deserves consideration as a constructional landform. Furthermore, the soil-forming and erosional properties of loess terranes also give them special geomorphic interest.

Loess and similar deposits form a cover 1–100 m thick over an estimated 10 percent of the land area (Pécsi, 1968). Thirty percent of the United States has an eolian cover (Ruhe, 1974, p. 152). In detailed studies, loess sheets in Europe and

Figure 14-12. The Sand Hills region of Nebraska. Stabilized dunes along Nebraska Highway 97, south of the Dismal River. (Photo: W. J. Wayne.)

Figure 14-13. Dunes in the Rub' al Khali, an erg covering 600,000 km² of southeastern Saudi Arabia. (Photo: Arabian American Oil Company.)

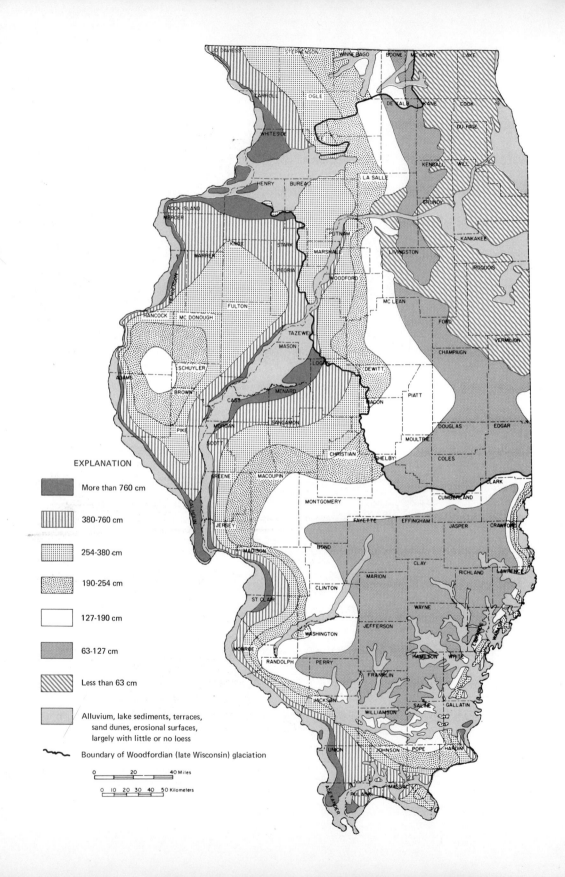

EXPLANATION

More than 760 cm

380-760 cm

254-380 cm

190-254 cm

127-190 cm

63-127 cm

Less than 63 cm

Alluvium, lake sediments, terraces,
sand dunes, erosional surfaces,
largely with little or no loess

Boundary of Woodfordian (late Wisconsin) glaciation

0   20   40 Miles

0 10 20 30 40 50 Kilometers

Figure 14-14. (*Opposite page.*) Loess thickness in Illinois (compare with Fig. 15-8). (From Willman and Frye, 1970, Plate 3.)

North America have been shown to consist of multiple depositional units, each correlative with a glacial age and separated by unconformities with soil profiles representing interglacial conditions. In Europe, archeologic sites are found in the buried soils. Loess deposits thicken exponentially (or as a hyperbolic function: see p. 74) toward source areas, usually outwash plains or the undulating fixed sand sheets that were probably barren during glacial ages (Fig. 14-14).

Loess is dominantly silt-size quartz but reflects the lithology of the source terranes and may be highly calcareous. Because of its fineness, carbonate content, and homogeneity, it consolidates readily and forms stable steep cliffs, often with a "case-hardened" surface of secondary calcium carbonate. The depositional surface is featureless, although parabolic dunes of loesslike silt and clay (**lunettes**) accumulate on the lee sides of dry lake beds in southern Australia (Hills, 1940; Stephens and Crocker, 1946). Most often, the uniform deposition results in topography that shows strong inheritance from preloess conditions (Lugn, 1935, p. 159). Loess erodes readily in humid climates and forms a characteristic high-density network of gullies with steep or vertical banks. The loess sheets associated with the last glaciation are relatively undissected and unweathered, but beyond the drift border of the last ice sheets, older loesses have much more developed erosional landscapes. Quaternary loess sheets can be dated by stratigraphic correlation with tephras and glacial drift and by paleomagnetic correlation with other Quaternary deposits. In the late Quaternary, radiocarbon dates on buried peat beds, gastropods, and artifacts are useful. Future quantitative studies of denudation rates on loess terranes are likely to be highly informative.

REFERENCES

BAGNOLD, R. A., 1941, *The physics of blown sand and desert dunes:* Methuen & Co., Ltd., London, 265 pp.

BALL, J., 1927, Problems of the Libyan desert: *Geog. Jour.*, v. 70, pp. 21–38, 105–28, 209–22.

———, 1933, Qattara depression of the Libyan desert and the possibilities of its utilization for power-production: *Geog. Jour.*, v. 82, pp. 289–314.

BERKEY, C. P., and MORRIS, F. K., 1927, *Geology of Mongolia: Natural history of central Asia*, v. 2, Am. Museum Nat. Hist., New York, 475 pp.

BLACKWELDER, E., 1934, Yardangs: *Geol. Soc. America Bull.*, v. 45, pp. 159–66.

COOPER, W. S., 1958, Coastal sand dunes of Oregon and Washington: *Geol. Soc. America Mem. 72*, 169 pp.

———, 1967, Coastal dunes of California: *Geol. Soc. America Mem. 104*, 131 pp.

HACK, J. T., 1941, Dunes of the western Navaho country: *Geog. Rev.*, v. 31, pp. 240–63.

HICKOX, C. F., JR., 1959, Formation of ventifacts in a moist, temperate climate: *Geol. Soc. America Bull.*, v. 70, pp. 1489–90.

HILLS, E. S., 1940, The lunette, a new landform of eolian origin: *Aust. Geographer*, v. 3, pp. 15–21.

HOLM, D. A., 1960, Desert geomorphology in the Arabian peninsula: *Science*, v. 132, pp. 1369–79.

INMAN, D. L., EWING, G. C., and CORLISS, J. B., 1966, Coastal sand dunes of Guerrero Negro, Baja California, Mexico: *Geol. Soc. America Bull.*, v. 77, pp. 787–802.

LUGN, A. L., 1935, Pleistocene geology of Nebraska: *Nebraska Geol. Survey*, Bull. 10, Ser. 2, 223 pp.

MURRAY, G. W., 1952, The water beneath the Egyptian western desert: *Geog. Jour.*, v. 118, pp. 443–52.

PÉCSI, M., 1968, LOESS, *in* FAIRBRIDGE, R. W., ed., *Encyclopedia of geomorphology:* Reinhold Book Corporation, New York, pp. 674–78.

PEEL, R. F., 1966, The landscape in aridity: *Inst. Brit. Geographers Trans.*, no. 38, pp. 1–23.

RUHE, R. V., 1974, *Geomorphology:* Houghton Mifflin Co., Boston, 246 pp.

SCHULTZ, C. B., and FRYE, J. C., eds., 1968, *Loess and related eolian deposits of the world:* Univ. of Nebraska Press, Lincoln, 355 pp.

SHARP, R. P., 1949, Pleistocene ventifacts east of the Big Horn Mountains, Wyoming: *Jour. Geology*, v. 57, pp. 175–95.

SMITH, H. T. U., 1965, Dune morphology and chronology in central and western Nebraska: *Jour. Geology*, v. 73, pp. 557–78.

STEPHENS, C. G., and CROCKER, R. L., 1946, Composition and genesis of lunettes: *Royal Soc. S. Australia Trans.*, v. 70, pp. 302–12.

VERSTAPPEN, H. T., and VAN ZUIDAM, R. A., 1970, Orbital photography and the geosciences —a geomorphological example from the central Sahara: *Geoforum*, v. 2, pp. 33–47.

WHITNEY, M. I., and DIETRICH, R. V., 1973, Ventifact sculpture by windblown dust: *Geol. Soc. America Bull.*, v. 84, pp. 2561–81.

WILLMAN, H. B., and FRYE, J. C., 1970, Pleistocene stratigraphy of Illinois: *Illinois Geol. Survey, Bull. 94*, 204 pp.

WILSON, I., 1972a, Sand waves: *New Scientist*, v. 53, pp. 634–37.

———, 1972b, Aeolian bedforms—their development and origins: *Sedimentology,* v. 19, pp. 173–210.

CHAPTER 15

Periglacial Morphogenesis

DEFINITIONS AND EXTENT

As originally defined and generally used early in this century, the term **periglacial** referred to the region in Europe that was peripheral to the margins of the Pleistocene glaciers where frost action was inferred to have been particularly active in the past. The traditional emphasis was on past climatic conditions now more equitable, but the term applies equally well to the tundra regions of today, in high latitudes primarily in the northern hemisphere. "Ideally, it is a zone of permanently- [preferred usage is 'perennially'] frozen subsoil (permafrost), seasonally-thawed topsoil (active layer), frequent changes of temperature across the freezing point, and an incomplete vegetation cover of herbaceous plants and dwarf species of trees, for example dwarf birch and willow [Sparks and West, 1972, p. 98]." The modern tendency is to consider the periglacial environment as near-glacial in the sense of either location or intensity of conditions. For practical reasons, the modern periglacial zone is regarded as coincident with the region of perennially frozen ground, or **permafrost** (Fig. 15-1) (Péwé, 1969, p. 2). Some synonyms include *cryergic* or *cryogenic* processes and morphogenetic regions.

No single climatic region corresponds to the periglacial zone. Low-latitude mountain tops are included, as well as the cold, dry continental interiors of Asia and North America and the maritime boreal coasts of Iceland, Canada, Scandinavia, and Alaska. In some periglacial regions, the mean annual temperature is close to 0°C, and the annual range varies only slightly because of maritime influences. In other regions, a short summer interval may be distinctly hot, with 20 hours or more of sunshine, but winters are dark and intensely cold and dry.

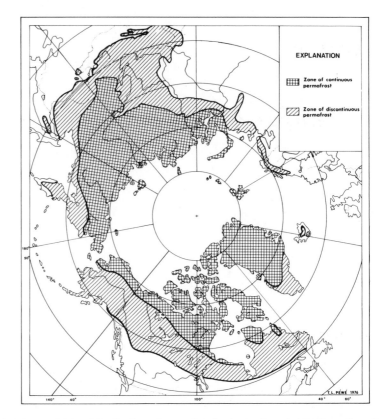

Figure 15-1.   Distribution of permafrost in the northern hemisphere. Compiled from many sources. Areas of sporadic permafrost are too small to show. (Courtesy T. L. Péwé.)

Neither can the former periglacial regions peripheral to the Pleistocene ice sheets be compared directly with the arctic tundras of today. Relict low-altitude periglacial soil structures have been discovered at latitudes at least as low as 39°N in the central United States (Wayne, 1967; Flemal, et al., 1973). Even if the periglacial interval was brief, its record at a latitude more than halfway toward the equator from the north pole should remind us not to extrapolate modern climatic conditions to the Pleistocene record. In the central plains of the United States at the edge of a Pleistocene ice sheet, the summer sun at noon was within 16° of the zenith. Days and nights must have been of the same length as present ones and not at all like the seasonal alternation of continuous daylight and darkness that characterizes the tundra regions north of the Arctic Circle.

The definitive feature of periglacial morphogenesis is intense freezing and seasonal thawing. Gelifluction (p. 172) and nivation (p. 355) are also characteristic processes. Some authors have included in the morphogenetic region all areas that experience seasonal frost, but this definition is too broad to be useful. By including in the definition the criterion of perennially frozen subsoil (permafrost), the periglacial morphogenetic region becomes much more coherent and identifiable (Péwé, 1969). Also, perennially frozen ground produces soil and sediment structures and landforms that remain after the climate changes, so relict periglacial conditions can be identified, and the affected areas can be shown on maps.

Permafrost now underlies approximately 20 percent of the world's land area, including about 85 percent of Alaska and about 50 percent of Canada and the Soviet Union (Fig. 15-1). All that is required to produce permafrost is a mean annual soil temperature below freezing. Because soil temperature at a depth of a meter or two closely approximates mean annual air temperature, the latter measure is often used to define the permafrost region. However, differences in soil moisture, porosity, vegetation cover, dissolved components, and ground-water pressure have marked effects on the thickness of the frozen layer, and subcategories of *sporadic, discontinuous,* and *continuous* permafrost are used. Some permafrost may have persisted since the last glaciation; in other areas, the permafrost has formed in postglacial time. As with glaciers, the permafrost limit seems to be shrinking during the last century. The maximum known depth of permafrost is about 1 km, but over wide areas of the arctic, a thickness of several hundred meters is commonplace (Bird, 1967, p. 40). Only 1 or 2 m at the surface is subject to seasonal melt. This is the **active layer.**

In spite of the vagueness of its definition, the periglacial morphogenetic region is impressive in size and geomorphic significance. The table on p. 308 lists 38 percent of the earth's land area under either a cold, treeless, snow climate or a boreal forest and snow climate. At least half of the combined areas of the two regions are underlain by permafrost. At various times during the Quaternary Period, recently enough so that some relict landforms display periglacial morphogenesis, perhaps as much as 40 percent of the land area has been subjected to periglacial conditions (Tricart, 1970, p. 28). In fact, Cotton (1963, p. 769) has suggested that most of the landforms in midlatitude regions, regarded by W. M. Davis and others of his time as the result of the "normal" processes of fluvial erosion, may be relict from past periglacial conditions. It would indeed be an odd trick of fate if the landforms of the northeastern United States and western Europe, which provided the observational evidence for the deduced "normal cycle of erosion," were actually relict periglacial landforms. Geomorphologists with field experience in present-day periglacial regions are steadily demonstrating the impact of former periglacial morphogenesis in regions of presently temperate climates. Both for its present and past importance, periglacial morphogenesis is one of the most important current topics of geomorphology.

PERIGLACIAL PROCESSES

The periglacial processes are characterized more by the criterion of degree or intensity than of kind. Weathering and soil formation are controlled by the intensity and frequency of freezing and thawing and by the generally frozen subsoil. Mass-wasting is dominated by solifluction of the active layer over permafrost (gelifluction).

Rock falls and rock glaciers are common in mountainous regions. Fluvial erosion is inhibited by the brevity of seasonal stream flow and the massive loads of sediment fed into streams by gelifluction. Wind becomes a relatively important agent of transport and deposition, with loess sheets creating major constructional landforms. In the review that follows, most of the processes have been described in previous chapters and will be only briefly noted. The emphasis in this chapter is on the ways the processes interact to produce a landscape recognizable as the result of periglacial morphogenesis.

Weathering

Freezing and Hydrofracturing.   In spite of continued uncertainty as to the exact process by which rocks are fractured by freezing and thawing of interstitial water (pp. 111–112), there is no doubt that regions now in the periglacial environment are characterized by great quantities of angular, fractured rock detritus (Fig. 15-2). The size of the fragments is largely a function of lithology and structure. Laboratory experiments suggest that repeated freezing and thawing of clay minerals can disaggregate mineral grains as small as 1 μm in diameter (McDowall, 1960). Water is a necessary but little-evaluated component in the process, but precipitation records for periglacial regions do not give much useful information about the availability of water within the ground. Certainly, dry rocks will not be disrupted by repeated

Figure 15-2.   Wright Valley, South Victoria Land, Antarctica. The mean annual temperature is about −20°C. No soil develops on the steep taluses. (Photo: C. Bull, Institute of Polar Studies, Ohio State University.)

temperature fluctuations through 0°C. Even for saturated rocks, the critical temperature for fracturing is probably well below 0°C. Air temperature fluctuations do not affect rocks to a depth of more than a few centimeters, so probably the annual temperature range is the significant factor in mechanical weathering by freezing (Cook and Raiche, 1962). Nevertheless, with all these qualifications and uncertainties, alternate freezing and thawing is generally considered the dominant weathering process in the periglacial zone (Flint, 1971, p. 267; Embleton and King, 1975, p. 4).

Frost heaving is the process of lifting surface particles by the formation of ice in soil. As ice nuclei form in soil, free water migrates toward these nuclei, which grow as veins or lenses of ice while adjacent sediments are dewatered (Everett, 1961). The grain size and pore size of sediments determine the rate of ice-crystal segregation and water movement. Silt is the optimal size of sediment, and frozen silt may contain up to 68 percent ice by volume (Bird, 1967, p. 43). Ice lenses as thick as 4 m have been reported from Alaska (Taber, 1943).

Needle ice consists of clusters of slender ice crystals up to several centimeters in length that form overnight in frosty wet soils. Although the phenomenon is much less impressive than the formation of segregated ice masses in perennially frozen ground, needle ice plays an important role in loosening surface soils and in lifting small stones to the surface of fine sediments. On hill slopes, frost heave and needle ice promote soil creep by lifting particles at right angles to the sloping surface, then permitting them to settle vertically on remelting (Figs. 8-6, 8-8).

Chemical Weathering. Freezing has been so closely associated with periglacial morphogenesis that other weathering processes have been slighted. Hydration, hydrolysis, and oxidation will occur whenever aerated water contacts minerals, but at low temperatures and with few microorganisms, chemical processes proceed only very slowly. Crusts of hydrated salts form on rocks at McMurdo Sound, Antarctica, and might be important to rock weathering there. In other coastal periglacial regions, where sea salts might collect by evaporation under low atmospheric humidity, similar weathering might also be expected.

An instructive illustration of weathering in a polar desert was provided by Kelly and Zumberge (1961). They analyzed a sequence of quartz diorite specimens from a single exposure at Marble Point, on McMurdo Sound, Antarctica. The samples ranged in degree of weathering from fresh rock to completely disaggregated, iron-stained grus. Although the authors assumed that a sequence of chemical changes would be illustrated by the samples, in reality no chemical weathering was found, except for oxidation of iron-rich biotite and accessory minerals. Bulk chemical and mineralogic composition did not change. No clay minerals formed. The disintegration was caused not by oxidation or hydration but by hydrofracturing and the subsequent crystallization of salts in pores and fractures. The site is exposed to sea spray during the brief summer, but no rain falls. This study provided substantial quantitative support for the general but previously unsubstantiated opinion that mechanical weathering predominates over chemical weathering in cold, dry climates. The question of limestone solution in cold regions was discussed in Chapter 7.

Mass-Wasting

Gelifluction.   Long before the term "periglacial" had been proposed, European geologists were aware that thick sheets of poorly sorted angular debris mantled their landscape beyond the limits of glaciation. The material was given quaint colloquial terms such as "head," or "coombe rock," and had been hypothesized to be the result of cold-climate, but not glacial, processes. During an expedition to the Falkland Islands, J. G. Andersson observed fluid sheets of debris moving down gentle slopes over perennially frozen ground and producing deposits similar to the "head" of European authors. He named the process **solifluction** (p. 172 and references there), later modified to **gelifluction** to specify flow over a frozen substrate. The process is generally accepted as being responsible for the deposits of "head," and, in fact, massive deposits of relict "head" are accepted as evidence of a former periglacial climate (Cotton and Te Punga, 1955).

Because of the perennially frozen substrate, mass-wasting is not a deep phenomenon in periglacial regions. Gelifluction occurs only in the active layer, rarely more than two meters thick. Resistant rock ledges are exposed to freeze and thaw as the active layer flows downhill, and they stand out as tors and low structurally controlled ridges on debris-covered slopes. Gelifluction occurs on very low slope angles, perhaps as low as 1° on rocks that break down to clay-rich debris (Fig. 15-3).

Figure 15-3.   Tundra landscape of gentle convex slopes above river-bank cliffs on the Colville River near Umiat, Alaska. The bentonitic shale is prone to gelifluction and other mass-wasting processes. (Photo: T. L. Marlar, U.S. Army Cold Regions Research and Engineering Lab.)

Figure 15-4. Ice-wedge polygons, northeastern Mackenzie River delta, Canada. Dark high-centered polygons are 10–20 m in diameter. (Photo: J. R. Mackay.)

On bouldery debris, gelifluction slopes may be as high as 20° (Tricart, 1970, pp. 105–6).

Along with the specific process of gelifluction, a variety of other mass movements are common features of periglacial morphogenesis. Snow and rock avalanches are frequent, especially during the springtime. Mud flows and debris flows break out of gelifluction lobes and sheets and are really only an extension of the gelifluction process. On clay-rich soils, debris flows in channels with levees have been described (Fig. 15-3) (Anderson, et al., 1969).

The annual cycle of freeze and thaw in the active layer produces large lateral stresses in the mixture of rock fragments, ice, and water. On level ground, slabs and cobbles may be lifted by frost heaving, but fine sediment fills in beneath them so that the layer becomes sorted, with the coarsest fragments at the surface. Even more commonly, cells of differential motion mutually interfere to form a variety of polygonal patterns (Fig. 15-4). The patterns may consist of either sorted or unsorted debris. On slopes steeper than 3 percent, polygonal patterns become attenuated downslope (Fig. 15-5). The driving forces that produce these and other patterns are freezing and thawing combined with gravity. The specific landforms are described in a subsequent section of this chapter.

The most prominent polygonal patterns are in areas of extremely low mean annual temperatures, probably averaging −6°C to −8°C. Permafrost is normally several hundred meters thick. In such regions, networks of **ice-wedge polygons** cover many thousands of square kilometers (Fig. 15-4). Each polygon, from 3–30 m in diameter or more, is outlined by shallow troughs, beneath each of which is a wedge-shaped foliated mass of ice, a meter or more broad at the top and extending downward from 3–7 m or more (Lachenbruch, 1962). Most authorities accept the contraction theory of ice-wedge polygon formation (Leffingwell, 1915; Lachenbruch, 1962) as the best explanation, although as noted in Chapter 6, the exact dynamics of fracture in freezing solids is by no means fully understood. As outlined by Lachenbruch (Fig. 15-6), the contraction theory is based on seasonal cracks that develop in totally frozen ground during winter. Ice and soil, like other solids, contract on cooling, and may explosively develop vertical cracks several millimeters wide and a meter or more in depth. In the spring, water from the active layer seeps down the crack and refreezes in the cold permafrost. During the following summer, warming and expansion (but not thawing) of the frozen layer causes sediments to be upturned and deformed

Figure 15-5.   Relict patterned ground and stone stripes on Snake River basalt, north of Glenns Ferry, Idaho (Malde, 1964). Plateau gradient is 2 percent southwestward; colluvium gradient is 20 percent southeastward. Light-colored areas are grassy; dark areas are basalt fragments. (Vertical aerial photo: U.S. Dept. of Agriculture.)

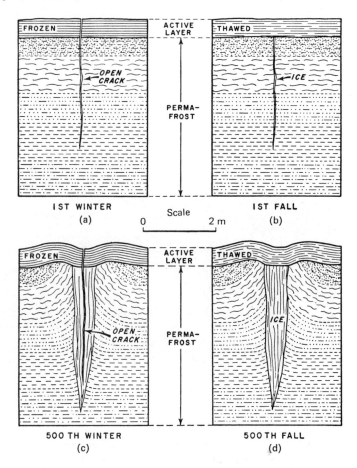

Figure 15-6. Schematic evolution of an ice wedge according to the contraction-crack theory. (From Lachenbruch, 1962, Fig. 1.)

against the ice wedge. Each winter the freeze-cracking is repeated along the initial plane of weakness, and each summer a new lamella of ice is introduced. After several centuries, large ice wedges have formed, their spacing and intersections outlining ice-wedge polygons. Because they can be replaced by later sediments as *ice-wedge casts*, ice wedges are among the few features that are preserved as valid criteria of former permafrost in regions not now periglacial. However, the identification of fossil ice wedges is difficult (Black, 1976).

Erosion and Transport Processes

Nivation. A subtle progression of processes can be observed beginning with weathering under a snow bank, to meltwater erosion from beneath a snow bank, to the downhill erosive creep of water-saturated or refrozen snow, to glacial erosion. The first three processes are grouped under the heading of **nivation,** or the erosive effects

Figure 15-7.   Nivation hollows on cryoplanation terraces cut into a greenstone ridge on Indian Mountain near Hughes, Alaska. (Reger and Péwé, 1976, Fig. 1. Photo: T. L. Péwé.)

associated with immobile patches of snow that persist into the summer season in the periglacial environment. Nivation, along with freezing and thawing and gelifluction, is one of the three groups of processes that characterize periglacial morphogenesis (St. Onge, 1969).

Snow patches that persist in sheltered positions on hillslopes have been observed to produce *nivation hollows* after only a few seasons. The snowmass produces a localized source of water around its edges. Some of the water percolates through the downhill edge of the snowbank and carries away any fine particles from the rock or soil beneath. Freezing and thawing may be intensified at the edges of the shrinking snowbank as well, although the snow insulates the rock and soil beneath it. Nivation hollows tend to be self-enlarging; a hollow once produced will catch more snow and shelter a larger snowbank in successive years (Fig. 15-7). Periglacial hill slopes commonly are lined with shallow nivation hollows, especially on the shaded and downwind side of ridge crests where snow accumulates and is shaded from sun for the longest time. Permafrost promotes widening rather than deepening of the hollows. In the conterminous United States, the favored location for nivation hollows is on the northeastern side of ridges, where snow is blown by westerly winds and is sheltered from sunlight in early summer.

The originator of the term *nivation* (Matthes, 1900) found no evidence that seasonal snow patches moved or abraded their beds. Subsequently, other researchers have either asserted or denied that slow creep in the snow associated with compaction, or actual slip of the snowbank over bedrock, contributed to bedrock erosion. A study of the nivation process in the Snowy Mountains of Australia (Costin, et al., 1964) proved conclusively that the downhill creep and slide of snow banks dragged basal rock fragments over bedrock and produced visible, although minor, abrasion marks on both the transported fragments and the subsurface. In a subsequent study (Costin, et al., 1973), steel rods notched to various depths were embedded in rock in nivation hollows, and their deformation or breakage under the force of moving snow was calibrated by laboratory tests. The pressure generated ranged from 160–3790 kPa (1.6–37.9 bars), equal to or greater than the shear stresses calculated for the base of valley glaciers (p. 369). Clearly, seasonal snow patches in nivation hollows, with floors that slope at angles of 10°–27°, do move downhill and abrade their floors with bed-load fragments. Because one of the definitive features of glaciers is evidence of motion (Chap. 16), proof of snow-patch abrasion makes the distinction between glacial erosion and nivation even less clear. In areas of possible former periglacial or glacial conditions, abraded surfaces have been cited as proof of glaciation. Such evidence may need to be reevaluated or at least used with great caution.

Fluvial Erosion.   Rivers in the periglacial zone have extremely complex hydraulic geometries. High-gradient mountain streams and large rivers on plains may continue to flow under ice cover throughout the winter. Even so, discharge is greatly decreased because of the lack of overland flow in winter. Small, low-gradient rivers may freeze solid and cease to flow entirely. Because the periglacial zone in general is semiarid or arid, river discharges are never high except during the melting season.

Springtime is the season of violent fluvial action. When meltwater begins to flow into headwater streams, they swell and burst their ice cover or flow over frozen river beds, progressively developing a flood wave that moves downstream. If the thaw season begins in the headwaters and progresses toward the river mouth, as it generally does in the larger south-to-north-flowing rivers of arctic North America and Siberia, huge ice jams and floods are seasonal events that are part of the streams' regimen. Briefly, the rivers move much larger loads and much coarser sediment than their mean annual discharges would suggest. Initially, the floods flow over basal ice on those rivers that were frozen solid, so bank erosion is much more extensive than channel deepening.

The discharge of small periglacial streams, especially mountain streams, follows a distinct daily cycle during the summer. Shortly after the hottest part of the day, the rivers reach peak discharge as overland runoff reaches the channels. Before dawn after a freezing night, runoff will have nearly ceased. Sediment transport and deposition goes through a daily cycle under such conditions, with alluvium deposited among shifting, braided channels. The sediment load delivered to streams also fluctuates with the daily cycle of gelifluction and other forms of mass-wasting. Most periglacial streams, other than mountain torrents, flow on aggraded beds. Mass-wasting seems to deliver sediment faster than periglacial rivers can move it to the sea. Aggradation results, but it is not yet clear how much of the aggradation is modern and how much is relict from Pleistocene times.

Eolian Transport.   Low precipitation, freeze-drying of exposed sediment, and strong winds combine to make the periglacial zone a region of strong eolian action. Clouds of sand and dust can be seen rising from the flood plains of rivers as the spring flood subsides (Fig. 15-8). Dunes are common features on the downwind banks of rivers. Sand sheets, less well sorted than dunes and sometimes stratified, are widespread in northern Europe and are attributed to windblown sediment settling on snow and being prevented from forming dunes. Very large areas of present and former periglacial regions are mantled with loess (p. 342; Fig. 14-14). Most of the plains of central Europe and Asia, and the Mississippi basin of North America, are loess mantled. The loess forms a blanket over the pre-existing landscape, not usually thick enough to form constructional landforms but thick enough to be the parent material for soils. Most of the world's wheat- and corn-producing regions are on plains of Pleistocene loess. In all cases, large outwash plains or fixed sand sheets can be shown to be the source of the windblown silt, because deposits thicken and grow coarser toward the source areas.

Figure 15-8.   Clouds of silt transported by wind from the Delta River flood plain, central Alaska. Compare with Figure 14-14. (Photo: U.S. Navy. Courtesy T. L. Péwé.)

PERIGLACIAL LANDFORMS

Major Landforms

The combination of intense frost action, permafrost, nivation, strong winds, incomplete vegetation cover, and seasonal flowing rivers imparts a strong morphogenetic imprint on landscapes. Periglacial plains are wastelands of patterned ground, rubble-strewn hillslopes, gelifluction lobes and sheets, and seasonal bogs or shallow lakes. Mountainous regions add rock glaciers, massive taluses, and a variety of terraced or benched slopes shaped by nivation and gelifluction. Many well-illustrated descriptions of periglacial scenery have been published (Muller, 1947; Hopkins, et al., 1955; Wahrhaftig, 1965; Bird, 1967; Ferrians, 1969; Péwé, 1975). Only the major landforms and some of the more spectacular minor forms are noted here. For more detailed descriptions of periglacial phenomena, recent books by Péwé (1969), Brown (1970), Tricart (1970), and Washburn (1973) are recommended, especially to American readers. An enormous literature in Russian, Polish, French, and the Scandinavian languages is readily available, as well.

The background of Figure 15-3 gives a good impression of the relief of periglacial plains on nonresistant rocks. Gelifluction causes mass-wasting on very low slopes, producing broad convex summits. Stream action is so brief and ineffective that small valleys cannot entrench but are choked by gelifluction debris. Large areas are covered with patterned ground.

Asymmetric Valleys.   Some periglacial valleys show notable asymmetry in cross-profile, even though structural control can be eliminated. The most typical condition is for north-facing slopes (in the northern hemisphere) to be steeper than the south-facing slopes. Where valleys trend north or south, the asymmetry disappears. The theory is that gelifluction is more intense on the valley side facing the sun for the longest time each day, and the shaded side stays colder, less prone to gelifluction, and steeper (Currey, 1964). Also, the copious colluvium from the south-facing slopes forces rivers to preferentially erode and steepen their north-facing valley walls. Relict valleys with such asymmetry are common in southern Indiana (latitude 38°N) adjacent to the Wisconsin glacial margin (Wayne, 1967). Tricart (1970, p. 137) suggested that near the southern limit of the arctic periglacial regions, permafrost, gelifluction, and gentle slopes might persist on cold, north-facing valley walls even as drier, warmer, nonpermafrost slopes maintain steeper gradients on the south-facing valley sides. The issue is obviously complex, but if a degree of asymmetry can be established in certain periglacial regions today, comparable asymmetry might be a useful criterion for establishing relict periglacial landforms. However, the low angle of the arctic summer sun, and the long arc of exposed hillsides, cannot be assumed for lower latitudes (p. 348).

Cryoplanation Terraces.   Flattened summit areas and benched hillslopes in Alaska were attributed to an assemblage of periglacial processes collectively termed **altiplanation** (Eakin, 1916, pp. 78–82). Current authors prefer the equivalent term, **cryopla-**

nation. By implication, the composite of processes could result in a regional periglacial surface of low local relief, not controlled by any base level, that would logically be called a "cryoplain" or an "altiplain." Although the term cryoplanation is frequently used in reference to processes in the literature of periglacial morphogenesis (Embleton and King, 1975), the deduced or implied penultimate landform of periglacial regions, the "cryoplain" or "altiplain," has never been seriously proposed as a climatic analog of a peneplain or a "pediplain" (pp. 301 and 302). Cryoplanation as a process includes a variety of distinct weathering, erosional, and depositional processes, and the term should be used in that sense with caution. However, **cryoplanation terrace** (or the Russian equivalent, **goletz terrace**) is a widely accepted term for the resulting landform (Fig. 15-7).

Many similar steplike periglacial terraces are of gelifluction lobes with steep, lobate, downhill faces or "risers" and somewhat flatter or gently sloping "treads." The steep faces are either of extremely coarse, angular, blocky debris, or sometimes turf. Flatter tread surfaces are poorly drained and of somewhat finer material. Cryoplanation terraces (Te Punga, 1956; Reger and Péwé, 1976) are cut surfaces, probably due to nivation, and are backed by steep headwalls or clusters of tors. Cryoplanation terraces are neither continuous over long horizontal distances nor of constant elevation, but they can be confused with river terraces or emerged shorelines, and periglacial geomorphologists should be aware of their abundance (Washburn, 1973; Embleton and King, 1975).

Block Fields, Rock Streams, and Rock Glaciers.   Masses of angular rock rubble on summits (**block fields** or **felsenmeer**) or streaming down mountainsides (**rock streams**) are striking features of modern periglacial landscapes, and their relict equivalents are also well known. Relict rock streams in the Appalachian Mountains consist of large, angular boulders and are now devoid of vegetation. They extend as much as 1000 meters downhill from a cliff or exposed ledge (Fig. 15-9). Although not now in motion, the Appalachian rock streams were almost certainly formed by periglacial mass-wasting (Clark, 1968; Potter and Moss, 1968). Gelifluction and frost creep (p. 173) are the likely processes, because although the rock streams now lack the water-retaining fine sediments in the upper few meters that would be required to maintain motion, the large blocks have interstitial finer sediments between them at depth. Hickory Run boulder field, one of the larger of the relict Appalachian rock streams, has a surface gradient of only 1° (Smith, 1953).

Rock glaciers were once thought to be moribund valley glaciers but are now recognized as distinct landforms, transitional between periglacial and glacial in origin. White (1971a, p. 44) defined rock glaciers as "accumulations of unsorted, till-like, coarse to fine rock debris, with an ice core and/or interstitial ice, having a glacier shape and spreading downvalley." Active rock glaciers are either tongue-shaped along valley axes or lobes on valley walls. They have steep frontal slopes, often with active taluses, and arcuate pressure ridges on their surfaces (Fig. 15-10).

If a mass of poorly sorted colluvium is sufficiently packed with ice within its pore spaces, the plastic deformation of the ice can produce downhill motion similar to but slower than glacier flow. The arbitrary difference between a rock glacier and a

Figure 15-9. Hickory Run boulder field, Carbon County, Pennsylvania. Boulders reach 2 m in length. (Photo: W. Bolles, Pennsylvania Dept. of Education.)

Figure 15-10. Vertical aerial view of the Arapaho (Colorado) rock glacier in late summer. Direction of motion (northeasterly) is toward the lower left of the picture. Principal lobe is about 200 m wide (White 1971b). (Photo: J. B. Benedict.)

glacier is that the former is mostly rock debris with interstitial ice, whereas the latter is mostly ice with included rock fragments. Several rock glaciers in the central Rocky Mountains are ice-cored and are transitional upslope to true valley glaciers (White, 1971a, 1971b, 1976; Potter, 1972). Larger rock glaciers in Alaska are ice-cemented rock fragments, similar to glaciers in all but the relative proportion of ice and rock (Wahrhaftig and Cox, 1959). Rock glaciers move forward by internal deformation and debris slides at their termini, resulting in a rolling or continuous-tread form of transport. Although they have not been proved to erode the rock surfaces over which they move, their movement is clearly transitional to glacial processes.

Minor Landforms

The landscape of periglacial regions is marked by a variety of patterns, many of which barely deserve the name "landform." Much of the terrain is depositional rather than erosional because of the contrast between intense mass-wasting and inefficient stream erosion. On the alluvial plains and other surfaces of low relief, permafrost overlain by the active layer contorts, erupts, or collapses the land surface into a maze of polygonal patterns, bogs or shallow lakes, ice-cored domes, and pits. Only the more striking forms are illustrated here.

Patterned Ground.   The relationship of ice-wedge polygons to permafrost has been described previously (p. 354). Less conspicuous polygons, outlined by open cracks or shallow depression but not underlain by ice wedges, are called **frost-crack polygons** (Washburn, et al., 1963). They are not diagnostic of permafrost and can form in any seasonally cold region if snow cover is thin during a cold interval. Some frost cracks may be preserved by windblown silt packing into them, but for the most part, the patterns are visible only during cold weather and are not preserved.

Polygons may be made of sorted or unsorted sediment and may have raised centers or raised rims (Fig. 15-4) (Washburn, 1973, and references therein). Generally, ice-wedge polygons are made of unsorted sediments. The contraction is compensated for by compressional deformation of frozen sediment adjacent to the crack around the polygon rather than by sorting and translocation of rock fragments within the polygonal prism of soil.

Sorted polygons involve more severe rearrangement of rock fragments, with coarser particles lifted to the surface or thrust outward to the edges of the polygons. A higher proportion of fine-grained sediment probably promotes sorting, because water is held longer without refreezing in fine sediments, and ice masses may segregate from the freezing soil to disrupt and heave the original stratification, if any. Sorted polygons may be preserved, because seasonal-flowing water permeates the coarse network of polygonal boundaries without eroding the fine sediments from the polygon cores. Sorted polygons are not diagnostic of the permafrost zone, although they are found there. They seem to be more abundant on wet, cold mountain plateaus and other places with intense freeze-thaw activity.

Sorted polygons, when occurring on slopes, are distorted laterally by gravity to **stone stripes** (Fig. 15-5), parallel segregations of coarse and fine sediment trending downhill. Like sorted polygons, stone stripes require active freeze-thaw processes but are not restricted to regions of permafrost. Once formed, the coarse stripes will channel runoff from the hillside. The percolating water probably erodes fine sediment from beneath the coarse debris and aids in lowering the stripes of coarse debris into trenches down hillsides. The excellent drainage provided by stone stripes enables hillslopes to remain stable at unusually steep angles, perhaps to 30°, without slumping or sliding.

The many bogs of subarctic periglacial regions have a striking ribbed pattern of turf stripes between elongate pools of water. These are the **string bogs** that cover a large proportion of the subarctic taiga or forested region. Permafrost is probably not involved in their formation, so they might be excluded from the periglacial morphogenetic region except for their widespread occurrence at the margins of the periglacial zone. Their present extent can be related to the extensive glacial derangement of rivers in formerly glaciated regions. Their origin is hypothesized to be due to lines of organic foam and debris that wash from the spruce forests and lodge as low dams across cold, slush-filled creeks or flooded areas (Thom, 1972). According to that hypothesis, string bogs form progressively from the downstream edge, each line of turf and grass forming at right angles to the slow flow of water, where it traps more organic debris until the next ridge forms above it.

Pingos.   Under certain hydrologic conditions, laccolithic intrusions (Chap. 3) of ice arch the overlying ground into a fractured dome or blister (Fig. 15-11) called a **pingo**. Pingos occur at the edges of alluvial plains or old lake beds and at the bases of long slopes. Probably the hydrostatic head in ground water is intensified by pressure

Figure 15-11.   Pingo, Wollaston Peninsula, Victoria Island, Northwest Territories, Canada. (Photo: A. L. Washburn.)

generated at the base of the permafrost or in unfrozen layers within permafrost. At some low point on an aquifer, pressure builds up sufficiently to arch the overlying permafrost and build up a lenticular mass of ice in the core of the mound. Other pingos are thought to be of a "closed-system" type, wherein the progressive inward and downward freezing of an old lake or bog generates pressure in the enclosed saturated area sufficient to drive pore water upward, where it freezes as a segregated mass in the core of a mound (Mackay, 1962; Holmes, et al., 1963). Fossil pingos, low circular mounds enclosing poorly drained basins, are reported from many areas of western Europe. A few similar forms in Illinois (Flemal, et al., 1973) and New Jersey (Wolfe, 1953) have been described, but they seem less common in North America than in Europe (Flemal, 1976). Like fossil ice wedges, fossil pingos are excellent evidence of former permafrost, and their presence near the southern limits of former glaciers gives important information about the former extent of the periglacial zone.

Thermokarst.   The global warming trend of the last two centuries has caused a reduction in area of permafrost regions. Some areas that were formerly underlain by continuous permafrost now have discontinuous or only sporadic occurrences. As permafrost melts, the reduction of total volume and the collapse of ground over former segregated ice masses produces a pitted karstlike terrain, to which the term **thermokarst** has been applied (Czudek and Demek, 1970). Many areas of permafrost are preserved only because of the insulating effects of forests or tundra vegetation. If land is cleared for roads, railways, airports, or pipelines, the collapse of permafrost can lead to destruction of the installations. An enormous literature of permafrost engineering has grown in recent years and promises to expand even more as transarctic oil pipelines are built to move petroleum southward from the Arctic coast.

The tundra is a delicate ecosystem that exhibits a distinct periglacial landform assemblage. The trans-Alaskan oil pipeline provoked violent controversy about its effects on tundra life, scenery, and the frozen substrate. It has been built, but with greater awareness of the regional environment than if it had been initially authorized without full examination of its impact. The tundra landscape is locally changed along the pipeline route, but we can hope that any lessons learned there can lead to better preservation and management of other tundra regions.

REFERENCES

ANDERSON, D. M., REYNOLDS, R. C., and BROWN, J., 1969, Bentonite debris flows in northern Alaska: *Science*, v. 164, pp. 173–74.

BIRD, J. B., 1967, *Physiography of Arctic Canada:* Johns Hopkins Press, Baltimore, 336 pp.

BLACK, R. F., 1976, Periglacial features indicative of permafrost: Ice and soil wedges: *Quaternary Res.*, v. 6, pp. 3–26.

BROWN, R. J. E., 1970, *Permafrost in Canada:* Univ. of Toronto Press, Toronto, 234 pp.

CLARK, G. M., 1968, Sorted patterned ground: New Appalachian localities south of the glacial border: *Science*, v. 161, pp. 355–56.

COOK, F. A., and RAICHE, V. G., 1962, Freeze-thaw cycles at Resolute, N.W.T.: *Geog. Bull.*, no. 18, pp. 64–78.

COSTIN, A. B., JENNINGS, J. N., BAUTOVICH, B. C., and WIMBUSH, D. J., 1973, Forces developed by snowpatch action, Mt. Twynam, Snowy Mountains, Australia: *Arctic and Alpine Res.*, v. 5, pp. 121–26.

COSTIN, A. B., JENNINGS, J. N., BLACK, H. P., and THOM, B. G., 1964, Snow action on Mt. Twynam, Snowy Mountains, Australia: *Jour. Glaciology*, v. 5, pp. 219–28.

COTTON, C. A., 1963, A new theory of the sculpture of middle-latitude landscapes: *New Zealand Jour. Geol. Geophys.*, v. 6, pp. 769–74.

COTTON, C. A., and TE PUNGA, M. T., 1955, Fossil gullies in the Wellington landscape: *New Zealand Geographer*, v. 2, pp. 72–75.

CURREY, D. R., 1964, A preliminary study of valley asymmetry in the Ogotoruk Creek area, northwestern Alaska: *Arctic*, v. 17, pp. 84–98.

CZUDEK, T., and DEMEK, J., 1970, Thermokarst in Siberia and its influence on the development of lowland relief: *Quaternary Res.*, v. 1, pp. 103–20.

EAKIN, H. M., 1916, The Yukon-Koyukuk region, Alaska: *U.S. Geol. Survey Bull. 631*, 88 pp.

EMBLETON, C., and KING, C. A. M., 1975, *Periglacial geomorphology*, 2d ed.: John Wiley & Sons, Inc., New York, 203 pp.

EVERETT, D. H., 1961, Thermodynamics of frost damage to porous solids: *Faraday Soc. Trans.*, v. 57, pp. 1541–51.

FERRIANS, O. J., KACHADOORIAN, R., and GREENE, G. W., 1969, Permafrost and related engineering problems in Alaska: *U.S. Geol. Survey Prof. Paper 678*, 37 pp.

FLEMAL, R. C., 1976, Pingos and pingo scars: Their characteristics, distribution, and utility in reconstructing former permafrost environments: *Quaternary Res.*, v. 6, pp. 37–53.

FLEMAL, R. C., HINKLEY, K. C., and HESLER, J. L., 1973, DeKalb mounds: A possible Pleistocene (Woodfordian) pingo field in north-central Illinois, *in* Black, R. F., Goldthwait, R. P., and Willman, H. B., eds., The Wisconsinan Stage: *Geol. Soc. America Mem. 136*, pp. 229–50.

FLINT, R. F., 1971, *Glacial and Quaternary geology:* John Wiley & Sons, Inc., New York, 892 pp.

HOLMES, G. W., FOSTER, H. L., and HOPKINS, D. M., 1963, Distribution and age of pingos in interior Alaska: *Permafrost Internat. Conf. Proc.*, NAS-NRC Pub. 1287, pp. 88–93.

HOPKINS, D. M., KARLSTROM, T. N. V., BLACK, R. F., WILLIAMS, J. R., PÉWÉ, T. L., FERNOLD, A. T., and MULLER, E. H., 1955, Permafrost and ground water in Alaska: *U.S. Geol. Survey Prof. Paper 264-F*, pp. 113–46.

KELLY, W. C., and ZUMBERGE, J. H., 1961, Weathering of a quartz diorite at Marble Point, McMurdo Sound, Antarctica: *Jour. Geology*, v. 69, pp. 433–46.

LACHENBRUCH, A. H., 1962, Mechanics of thermal contraction cracks and ice-wedge polygons in permafrost: *Geol. Soc. America Spec. Paper 70*, 69 pp.

LEFFINGWELL, E. deK., 1915, Ground-ice wedges; the dominant form of ground-ice on the north coast of Alaska: *Jour. Geology*, v. 23, pp. 635–54.

MACKAY, J. R., 1962, Pingos of the Pleistocene Mackenzie delta area: *Geog. Bull.*, v. 18, pp. 21–63.

MALDE, H. E., 1964, Patterned ground in the western Snake River Plain, Idaho, and its possible cold-climate origin: *Geol. Soc. America Bull.*, v. 75, pp. 191–207.

MATTHES, F. E., 1900, Glacial sculpture of the Bighorn Mountains, Wyoming: *U.S. Geol. Survey*, 21st Ann. Rept., pt. 2, pp. 167–90.

McDowall, I. C., 1960, Particle size reduction of clay minerals by freezing and thawing: *New Zealand Jour. Geol. Geophys.*, v. 3, pp. 337–43.

Muller, S. W., 1947, *Permafrost or permanently frozen ground and related engineering problems:* Edwards Bros., Ann Arbor, Mich., 231 pp.

Péwé, T. L., ed., 1969, *The periglacial environment, past and present:* McGill-Queen's Univ. Press, Montreal, 487 pp.

———, 1975, Quaternary geology of Alaska: *U.S. Geol. Survey Prof. Paper 835*, 145 pp.

Potter, N., Jr., 1972, Ice-cored rock glacier, Galena Creek, northern Absaroka Mountains, Wyoming: *Geol. Soc. America Bull.*, v. 83, pp. 3025–57.

Potter, N., Jr., and Moss, J. H., 1968, Origin of the Blue Rocks block field and adjacent deposits, Berks County, Pennsylvania: *Geol. Soc. America Bull.*, v. 79, pp. 255–62.

Reger, R. D., and Péwé, T. L., 1976, Cryoplanation terraces: Indicators of a permafrost environment: *Quaternary Res.*, v. 6, pp. 99–109.

St. Onge, D. A., 1969, Nivation landforms: *Canada Geol. Survey Paper 69–30*, 12 pp.

Smith, H. T. U., 1953, Hickory Run boulder field, Carbon County, Pennsylvania: *Am. Jour. Sci.*, v. 251, pp. 625–42.

Sparks, B. W., and West, R. G., 1972, *The ice age in Britain:* Methuen Co., Ltd., London, 302 pp.

Taber, S., 1943, Perennially frozen ground in Alaska: Its origin and history: *Geol. Soc. America Bull.*, v. 54, pp. 1433–548.

Te Punga, M. T., 1956, Altiplanation terraces in southern England: *Biul. Peryglac.*, v. 4, pp. 331–38.

Thom, B. G., 1972, Role of spring thaw in string-bog genesis: *Arctic*, v. 25, pp. 237–39.

Tricart, J., 1970, *Geomorphology of cold environments* (transl. by E. Watson): Macmillan and Co., Ltd., London, 320 pp.

Wahrhaftig, C., 1965, Physiographic divisions of Alaska: *U.S. Geol. Survey Prof. Paper 482*, 52 pp.

Wahrhaftig, C., and Cox, A., 1959, Rock glaciers in the Alaska Range: *Geol. Soc. America Bull.*, v. 70, pp. 383–436.

Washburn, A. L., 1973, *Periglacial processes and environments:* St. Martin's Press, New York, 320 pp.

Washburn, A. L., Smith, D. D., and Goddard, R. H., 1963, Frost cracking in a middle-latitude climate: *Biul. Peryglac.*, v. 12, pp. 175–89.

Wayne, W. J., 1967, Periglacial features and climatic gradient in Illinois, Indiana, and western Ohio, east-central United States, *in* Cushing, E. J., and Wright, H. E., Jr., eds., *Quaternary paleoecology:* Yale Univ. Press, New Haven, Conn., pp. 393–414.

White, S. E., 1971a, Rock glacier studies in the Colorado Front Range, 1961 to 1968: *Arctic and Alpine Res.*, v. 3, pp. 43–64.

———, 1971b, Debris falls at the front of Arapaho rock glacier, Colorado Front Range, U.S.A.: *Geograf. Annaler*, v. 53, ser. A, pp. 86–91.

———, 1976, Rock glaciers and block fields, review and new data: *Quaternary Res.*, v. 6, pp. 77–97.

Wolfe, P. E., 1953, Periglacial frost-thaw basins in New Jersey: *Jour. Geology*, v. 61, pp. 133–41.

CHAPTER 16

Glaciers as Morphogenetic Agents and Landforms

INTRODUCTION

Glaciers are masses of ice (with included rock debris and air) on land, which show evidence of present or past internal deformation and motion. Water in the solid state has thermal and physical properties very different from those in the liquid state. The transition from one state to the other is abrupt. Therefore, the morphogenesis of landscapes under glaciers is very different from the processes and forms dominated by liquid water, and the boundaries between the glacial morphogenetic region and other regions are unusually sharp.

Of all the water near the surface of the earth, only about 2 percent is on the land, in the solid state as glacier ice (Fig. 5-5). Even this small fraction is sufficient to cover entirely one continent (Antarctica) and most of the largest island (Greenland) with ice to an average thickness of 2.5 km over Antarctica and 1.5 km over Greenland. At the present time, about 10 percent of the earth's land area is ice covered. An additional 20 percent has been ice covered repeatedly during the glaciations of the Pleistocene Epoch, much of it as recently as 15,000 years ago. Glacial morphogenesis has been the dominant factor in shaping the present landscape of North America northward of the Ohio and Missouri Rivers and of Eurasia northward of a line from Dublin eastward through Berlin to Moscow and beyond the Urals (Fig. 16-1). In addition, mountains that carry glaciers and year-round snowfields today have been glaciated to an altitude 1000–1500 m lower than the present snowline.

Glaciers, especially continent-sized ice sheets, pose unique problems for geomorphic description. In one sense, the surface of an ice sheet is a constructional landform, built of layers of snow compressed to ice. However, ice is an easily deformed

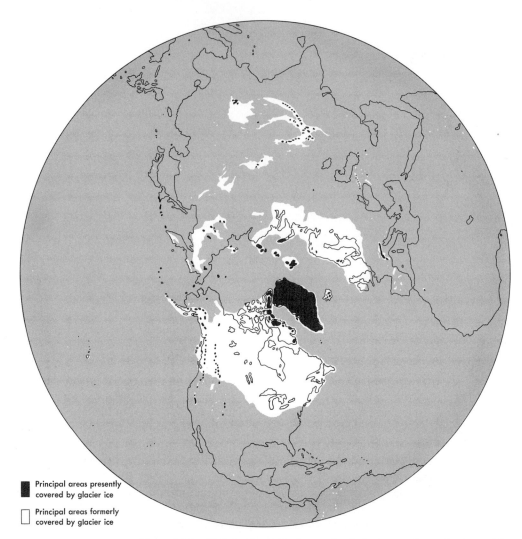

Figure 16-1. Existing large glaciers and principal areas formerly covered by glacier ice, northern hemisphere. (Redrawn from Flint, 1971, Figs. 4-8 and 4-9.)

plastic solid and assumes a dome-shaped or elliptical cross section (Figs. 16-8, 16-9, and 16-11) that must be defined as being structurally controlled. Erosion plays only a minor role in shaping the surface of a glacier, yet glaciers are powerful eroding agents of other rocks. An estimated 7 percent of all modern erosion is being done by glaciers (p. 197), which makes them second only to rivers in erosional effectiveness, although it is a very poor second. Therefore, we need to study ice first as a land-forming structure; next as a landscape with landforms shaped by such odd geomorphic processes as rheid flow, brittle fracture, and melting; and finally as a geomorphic agent that is shaping and has shaped many terrains.

Three distinct scientific disciplines are involved in glacial morphogenesis. The first is **glaciology**, the science of ice. Next is **glacier morphology**, the description and explanation of ice surfaces or "ice-scapes." These two subjects are covered in this chapter. The third scientific discipline is **glacial geomorphology**, which describes and explains the geomorphic work of glaciers, past and present. It is a large enough subject to be considered separately in Chapter 17. No geomorphologist can appreciate the work of ancient glaciers in shaping some of the world's most dramatic landscapes unless he or she understands the physical and thermal properties of ice. But all the theoretical studies in the world cannot replace a walk onto the surface of a modern glacier for the comprehension of glacial morphogenesis.

GLACIOLOGY

Glaciology is the scientific study of ice. It includes the study of ice crystals in high clouds, hail, and snow; frozen lake, river, and ocean water; and glacier ice. Glaciologists are also meteorologists, physicists, and geologists. Ice can be studied either as an easily deformed crystalline solid with a structure to be analyzed by metallurgical techniques or as a geomorphic agent that erodes valleys and transports massive loads of rocky debris. What we know about glaciers has been learned from a wide field of experiment and observation (Glen, 1974, 1975).

Flow in Glaciers

Plastic Flow.   Naturally formed ice is generally quite pure, with only minor amounts of dissolved salts and gasses. Ice behaves very much as any polycrystalline metal a few tens of degrees Celsius below its melting point. It deforms under shear stresses in somewhat the same way that iron deforms when it is heated to a bright red color. The stress required to deform a single crystal is very low; under a low sustained pressure, an ice crystal can be drawn out into a thin ribbon in a few days (Fig. 16-2) (Glen, 1952). Polycrystalline ice deforms in the same manner but only about 1 percent as fast. A mass of polycrystalline ice subjected to a shear stress of 160 kPa (1.6 bar) deforms by an amount equal to its original length in a year (Fig. 16-3). At only slightly larger stresses, the strain rate becomes very rapid. By assuming that ice is a perfect plastic that does not deform at all at stresses less than 100 kPa (1 bar), but deforms infinitely rapidly at 100 kPa so that no additional shear stress can be applied to it, theoretical models of ice flow have been developed that closely approximate the flow of valley glaciers and ice caps. Several lines of evidence suggest that shear stresses at the base of glaciers are never far from the range of 50–150 kPa (0.5–1.5 bars), so the simplifying assumption of perfect plasticity is appropriate.

At shear stresses greater than 50 kPa, the plastic deformation of ice shown in Figure 16-3 fits an equation of the form;

$$\dot{\gamma} = a\tau^n$$

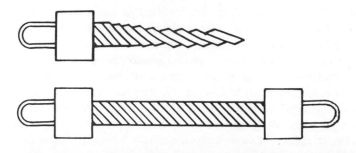

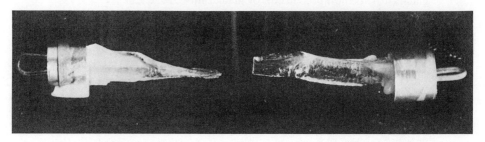

Figure 16-2. Single crystal of ice, 13 mm² cross section, c optical axis 45° to specimen axis, deformed by sustained tension of 0.4 kg. (Photo: J. W. Glen, reproduced from the Journal of Glaciology by permission of the International Glaciological Society.)

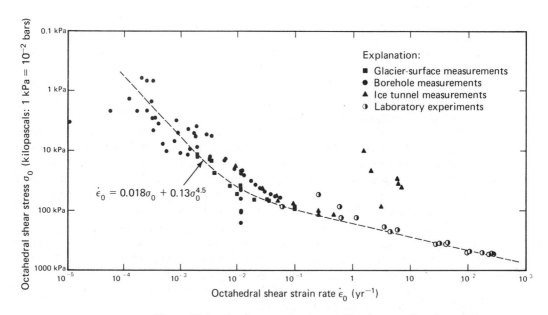

Figure 16-3. Strain rate of polycrystalline ice as a function of shear stress. Deformation becomes rapid at stresses greater than 100 kPa (1 bar). (Meier, 1960, Fig. 40, compiled from numerous sources.)

where $\dot{\gamma}$ is the strain rate, or rate of permanent deformation measured in percent change of length per time, τ is the shear stress and a and n are constants. Experimental values of n for polycrystalline ice vary between 1.9 and 4.5, with a mean of about 3. The constant a, but not n, is strongly temperature-dependent, so that for a given shear stress, the strain rate is only one-tenth as great in ice at $-22°C$ as in ice at $0°C$ (Kamb, 1964; Paterson, 1969, p. 83). A more complex form of the stress-strain equation is shown in Figure 16-3. It is important to note that confining, or hydrostatic, pressure does not significantly affect the strain rate of ice. Only shear, or directional, stress causes plastic deformation. Confining pressure does not make the ice at the bottom of a glacier any more "plastic" than the ice at the surface. Many serious errors in interpreting glacier flow have been made because of this misconception.

On the basis of the equations of plastic flow, it can be shown that a slab of ice resting on an inclined rock surface will deform downslope as a solid, without any melting and refreezing. The rate of movement at the ice surface will be proportional to the thickness and surface slope of the ice (Paterson, 1969, p. 78). The flow is laminar; theory and observation agree that the motion is most rapid at the surface and progressively less rapid with depth (Fig. 16-4). Also, theory and observation agree that flow is controlled by the surface slope of the glacier, not by the slope of the rock floor beneath it, as long as the ice thickness is significantly greater than the bedrock relief. It is quite possible for basal ice to move into and out of an ice-filled basin or over an obstruction, controlled only by the slope on the surface of the ice above the basin or obstruction.

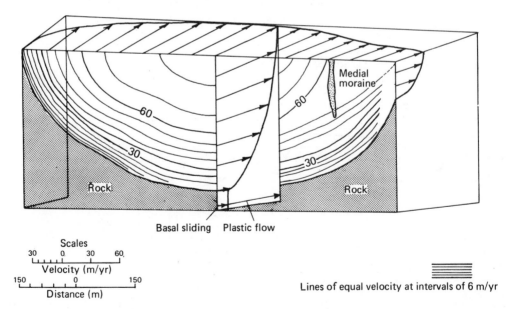

Figure 16-4. Distribution of velocity within the Saskatchewan Glacier, Alberta, Canada, at a measured cross section. About 30 percent of the maximum surface motion is due to basal sliding. (From Meier, 1960, Fig. 42.)

Theoretical models that assume ice to be a perfectly plastic solid with a yield stress of 100 kPa (1 bar) give excellent predictions of the shape, thickness, and rate of flow of glaciers, but the results are not perfect. The obvious reason is that real glaciers are complex masses of ice, with internal temperature gradients and impurities, that move over or between rock surfaces along discontinuities that cannot be modeled theoretically. Also, during movement, deforming ice crystals in a polycrystalline mass grow and re-orient themselves in the stress field by a process of solid-state recrystallization. Thus the grain size and orientation, which greatly affect the rate of deformation, change during deformation. Ice under high stress forms small, interlocking grains, but if the stress is reduced, these small grains "anneal" to form oriented grains as wide as 10 cm.

Regelation Slip and Basal Sliding. If ice is very close to its melting temperature, an additional form of flow can occur, called **regelation slip**. This is not a solid state phenomenon but involves pressure-dependent melting and freezing (Fig. 6-5). Consider a glacier flowing around a rock obstruction at its base, with the ice close to the melting temperature for the appropriate confining pressure. On the upglacier side of the obstruction, the ice presses against rock, locally increasing pressure, depressing the melting temperature, and producing water *at the same temperature as the surrounding ice*. The water migrates along grain boundaries within the ice or over the rock surface to regions of less pressure, especially on the downglacier side of the obstruction where ice is being drawn away from the rock. Here the water refreezes as bubble-free *regelation ice* (Fig. 16-5). Regelation is essentially an isothermal process. The only thermal gradients are those produced by the pressure gradients around the obstruction.

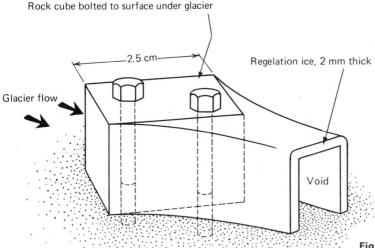

Figure 16-5. Experimental demonstration of the formation of regelation ice beneath a temperate glacier. (From Peterson, 1970.)

Mathematical models of regulation slip are imperfect, because it is essentially the result of boundary conditions between unlike materials and phases. The controlling factor, assuming that the temperature is at or very near the melting point, is the bed roughness, or size of the limiting obstruction (Kamb, 1964, p. 358). Large obstructions can be passed by solid-state plastic deformation, but smaller obstructions are more efficiently passed by regulation slip. One theory suggests that the controlling wavelength of bedrock obstruction, which would permit the optimal contributions by both plastic deformation and regulation slip, is about 0.5 m (Kamb and LaChapelle, 1964). However, obstructions of this size are not common on glaciated rock surfaces. Streamlined glaciated landforms about ten times as large are common and are thought to be stable forms under moving ice, somewhat like ripple marks and dunes under moving water and air. The lack of agreement between theory and observation illustrates the imperfections of present understanding of regulation slip.

Regulation slip produces a film of water locally at the base of a glacier. If this film becomes continuous, perhaps because of geothermal or frictional heat, the entire mass of the glacier can move forward or downslope along the rock surface by gross **basal sliding**, in addition to moving by internal plastic deformation and regulation slip. As shown by borehole measurements (Fig. 16-4), basal sliding (including regulation slip) can account for 50 percent or more of the total observed surface motion of valley glaciers. At the extreme of 100 percent basal sliding, the movement becomes similar to a rock slide or slump along a basal surface (Chap. 8).

Brittle Fracture; Faults and Crevasses.   In their thin downstream or marginal regions, glaciers also "flow" by movement along discrete shear planes such as thrust faults. Pressure of moving ice against marginal ice that is either frozen fast to rock, or too full of rock debris to move, or blocked by a large obstruction, causes a thrust fault to form, dipping upglacier. Actively moving ice shears forward and over the stagnant basal portion (Schytt, 1956). Thrust faults of this sort may reach entirely through the glaciers and are an important means by which rock debris from the base of the ice is brought upward to the surface.

Brittle failure also contributes to surface topography. Wherever a glacier moves over an obstruction at shallow depths or drops down a buried cliff, the surface ice is fractured by a maze of **crevasses**, mostly vertical and extending tens of meters deep (Figs. 16-6, 16-12). These are tension cracks that heal at the depth where the stresses can be compensated by plastic flow. The deepest known crevasses, in the cold brittle ice of Antarctica, are about 50 m deep. Crevasses often become bridged by drifting snow and are one of the major dangers for explorers. A prominent arcuate crevasse, called the **bergschrund**, is at the extreme head of most valley glaciers.

Surges.   Some glaciers are known to move suddenly forward at their termini, advancing at a rate of kilometers per year instead of the average rate of a few meters per year. These **surges** rarely last more than a few months to a year, but they create a chaotic pattern of crevasses on the glaciers. They were first reported on a group of Alaskan glaciers early in this century (Tarr, 1909) and were attributed to an excessive

Figure 16-6. Yentna Glacier, Alaska Range. Late summer view; equilibrium line is visible toward the left rear. Rhythmic ridges on the left are *ogives*, formed by seasonal melting as the glacier descends a buried ridge. Lateral moraines of tributary glaciers become medial moraines in foreground, but maintain their individual identity. (Photo: Austin Post, U.S. Geological Survey.)

accumulation of snow from avalanches produced by an earthquake in 1899. This idea has not been supported by subsequent observations, but ever since the major Alaskan earthquake of 1964, a close watch has been kept on the glaciers of the Alaska Range in hopes of observing the initial phases of surging.

Surging has been associated with the theory of *kinematic waves*, local compressional bulges that can move through glaciers at four to five times the velocity of the glacier movement (Kamb, 1964; Paterson, 1969, pp. 198–212). Kinematic waves originate in the upper part of valley glaciers as a local thickening and migrate downglacier, fracturing the surface ice as they pass through. However, surges seem to involve velocities much greater than kinematic wave theory and result in a thinning of the ablation area, not thickening. Some catastrophic internal instability of flow is implied by surging, such as a sudden increase in water at the base so that the glacier becomes "unstuck" from its bed.

A growing hypothesis is that an ice sheet might surge only along a restricted segment of its perimeter, independent of the classic theory of radial-spreading flow. If so, the glacier margin could rapidly advance in one region while shrinking elsewhere (Wright, 1973). The possibility has important implications for Pleistocene chronology (Chap. 17).

Flow Lines within Glaciers.   Glaciers originate in regions where snow accumulation exceeds loss, and they flow outward or downward to regions where losses exceed accumulation (Fig. 16-7). The *terminus*, or downglacier extremity, represents the line where losses by all causes (melting, sublimation, erosion, and calving into water—collectively termed **ablation**) equal the rate at which ice can be supplied by accumulation and forward motion. In the *zone of accumulation*, vectors of particle motion are downward into the ice mass as each year's snowfall adds a new surface layer to the glacier. In the *zone of ablation*, vectors of particle motions point toward the ice surface, because a surface layer is annually removed, exposing progressively deeper ice (Fig. 16-7). At the cross section of the glacier where upglacier net accumulation and downglacier net ablation are in exact balance, internal flow vectors are parallel to the glacier surface. The equilibrium cross section of a glacier is located at the **equilibrium line**, or downglacier edge of net annual accumulation. Theoretical models of glacier flow fit observations at the equilibrium line especially well (Andrews, 1975).

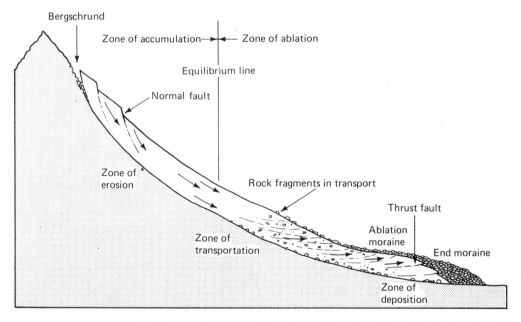

Figure 16-7.   Trajectories of particle motion in a valley glacier. The large area of the zone of accumulation is not well shown in this longitudinal cross section.

Flow lines in a circular ice cap or ice sheet are less well known. The ice must spread radially in plan view from the area of accumulation, as well as outward along any cross section. However, the general flow lines (Fig. 16-8) are probably much as inferred by early workers: downward near the center and then outward in the marginal zone of ablation. Obstructions such as enclosing mountain ranges obviously complicate the flow patterns. Paterson (1969, Chap. 9) reviewed the theories of flow in ice sheets and concluded that the horizontal component of ice motion in the central part of a large ice cap is very low. For an ice sheet 2000 km wide, with a central thickness of 4700 m, a parabolic profile, and net accumulation rate of 15 cm of ice per year, Paterson calculated that it would take a particle of ice 75,000 years to travel to the edge from a starting point 50 km from the center. About 60 percent of the time would be spent in moving the first 250 km outward from the starting point. The dimensions and rates are reasonably appropriate for the existing Antarctic and Greenland ice sheets and for the former North American and Scandinavian ice sheets.

For various reasons, excluding hydrostatic pressure, shear in an ice cap should be most rapid near the base. The highest temperatures should be there because of the poor conductivity of ice and the geothermal heat flow. Ice at higher temperatures deforms more rapidly. Also, the greatest shear stress is at the base of the ice, although this factor does not vary greatly. If any regelation slip occurs, it too will be at the base. A possible contribution of frictional heat must also be considered. Therefore, reasonable models of flow in ice caps are obtained by assuming that all the shear takes place in a basal layer only a few meters thick, and the overlying ice is passively transported. Under such assumptions, flow rate is proportional only to the surface

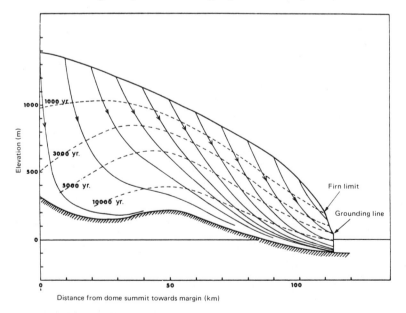

Figure 16-8. Particle paths and calculated ages of ice in a cross section of Law Dome, a small independent ice dome on the edge of the Antarctic ice sheet in Wilkes Land (111°E longitude). (From Budd and Morgan, 1973, Fig. 4.)

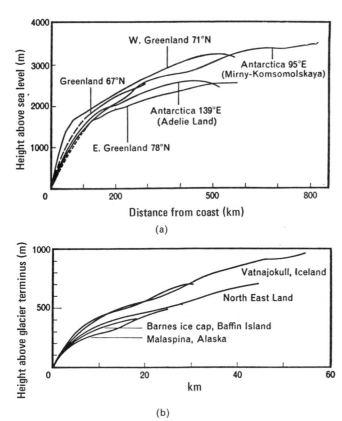

Figure 16-9. Surface profiles of (*a*) the ice sheets of Greenland and Antarctica, and (*b*) smaller ice sheets and ice caps. In both figures, bedrock relief is slight relative to ice thickness. (Robin, 1964, Fig. 5.)

slope of the ice, which develops an elliptical or parabolic profile (Figs. 16-9, 16-11). However, none of the theoretical models fit the central or near-margin parts of any ice sheet very well. Neither do they account for surges or other disproportionate rates of flow along certain radii while the remainder of the ice sheet remains in equilibrium.

Surface Flow in Valley Glaciers. Among the oldest observations of glacier flow were reports that lines of stakes straight across valley glaciers would bulge progressively downvalley year after year, proving that the center part of the ice surface moved faster than the margins. Many subsequent experiments have confirmed the fact (Fig. 16-4). The mean velocity of the surface ice is within a few percent of the mean velocity of the entire glacier cross section (Paterson, 1969, p. 109). As a useful result of this approximation, glaciologists can estimate the annual discharge of ice through the equilibrium-line cross section of a glacier by measuring surface velocities only. It is not necessary to drill boreholes and measure velocities within the ice. In this respect especially, the laminar flow of a glacier is very different from the turbulent flow of a river, for which an entire velocity cross section must be measured to determine mean discharge. Shear stress, or the lateral rate of change of velocity, is highest near the edges of valley glaciers. Most crevasses are there.

Thermal Classification of Glaciers

From the foregoing discussion of ice deformation and glacier flow, two categories of glaciers can be visualized. One category is of glaciers that are cold from top to bottom, with the entire thickness of ice well below the pressure-melting temperature. The other category is of glaciers that are at or very close to the pressure-melting temperature, so that small pressure variations can produce melting and regelation slip. Actually, this thermal classification is better applied to parts of glaciers rather than entire valley glaciers or ice caps, because the thermal status of one part of a glacier may be different from that of another part. A discussion of the thermal status of glaciers and parts of glaciers uses terminology first proposed over 40 years ago (Lagally, 1932; Ahlmann, 1933, 1935), but the examples and conclusions are much more recent. The thermal condition of glaciers has a very important bearing on both their form and their ability to shape landscapes.

Cold, Polar, or Dry-Base Glaciers. A 1387-meter borehole through the Greenland ice sheet at Camp Century, about 100 km from the west edge of the ice sheet, was completed in 1966 (Figs. 16-10, 16-11). At the time, it was the deepest borehole made in ice and the first to penetrate entirely through a polar ice cap. The temperature profile illustrates well the characteristics of a **cold**, or **polar**, glacier. The ice temperature at a depth of 10 m was −24°C, very close to the mean annual air temperature.

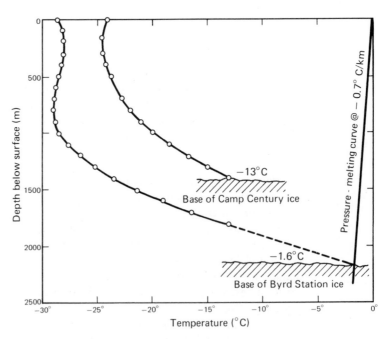

Figure 16-10. Temperature profiles of boreholes at Camp Century, Greenland (Weertman, 1968) and Byrd Station, Antarctica (Gow, et al., 1968).

A minimum temperature of $-24.6°C$ was reached at a depth of 154 m; then the temperature rose steadily to $-13.0°C$ at the base of the ice. The entire thickness of ice at the Camp Century borehole is far below the pressure-melting temperature. No water phase can exist, regelation slip cannot occur, and the rather slow surface motion of only 3.3 m/yr must be due entirely to plastic flow and minor brittle fracturing.

Cold, or polar, glaciers are also called **dry-base** glaciers, because the basal ice must be frozen firmly to the rock beneath. Experiments demonstrate that the surface bond between silicate minerals and ice is considerably stronger than the yield strength of ice, so that any motion of a cold glacier must be by internal plastic flow. Just one consequence of this fact is that cold glaciers cannot be expected to produce striations and grooves by sliding over bedrock. Most of the motion is presumed to be concentrated in the basal, relatively "warm" ice, however, and differential shear in the basal zone might break off rock projections or fracture larger rock fragments in transport.

Temperate, or Wet-Base, Glaciers.   Drawn in Figure 16-10 is the theoretical graph of the temperature of ice at the pressure-melting point with depth in a glacier. Melting temperature changes with depth in a glacier at about $-0.7°C/km$ (see also Fig. 6-5). If the mean annual temperature of the accumulating ice is close to $0°C$, then with progressive burial, pressure melts a minute amount of ice, and the ice and water phases coexist at progressively lower temperatures with depth. This is the definition of a **temperate**, or **wet-base** glacier. Regelation slip and basal sliding are probable, and flow rates are likely to be at least twice as fast as in polar glaciers. Striated, grooved basal rock surfaces are possible, as are numerous other significant forms of glacier transport and erosion (Chap. 17), because the ice slides over rock or compact sediment on a film of water.

An important characteristic of a temperate glacier is that the temperature at the base is less than at the surface. No geothermal heat can flow upward as it does in a polar glacier. The geothermal heat that reaches the base of a temperate glacier cannot be conducted upward but must melt ice. This geothermal heat, plus an increment of heat generated by sliding friction that is also trapped at the base of the ice, assures that a temperate glacier must move on a wet base (Weertman, 1966). The nature of the water film is not certain. It is probably a very thin film, only 1 or 2 mm in thickness, but it may fill pockets in the ice in the lee of rock obstructions or gather in deeper basins. Certainly, permafrost cannot exist under a temperate glacier (Werenskiold, 1953).

Intermediate Categories.   Ahlmann (1933, 1935) subdivided polar glaciers (*arctic* glaciers in his original terminology) into two classes, *sub-polar* and *high-polar*. Although both are cold at depth, summer melting on the sub-polar glacier might produce a surface layer 10–20 m deep, which is at the melting temperature. Ahlmann's high-polar glacier would be typified by the Greenland icecap at Camp Century (Fig. 16-10) in which not even a surficial layer of snow melts in the summer.

Lagally (1932), in a classification otherwise almost identical to Ahlmann's polar and temperate glaciers, proposed an *intermediate* type that would be cold or polar at the top but be at the pressure-melting temperature near the base. The intermediate category of Lagally was accepted as an interesting theoretical possibility but was generally overlooked until 1968, when the first borehole was drilled to bedrock at Byrd Station in Antarctica, (Fig. 16-10) (Gow, et al., 1968). At a depth of 2164 m, the bottom of the ice was reached, but at the basal contact was a film of water at least 1 mm in thickness, under enough pressure to rise 50 m up the drill hole and displace the antifreeze solution that had been used to keep the hole open. At the base of the ice sheet, the calculated melting temperature is $-1.6°C$. The surface temperature is $-28.8°C$. The steeper-than-average thermal gradient in the lower half of the Byrd Station borehole suggests a geothermal heat flow about 50 percent greater than the world average. Because West Antarctica is an ice-covered volcanic archipelago, the higher heat flow is reasonable. The significance of the borehole data to the history and nature of flow of the Antarctic ice sheet has yet to be fully evaluated. Subsequent radar depth soundings of the Antarctic ice sheet have indicated that large areas of the basal ice are wet (Oswald and Robin, 1973). The effectiveness of the wet-base condition is demonstrated by the fact that the Byrd Station drill bit was unable to obtain a sample of the sub-ice material, because the ice moved laterally several centimeters per day and constantly pinched the drilling tool in the borehole. The substrate may be either rock or saturated, clay-rich basal sediment.

GEOMORPHOLOGY OF GLACIER SURFACES

The simplest morphologic subdivision of glaciers is to distinguish those that flow between confining rock walls, or **valley glaciers**, and those that flow unconfined by virtue of their great thickness, the **ice caps** and **ice sheets**. Although the morphology of all glaciers is controlled by the rheology of ice, important differences are to be noted whether flow is confined or unconfined.

Valley Glaciers

A genetic sequence can be recognized in mountainous terrains, progressing from seasonal snow in nivation hollows, to permanent ice fields, to valley glaciers. The essential component for any glacier is an area in which annual accumulation exceeds all ablation losses. As the accumulating snow layers thicken and compact, the yield stress of ice is exceeded, and glacier flow begins.

The conversion of snow to glacier ice follows a regular sequence. Fresh snow is full of entrapped air and may have a bulk specific gravity even lower than 0.1. Snowflakes readily sublime, so old snowflakes lose their frilly margins and become more globular. Meltwater often freezes onto snowflake nuclei, so aging also tends to increase the size of the ice granules. Old snow, such as remains on sheltered mountain slopes in early summer, has the texture of very coarse sand. This is the late-season *corn snow* or *buckshot snow*, familiar to avid skiers. The loose granular mass is about one-half ice and one-half entrapped air and has a bulk specific gravity of about 0.5.

Snow that has survived a summer melting season is called **firn** in German or **névé** in French. Both terms are widely used in English. Firn, or névé, is an intermediate step in the conversion of snow to glacier ice. It is granular and loose unless it has formed a crust. It represents the net positive balance between winter accumulation and summer losses.

As successive annual layers accumulate, the deep firn is compacted. Individual ice grains freeze together. By definition, when the grains are frozen together so that the mass is impermeable to air, firn becomes glacier ice. The bulk specific gravity is usually about 0.8 by this stage of consolidation. Most glacier ice is a polycrystalline mass of frozen water plus a variable amount of air. Other components are dust and rock fragments that have fallen, washed, or been blown onto the ice surface, and rock that has been eroded from beneath the glacier. The bulk specific gravity of glacier ice ranges from about 0.8 to 0.9, which is approximately the specific gravity of pure, gas-free ice.

In a mountainous terrain, the colder temperatures and frequently higher precipitation at higher altitudes produce ice fields at the heads of glaciers. This is the zone of accumulation (Fig. 16-7), and as a rule of thumb, about 70 percent of the surface area of a valley glacier is in this zone. The zone of accumulation is separated from the zone of ablation by the equilibrium line. The annual discharge of ice through the cross section of the equilibrium line is an important measure of the glacier's regimen. The equilibrium line is approximated by the **firn limit**, or the lowest line of surviving accumulation at the end of summer, and by the annual **snow line** on the glacier and adjacent mountain slopes (Müller, 1962).

Downglacier from the equilibrium line, the glacier has net annual loss of ice. Melting is the major factor, but sublimation, wind ablation, and iceberg calving into lakes or the ocean also contribute. A glacier is in equilibrium only if annual discharge through the equilibrium line is equal to the net annual accumulation upglacier and the net annual ablation downglacier. Otherwise, it will grow or shrink. Ablation is at a maximum toward the downglacier terminus, and the terminus itself marks the line where ablation is complete. Some valley glaciers have steep ice faces at their termini, but others end deeply buried in transported rock debris, so that the exact location of the terminus is uncertain.

Many large valley glaciers are fed by more than one *tributary glacier* (Fig. 16-6). The surfaces of merging glaciers join at the same level, because ice is easily deformed and responds to the lateral stresses imposed by an entering tributary. However, erosive power is not proportional to thickness, and even though a small tributary may merge with a larger glacier at a common surface level, the bedrock floor under the smaller glacier is likely to be far above the axis of the floor under the larger glacier (p. 391). Tributary glaciers do not turbulently mix with the trunk glacier, as rivers do. Because of the nature of flow in ice, the tributary glacier retains its identity as a separate stream or surface ribbon of ice (Fig. 16-6), although it may become progressively attenuated or enfolded into the trunk glacier.

Glaciers have much higher surface gradients than rivers, commonly averaging 10 percent (6°). On ice cliffs or ice falls steeper than 45°, ice avalanches are nearly continuous. Ledges buried beneath the ice are reflected by crevasse fields (*séracs*) on the ice surface (Fig. 16-6). The long profile of a glacier surface is likely to be marked by level areas, steep drops, and even local reversals of slope, much in contrast to the surface gradient of a river.

On the surface of a temperate valley glacier in the ablation zone, a variety of ephemeral landforms develop due to melting, collapse of ice tunnels, and erosion by streams flowing on, in, or along the glacier. Many features are analogous with karst features and may be collectively called **glacier karst**. **Thaw lakes** and swallow holes (**moulins**) mark the ice surface. Segregations of debris-rich ice stand as ice-cored mounds or ridges above the debris-covered surface. Lakes may drain and then abruptly form again when a crevasse or moulin opens at depth, and then clay-rich sediments seal it closed again. Ice tunnels within or at the base of the glacier discharge sediment-laden water from the terminus.

Ice Caps and Ice Sheets

Unconfined glaciers are so large that they form their own orographic precipitation pattern. The accumulation zone is in the central region of the dome, and the ablation zone is on the periphery, below the equilibrium line. The regimen of an ice cap is very sensitive to slight climatic changes, because a slight raising or lowering of the equilibrium line on a convex-skyward dome can have an enormous effect on the relative areas of accumulation and ablation. Numerous small or moderate-sized ice caps on Iceland, Scandinavia, and the Canadian Arctic islands could not reform today if they once fell below the firn limit; sustaining orographic precipitation is a consequence of the ice caps themselves.

The radial profiles of continent-sized ice sheets and the somewhat smaller, but otherwise similar, ice caps differ very little from each other (Fig. 16-9). Theories of radial flow that assume shear stresses of only 20–30 kPa (0.2–0.3 bars) at the horizontal base of ice sheets, and either plastic flow concentrated just above the base or basal sliding, predict parabolic or elliptical radial profiles very similar to those actually observed (Robin, 1964). The equations of the predicted profiles are not very sensitive to either internal temperature or rate of accumulation, two factors that strongly affect the steepness of valley glaciers. Note the considerable vertical exaggeration on Figures 16-9 and 16-11. At true scale, the thickness of the ice could barely be shown by a slight thickening of the horizontal base lines. A published radar

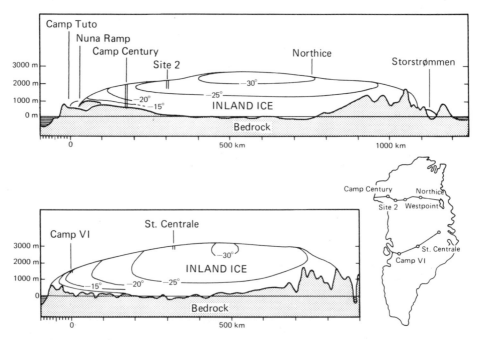

Figure 16-11. Profiles and estimated temperatures of the Greenland ice sheet (the Inland Ice). Vertical exaggeration 50 : 1. (Weidick, 1975, Fig. 4.)

profile of the North Greenland ice sheet (Walker, et al., 1968) gives an excellent idea of the true-scale surface relief, bottom relief, and thickness of a large ice sheet.

Most of the Greenland ice sheet perimeter is a series of broad, low ramps. Along some parts of the perimeter, where the terminus is on land, the glacier ends as a vertical ice cliff as much as 100 m in height. The cliffs may be due to a combination of factors, such as low temperatures, compass orientation relative to a low sun angle that causes sublimation directly from the cliff face, and cold, brittle ice almost free of enclosed sediment except at the base. The Greenland ice sheet (Known as the *Inland Ice*) rests on a basin-shaped interior plateau surrounded by mountain ranges (Fig 16-11). Around much of its margin, the ice moves between exposed or slightly buried mountains peaks as **outlet glaciers**. These have many of the geomorphic properties of valley glaciers. At the head of each outlet glacier is an area of shallow concavity sloping toward the outlet, with many wide, deep crevasses. This is the "drawdown" area, or "ice-shed" of that particular outlet glacier.

Much of the perimeter of the Antarctic ice sheet is also a great cliff, but that cliff ends at a **grounding line** in the ocean (Fig. 16-8) from which enormous tabular icebergs calve off. The surface of the Antarctic ice sheet is a monotonous dome with only minor areas of exposed bedrock, mostly near the coast. Flow lines define ice divides and drainage basins, but surface gradients are very low, and little relief is apparent. A few **nunataks**, peaks surrounded by ice, rise above the surface (Fig. 16-12).

Figure 16-12.   South-facing slope of Mount Takahe, a volcanic nunatak in Marie Byrd Land, West Antarctica. Other nunataks in distance. Ice flow from cirques is southward (left); continental ice sheet flows northward (right). Distance from ice-sheet surface to caldera rim (extreme right) is about 10 km. (Photo: U.S. Navy for U.S. Geological Survey, TMA 1719 F-33 Exp. 161.)

REFERENCES

AHLMANN, H. W., 1933, Scientific results of the Swedish-Norwegian Arctic Expedition in the summer of 1931, Part VIII, Glaciology: *Geograf. Annaler*, v. 15, pp. 161–216, 261–95.

———, 1935, Contribution to the physics of glaciers: *Geog. Jour.*, v. 86, pp. 97–113.

ANDREWS, J. T., 1975, *Glacial systems:* Duxbury Press, North Scituate, Mass. (a division of the Wadsworth Publishing Co., Belmont, Cal.), 191 pp.

BUDD, W. F., and MORGAN, V. I., 1973, Isotope measurements as indicators of ice flow and paleo-climates, *in* van Zinderen Bakker, E. M., ed., *Paleoecology of Africa and the surrounding islands and Antarctica*, A. A. Balkema, Cape Town, v. 8, pp. 5–22.

FLINT, R. F., 1971, *Glacial and Quaternary geology:* John Wiley & Sons, Inc., New York, 892 pp.

GLEN, J. W., 1952, Experiments on the deformation of ice: *Jour. Glaciology*, v. 2, pp. 111–14.

———, 1974, Physics of ice: *U.S. Army Corps of Engineers, Cold Regions Res. and Eng. Lab., Cold Regions Sci. and Eng. Mono. II-C2a*, 81 pp.

———, 1975, Mechanics of ice: *U.S. Army Corps of Engineers, Cold Regions Res. and Eng. Lab., Cold Regions Sci. and Eng. Mono. II-C2b*, 43 pp.

GOW, A. J., UEDA, H. T., and GARFIELD, D. E., 1968, Antarctic ice sheet: Preliminary results of first core hole to bedrock: *Science*, v. 161, pp. 1011–13.

KAMB, B., 1964, Glacier physics: *Science*, v. 146, pp. 353–65.

KAMB, B., and LaCHAPELLE, E., 1964, Direct observations of the mechanism of glacier sliding over bedrock: *Jour. Glaciology*, v. 5, pp. 159–72.

LAGALLY, M., 1932, Zur Thermodynamik der Gletscher: *Zeitschr. für Gletscherkunde*, v. 20, pp. 199–214.

MEIER, M. F., 1960, Mode of flow of Saskatchewan Glacier, Alberta, Canada: *U.S. Geol. Survey Prof. Paper 351*, 70 pp.

MÜLLER, F., 1962, Zonation in the accumulation areas of the glaciers of Axel Heiberg Is., N.W.T., Canada: *Jour. Glaciology*, v. 4, pp. 302–11.

OSWALD, G. K. A., and ROBIN, G. DEQ., 1973, Lakes beneath the Antarctic ice sheet: *Nature*, v. 245, pp. 251–54.

PATERSON, W. S. B., 1969, *The physics of glaciers:* Pergamon Press, Elmsford, N.Y., 250 pp.

PETERSON, D. N., 1970, Glaciological investigations on the Casement Glacier, southeast Alaska: *Ohio State Univ. Inst. of Polar Studies, Rept. no. 36*, 161 pp.

ROBIN, G. DEQ., 1964, Glaciology: *Endeavour*, v. 23, pp. 102–7.

SCHYTT, V., 1956, Lateral drainage channels along the northern side of the Moltke Glacier, northwest Greenland: *Geograf. Annaler*, v. 38, pp. 64–77.

TARR, R. S., 1909, The Yakutat Bay region, Alaska: *U.S. Geol. Survey Prof. Paper 64*, 183 pp.

WALKER, J. W., PEARCE, D. C., and ZANELLA, A. H., 1968, Airborne radar sounding of the Greenland ice cap: Flight 1: *Geol. Soc. America Bull.*, v. 79, pp. 1639–46.

WEERTMAN, J., 1966, Effect of a basal water layer on the dimensions of ice sheets: *Jour. Glaciology*, v. 6, pp. 191–207.

———, 1968, Comparison between measured and theoretical temperature profiles of the Camp Century, Greenland, borehole: *Jour. Geophys. Res.*, v. 73, pp. 2691–700.

WEIDICK, A., 1975, Review of Quaternary investigations in Greenland: *Grønlands Geologiske Undersøgelse Miscell. Papers*, no. 145; distributed as *Ohio State Univ. Inst. of Polar Studies, Rept. no. 55*, 161 pp.

WERENSKIOLD, W., 1953, Extent of frozen ground under the sea bottom and glacier beds: *Jour. Glaciology*, v. 2, pp. 197–200.

WRIGHT, H. E., JR., 1973, Tunnel valleys, glacial surges, and subglacial hydrology of the Superior Lobe, Minnesota, *in* Black, R. F., Willman, H. B., and Goldthwait, R. P., eds., The Wisconsinan Stage: *Geol. Soc. America Mem. 136*, pp. 251–76.

CHAPTER 17

Glacial Morphogenesis

The glacial climatic zone is defined by the combination of temperature and precipitation that permits glaciers to cover the rocky landscape. The glacial morphogenetic region is sharply circumscribed by the edge of a glacier. This is one of the few examples that can be cited where geomorphic processes and forms change abruptly at a sharp boundary. Only at a coast is there such an equally abrupt process-controlled change in landforms.

The difficulty of describing glacial morphogenesis is that most of the processes go on beneath a layer of ice, which is properly regarded as a common rock type at the earth's surface. Therefore, strictly speaking, glacial morphogenesis is a subsurface activity, and the true glacier landscape is on top of the ice, not beneath it. These are the forms described in the previous chapter. This chapter considers the geomorphic work of glaciers in shaping the surface morphology of other rocks. However, these are not landscapes until they are exposed by the disappearance of the glaciers. In this sense, all glaciated landscapes are relict even if they were exposed by the previous year's retreat of a glacier terminus.

Because of the effectiveness of glacial erosion, transport, and deposition, and of the associated work of water and wind, glaciated landforms preserve their morphogenetic origins long after glaciers cease to shape them. In very large areas of middle-latitude Europe and North America, the landforms of glacial origin totally dominate the few landforms developed in postglacial time, even though the areas have been deglaciated for as long as 18,000 years. Much of the total postglacial landscape evolution may have been produced during a brief phase of periglacial morphogenesis immediately after deglaciation (p. 349), but the impact of late-glacial processes has yet to be firmly established. Clearly, glacial processes are able to shape landscapes so powerfully that a much longer period of fluvial erosion is required to remove the evidence of glaciation. The oldest Pleistocene glacial deposits of the central United States, although deeply weathered and dissected, still control the regional topography although they are over 2 million years old (p. 407).

GLACIAL EROSION

Processes

Abrasion. Table 17-1 lists the hardness, by the standard Mohs' scale, of ice at various temperatures. From the table, it is clear that some soft minerals or weathered rock surfaces can be abraded if very cold ice slides over them. However, the very low yield stress of ice is more likely to cause plastic deformation of the ice pressing against a rock surface rather than abrasion of the rock. Ice grains blown by cold winds might be expected to polish rock surfaces as blowing sand does, but cold ice probably shatters before it significantly abrades rock surfaces. Therefore, ice alone, even when very cold, is probably not an effective abrasive. (However, press reports have described Soviet optical works using very cold ice as a polishing and grinding abrasive for large astronomical telescope mirrors.)

TABLE 17-1. BRITTLE HARDNESS OF ICE AT VARIOUS TEMPERATURES (MOHS' SCALE)

Temperature	Hardness
0°C	1.5
−15°C	2–3
−30°C	3–4
−40°C	4
−78.5°C	6 (hardness of orthoclase)

SOURCE: Shumskii, 1964, pp. 39–40.

The abraded scratches and grooves that characterize glaciated rock surfaces are undoubtably caused by rock fragments in the ice rather than by the ice itself (Boulton, 1974, p. 43). If the glacier is sliding over its base on a water film, any fragment larger than the thickness of the water film will be rolled or dragged against the rock surface. Glaciated surfaces show a range of abraded microrelief forms from hairline *striations* to *grooves* a meter or more in depth and width. Careful study of glacial striations and grooves can sometimes reveal the direction of ice motion, as when a piece of rock is chipped off and dragged downglacier over the surface, or when a resistant projection protects less resistant rock in its lee from abrasion (Fig. 17-1). Regelation slip (p. 372) must be an effective process of abrasion, because if a sand grain is embedded in basal ice, it will move into contact with the rock surface as the enclosing ice melts. Abrasion is ineffective beneath polar glaciers (Boulton, 1974, p. 83).

Figure 17-1. Glacially abraded surface of limestone, with a protruding chert concretion, a streamlined tail of limestone, and a shallow marginal moat. (Photo: A. Meike.)

Figure 17-2. Glacially scoured metamorphic rock suggesting a highly plastic abrasive medium, near Portør, Norway, on the Skagerrak coast. Match box is 5.4 cm long. (Photo: Just Gjessing.)

Some polished and striated rock surfaces show patterns of flow that diverge from the regional direction in minute, complex ways (Fig. 17-2). Local reversals of flow direction that give eddylike patterns, helical flow along shallow troughs, and potholes suggest a highly plastic abrasive substance (Gjessing, 1966). Very likely, the base of some temperate and intermediate glaciers (pp. 379 and 380) is underlain by a sheared mass of water-saturated, probably clay-rich, sediment under high confining pressure and shearing stress. The sediment moves generally in the same direction as the overlying ice but sensitively responds to local stress gradients around obstructions. Such saturated basal sediment has been exposed by rapid retreat of some glacier termini and has been observed in tunnels dug along the base of valley glaciers. It also makes constructional landforms when it squeezes up into crevasses near the margin of glaciers.

Not only bedrock surfaces are abraded; the rock fragments carried in glaciers also are subjected to abrasion. A significant proportion of glacially transported pebbles and cobbles have planocurved *facets* on one or more sides, showing that they were held in contact with other rocks as they ground past each other. Two of the strongest proofs that a region has been glaciated by a wet-base glacier are (1) a regional pattern of striated and grooved bedrock surfaces, and (2) abraded stones of exotic origins in the detritus overlying the surfaces.

Plucking.   During regelation slip, water migrates to the downglacier side of a sub-ice obstruction and refreezes (p. 372). Fractured rock can be incorporated into the regelation layer and carried away by the glacier. This process is called glacial **plucking**. It is not clear whether regelation ice has enough tensile strength to actually pull away lee-side pieces of rock or whether pieces already loosened by the stress of the moving ice are simply frozen in and carried away (Boulton, 1974, pp. 69–75). Whatever the process, plucking is much more effective than abrasion on fractured rocks. Jahns (1943) reconstructed the sheeting joints on crystalline metamorphosed rocks near Boston, Massachusetts (p. 106) to show that plucking had removed several times the volume of rock from the lee sides of hills than abrasion had removed from the upglacier, or **stoss** (German for *push*) sides.

The combination of abrasion and plucking exploits structural details of rocks, especially joints. A variety of streamlined landforms, such as *stoss-and-lee topography*, *roches moutonnées*, and *crag-and-tail* hills are variants of the combined abrasion and plucking process. They may form equally well under ice sheets or valley glaciers, but a wet-base or temperate thermal status is necessary. All are small hills with abraded upglacier sides, but the crag-and-tail variant has a streamlined tail of compact sediment or erodible bedrock sheltered behind a more resistant crag (Fig. 17-1), whereas the other two forms have steepened, plucked lee faces.

Erosional and Residual Landforms

Troughs.   The erosional landforms of valley glaciers rank with the most spectacular scenery on earth. The European Alps are so thoroughly dissected by valley glaciers that the entire assemblage is sometimes simply referred to as *alpine topography*. The

Figure 17-3.   View northeastward over Mount Cook (3800 m) up the Tasman Glacier, New Zealand. Horns and arêtes are prominent. (Photo: New Zealand National Publicity Studios.)

fundamental landform conponent is the **glacial trough**, the channel of a former valley glacier. The cross profile of a glacial trough is a catenary curve, not unlike the cross profile of a river channel but quite different from the profile of an entire river valley. Concave slopes dominate, rising from the valley floor to steep cliffs or stepped-back, ice-abraded shoulders (Figs. 16-6, 17-3, 17-10, and 18-6). In long profile, glacial troughs may be a series of basins separated by cross-valley ledges of more resistant rocks (Gutenberg, et al., 1956). If not filled with sediment, the basins form chains of lakes, each with a waterfall over the downstream ledge to the next lake. These delightful chains are called **paternoster lakes**, from the resemblance to shining beads on a rosary.

Trough Lakes and Fiords.   In a glacial trough that has either a high rock bar or a massive end moraine blocking its lower end, a **trough lake** will drown the valley floor. Trough lakes are usually long, steep-sided, deep, and cold, with a minimal inflow of sediment because of the clean abraded valley walls and the upstream settling basins on most tributary streams. The Swiss lakes, the Como region of northern Italy, the Lake District of England, the Finger Lakes of central New York (Fig. 17-4), and many other regions of large trough lakes have become justly famous as scenic tourist areas. Hydroelectric projects make use of the large storage volumes and abrupt gradients associated with trough lakes in mountainous regions.

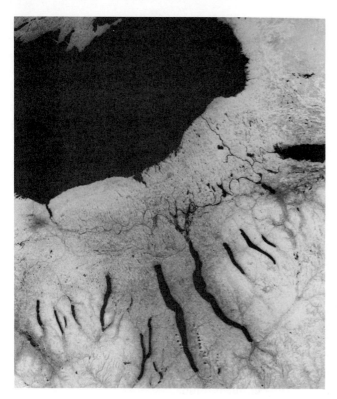

Figure 17-4. The Finger Lakes of central New York. All the lakes drain north to Lake Ontario (top) via the Oswego River. Cayuga Lake, the easternmost of the two longest lakes, is 61 km long. A light snow cover enhances the relief. (NASA ERTS E-1243-15244.)

Fiords are glacial troughs that have been entered by the sea (Fig. 20-2). Their form is no different from that of any other trough except that near the outer coast, many fiords shoal rapidly to only a fraction of the depth between more restricting walls farther inland, where multiple basins may be 1 km or more in depth. In some fiords, the bar at the mouth is an end moraine, but usually it is bedrock and is attributed to a reduction in intensity of vertical erosion as the valley glacier either began to float (Crary, 1966) or spread out as a piedmont lobe on the land that would have been exposed by the worldwide lowering of sea level during glacial intervals.

The side walls of glacial troughs are straightened by abrasion and plucking. The interlocking spurs that interrupt views along a river valley are cut away, giving a series of **ice-faceted spurs** that align between lateral valleys. Where tributary glaciers joined the trunk ice stream, smaller glacial troughs form, but their floors are usually well above the floor of the main valley (p. 382). Waterfalls cascade down the main valley wall from these **hanging valleys**. The accordance of valley floors, so important to Playfair's law, is absent in alpine glacial scenery.

Cirques.   At the head of each trough, in the accumulation area of the glacier that eroded the valley, are steep-headed, semicircular basins, or **cirques**. The bowl shape is a consequence of intense frost action and nivation processes adjacent to the ice field plus the flow lines in the zone of accumulation (Fig. 16-7) that carry surface rock debris downward toward the base of the glacier, where the bed is heavily abraded. A cirque has much the shape of a human heel-print in sand and is often "down-at-the-heel," that is, more deeply eroded as a basin behind an abraded rock bar at the lip of the cirque. A cirque lake, or **tarn**, dammed by the rock bar, occupies the basin in many cirques.

Cirques are a unique and diagnostic feature of alpine glaciation and have been given many local names, indicating their recognition as landforms. In addition to cirque (French), other terms such as *cwm* (Welsh), *corrie* (Scottish), *kar* (German), *botn* (Norwegian), and *nisch* (Swedish) are encountered in the geomorphic literature. The association of active cirques with the accumulation areas of modern valley glaciers led to the recognition of similar relict forms very early in the history of geomorphology. By nature of their position high on mountain slopes, many cirques that fed Pleistocene valley glaciers still carry small cirque glaciers or greatly reduced valley glaciers in them. Many cirques are compound, with a nested series of progressively smaller and higher forms representing stages in the reduction of icefield size in late-glacial and postglacial time.

Cols, Arêtes, and Horns.   If a plateau or mountain range is lightly but incompletely dissected by glacial troughs and cirques, the former upland surfaces remain as remnants between the steep, concave glaciated forms (Fig. 17-5). This condition is well shown in the central Rocky Mountains and the Big Horn Mountains, where it has been described as a **scalloped upland**, or **biscuit-board** topography. The latter term describes the landscape well, by analogy to a thick layer of biscuit dough on a cutting board, with arcuate pieces removed by cirque erosion.

It is not hard to deduce the sequential development of a scalloped upland into a **fretted upland**, completely dissected by cirques. As two steep-sided cirques intersect, U-shaped notches cut the divide between their headwalls. Such a gap is called a **col** in alpine terminology. A saw-toothed divide consisting of a series of cols and inter-

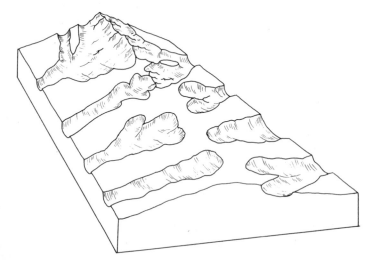

Figure 17-5.   A scalloped upland on a broad, gentle divide (foreground) evolving by intersection of opposing cirques into a fretted upland of arêtes, horns, and cols at higher altitudes (background).

vening ridge segments is an **arête**. Where three or more cirques intersect, a higher pyramidal or triangular-faceted peak may rise above the general level of the arête. These peaks are the **horns** (or *Matterhorn peaks*, after the type example) of alpine scenery. A summit area of arêtes, cols, and horns represents the complete destruction of the preglacial surface (Fig. 17-3), and is analogous to the mature dissection of a regional fluvial landscape. The most prominent horns are at the intersection of three or more radiating arêtes.

Through Valleys and Breached Divides.   If mountain glaciation increases in intensity until a mountain ice cap develops, former regional divides may be breached by glacial erosion, and major diversions of postglacial stream flow result. The Brooks Range of Alaska was glaciated more intensely on the southern, snowier side of the regional divide, whereas only short valley or piedmont glaciers formed on the arid northern Arctic slope. Headward erosion by the south-flowing glaciers breached the preglacial regional divide in several places, and major ice streams that originally flowed southward on relatively gentle profiles were diverted down the steeper northward troughs. Breached divides of this sort become strongly abraded and low. In the Brooks Range, the divide at Anaktuvuk Pass, between the Yukon and Arctic drainage, is only slightly over 600 m above sea level and is on a swampy valley floor among moraines (Porter, 1966).

Similar breached divides in the glaciated northern Appalachian Mountains were produced by multiple advances of the Pleistocene ice sheets that overrode the preglacial divides between the St. Lawrence–Great Lakes drainage and the southerly draining Susquehanna, Delaware, and Ohio Rivers. Powerful erosion through constricted valleys at the base of the ice sheets produced the Finger Lakes troughs (Fig. 17-4) on the north side of the divide but left flat-gradient, swampy **through valleys** between north- and south-flowing tributaries (Tarr, 1905). The present divide is inferred to be close to the preglacial position in most areas, but valley forms have been modified into a series of intersecting troughs "like the intersecting passages of some unroofed ants' nest" (Clayton, 1965, p. 51; see also Linton, 1963; Coates, 1966). Undoubtedly, much of the erosion of the through valleys can be attributed to meltwater overflowing into the southerly drainage from large ice-dammed lakes north of the divide at times when the ice margin blocked the St. Lawrence lowland. Each ice advance must have been preceded by an intense period of fluvial erosion at the overtopped divides. Periglacial processes may have helped prepare the terrane for glacial scour, too.

Regional Denudation

Valley Glaciers.   Table 12-3 presents some evidence that the regional denudation rate in mountains with glaciers approaches 1 mm/yr, or 1 m/1,000 yr, the most rapid of any morphogenetic region. The combination of high relief, steep slopes, intense nivation, and glaciers must indeed have a powerful impact on landscapes. The share of the work borne by glaciers alone is hard to evaluate, because glaciers are just one aspect of geomorphic processes in high snowy mountains. It seems certain that alpine glaciation, in which most of the upland is buried under a mountain ice cap from which valley glaciers radiate, must be one of the most rapid means of destroying mountains. Deductions concerning an alpine "cycle of erosion" were never satisfactory, because after the "fretted upland" stage of mature dissection, any further erosion must lower the ice fields toward or below the regional equilibrium line, reduce the accumulation areas, and thus reduce the effectiveness of glacial processes before a deduced endform can be reached. The **gipfelflur**, or "sea of peaks" all at about the same elevation, is a well-known feature of alpine scenery (Fig. 17-3), but whether the horns that rise above mountain ice fields represent remnants of an ancient erosion surface or represent the mean level of intersection of adjacent cirques and troughs cannot be determined. Also, in the Holocene Epoch (a postglacial or interglacial interval), glaciers all over the earth have greatly receded, and the high denudation rates of mountainous regions are at least partly due to the many oversteepened slopes relict from glacial erosion and deposition but now exposed to mass-wasting and fluvial erosion (Church and Ryder, 1972). Some authors have proposed a hypothesis of glacial protection, in which glaciation preserves a landscape by burying it and stopping fluvial erosion. That hypothesis might be considered for high-polar, or dry-base, ice sheets or ice caps (Andrews and LeMasurier, 1973, p. 80) but is hardly tenable for valley glaciers.

Ice Caps and Ice Sheets.   An extreme range of views can be found concerning the effectiveness of large ice sheets in denuding landscapes. Some authors (White, 1972) argue that the Laurentide ice sheets of eastern North America completely stripped the Paleozoic and younger sedimentary rocks off large areas of the Canadian shield. Other authors (Ambrose, 1964) have noted that the buried relief of the crystalline shield, where it is unconformably overlain by lower Paleozoic or Cretaceous sedimentary rocks, is very similar to the local relief on the shield today. By this line of reasoning, multiple Pleistocene glaciations have done little more than to remove the weathered surface layers and exploit structural weaknesses by glacial plucking (Prest, 1971, p. 697). The view of minimal regional denudation by ice sheets is supported by the rather small volume of glacial deposits on the periphery of the Laurentide ice sheet and by the fact that most of the glacial sediment was derived from relatively near the ice margin, especially on the southern quadrant in the United States (Gravenor, 1975).

Discoveries of deeply weathered kaolinitic rock in construction sites at Boston and of bauxite pebbles in glacial deposits on Martha's Vineyard, Massachusetts (Kaye, 1967a, 1967b) support the inference that in preglacial times, possibly in the late Tertiary Period, subtropical or tropical weathering affected the northeastern United States. Such weathering is known to affect rocks to depths of 100 m or more (Fig. 6-10), so it is possible that glacial erosion has accomplished little more than to remove the previously weathered rock from northeastern North America. The very large crystalline **erratics** derived from the Precambrian shield, but widely scattered in southern Canada and the glaciated United States east of the Rocky Mountains, need not represent erosion into fresh, unweathered rock; they could very likely be corestones (Fig. 11-3) from the preglacial weathered layer (Feininger, 1971). The control of stoss-and-lee landforms around Boston by sheeting joints, referred to earlier in this chapter, is also consistent with the hypothesis that glacial erosion by the Laurentide ice sheets removed much weathered rock but did not fundamentally reshape the bedrock relief. The nearly total lack of rock debris in the tabular icebergs derived from Antarctica suggests that glacial erosion is now inactive there (Warnke, 1970). However, the thick deposits of glacier-derived submarine sediments around Antarctica suggest that at some time, probably in the late Cenozoic Era, the Antarctic ice sheet eroded a considerable mass off the continent.

GLACIAL TRANSPORT AND DEPOSITION

Glaciers carry rock debris on, in, and beneath the ice. Valley glaciers probably derive most of their sediment loads from intense weathering and mass-wasting on adjacent slopes. An unknown amount is actually eroded by the moving glacier. Flow lines in a valley glacier carry surface debris downward into the glacier in the zone of accumulation and upward to the surface in the zone of ablation, so debris derived by mass-wasting or by subglacial erosion can be thoroughly mixed. Rock debris that composes lateral and medial moraines seems to stay fairly high in the glacier cross section, however (Fig. 16-6).

Ice sheets that bury landscapes cannot gain sediment by mass-wasting, because nothing is exposed above the ice. The Antarctic and Greenland ice sheets are notably free of any sediment except minute amounts of windblown dust, except very near their bases. Neither ice sheet appears to be transporting much rock debris, but they may not be good models for the former ice sheets of Europe and North America, which expanded and shrank repeatedly during the Pleistocene Epoch.

Several traditional terms have been formalized to describe rock debris that is both in transport on modern glaciers and is inferred to have been transported by former glaciers. The oldest word is **moraine**, used to describe the hummocky or ridged depositional landscapes on the margins of alpine glaciers and also to describe the rock debris in and on glaciers today. Some confusion results if the word moraine is used to describe both the material and the landforms, so only the latter meaning is used here. Moraine, then, is simply a collective term for depositional landforms of direct glacial origin. A classification of moraine and related glaciofluvial forms is outlined on the following pages.

For centuries before the glacial theory of Agassiz (1840), the poorly sorted, unconsolidated surficial deposits of northern Europe were supposed to have been the result of the biblical flood of Noah's time. In Germany, the term *diluvium* was widely used. In the British Isles, the same material was called *drift*, a term also used to describe marine sediment in transport along coasts. With the acceptance of the glacial theory to explain these deposits, **glacial drift** (or simply **drift** if no ambiguity is possible) has come to be the general English term for all sediment transported and deposited by glaciers or associated rivers. It is important not to confuse landforms and materials when studying glacial morphogenesis, so *drift* is used consistently to describe glacier-transported sediment, and *moraine* is used as a general term for glacier-depositional landforms.[1] The value of the distinction should become apparent in the following pages.

Glacial Drift

The following brief glossary defines some of the more common terms for categories of glacial drift. Details of composition, sorting, stratification, and genesis are left to specialized texts (Flint, 1971; Price, 1973; Embleton and King, 1975; Jopling and McDonald, 1975; Sugden and John, 1977), but this vocabulary is necessary to appreciate the genesis and morphology of morainic and related glaciofluvial landscapes.

1. **Till**   Nonsorted, nonstratified glacial drift, deposited directly from ice without reworking by meltwater. Called *boulder clay* by some British workers.
   a. *Basal*, or *lodgment*, till—Compact, tough, dense till, often clay-rich, deposited at the base of a temperate or wet-base glacier. Stones may show preferred orientation parallel to former flow.
   b. *Ablation* till—Less consolidated than basal till, often lacking the finer grain sizes. Carried on the surface of melting ice and "let down" without being overridden. Also used for the cover of debris on the ablation zone of modern glaciers, although "ablation moraine" is also applied to the sheetlike form of that debris.
2. **Ice-contact stratified drift**   Drift modified by meltwater during or after deposition. Considerable stratification and sorting is usual. Although water-modified, the large range of sizes, chaotic sedimentary structures, slump structures, faults, surface topography, and inclusions of till all confirm an origin in contact with, or in close proximity to, melting ice.
3. **Outwash**   Alluvium deposited by meltwater not in close proximity to melting ice. Sorting and stratification are more regular than in ice-contact stratified drift. Similar to other coarse alluvium, except as influenced by the extreme variability of discharge from glacial sources.
4. **Lacustrine sediment**   Lake-bottom deposits, usually fine grained, formed in settling basins on or near melting ice. Often forms silt and clay plains of broad extent, with distinctive erosional and soil-forming properties. Sometimes **varved**, or composed of rhythmic silt-clay couplets each representing an annual cycle of deposition.
5. **Eolian sediment**   Sand dunes, sand sheets, and loess derived from other types of glacial drift. Genetically undifferentiated from other eolian sediments except by regional and temporal association with other drift.

[1]See the discussion on p. 5 concerning the use of the terms *terrane* and *terrain*.

6. **Glacial marine sediment**   Clay-rich, but often stony, marine deposits on shelves and deep-ocean floor. Characterized by angular, mechanically crushed fragments and anomalous coarse particles embedded in fine mud. Coarse fraction attributed to ice rafting.

Moraines and Other Constructional Drift Landforms

Some moraines and constructional glaciofluvial landforms are genetically related to a specific kind of drift. Examples are *till plains*, which by definition are low-relief areas of eroded or deposited till; *eskers*, which by origin in an ice tunnel or channel must be made of ice-contact stratified drift; or *outwash plains*, which are depositional plains of outwash. Other terms for constructional glacial landforms do not indicate the nature of the drift from which they are built. *End moraines* range in composition from entirely till to entirely ice-contact stratified drift: Almost all moraines contain some mixture of these materials.

In an attempt to sort out the nomenclature of moraines, Prest (1968) proposed a classification based on the orientation of the landforms with respect to ice movement (Table 17-2). The classification is intended for landforms built by continental ice sheets but includes forms built by valley glaciers as well. The scheme is vaguely similar to the genetic classification of sand dunes (Fig. 14-9) in that longitudinal (parallel to ice flow), transverse (perpendicular to ice flow), and nonoriented forms are included. The transverse forms are mostly ice-marginal deposits, built during the last recession of the ice margin, but include some forms that are shaped beneath

TABLE 17-2.   CLASSIFICATION OF MORAINE*

Transverse to Ice-flow (*Controlled Deposition*)	*Parallel to Ice-flow* (*Controlled Deposition*)	*Nonoriented* (*Uncontrolled Deposition*)
Ground moraine—corrugated. End moraine; includes terminal, recessional, and push moraine. Ice-thrust moraine. Ribbed moraine. De Geer moraine; includes cross-valley moraine. Interlobate and kame moraine. Linear ice-block ridge; includes ice-pressed and ice-slump ridge, and crevasse-filling.	Ground moraine—fluted and drumlinized; includes drumlin, drumlinoid ridge, crag-and-tail hill. Marginal and medial moraine. Interlobate and kame moraine. Linear ice-block ridge; includes ice-pressed and ice-slump ridge, and crevasse-filling.	Ground moraine—hummocky (low relief); includes some ablation moraine. Disintegration moraine— hummocky and/or pitted (mainly high relief); includes dead-ice, stagnation, collapse, and ablation moraine. Interlobate and kame moraine. Irregular ice-block ridge and rim ridge; includes ice-pressed and ice-slump ridge, and ablation slide moraine.

Source: Prest, 1968.

*Included are landforms constructed from till or other drift that was deposited directly from, or in association with, glacier ice.

ice during ice advance. The forms that parallel ice flow, by contrast, are nearly all produced by actively moving ice, although some form in stagnant, melting ice along channels and crevasses parallel to former flow directions. The nonoriented forms are commonly built of a mixture of ablation till and ice-contact stratified drift, during a time of regional melting with insignificant ice motion. Longitudinal forms are likely to predate transverse and nonoriented forms in the same region. For instance, younger end moraines or kame moraines may be banked against, or draped over, drumlins and fluted ground moraines. Some of the morainic terms apply equally well to transverse, longitudinal, and nonoriented forms. For instance, drift pressed into the base of a crevassed and melting ice sheet may produce ridges oriented in several direction or produce a nonoriented maze of ridges and mounds. Drift is a notoriously mixed sediment; within a single moraine (Fig. 17-6), deltaic structures, lodgment till, and deformed ice-contact stratified drift may be found together.

Regional patterns are helpful in interpreting glacial landforms, because the terrane over which the glacier moved and the ablation history largely determine the regional character of both the drift and the landforms. Many morainic forms can be assembled in a genetic sequence that was called a *glacial series* (Fig. 17-6) by early workers (Penck and Brückner, 1909; von Engeln, 1921). The principle is a good one. At any one time during the uncovering of a landscape by ice, an end moraine may be forming at the ice margin, fronted by an outwash plain. Ice tunnels and crevasses behind the moraine may be the sites of eskers, crevasse fillings, and kames. Meltwater may be traced through an eroded channel across the end moraine to a fan or delta that coalesces with the outwash plain in front of it. The ice-contact, or *proximal*, part of the end moraine is of till or ice-contact stratified drift. On the *distal* or outwash side of the end moraine, the drift is mostly outwash, with inclusions of slumped ice-contact stratified drift or till. The entire assemblage represents contemporary events and, when understood as a unit, gives a good concept of the activity along a receding glacier terminus.

Morainic Landforms Constructed Primarily of Till.   Many of the ice-parallel morainic landforms are streamlined, and when developed in clusters give a vivid impression of ice motion. Similar erosional forms, such as roches moutonées and crag-and-tail hills have been described previously. Best known of the streamline ice-molded forms are **drumlins** (Muller, 1974). These are elliptical or ovoid hills, blunt on the upglacier end, with an elongate downglacier tail (Fig. 17-7). They occur in clusters, most frequently in areas of clay-rich basal till. Internal composition is not a criterion for calling a streamlined hill a drumlin, and known examples range from those that are nearly all rock with a thin veneer of till to those that are entirely till with no rock core. Classic examples are to be seen in northern Ireland, in New Brunswick, near Boston, Massachusetts, in southern Ontario and central New York, and in southern Wisconsin. In each region, hundreds or thousands of drumlins give the landscape a distinct "grain," or linear trend. The drumlins tend to be of comparable size and shape within each region and show evidence of harmonic spacing (Reed, et al., 1962). Table 17-2 lists as variants such forms as fluted (ridged) ground moraine and elongate "drumlinoid" ridges.

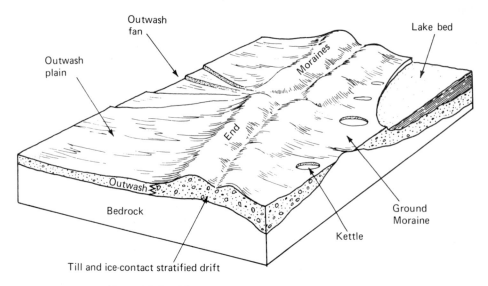

Figure 17-6. The concept of a glacial series. A group of morainic and related glaciofluvial landforms are either synchronous or sequential in time and are genetically related.

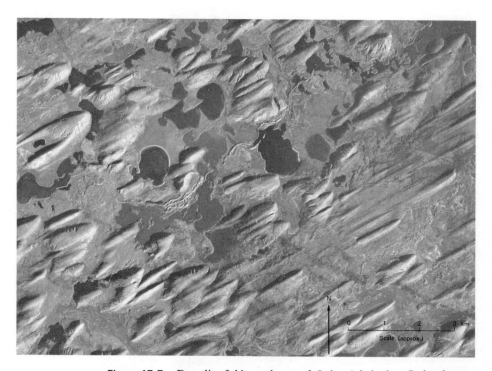

Figure 17-7. Drumlin field southeast of Lake Athabaska, Saskatchewan, Canada. Ice flowed from the northeast. Also shown are numerous kettle lakes and a system of anastomosing eskers. (Photo: Surveys and Mapping Branch, Dept. of Energy, Mines, and Resources, Canada.)

A theory of "dilatant flow" within water-saturated basal till has been proposed to explain drumlin formation (Smalley and Unwin, 1968). A plastic mixture of rock debris, ice, and water, similar to the hypothesized plastic scouring substance beneath wet-base glaciers (p. 379), could flow as a viscous fluid as long as it retains water. If, however, an increased pressure region upglacier of a bedrock obstruction or even of a large stone within the mass were to cause local dewatering, the mixture would stiffen and lodge. The initial obstruction would be enlarged by the mass of rigidified till and continue to grow to some equilibrium streamlined form or until another obstruction in the vicinity caused a similar response. The theory has the advantage of explaining the size, shape, compositional variation, and spatial distribution of drumlins. The process requires a wet-base glacier, which is also consistent with the usual abundance of ice-contact stratified drift in drumlin fields.

Glacial Landforms Constructed Primarily of Ice-Contact Stratified Drift.   A great variety of distinctive landforms are made of ice-contact stratified drift. Oriented and nonoriented forms are both common. The terminology is complicated, because various terms can be applied to either single landforms or groups of forms. Some landforms are parts of moraines; many are not.

Figure 17-8.   Internal structure of a kame near West Durham, Maine. Coarse gravel foreset beds at bottom were deposited by a prograding delta among ice blocks. Overlying glaciofluvial sand and gravel were probably deposited by an aggrading meltwater river.

Kame is representative of the complexities of this category of landforms. By derivation, a kame is an isolated hill or mound of stratified drift deposited in an opening within or between ice blocks (Fig. 17-8). If the opening was between ice and a valley wall, the resulting form can be called a **kame terrace**. **Kame deltas** are flat-topped hills with an ice-contact proximal face and a semicircular plan view, deposited in standing water at an ice margin. A **moulin kame** is believed to form under stagnant ice at the base of a *moulin* or glacier-karst swallow hole (p. 382). It is clear that a variety of environments adjacent to melting glaciers can produce kames.

Another important landform of ice-contact stratified drift is the **esker** (Fig. 17-7). An esker is a sinuous or meandering ridge of stratified sand and gravel, commonly with very well-rounded pebbles and cobbles. Eskers form in open channels on glaciers and in tunnels in or beneath them. If large quantities of sediment-laden meltwater are discharging through a karstlike hydrologic system within a glacier, the sediment may be deposited on the floors of various tunnels or channels while the ice roof or walls are widened or raised by melting. Eskers show sinuosities comparable to the dimensions of rivers and clearly show by internal structures that they are water-laid. However, they cross bedrock ridges, rising and falling as they are traced across moraine landscapes. Confined flow must be invoked to permit them to be deposited on uphill gradients. The esker systems of northeastern United States, eastern Canada, and southern Finland and Sweden are noteworthy. Most trend broadly perpendicular to major end moraines and parallel to former ice flow, probably because water flowed under hydrostatic head toward the ice margin.

A **kettle** is a basin of nondeposition within morainic terrain formed where an ice block was buried by till or ice-contact stratified drift and was then melted. Many areas of ground moraine are chaotic mazes of kettles among kames and **crevasse fillings**, the latter representing castings made in ice molds which then melted. Extremely complex internal structures, representing multiple episodes of melting and drift deposition, characterize many areas of kames and kettles (Fig. 17-9). Small kettles develop angle-of-repose side slopes and circular outlines very similar to those of collapse dolines (Fig. 7-9). Larger ones are complexly segmented (Fig. 17-7). Kettle lakes are favorite scenic attractions and a fisherman's paradise. Thoreau's Walden Pond may be the best known.

Glacial Landforms Constructed Primarily of Outwash. Meltwater alluvium aggrades the floors of glaciated valleys downstream from melting glaciers. These **valley trains** (Fig. 17-10) have high gradients and braided patterns appropriate for their massive bed loads. On the margin of an ice sheet, broad **outwash plains** form from confluent alluvial fans that emerge from multiple meltwater valleys through end moraines, or the plains grade directly onto the buried ice margin. The Icelandic term **sandur** (plural: sandr) is sometimes applied. The Icelandic sandr are notable for major episodic floods that inundate hundreds of square kilometers during events called *jökulhlaups* ("glacier bursts"), when ice dams break and release masses of impounded water from the margins of mountain ice caps. Active valley trains and outwash plains are excellent source areas for loess.

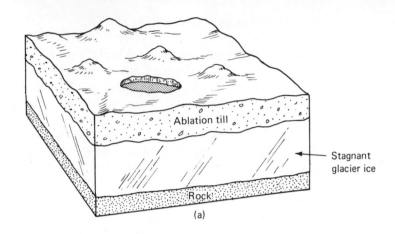

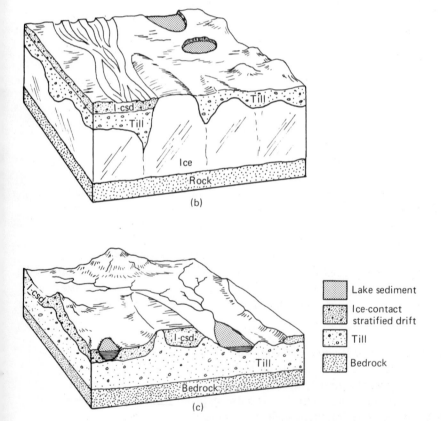

Figure 17-9. Hypothetical sequence of events in the evolution of a flat-topped kame. (*a*) Cover of ablation till over stagnant ice. (*b*) Till slumps and washes into deepening crevasses, exposing ice on a chaotic ablation surface. (*c*) Glaciofluvial sediment washed into a depression when ice melted, later became the crest of a kame, surrounded by a rim of till. Other ice-contact stratified drift deposits are slumped and tilted by final melting.

Figure 17-10. View upstream over the braided valley train of the Tasman River, South Island, New Zealand. Tasman Glacier, mostly covered with ablation drift, in right distance. Mount Cook is high peak on the left. (Photo: V. C. Brown & Son, Photographers.)

INDIRECT MORPHOGENETIC EFFECTS OF GLACIATION

Postglacial Isostatic Uplift

Sound theory and abundant evidence (Fig. 17-11) support the conclusion that ice caps larger than 50 km in diameter isostatically depress the earth's crust (p. 23). The response time of the earth to the removal of an ice sheet is slow enough so that, following deglaciation, the land surface continues to rise for a few thousand years at an exponentially decreasing rate until isostatic adjustment is complete. The Laurentide and Scandinavian ice sheets may have been 3000 m thick at their maximum (Fig. 17-12), so the potential isostatic recovery is an impressive 1000 m. However, depending on the rate and nature of deglaciation, much of the recovery may have been completed prior to the land being uncovered by ice (Andrews, 1970). Uplift continues today at a maximum known rate of 9 mm/yr near the north end of the Baltic Sea (Fig. 17-13).

Figure 17-11. Emerged Holocene beach ridges on Östergarnsholm, eastern Gotland, Sweden. The contemporary uplift rate is about 2 mm/yr (Fig. 17-13). (Photographer and copyright holder: Arne Philip, Visby, Sweden. Courtesy IGCP Project Ecostratigraphy.)

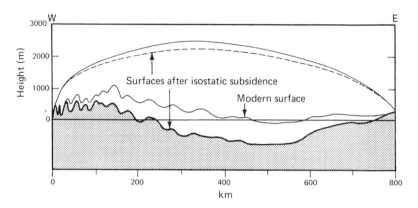

Figure 17-12. Estimated surface profile of the Fennoscandian ice sheet along a line eastward from Trondheim, Norway, to Helsinki, Finland. Note the very large area of ice that was grounded below sea level in the Baltic basin. (Robin, 1964, Fig. 7.)

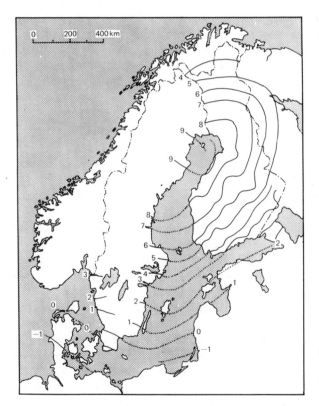

Figure 17-13. Contemporary emergence rates in the Baltic basin and vicinity (mm/yr). Emergence is relative to sea level, which is also rising 1 mm/yr or more. (Flint, 1971, Fig. 13-2.)

The most obvious geomorphic effects of postglacial isostatic uplift are to be found on coasts, where emergence so dominates that no shoreline has a chance to develop large erosional or depositional forms before it is uplifted beyond the reach of waves (Blake, 1975). River gradients must be affected by the tilting associated with uplift, and lakes or swamps may have formed or drained, but the extreme deranged drainage patterns of glaciated regions make the subtleties of tilting hard to recognize. Rivers flowing away from the center of uplift should show simultaneous rejuvenation as their gradients are increased by tilting; rivers flowing toward the center of maximum uplift, as do many of the rivers that drain into Hudson Bay, should be *defeated* by uplift and loss of gradient. The differential tilt on postglacial shorelines is typically of the order of 1 m/km, or 0.1 percent, which is comparable to the gradient of many river segments, so the effect on stream hydrology cannot be neglected.

The Great Lakes of North America form a natural laboratory for studying postglacial uplift and tilting. Abandoned shorelines and contemporary water-level gages prove that the lakes are deepening on their southern shores and shoaling on the northern. For instance, Lake Michigan and Lake Huron now form a continous water level through the very deep Straits of Mackinac. The northeastern parts of the combined lakes are now rising about 30 cm/century relative to their southwestern parts. Furthermore, the outlet of the combined lakes near Detroit is rising at about 7.5 cm/century relative to a former glacial overflow at Chicago. The discharge of Lake Michigan will be diverted naturally from the St. Lawrence system to the Mississippi in the next 3200 years if control dams in the Chicago Sanitary and Ship Canal do not prevent the changes (Clark and Persoage, 1970, p. 632). The magnitude of the potential change suggests that widespread regional drainage diversions are possible by postglacial uplift.

Glacial Eustatic Fluctuations of Sea Level

The volume of water abstracted from the sea at times of ice-sheet expansion and returned to the sea during interglacial intervals of the Pleistocene Epoch has had a worldwide, or **eustatic**, effect on morphogenesis. Estimates of the maximum lowering of sea level during a full glacial interval, as compared to today, vary from 80–140 m. The lowering was enough to lay bare about one-half of the continental shelf area so that all rivers had to extend and deepen their valleys to the low temporary base level. At present, all rivers enter the sea either in estuaries or across thick deltaic fills. Both features are a direct consequence of glacial eustatic sea-level fluctuations but are so much a part of the Quaternary fluvial geomorphic system that we find it hard to deduce what river mouths would be like if sea level had been stable for the last few million years. The nickpoints produced during glacial low sea level have migrated upstream various distances and are generally buried under postglacial alluvium, so they are not part of the modern valley morphology. However, it is a useful mental exercise to deduce the hypothetical morphology of any major river valley during glaciation, with the assumption that sea level was 100 m lower than present (compare with Fig. 10-1). During glacial intervals, large paired terraces must have flanked the gorgelike lower valleys of many rivers.

Glacial eustatic fluctuations of sea level have had an even greater effect on coastal morphology than postglacial uplift, which affected only the deglaciated areas. The rising sea level of the last 15,000 years has drowned all coasts except those with rapid glacial-isostatic or other tectonic uplift. During the most rapid phase of deglaciation, from about 10,000–7000 years ago, sea level may have risen at a rate of 10 mm/yr. The consequences of this submergence to coastal morphology are considered at length in Chapters 19 and 20. It can be noted here that the sea now appears to be only a few meters below its level of the last interglacial, about 120,000–140,000 years ago. Many coastal landforms that on first glance seem to be in equilibrium with modern processes have been shown to be relict from the last interglacial and only recently reoccupied by the sea. Many morphologic features of coastal plains and lower river valleys may be equally suspect of having been inherited from an earlier time.

CHRONOLOGY OF LATE CENOZOIC GLACIATION

To appreciate the impact of glacial morphogenesis on the earth, a review of the chronology of glaciations is required. Most of the earth's landforms have taken shape during late Cenozoic time, when ice sheets had formed in Antarctica and possibly elsewhere in mountainous regions.

Evidence of glaciers in West Antarctica can be traced back at least 27 million years. Hyaloclastites (p. 65) of that age are exposed above the ice plateau of Marie Byrd Land but were erupted under ice. Thus an ice cap probably covered the volcanic archipelago of West Antarctica as early as Late Oligocene time and perhaps earlier (LeMasurier, 1972). By 25–30 million years ago, Australia had crossed 50°S latitude as it moved toward the equator, the Circum-Antarctic Current had developed (Kennett, et al., 1974), and the large East Antarctic plateau was probably ice-covered. At that time, the modern global atmospheric and oceanic circulation patterns developed. Our present climatic morphogenetic regions are no older than late Oligocene or early Miocene time. By 9 or 10 million yerars ago, valley glaciers or mountain ice caps were forming in the western Cordillera of North America (Denton and Armstrong, 1969). The Greenland ice sheet may have formed 3 million years ago in response to the emergence of the isthmus of Panama and the development of the Gulf Stream (Hopkins, 1972).

The chronology of the late Cenozoic glaciations of North America and Europe is not established, although rapid progress is now being made by fission-track and potassium/argon dating on volcanic tephras that are interstratified with drift (p. 80). In particular, several fine-grained tephras known collectively as "Pearlette Ash" are widespread on the Great Plains of the United States from South Dakota to Texas. Formerly thought to be a single ash bed of late Kansan or early Yarmouth age (Flint, 1971, pp. 549–51), the tephras at numerous localities give ages that cluster in three groups. The oldest dates of about 1.9 million years are from tephra that presumably is younger than the Nebraskan glaciation, the oldest of the named glaciations in central North America. Other ash beds with ages of 1.2 million years and 0.6 million years probably bracket the second oldest, or Kansan, glaciation. All three groups of ashes predate the Illinoian glaciation (Naeser, et al., 1973; Boell-storff, 1974). Early Pleistocene outwash terraces in the Rhine valley of Europe are also overlain by volcanic ash about 0.6 million years old.

The radiometric dates summarized above demonstrate that substantial regions experienced glacial morphogenetic conditions well before the Pleistocene Epoch (the traditional "Ice Age") began, about 1.8 million years ago. The traditional subdivision of the Pleistocene Epoch into four glacial ages with three intervening interglacial ages will certainly be greatly revised in the near future.

A major glaciation, in many regions the most extensive of all, is called the Illinoian in North America and the Riss or Saale in Europe. It began less than 0.5 million years ago and consisted of several advances separated by local retreats but is not yet adequately dated. The next youngest recognized event in the glacial-inter-glacial sequence is a time when sea level was a few meters above present, spanning or recurring during the interval from at least 140,000 to 120,000 years ago. Numerous coral reefs grew above present sea level at that time, and the eroded bases of these reefs are the foundation of modern reefs (Chap. 19). In areas of strong tectonic uplift, prominent coastal terraces record this interglacial interval (Fig. 2-6).

There are numerous botanical, pedological, and geological lines of evidence that the last interglacial, called the Sangamon in North America and the Riss/Würm or Eem in Europe, was somewhat warmer than present. Deciduous forests extended into what is now the conifer zone north of the Great Lakes, for instance. The required warmth is consistent with less glacier volume than today and with a sea level a few meters higher than present. Therefore, part or all of the interval of 140,000–120,000 years ago, dated primarily from emerged coral reefs, is widely and confidently cor-related with the last interglacial.

Sometime slightly less than 120,000 years ago, the earth's climate cooled sharply, and the last major glaciation began (Fig. 17-14). Sea level fell in a series of pulses, separated by brief intervals of high sea level that were nevertheless well below present sea level. The entire interval from about 115,000 to about 10,000 or 6000 years ago is now called the Wisconsin glaciation in North America and the Weichsel or Würm glaciation in Europe. It consisted of short periods of ice advance, called **stadials**, separated by periods of retreat but colder than an interglacial, called **interstadials**. The interstadials are dated by coral reef growth that produced terraces on rising orogenic coasts (Figs. 2-6, 17-14) and in glaciated regions by radiocarbon ages of organic deposits between the younger drifts. Each stadial seems to have been more severe than the previous, culminating in the maximum advance of the entire glacia-tion only about 15,000–18,000 years ago. The outermost moraines of the last glaciation around the periphery of the Laurentide and Scandinavian ice sheets all fall within that time range.

The worldwide cooling during the Wisconsin glacial maximum was not extreme. The average change in summer sea-surface temperature 18,000 years ago was only $-2.3°C$ (CLIMAP, 1976; Gates, 1976). However, the North Atlantic surface water was as much as 18°C colder, the western North Pacific was 10°C colder, and the equatorial Pacific was 6°C cooler than present summer temperatures (Fig. 17-15).

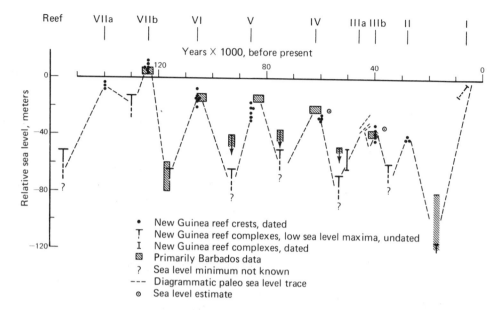

Figure 17-14. The last 140,000 years of glacier-controlled sea-level fluctuations, as inferred from the ages and altitudes of coral-reef terraces on tectonic coasts (Fig. 2-6). An assumption of constant rates of tectonic uplift was used to calculate the amplitudes of the oscillations. (Bloom, et al., 1974, Fig. 5.)

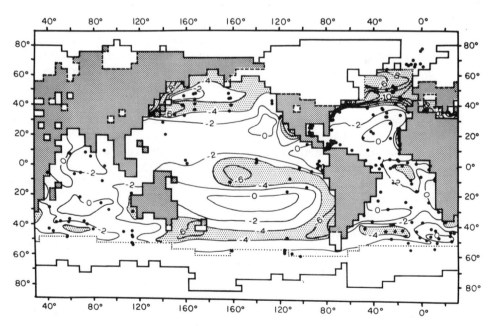

Figure 17-15. Difference (°C) between summer sea-surface temperatures of 18,000 years ago and today. (CLIMAP, 1976, Fig. 2; © 1976, American Association for the Advancement of Science.)

409

The extreme cooling of the North Atlantic correlates well with the morphogenetic evidence of glacial and periglacial conditions in eastern North America and north-western Europe (p. 421).

The marginal retreat of the Wisconsin/Weichsel ice sheets was dramatic. By 11,000 years ago, the ice margins were north of the Great Lakes and in the Baltic Sea. The Gulf Stream had diverted back toward northern Europe from its glacial-maximum track toward Portugal. By 9000 years ago, the Scandinavian ice sheet was separating into small highland residuals in northern Sweden, and the sea was entering Hudson Bay, calving the thickest part of the complex Laurentide ice sheet and radically changing the flow directions of the ice that remained on the highlands east and west of the Bay. By 6500 years ago, the remnant Laurentide ice sheets were gone, sea level was within a few meters of its present level, and the climate was approximately as today.

In the past 6000 years there have been several climatic oscillations that have affected geomorphic and pedologic processes, but they are not comparable to the glacial–interglacial changes. Perhaps a temperature range of 1°C has been involved in the middle latitudes, in contrast to the 8° to 10°C difference between glacial and postglacial climates in the same regions. Between 6000 and 3000 years ago, most areas were warmer and drier than today. This was the **Hypsithermal Interval** (Deevey and Flint, 1957). Most of the cirque glaciers in the United States and Canadian Rockies and the European Alps were gone, as were most of the small mountain ice caps in Norway. The valley glaciers along the Alaskan coast retreated far up their valleys. Then, in a cooling trend with a series of minor oscillations, the climate approached present conditions, and valley glaciers expanded somewhat. The last century has seen a dramatic retreat of valley glaciers from the moraines built during the last times of expansion between 1650 and 1850 A.D. Systematic geomorphic study of glaciers has been done almost entirely during an interval of glacier retreat and negative glacier budgets. We know very little about the events that accompany glacial advances (except when caused by surges).

The causes of glaciations and of climatic change in general are unknown. Certain periodicity, as in Figures 17-14 and 18-4, suggests that planetary orbital parameters might be involved. Certainly, glaciation in middle latitudes was accompanied by lower temperatures over the entire earth (Fig. 17-15). As is discussed in the next chapter, there is very little information about climatic change in the low latitudes or nonglaciated regions.

The boundaries of glaciated regions can be determined to within a kilometer or even a few meters, so it has been relatively easy to define the glacial morphogenetic region. Not at all as simple are the tasks of defining the former extent of the peri-glacial zone and of identifying the effects of alternating or changing morphogenetic regions outside of the glacial limits. There is abundant evidence that climates have changed during the late Cenozoic Era while the modern landscape has taken shape. We use the glacial chronology, as imperfect as it is, as a framework for the chronology of morphogenetic change in other regions, but many mistakes have been made, and many assumptions have been proved false.

REFERENCES

AGASSIZ, L., 1840, *Études sur les glaciers:* privately publ., Neuchâtel, 346 pp.

AMBROSE, J. W., 1964, Exhumed paleoplains of the Precambrian shield of North America: *Am. Jour. Sci.*, v. 262, pp. 817–57.

ANDREWS, J. T., 1970, Geomorphological study of post-glacial uplift with particular reference to Arctic Canada: *Inst. Brit. Geographers Spec. Pub. 2*, 156 pp.

ANDREWS, J. T., and LeMASURIER, W. E., 1973, Rates of Quaternary glacial erosion and corrie formation, Marie Byrd Land, Antarctica: *Geology*, v. 1, pp. 75–80.

BLAKE, W., JR., 1975, Radiocarbon age determinations and postglacial emergence at Cape Storm, southern Ellesmere Island, Arctic Canada: *Geograf. Annaler*, v. 57, ser. A, pp. 1–71.

BLOOM, A. L., BROECKER, W. S., CHAPPELL, J. M. A., MATTHEWS, R. K., and MESOLELLA, K. J., 1974, Quaternary sea level fluctuations on a tectonic coast: New ^{230}Th/^{234}U dates from the Huon Peninsula, New Guinea: *Quaternary Res.*, v. 4, pp. 185–205.

BOELLSTORFF, J., 1974, Fission-track ages of Pearlette family ash beds: Comment: *Geology*, v. 2, p. 21.

BOULTON, G. S., 1974, Processes and patterns of glacial erosion, *in* Coates, D. R., ed., *Glacial geomorphology:* State Univ. of New York Publications in Geomorphology, Binghamton, N.Y., pp. 41–87.

CHURCH, M., and RYDER, J. M., 1972, Paraglacial sedimentation: A consideration of fluvial processes conditioned by glaciation: *Geol. Soc. America Bull.*, v. 83, pp. 3059–72.

CLARK, R. H., and PERSOAGE, N. P., 1970, Some implications of crustal movement in engineering planning: *Canadian Jour. Earth Sci.*, v. 70, pp. 628–33.

CLAYTON, K. M., 1965, Glacial erosion in the Finger Lakes region (New York State, U.S.A.): *Zeitschr. für Geomorph.*, v. 9, pp. 50–62 (discussion, v. 10, pp. 475–77).

CLIMAP PROJECT MEMBERS, 1976, Surface of the ice-age earth: *Science*, v. 191, pp. 1131–37.

COATES, D. R., 1966, Discussion of K. M. Clayton, "Glacial erosion in the Finger Lakes region (New York State, U.S.A.)": *Zeitschr. für Geomorph.*, v. 10, pp. 469–74.

CRARY, A. P., 1966, Mechanism for fiord formation indicated by studies of an ice-covered inlet: *Geol. Soc. America Bull.*, v. 77, pp. 911–29.

DEEVEY, E. S., and FLINT, R. F., 1957, Postglacial Hypsithermal interval: *Science*, v. 125, pp. 182–84.

DENTON, G. H., and ARMSTRONG, R. L., 1969, Miocene-Pliocene glaciations in southern Alaska: *Am. Jour. Sci.*, v. 267, pp. 1121–42.

EMBLETON, C., and KING, C. A. M., 1975, *Glacial geomorphology:* John Wiley & Sons, Inc., New York, 573 p.

VON ENGELN, O. D., 1921, The Tully glacial series: *N.Y. State Museum Bull. 227–228*, pp. 39–62.

FEININGER, T., 1971, Chemical weathering and glacial erosion of crystalline rocks and the origin of till: *U.S. Geol. Survey Prof. Paper 750-C*, pp. 65–81.

FLINT, R. F., 1971, *Glacial and Quaternary geology:* John Wiley & Sons, Inc., New York, 892 pp.

GATES, W. L., 1976, Modeling the ice-age climate: *Science*, v. 191, pp. 1138–44.

GJESSING, J., 1966, On "plastic scouring" and "subglacial erosion": *Norsk Geografisk Tidsskrift*, v. 20, pp. 1–37.

GRAVENOR, C. P., 1975, Erosion by continental ice sheets: *Am. Jour. Sci.*, v. 275, pp. 594–604.

GUTENBERG, B., BUWALDA, J. P., and SHARP, R. P., 1956, Seismic explorations on the floor of Yosemite Valley, California: *Geol. Soc. America Bull.*, v. 67, pp. 1051–78.

HOPKINS, D. M., 1972, Changes in oceanic circulation and late Cenozoic cold climates [Abs.]: *Internat. Geol. Congress, 24th, Montreal, 1972, Abstract Vol.*, p. 370.

JAHNS, R. H., 1943, Sheet structure in granites: Its origin and use as a measure of glacial erosion in New England: *Jour. Geol.*, v. 51, pp. 71–98.

JOPLING, A. V., and McDONALD, B. C., eds., 1975, Glaciofluvial and glaciolacustrine sedimentation: *Soc. Econ. Paleontologists and Mineralogists Spec. Pub. 23*, 320 pp.

KAYE, C. A., 1967a, Fossiliferous bauxite in glacial drift, Martha's Vineyard, Massachusetts: *Science*, v. 157, pp. 1035–37.

———, 1967b, Kaolinization of bedrock of the Boston, Massachusetts, area: *U.S. Geol. Survey Prof. Paper 575-C*, pp. 165–72.

KENNETT, J. P., HOUTZ, R. E., ANDREWS, P. B., EDWARDS, A. R., GOSTIN, V. A., HAJOS, M., HAMPTON, M. A., JENKINS, D. G., MARGOLIS, S. V., OVENSHINE, A. T., and PERCH-NIELSON, K., 1974, Development of the Circum-Antarctic Current: *Science*, v. 186, pp. 144–47.

LeMASURIER, W. E., 1972, Volcanic record of Antarctic glacial history: Implications with regard to Cenozoic sea levels, *in* Price, R. J., and Sugden, D. E., eds., Polar geomorphology: *Inst. Brit. Geographers Spec. Pub. no. 4*, pp. 59–74.

LINTON, D. L., 1963, The forms of glacial erosion: *Inst. Brit. Geographers Trans.*, v. 33, pp. 1–28.

MULLER, E. H., 1974, Origins of drumlins, *in* Coates, D. R., ed., *Glacial geomorphology*: State Univ. of New York Publications in Geomorphology, Binghamton, N.Y., pp. 187–204.

NAESER, C. W., IZETT, G. A., and WILCOX, R. E., 1973, Zircon fission-track ages of Pearlette family ash beds in Meade County, Kansas: *Geology*, v. 1, pp. 187–89.

PENCK, A., and BRÜCKNER, E., 1909, *Die Alpen im Eiszeitalter:* Tauchnitz, Leipzig, 3 v., 1199 pp.

PORTER, S. C., 1966, Pleistocene geology of Anaktuvuk Pass, Central Brooks Range: *Arctic Inst. of North America Tech. Paper 18*, 100 pp.

PREST, V. K., 1968, Nomenclature of moraines and ice-flow features as applied to the glacial map of Canada: *Canada Geol. Survey Paper 67-57*, 32 pp.

———, 1971, Quaternary geology of Canada, *in* Geology and economic minerals of Canada, *Economic Geology Rept. no. 1*, 5th ed.: Dept. Energy, Mines and Resources, Ottawa, Canada, pp. 676–764.

PRICE, R. J., 1973, *Glacial and fluvioglacial landforms:* Oliver and Boyd, Edinburgh, 242 pp.

REED, B., GALVIN, C. J., and MILLER, J. P., 1962, Some aspects of drumlin geometry: *Am. Jour. Sci.*, v. 260, pp. 200–210.

ROBIN, G. DEQ., 1964, Glaciology: *Endeavour*, v. 23, pp. 102–7.

SHUMSKII, P. A., 1964, *Principles of structural glaciology* (transl. by D. Kraus): Dover Publications, Inc., New York, 497 pp.

SMALLEY, I. J., and UNWIN, D. J., 1968, Formation and shape of drumlins and their distribution and orientation in drumlin fields: *Jour. Glaciology*, v. 7, pp. 377–90.

SUGDEN, D. E., and JOHN, B. S., 1976, *Glaciers and landscape:* John Wiley & Sons, Inc., New York (a Halsted Press Book), 384 pp.

TARR, R. S., 1905, Drainage features of central New York: *Geol. Soc. America Bull.*, v. 16, pp. 229–42.

WARNKE, D. A., 1970, Glacial erosion, ice rafting, and glacial-marine sediments: Antarctica and the Southern Ocean: *Am. Jour. Sci.*, v. 269, pp. 276–94.

WHITE, W. A., 1972, Deep erosion by continental ice sheets: *Geol. Soc. America Bull.*, v. 83, pp. 1037–56.

CHAPTER 18

Alternating Quaternary Morphogenetic Systems

THE CONCEPT OF ALTERNATING QUATERNARY MORPHOGENETIC SYSTEMS

In the previous five chapters, it has not been possible to describe morphogenetic systems without reference to examples of relict Quaternary landforms. Glacial moraines of several ages are 1000 m or more lower on mountains than any valley glaciers now reach, or 2000 km nearer the equator than the latitude of any present ice sheet. Deserts are dissected by river networks in which no water has flowed in human history; longitudinal dune systems are now soil- and grass-covered. Any attempt to classify morphogenetic regions is immediately confronted by the need to separate presently evolving landforms from those relict from earlier Quaternary time. The problems are so difficult that only a start has been made in solving them, and numerous contradictions and inconsistencies among the morphogenetic models and examples are cited in this chapter. All that can be done is to outline the general concepts of Quaternary climatic change and to illustrate by selected examples the impact of alternating morphogenetic systems on the present landscape. Relicts of Tertiary or pre-Cenozoic morphogenetic systems, claimed for tectonically inactive regions especially in Africa and Australia, may be due either to latitudinal shifts in the positions of the continents (Frakes and Kemp, 1972; Kennett, et al., 1974) or to pre-Quaternary climatic changes. If there are difficulties in recognizing Quaternary changes of climate and their impact on landscapes, the problems of even older events and forms are at present insuperable. However, there is real promise that eventually the positions of the continents in pre-Quaternary times will be worked out with sufficient accuracy so that the effects of plate tectonics and climatic change can be distinguished.

Although the causes of climatic change are unknown, it is clear that the earth has experienced a series of climatic alternations, each of comparable magnitude, during at least the last 2 million years. The chronology of these events insofar as it is now known was reviewed at the end of Chapter 17. The most obvious effect of the changes was the repeated expansion of the glacial climates on the larger and more humid parts of the northern-hemisphere continents, with a corresponding expansion of the periglacial morphogenetic region as well. Alternating with the glacial episodes have been a series of interglacial (and "interperiglacial") intervals with morphogenetic regions approximating their present distribution. Within the last glaciation and presumably during earlier ones as well were a series of lesser climatic fluctuations (Fig. 17-14) more nearly glacial than interglacial in character. Quite possibly, interglacial intervals may also have had lesser climatic fluctuations all more nearly interglacial than glacial in character. Certainly the Holocene has been marked by such lesser fluctuations, some of which have been called "little ice ages," or "neoglaciations."

In lower latitudes, especially in the seasonally wet-and-dry climatic zones that border the great arid regions, Quaternary climatic change was expressed more as changes in effective precipitation than as temperature changes. Although we have abundant reasons to believe that the climatic changes of the Quaternary Period were worldwide, the resulting changes in morphogenesis varied from region to region. We know more about the strongly contrasting glacial-interglacial morphogenetic systems than we know about any of the others, and in the following sections, the chronology of these contrasting systems is used as a temporal framework within which other alternating morphogenetic systems can be placed.

Temperature Changes in Late Quaternary Time

During the last glaciation, approximately 15,000–20,000 years ago, the present morphogenetic systems were strongly displaced (Figs. 17-15, 18-1, and 18-2). Mean annual temperature was lowered 8°–10°C in the middle-latitude continental interiors and a few degrees in the surface waters of the tropical oceans. The cooling caused the greatest latitudinal shift of climatic zones in the middle latitudes of the northern hemisphere and only slight displacement of tropical zones. According to one model (Fig. 18-1), the zones of tropical forest and humid savanna actually expanded slightly as a result of decreased evaporation and cooler mean temperature. Another model (Fig. 18-2) does not show an expansion of the tropical humid regions. Neither of Figures 18-1 and 18-2 was intended as anything more than an idealized meridianal transect of average climatic conditions. As such, both are still very useful. However, Figure 17-15 demonstrates that during the last glacial maximum, the temperature changes were not the same on all meridians but were very complex. We assume, but have little proof, that the cooling associated with the previous Pleistocene glaciations was similar.

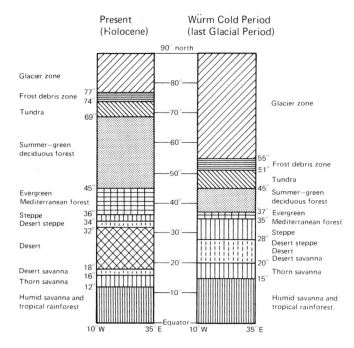

Figure 18-1. The present climatic zones along a meridianal belt from the equator to the north pole through Europe, compared with the hypothetical climatic zones during the last glaciation. (Büdel, 1957, Fig. 1.)

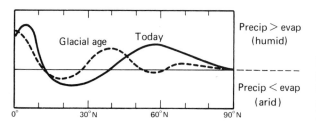

Figure 18-2. Model of the effective precipitation in the northern hemisphere during a glaciation and today (or during an interglacial interval). (From Flohn, 1953, Fig. 1.)

Precipitation Changes in Late Quaternary Time

The worldwide cooling of the last glaciation is well established, but the corresponding effects on precipitation are not at all clear. Most authorities reason that overall cooling must have reduced total evaporation and therefore total precipitation; values commonly suggested are about 80 percent of the present values shown by Figure 5-5 (Flint, 1971, p. 417). The former notion that glaciations were also times of increased precipitation ("pluvials") is very likely wrong. As suggested by Figure 18-2, the cool, humid, continental regions now in the westerly storm track of latitudes 45°–60° north were distinctly drier during glacial ages.

Part of the evidence for inferring climates during glacial ages is morphogenetic. At present, the snow lines on maritime west coasts rise to the east as the available moisture is lost. The altitude of glacial-age cirque floors, although 1000–1500 m lower than the present snow line, also rise inland at about the same rate as the modern snow line [Fig. 18-3(*a*)]. The parallelism of the two lines suggests that the cause of the glacial-age lowering of snow line was cooler temperature rather than increased precipitation (Flint, 1971, pp. 63–73). If precipitation had decreased during glaciation, coastal cirques would be lowered by an amount proportional to the cooling, but cirque-floor altitudes would again increase rapidly inland and converge with the modern snow line [Fig. 18-3(*b*)]. Such convergence has been found in the Cascade Range of Washington, where west-wind moisture was blocked by a large glacier lobe in the Puget Sound lowland (Porter, 1964).

Even though temperature and total precipitation may have decreased during a glacial age, the *effective* precipitation may have increased because of reduced evapo-

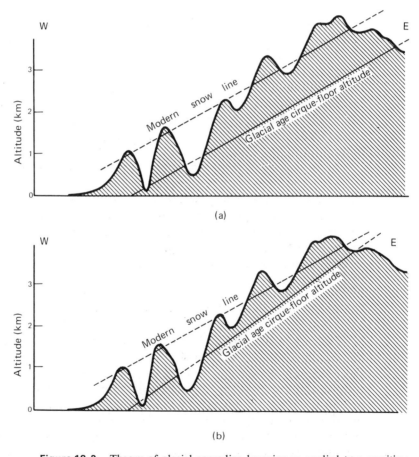

(a)

(b)

Figure 18-3. Theory of glacial snow-line lowering as applied to a maritime west coast mountain range. In the most usual condition, (*a*), the snow line of glacial time (as indicated by cirques) was lowered approximately parallel to the modern snow line. If inland convergence were found, (*b*), cooling and either a reduction of precipitation inland or an increase in precipitation near the coast could be inferred.

ration. In particular, the large areas of subtropical aridity (Fig. 13-1) that characterize the modern planetary circulation may have been reduced in size and displaced toward the equator (Figs. 18-1, 18-2). The Mediterranean forest and steppe regions on the poleward sides of the arid belts were widened as winter rains brought moisture to areas such as the northern Sahara and the southern edge of the Australian desert. The shifts on the equatorial side of the arid regions are still uncertain. Aridity may have displaced part of the savanna region, or the humid tropics may have stayed approximately fixed and simply compressed the dry zones. Both hypotheses are illustrated in Figures 18-1 and 18-2. Case studies from Brazil, Africa, and Australia, presented later in this chapter, illustrate the difficulties of interpretation.

Morphogenetic Effects of Late Quaternary Climatic Change

To illustrate the hypothetical morphogenetic effects of the climatic changes during a glacial age, Figure 6-10 should be compared with Figures 17-15, 18-1, and 18-2. In Chapter 6, the zonal or latitudinal changes in weathering processes were described with climate as a major factor. If the climatic zones were to shift, then the nature and intensity of weathering and soil-forming processes must also change. Similar deductions can be made for mass-wasting and the fluvial morphogenetic system.

As an indirect effect of glaciations, sea level repeatedly fell at least 100 m (Fig. 17-14). This had the effect of rejuvenating all major rivers and steepening both their longitudinal and cross-valley profiles in coastal regions. Also, in regions with wide continental shelves, present maritime climates became more continental as the shoreline withdrew 100 km or more seaward of its present position. Superimposed on the colder and possibly drier glacial-age climates, the increased continentality probably contributed to the morphogenetic contrasts of these regions.

Tertiary and Early Quaternary Morphogenetic Changes

Evidence in the northeastern United States of preglacial deep weathering of a sort usually associated with tropical or subtropical climates (p. 394) hints at morphogenetic changes far greater than the alternating systems of Quaternary glacial-interglacial times. However, the older relict forms and deposits are so rare that they are only recorded, not organized into a morphogenetic chronology. In a similar way, an interval of mid-Cenozoic lateritization over much of Australia (Hays, 1971; Mulcahy, 1971; Mabbutt, 1971) is the starting point for most geomorphic studies of that continent. The laterite is widespread in regions that are now either too dry or too cool for comparable weathering. Perhaps the late Cenozoic displacement of continents by lithospheric plate motions (p. 407) has disrupted former climatic patterns and intensified latitudinal climatic gradients. Possibly the alternating Quaternary morphogenetic systems are just one result of the tectonic-produced pattern of continent orientation that developed progressively during the late Cenozoic Era.

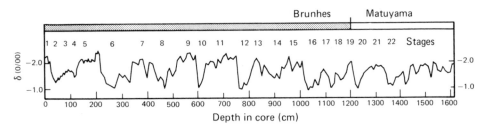

Figure 18-4. Oxygen-isotope ratios from a planktonic foraminiferal species in a tropical Pacific deep-sea sediment core. Delta (δ) is related to ice volumes on land in such a way that isotope stage 1 (uppermost 22 cm of core) represents the Holocene, stage 2 is the last full glacial interval (from 13,000–32,000 years ago), and so on. The Brunhes-Matuyama paleomagnetic reversal, which occurred 700,000 years ago, is recorded at a depth of 1200 cm in the core. Isotope stages 1–5 are similar to the record of sea-level changes inferred from emerged coral reefs (Fig. 17-14). (Shackleton and Opdyke, 1973, Fig. 9.)

Interpretation is complicated by the effects of biological evolutionary changes, especially the first appearance of grasses (pp. 7 and 323).

For at least the last 800,000 years, there is reasonable evidence for alternating climatic morphogenesis, recorded in deep-sea stratigraphy. The isotopic chemistry of pelagic and benthonic foraminifera is a record of changes in ice volumes and, to a lesser extent, oceanic surface temperatures (Fig. 18-4). Deep-sea cores that record these changes are dated by radioactive decay or by reference to paleomagnetic reversals with an assumption of constant sedimentation rate throughout the length of each core (Shackleton and Opdyke, 1973, 1976). At least 22 "stages," each representing a significant change in terrestrial ice volumes and therefore of temperature, are recorded in the last 800,000 years. Some changes are of interglacial-glacial status, whereas other "stages" represent glacial stadia or interstadials. The cyclic, although complex, oscillations that have taken place on a scale of 10^5 years should warn us to use great caution in interpreting any landscape as the result of a single morphogenetic system. The amplitudes of the larger climatic oscillations shown by Figure 18-4 are equivalent to the contrast between full glacial and interglacial morphogenetic conditions on the land.

EXAMPLES OF ALTERNATING QUATERNARY MORPHOGENETIC SYSTEMS

The magnitude and chronology of Quaternary climatic changes are the subjects of active research and are certain to be revised. However, the concept forms a framework for a series of examples that illustrate the effects of alternating morphogenetic systems on a variety of landscapes. Inconsistencies and outright contradictions will be discovered by comparing the various examples, yet they illustrate the powerful impact of Quaternary climatic change on landscapes. Eventually a theory must be developed that can mutually reconcile all of these and many other examples.

Alternating Glacial (and Periglacial)
and Interglacial Systems

Upper Mississippi Basin.   With its major tributaries, the Missouri and Ohio Rivers, the Mississippi carried meltwater from almost the entire southern quadrant of the Laurentide ice sheets during at least the last two glaciations (Fig. 16-1). The Missouri River was created as an ice-marginal stream from segments of rivers that formerly flowed northeastward down the regional slope of the Great Plains (Fig. 18-5). Much of the ancestral drainage formerly went toward Hudson Bay (Flint, 1971, pp. 232–35). The drainage diversions occurred prior to the Wisconsin glaciation. Probably each earlier glaciation contributed to the changes.

The Ohio River had a comparable but older origin. In preglacial time, the western Appalachian drainage probably flowed northwest but was directed more nearly westward by one of the first Pleistocene ice sheets (Tight, 1903; Teller, 1973). A buried valley comparable in size to the present Ohio River valley has been traced westward across Ohio, Indiana, and Illinois. The valley of the ancestral river, named the Teays Valley, has been nearly buried by several subsequent drifts (Fig. 18-5).

During each glacier advance southward across the central lowlands of the United States, north-flowing rivers were blocked by ice and diverted southward to the Mississippi system. A sequence of events can be deduced, with an initial phase of high meltwater discharge but primarily suspended load, because the coarser drift was trapped in ice-marginal lakes. Then, as the ice margin crossed the divide between northward-flowing and Mississippi drainage, outwash poured down the Mississippi. The previously eroded deep trench, cut to a glacially lowered sea level (Fig. 10-1), then began to aggrade. Later, when the ice margin had again retreated north of the ice-marginal lakes on the regional divide, only water and suspended load were carried by the Mississippi. The modern meandering channel developed on fine-grained, flood-plain alluvium. Postglacial uplift has aided the upstream reaches to entrench

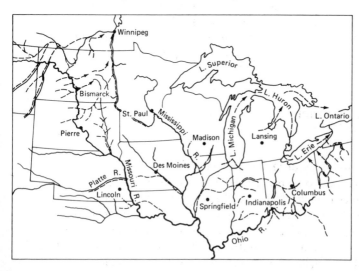

Figure 18-5. Preglacial drainage networks in the upper Mississippi drainage basin. Present rivers are shown in bold lines.

and leave outwash terraces on valley sides, but the downstream part of the river has aggraded a modern flood plain to rising sea level, deeply burying the outwash sand and gravel of glacial times. The alternation of glacial and interglacial hydrologic regimes has left flights of loess-capped terraces and many buried channels throughout the Mississippi watershed. In general, the regional relief has been decreased by thick sheets of glacial drift with a subdued morainal topography.

Southward from the maximum advances of the Pleistocene ice margins, tundra vegetation extended at least 300 km on the Appalachian Plateau (Maxwell and Davis, 1972). Periglacial patterned ground is known from many regions of central and east-central United States (p. 348). Periglacial morphogenesis must have influenced the development of the Mississippi Basin, but the effects have not been separated from the massive effects of glaciation.

Northwestern Europe.   The present climate of Europe grades from cool maritime forests on the Atlantic seaboard (1) eastward to continental steppe grasslands in Russia and Poland, and (2) southward to Mediterranean subhumid conditions. During each glaciation, most of nonglaciated Europe north of the Alps became a periglacial region. The maritime influence was greatly decreased by the width of the exposed continental shelf and by the southward diversion of the Gulf Stream (Fig. 17-15), which then did not affect the European climate north of Spain (Ruddiman and McIntyre, 1973). The Alpine ice cap that formed synchronously with each expanding Scandinavian ice sheet blocked warm air and moisture from the south.

Glacial-interglacial, alternating climatic morphogenesis seems to have followed a repeated pattern (West, 1968; van der Hammen, et al., 1971). With the onset of glaciation, the climate grew cold, but initially remained fairly humid in western Europe. Intense gelifluction characterized the wet, cold, periglacial interval. Then, at each of the glacial maxima, the climate became cold and dry, possibly because pack ice and lower sea level blocked the northward flow of the warming Gulf Stream. Thick loess deposits mantled frost-churned residual soils and the gelifluction deposits of the wetter phase. As the ice margins retreated again, the climate warmed and became more humid, with minor fluctuations. The pollen stratigraphy in western European bogs and lakes records a late-glacial tundra environment changing to birch and pine forest about 8500 years ago. The climate then was cold and rather dry. About 5000 years ago, the climate became maritime and moist and perhaps 2°C warmer than the average of today. This was the Hypsithermal Interval in Europe. Heavy deciduous forests covered northwestern Europe, and peat bogs were widespread on uplands and lowlands. About 3000 years ago, the forests began to degenerate, perhaps owing to a slightly cooler, drier climate but more likely owing to the beginning of Neolithic agriculture. People have strongly affected subsequent morphogenesis in western Europe by cutting the forests and planting crops.

The highlands of Scandinavia and the Alps support small ice caps and valley glaciers today, although all have reformed after the Hypsithermal Interval and are not the remnants of the last glaciation. The drift plains of northern Europe have low relief, dissected by abandoned ice-marginal valleys. Loess is abundant, especially

Figure 18-6. Pindaunde Valley, a glaciated trough with rock-basin lakes on the flank of Mount Wilhelm, Papua New Guinea. The building at the lower end of the lakes is at an altitude of 3500 m. Glaciated landforms extend down to 3200 m, below which steep V-shaped valleys are covered with tropical rainforest. (Photo: G. S. Hope.)

in the intermontane basins of eastern Europe. The cereal-producing steppe lands of Austria, Poland, Hungary, and the Ukranian SSR are all loess mantled, with silt derived from large outwash rivers such as the Danube, Dneiper, and Don.

South of the glacial limit, western European landscapes are dominated by the multiple episodes of periglacial morphogenesis. Slopes are mantled with thick gelifluction sheets. Valleys are heavily aggraded. Cave stratigraphy and rock paintings prove that Paleolithic people hunted the tundra animals during the last glacial age and witnessed the periglacial processes.

High-Altitude New Guinea.   One of the most striking results of alternating glacial and interglacial morphogenetic systems is the landscape of high-altitude equatorial regions. On the island of New Guinea, 3°–6° south of the equator, only three of the highest mountain peaks, all above 4600 m in height, now support glaciers (Verstappen, 1964). From an undetermined time prior to 22,000 years ago until as recently as 9000 years ago, mountain ice caps or valley glaciers shaped at least 20 mountains, covering a total area of 5000–6000 km². The firn limit was about 1000 m lower than present, and the tree line, as well as vegetational zones in the lower rainforests, were comparably lowered (Galloway, et al., 1973; Hope and Peterson, 1975; Bowler, et al., 1976). The result is that areas now in high-altitude grasslands, just above the tree line, have U-shaped glaciated troughs, paternoster lakes, hanging valleys, and other typically alpine landforms (Fig. 18-6). Moraine ridges cross valley floors and dam lakes. Soils are shallow. Weakly developed cirques dissect mountain ridges into arêtes. Areas between 3200 and 2200 m, now in montane rainforest, were covered by alpine grasslands until about 10,000 years ago. Valley forms change from broad, open troughs in the deglaciated highlands to steep, deep gorges across the forested lower mountain ridges. Entire assemblages of landforms, each appropriate to a modern climatic and vegetational zone on the mountain sides, are found relict about 1000 m lower. The entire shift is consistent with a climatic cooling of about 5°–6°C, with no significant change in precipitation (Löffler, 1972). The climatic interpretation is complicated by the influence of human agriculture for the last several thousand years but is nevertheless very similar to the history of montane equatorial regions in Africa and South America.

Alternating Pluvial and Nonpluvial Systems

The term **pluvial** implies greater precipitation but in fact is more accurately defined by an excess of precipitation over evaporation, whether the climatic cause is greater precipitation, reduced evaporation, more effective seasonal distribution of precipitation, or any combination of these. Many regions now arid or semiarid show evidence of former pluvial conditions. In some of the regions, times of pluvial morphogenesis correlate with glacial ages; in others, the glacial ages were nonpluvial.

Figure 18-2 summarizes the hypothesis that during a glacial age, pluvial conditions prevailed in the zones between latitudes 25° and 35°, where there is widespread aridity or semiaridity today (Fig. 13-1). Three examples of alternating pluvial–

nonpluvial morphogenesis are offered to demonstrate that in this latitudinal zone, pluvial and glacial ages are correlative. Subsequently, three other examples of relict pluvial morphogenesis demonstrate a correlation with interglacial ages. The latter three examples are from tropical regions, not from the extra-tropical latitudinal belts.

Southwestern United States.   About 140 closed tectonic basins mark the semiarid or arid Basin and Range province of southwestern United States (Fig. 2-4). Most of them contained lakes during the last glaciation, ranging from shallow playas to Lake Bonneville, which was more than 330 m deep at its maximum. A few of the lakes were fed by meltwater from valley glaciers on the adjacent ranges, but most of them developed only because of a more favorable precipitation/evaporation ratio. Various lines of evidence (Flint, 1971, pp. 442–51 and references therein) suggest that the pluvial lakes resulted from precipitation about 80 percent greater than now, evaporation about 30 percent less, and mean annual temperature 5°–8°C cooler. Cloud cover and seasonality during the pluvial intervals are unknown.

In two pluvial lake basins, end moraines that were built at the mouths of valleys by glaciers from adjacent mountain ranges are cliffed by high-level lake shorelines. Outwash correlated with recessional moraines built deltaic terraces at the level of the pluvial shorelines as well. In only these two lakes, pluvial Lake Russell in the Mono Basin, and pluvial Lake Bonneville in the Great Salt Lake Basin, is it possible to prove by such morphologic evidence that glaciation and pluvial conditions were synchronous. However, a growing number of radiocarbon-dated sediment cores from various playas prove that they were pluvial during at least the last glaciation (Flint, 1971, pp. 431–33 and 449–50) and probably during earlier glacial intervals as well.

The most obvious morphologic effects of a pluvial climate are the shoreline features of the former lakes. Shore platforms, wave-cut cliffs, spits, bars, and deltas, all trace the ancient shorelines (Fig. 18-7). River terraces are graded to former lake levels. Basin floors are blanketed with lacustrine sediments, including salt layers and buried soils that record nonpluvial conditions. Sand dunes are associated with ancient shorelines, as well as being built on the dry playas today. Periods of soil formation and valley incision with presumed more effective vegetation cover have alternated with periods of mass-wasting and aggradation, presumably during nonpluvial times, but the validity of Huntington's principle (p. 248) has yet to be established in the region. Several lesser cycles of arroyo entrenchment and alluviation have occurred within postpluvial time, so an exact correlation with the pluvial–nonpluvial cycle is not likely.

An unsolved problem of morphogenesis in the southwestern United States concerns the role of pluvial climates in pediment development (p. 321). Was pediment cutting accelerated or inhibited by the pluvial morphogenetic system? Were pediments dissected by more effective precipitation during pluvial times? Alternating pluvial–nonpluvial climates must be significant to pediment formation, although no one has yet been able to demonstrate the significance. An even more challenging hypothesis is that the pediments of the Mohave Desert, California, developed in a

Figure 18-7. View southeast over Payson, Utah. Large gravel spit in foreground was built northward by longshore currents in a succession of pluvial lakes in the Great Salt Lake Basin. Benches were eroded during falling lake levels. (Photo: H. J. Bissell.)

semiarid pre-Pliocene climate and have subsequently been exhumed and only slightly modified (Oberlander, 1974).

The Mediterranean Basin.   The Mediterranean climate and morphogenetic system is characterized by summer drought and winter rains. Floods and droughts are both tragically frequent. Mountainous regions rank high in denudation rates (Table 12-3). Geomorphic, archeologic, and botanic evidence prove that the north coast of Africa and the south coast of Europe were more heavily forested during the last glaciation, even though total precipitation was probably less than now (Flint, 1971, p. 619). The present forest zone of Europe was compressed and shifted southward during glacial times. Pluvial conditions were the result of cooler temperatures and possibly more effective precipitation rather than of increased precipitation (Farrand, 1971).

Mediterranean river valleys are characterized by alluvial terraces into which the present stream channels are incised (Vita-Finzi, 1969). The alluvium is stratigraphically younger than shorelines of the last interglacial, forms fans and terraces that were graded to a sea level lower than present, contains floral remains of cooler temperature, and in places can be traced upstream into periglacial deposits. All this evidence proves that valley aggradation occurred during the last glacial age, probably during the interval of ice-sheet expansion and falling sea level. Intensified frost weathering on highlands and at least seasonally more intense rain have been claimed as causes of the aggradation (Vita-Finzi, 1969, p. 100). Calcareous tufa deposits around springs suggest correlative higher ground-water levels, also the result of more effective precipitation.

About 9000 years ago, the alluvial fans and flood plains of the Mediterranean coast began to be dissected. While the lower reaches of rivers became estuaries, their upstream portions were entrenching as postglacial streams established more gentle gradients. Human intervention in the form of dams and soil-conservation systems became a factor in subsequent valley development. Erosion was inhibited by check dams, but by about 1000 years ago, a new cycle of aggradation had begun to bury classical Roman dams and buildings. Finally, for the last five centuries, erosion has again dominated.

The correlation of aggradational phases in Mediterranean valleys with the cooler and pluvial times of the last glaciation and of the "little ice age" of post-Roman times in Europe contradicts Huntington's principle (p. 248). However, if vegetational cover was incomplete in the Mediterranean basin, owing either to climatic or human factors, increased precipitation could cause valley aggradation instead of landscape stabilization and valley incision. The last pluvial was correlative with glaciation in northern Europe and local periglacial conditions. It was a time of cooler temperature, decreased total precipitation but more effective precipitation/evaporation ratio, heavier forest cover, alluviation in river valleys, and lower sea level. Clearly, the modern Mediterranean landscape is the result of extremely complex alternating morphogenetic systems with the further complication of many thousands of years of human impact.

Southern Australia.   A series of pluvial lakes formed in the alluvial plains of the semiarid Murray Basin in southeastern Australia. Sediment cores from some of these lakes have been radiocarbon-dated to give a pluvial chronology for the last 40,000 years, which is broadly similar throughout the region (Bowler, et al., 1976, especially Fig. 9-1). Until about 17,000 years ago, the lakes contained more water than now, the effect of a favorable precipitation/evaporation ratio. Many of the lakes dried completely about 17,000 years ago. Those in more arid regions have not held water for more than brief seasonal intervals since. Those in somewhat more humid regions record weak pluvial periods between 10,000 and 5000 years ago and again during the last 2000 years. The area is at about 35°S, a latitude equivalent to similar lakes in the southwestern United States. Until 17,000 years ago, the correlation with glacial and pluvial events in the northern hemisphere is excellent, strongly supporting theories that in the southern as well as northern middle latitudes, pluvial intervals correlate with glaciations. For the last 17,000 years, however, the Australian lakes record brief pluvial periods that do not correlate with the southwestern United States record. As temperature rose in postglacial time, the United States lakes dried up, but the Australian lakes recovered again between 10,000 and 5000 years ago. An unexplained increase in Australian precipitation must be invoked, possibly related to the rising sea-surface temperatures and changes in atmospheric circulation. Bowler, et al. (1976) maintained that the pluvial intervals were primarily due to reduced evaporation, but Dury (1973) gave hydraulic-geometry arguments for at least 50 percent greater precipitation during times of higher lake levels.

Major changes in fluvial morphogenesis in southeastern Australia are correlated with pluvial–nonpluvial conditions. In the last 45,000 years, three cycles of fluvial morphogenesis occurred on the Murray River and its tributaries. Each began with pluvial high discharge and intense aggradation by large rivers and ended with soil formation and smaller channel networks. The oldest cycle is poorly dated. The second of the three pluvial events is dated at between 30,000 and 25,000 years ago. The youngest of the events dates from 25,000 or 20,000 years ago to about 16,000 or 13,000 years ago (Bowler, et al., 1976, p. 381). During similar cyclic changes, the Murrumbidgee River channels may have carried a discharge five times that of the modern river (Schumm, 1968, p. 27). Meander wavelengths, meander radii, and channel widths of the ancestral fluvial systems are much larger than the equivalent dimensions of modern streams. The modern cycle of channel development has continued for the last 13,000 years, when discharge and channel dimensions have been roughly equivalent to those of the modern rivers. A decrease in discharge of three- to fivefold since 16,000 years ago can be inferred. Clearly, the pluvial morphogenetic interval for the southeastern Australian lakes and rivers is correlative with the last glaciation in the northern hemisphere, although postpluvial conditions in Australia began 6000 years before postglacial time (as defined in Europe). Subsequently, the pluvial conditions have recurred without obvious northern hemisphere correlatives. Earlier pluvial stages, beyond the 40,000-year range of radiocarbon dating, are also recognized in the Australian sites.

Southeastern Brazil.   The best example of alternating climatic morphogenesis, in which pluvials correlate with interglacial or Holocene conditions in higher latitudes, is in southeastern Brazil (Bigarella, et al.,1965). The landscapes of the presently humid, forested, southeastern coast of the country have steplike valley cross-profiles of nonstructural benches alternating with steep slopes. Traced into more arid northeastern Brazil, each bench becomes part of a regional pediment or "pediplain" separated from the next lower pediment by an escarpment.

Three cycles of alternating semiarid sediment erosion and humid valley incision are recognized (Fig. 18-8). Within each, minor fluctuations toward the opposite extreme are recognized. During the semiarid half of each cycle, pediments were cut across uplands, and coarse alluvium was transported down high-gradient braided alluvial channels to the sea, which was lower than present, thus establishing the basic correlation of nonpluvial and glacial conditions. During the humid half of each cycle, pediments and alluvial surfaces alike were forested and deeply weathered. The surface stabilization and higher stream discharge allowed rivers to entrench their upper valleys, even though their mouths were graded to a higher sea level. Minor semiarid episodes during humid intervals promoted scarp retreat and valley widening, and minor humid episodes during semiarid intervals caused minor valley entrenchment.

Because coarse alluvium and steep pediments are graded to lower sea level, and because humid valley incision is graded to a sea level comparable to today's, the basic pluvial = interglacial correlation is established in the Brazilian alternating mor-

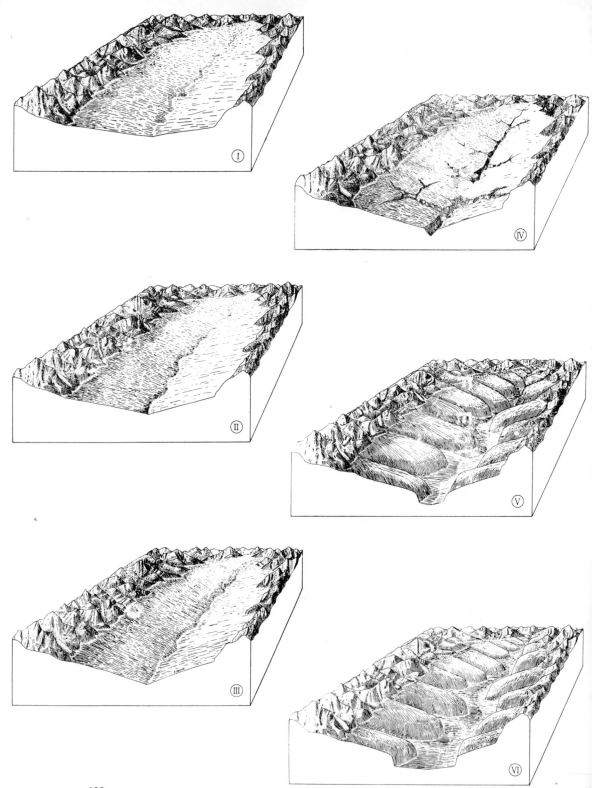

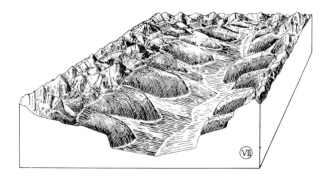

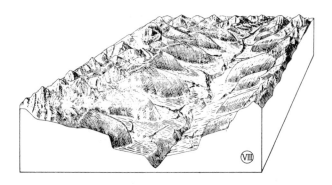

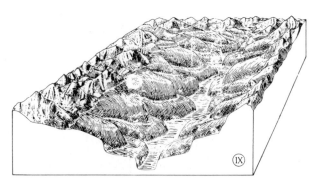

Figure 18-8. Hypothesis of slope evolution in southeastern Brazil under alternating semiarid and humid conditions. I–III: semiarid pediments lowered by slight changes in local base level and minor climatic changes. IV: entrenchment in a humid climate. V: valley widening and alluviation by minor climatic changes within a humid phase. VI: renewed pedimentation under a semiarid climate. VII–IX: repetition of stages III–V. (Bigarella, et al., 1965, Figs. 6, 7, and 8.)

phogenetic systems. Further proof has been provided by a radiocarbon date of about 14,000 years from organic debris in coarse terrace alluvium on one of the rivers (Pflug, 1969). In the Rio Dôce valley, after an earlier stage of deep erosion, alluvium was aggraded to so high a level that tributary streams were ponded against it as lakes. At that time, about 14,000 years ago, the uplands were providing large amounts of feldspar-rich, mechanically weathered sand and gravel. Subsequently, the river has again entrenched the alluvium as deep as 20 m, but the dammed lakes in the lower tributaries are still held about 20 m above the stream by the alluvium. The entire region is under tropical rainforest today.

East Africa. Supposed evidence of higher level lakes in the rift valleys of eastern Africa (Fig. 3-17) led to the early concept of "pluvials" that were intuitively correlated with the glacial sequence of the Alps. No dating methods or means of correlation were then available. Subsequently, some of the "lake beds" proved to be volcanic tephra, others are alluvium, and none can be proved to be related to glaciation. All of the names given to the alleged pluvials have been abandoned. Recent studies (Butzer, et al., 1972; Street and Grove, 1976) by radiometric dating on both lacustrine sediments and volcanic rocks definitely show major climatic changes, but the supposed correlation with glacial events is disproved (Fig. 18-9). Key eastern African lakes were at high levels sometime prior to 21,000 years ago but were at low levels and high salinity until about 10,000 years ago, nearly at the end of the last glaciation. Then, for 2000 years, they were at maximum levels, and several of them overflowed into adjacent tectonic basins. Some of them have had several moderately high levels during the Holocene, but as with the Australian lakes, no climatic reason for these fluctuations has been hypothesized. In the eastern African lakes, the major pluvial interval was neither glacial nor interglacial but rather at the transition from glacial to

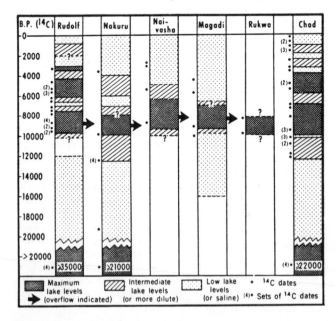

Figure 18-9. History of lake levels in eastern and central Africa. All lakes presently lack outlets. (Butzer, et al., 1972, Fig. 4; © 1972, American Association for the Advancement of Science.)

interglacial conditions. To further complicate the record, the Nile River, which has its source in the eastern African highlands, had much greater discharge than now between about 17,000 and 5000 years ago. The evidence of greater river discharge before 12,000 years ago is directly contradictory to the evidence of low lake levels then, although since 12,000 years ago, the correlation of river discharge and lake levels is fairly good. Certainly the erosional and depositional morphology of eastern Africa has been shaped by alternating Quaternary morphogenetic systems, but the record is greatly complicated by continuing tectonic movements (p. 47). River captures and diversions on the warping eastern African plateau may have affected discharges, and even the climate of the region may have been changed by tectonic uplift.

India.   As a final example of low-latitude, alternating pluvial and interpluvial systems, the few available studies from India can be summarized. River valleys in the Deccan region of northwestern peninsular India record successive Quaternary episodes of (1) deep erosion, (2) deposition of "older alluvium" tens of meters in thickness, with the development of alluvial plains several kilometers broad, (3) entrenchment and terrace cutting into the older alluvium, and (4) deposition of Holocene silty alluvium. The older alluvium has lenses of cemented gravel in a silt or sand matrix, which suggests deposition as alluvial fans or channel and point-bar deposits. It contains artifacts that formerly were assigned to the Middle Pleistocene, predating the last glaciation, but the age of the artifacts is now doubted, and the older alluvium may be only 33,000 years old (Rajaguru, 1969). It is correlated with glacial-age cooler air temperatures and ocean-surface temperatures, a weakening of the Indian monsoon system, and reduced total precipitation. During glacial times, semiarid erosion processes seem to have dominated, although unlike Brazil, a correlation with lower sea level has not been established.

In the Thar desert, just north of the Deccan Plateau, an episode of Pleistocene dune building choked former river valleys. The arid phase ended abruptly about 10,000 years ago, when lakes formed behind the dunes. The lakes have gradually shrunk during the Holocene but were fresh until about 3000 years ago. Now they are saline (Singh, et al., 1972; Singh, et al., 1974). Clearly, the Thar desert was drier during at least the late stages of the last glacial age and then abruptly more humid. The chronology is somewhat similar to that of the eastern African rift lakes (Fig. 18-9). Other evidence of dune growth and soil formation in western India also supports the theory that glacial ages were nonpluvial in the Indian subcontinent (Verstappen, 1970).

SUMMARY

The examples that have been presented generally support the hypotheses of Quaternary climatic change approximately as shown by Figures 18-1 and 18-2. There is some indication that rapid geomorphic change occurred during the transition from glacial to interglacial or pluvial to nonpluvial rather than during the peak of either extreme. It is important to remember that the time scale of these climatic changes is on the order of 100,000 years, with lesser oscillations on the order of 20,000 years each. These intervals are long enough to put distinctive morphogenetic imprints on landscapes but far too short to shape any region totally. Two possible exceptions are the humid equatorial tropics and the continental interior desert regions, which may have felt the least impact of Quaternary alternating morphogenetic systems. Perhaps this is one reason why the oldest of the earth's landscapes remain in the arid parts of Australia and Africa. The humid tropics, even though buffered from Quaternary climatic change by an excess of heat and moisture, could not escape the geomorphic impact of sea-level fluctuations. Arid regions with internal drainage may have been unaffected even by this otherwise worldwide morphogenetic factor.

REFERENCES

BIGARELLA, J. J., MOUSINHO, M. R., and DA SILVA, J. X., 1965, *Processes and environments of the Brazilian Quaternary:* Universidade de Paraná, Curitiba, Brasil, 69 pp.

BOWLER, J. M., HOPE, G. S., JENNINGS, J. N., SINGH, G., and WALKER, D., 1976, Late Quaternary climates of Australia and New Guinea: *Quaternary Res.*, v. 6, pp. 359–94.

BÜDEL, J. K., 1957, The ice age in the tropics: *Universitas* (Quarterly English Language Ed.), v. 1, pp. 183–91.

BUTZER, K. W., ISAAC, G. L., RICHARDSON, J. L., and WASHBOURN-KAMAU, C., 1972, Radiocarbon dating of East African lake levels: *Science*, v. 175, pp. 1069–76.

DURY, G. H., 1973, Paleohydrologic implications of some pluvial lakes in northwestern New South Wales, Australia: *Geol. Soc. America Bull.*, v. 84, pp. 3663–76.

FARRAND, W. R., 1971, Late Quaternary paleoclimates of the eastern Mediterranean area, *in* Turekian, K. K., ed., *Late Cenozoic glacial ages:* Yale Univ. Press, New Haven, Conn. pp. 529–64.

FLINT, R. F., 1971, *Glacial and Quaternary geology:* John Wiley & Sons, Inc., New York, 892 pp.

FLOHN, H., 1953, Studien über die atmosphärische Zirkulation in der letzten Eiszeit: *Erdkunde*, v. 7, pp. 266–75.

FRAKES, L. A., and KEMP, E. M., 1972, Influence of continental positions on early Tertiary climates: *Nature*, v. 240, pp. 97–100.

GALLOWAY, R. W., HOPE, G. S., LÖFFLER, E., and PETERSON, J. A., 1973, Late Quaternary glaciation and periglacial phenomena in Australia and New Guinea, *in* van Zinderen Bakker, E. M., ed., *Paleoecology of Africa and the surrounding islands and Antarctica*, A. A. Balkema, Cape Town, v. 8, pp. 125–38.

VAN DER HAMMEN, T., WIJMSTRA, T. A., and ZAGWIJN, W. H., 1971, The floral record of the late Cenozoic of Europe, *in* Turekian, K. K., ed., *Late Cenozoic glacial ages:* Yale Univ. Press, New Haven, Conn. pp. 391–424.

HAYS, J., 1971, Land surfaces and laterites in the north of the Northern Territory, *in* Jennings, J. H., and Mabbutt, J. A., eds., *Landform studies from Australia and New Guinea:* Australian National Univ. Press, Canberra, pp. 182–210.

HOPE, G. S., and PETERSON, J. A., 1975, Glaciation and vegetation in the high New Guinea mountains: *New Zealand Royal Soc. Bull. 13*, pp. 155–62.

KENNETT, J. P., HOUTZ, R. E., ANDREWS, P. B., EDWARDS, A. R., GOSTIN, V. A, HAJOS, M., HAMPTON, M. A., JENKINS, D. G., MARGOLIS, S. V., OVENSHINE, A. T., and PERCH-NIELSON, K., 1974, Development of the Circum-Antarctic Current: *Science*, v. 186, pp. 144–47.

LÖFFLER, E., 1972, Pleistocene glaciation in Papua and New Guinea: *Zeitschr. für Geomorph.*, Supp. v. 13, pp. 32–58.

MABBUTT, J. A., 1971, Denudation chronology in central Australia, *in* Jennings, J. N., and Mabbutt, J. A., eds., *Landform studies from Australia and New Guinea:* Australian National Univ. Press, Canberra, pp. 144–81.

MAXWELL, J. A., and DAVIS, M. B., 1972, Pollen evidence of Pleistocene and Holocene vegetation on the Allegheny Plateau, Maryland: *Quaternary Res.*, v. 2, pp. 506–30.

MULCAHY, M. J., 1971, Landscapes, laterites and soils in southwestern Australia, *in* Jennings, J. N., and Mabbutt, J. A., eds., *Landform studies from Australia and New Guinea:* Australian National Univ. Press, Canberra, pp. 211–30.

OBERLANDER, T. M., 1974, Landscape inheritance and the pediment problem in the Mohave Desert of southern California: *Am. Jour. Sci.*, v. 274, pp. 849–75.

PFLUG, R., 1969, Quaternary lakes of eastern Brazil: *Photogrammetria*, v. 24, pp. 29–35.

PORTER, S. C., 1964, Composite Pleistocene snow line of Olympic Mountains and Cascade Range, Washington: *Geol. Soc. America Bull.*, v. 75, pp. 477–82.

RAJAGURU, S. N., 1969, On the late Pleistocene of the Deccan, India: *Quaternaria*, v. 11, pp. 241–53.

RUDDIMAN, W. F., and McINTYRE, A., 1973, Time-transgressive deglacial retreat of polar waters from the North Atlantic: *Quaternary Res.*, v. 3, pp. 117–30.

SCHUMM, S. A., 1968, River adjustment to altered hydrologic regimen—Murrumbidgee River and paleochannels, Australia: *U.S. Geol. Survey Prof. Paper 598*, 65 pp.

SHACKLETON, N. J., and OPDYKE, N. D., 1973, Oxygen isotope and paleomagnetic stratigraphy of equatorial Pacific core V28–238: Oxygen isotope temperatures and ice volumes on a 10^5 year and 10^6 year scale: *Quaternary Res.*, v. 3, pp. 39–55.

———, 1976, Oxygen-isotope and paleomagnetic stratigraphy of Pacific core V28–239 late Pliocene to latest Pleistocene, *in* Cline, R. M., and Hays, J. D., eds., Investigations of late Quaternary paleoceanography and paleoclimatology: *Geol. Soc. America Mem. 145*, pp. 449–64.

SINGH, G., JOSHI, R. D., CHOPRA, S. K., and SINGH, A. B., 1974, Late Quaternary history of vegetation and climate of the Rajasthan Desert, India: *Royal Soc. London Philos. Trans.*, ser. B, v. 267, pp. 467–501.

SINGH, G., JOSHI, R. D., and SINGH, A. B., 1972, Stratigraphic and radiocarbon evidence for the age and development of three salt lake deposits in Rajasthan, India: *Quaternary Res.*, v. 2, pp. 496–505.

STREET, F. A., and GROVE, A. T., 1976, Environmental and climatic implications of late Quaternary lake-level fluctuations in Africa: *Nature*, v. 261, pp. 385–89.

TELLER, J. T., 1973, Preglacial (Teays) and early glacial drainage in the Cincinnati area, Ohio, Kentucky, and Indiana: *Geol. Soc. America Bull.*, v. 84, pp. 3677–88.

TIGHT, W. G., 1903, Drainage modifications in southeastern Ohio and adjacent parts of West Virginia and Kentucky: *U.S. Geol. Survey Prof. Paper 13*, 111 pp.

VERSTAPPEN, H. T., 1964, Geomorphology of the Star Mountains: *Nova Guinea, Geology*, v. 5, pp. 101–58.

————, 1970, Eolian geomorphology of the Thar desert and paleoclimates: *Zeitschr. für Geomorph.*, Supp. v. 10, pp. 104–20.

VITA-FINZI, C., 1969, *The Mediterranean valleys: Geological changes in historical times:* The University Press, Cambridge, U.K., 140 pp.

WEST, R. G., 1968, *Pleistocene geology and biology, with especial reference to the British Isles:* John Wiley & Sons, Inc., New York, 377 pp.

PART IV

Coastal and Submarine Geomorphology

The fluid that covers 71 percent of the earth's rocky crust is sea water, not air. Weathering, mass-wasting, and erosion proceed in very different ways under water than under air. The sea floor is mantled with a thin layer of sediment that settles out of the overlying water mass, vaguely analogous to the loess mantle that covers large areas of subaerial landscape. The fine sediment, both biogenic and terrigenous, that covers most of the sea floor is in places overwhelmed by many thousands of meters of coarser sediment derived from subaerial erosion. Most of the coarse sediment stays near the continental margins, but some is carried by subaqueous mass-wasting into the deepest ocean basins.

Most ocean-floor geomorphologists suppose that submarine topography is more constructional than destructional. Volcanic and tectonic mountains and hills rise above depositional abyssal plains that are almost perfectly flat. So little of the ocean floor has even been seen that a systematic geomorphology cannot hope to be written. Our information comes primarily from precision seismic depth recorders that trace profiles of the ocean floor several kilometers below ships. By analogy, try to imagine a systematic geomorphology of subaerial landscapes based entirely on profiles and samples collected from aircraft that hovered above a continuous cloud cover!

The systematic geomorphology of coasts involves a localized set of processes and landforms encountered neither in the subaerial nor in the submarine landscape. Energy is fed into the coastal geomorphic system primarily by surface ocean waves and is expended over an area that is narrow in height and width relative to its great length. Unless the relative level of land and sea has changed, all coastal landforms are in a zone about 10 m above and below mean sea level and a few kilometers inland and seaward of the average shoreline position. Yet within this narrow, almost linear zone, a complex range of landforms are eroded and built, organisms evolve special forms of adaptation, and human activity is focused. By United Nations estimates, two-thirds of the world's people live in the 10 percent of land defined as the coastal zone. Food resources, commerce, and recreation are critically dependent on the dynamic landforms of coasts. Coastal scenery has been the subject of great artistic expression and massive engineering projects.

A peculiar and important aspect of coastal geomorphology is that the narrow vertical operating range of coastal processes ensures that small relative changes of land and sea level will produce relict coastal landforms above or below those that are now forming. Considering the frequency and magnitude of Quaternary sea-level fluctuations, it is hardly surprising that modern coastal morphology is highly complex. Relict and active landforms are intimately mixed and frequently confused. Subaerial processes interfere with or contribute to coastal geomorphic change as well.

The uppermost 100 m of sea water is a zone of enormous biologic productivity. In certain climatic regions, organisms build major shallow-water or coastal landforms. Entire island archipelagos are emerged coral reefs, uplifted by tectonic movements or relict from times of higher Quaternary sea levels. There is no subaerial analog for these huge biogenic constructional landforms.

Climatic morphogenesis is not a major topic in coastal geomorphology. Certainly coral reefs and mangrove swamps are tropical, and ice-pushed beach ridges are high-latitude landforms. However, long-period surface waves transport energy across entire oceans with minimal loss, diffusing climatic morphogenetic gradients. The degree of exposure to surface waves is more significant to coastal geomorphology than the local climate.

In Chapters 19 and 20, the special processes and landforms of coastal regions are systematically considered. The deep-sea topography, although of great extent and geomorphic interest, is so poorly known that only the most significant processes and the regional landforms at the scale of geomorphic provinces are outlined in Chapter 21.

CHAPTER 19

Shore-Zone Processes
and Landforms

The **shoreline** is the line of demarcation between land and water. It fluctuates from moment to moment, influenced by waves and tides. The *shore zone*, or simply **shore**, is the zone affected by wave action. It is conveniently subdivided into: (1) the **offshore**, the shallow bottom seaward of the breaking waves; (2) the **nearshore**, between low-tide level and the breaker zone; (3) the **foreshore**, which extends from low-tide level to the limit of high-tide, storm-wave effects; and (4) the **backshore**, from the limit of recent or frequent storm waves landward to the base of a cliff, dune, or vegetated beach ridge. These definitions are not precise: "Offshore" oil is sought on the continental shelves to a depth of 200 m or more. However, in nautical use, the offshore zone is a place of hazardous navigation in contrast to the "open" sea. Lakes and ponds as well as oceans have shores, but the more general word **coast** applies only to the geographic region adjacent to an oceanic shore.

ENERGETICS OF SHORE-ZONE PROCESSES

Wind Waves

Most of the energy expended in the shore zone is transmitted there by wind-generated waves on the water surface. It is easy to visualize how a shearing stress by wind over water produces mass transport in the surface water layer, but why the water surface should assume a wave form is not simple to explain. One theory emphasizes the inherent gustiness of wind; another theory explains waves as a result of slight pressure differences over small ripples. The obvious phenomenon of waves being kicked up by a wind is in fact an elusive and complex process (Kinsman, 1965).

437

In an area of strong winds and storminess over the ocean, the water surface is thrown into a confused mass of waves that intersect in peaks and troughs. Smaller waves and ripples run up the backs of larger waves. When "a **sea** is running," it is best for ships to avoid the area. Waves as high as 34 m have been reported, but accurate observation of such waves is understandably difficult. The height of waves generated by wind depends on wind *speed*, the *duration* of wind from one direction, and the *fetch*, or length of water surface over which the wind blows. For example, a near gale (wind speeds to 17 m/sec) must blow for 6–12 hours over a fetch of 40–140 km to produce waves 50–75 percent as high as the theoretical maximum height of 6 m. Tables of predicted wave heights for various combinations of wind speed, duration, and fetch are available to mariners.

Radiating outward from the generating area of a sea, waves with the longest wavelengths advance most rapidly. Steep waves, with large height/length ratios, decay rapidly, but long low waves radiate for thousands of kilometers across oceans with little energy loss (Fig. 19-1). The regular pattern of smooth, rounded waves that characterizes the surface of the ocean during fair weather is called **swell**. Swell is usually composed of many wave trains of different wavelengths, moving outward from more than one generating area. From the air, swell patterns look like a grid of intersecting lines. Swell approaches a shore as a complex of wave trains that combine to produce alternately a succession of higher waves and a succession of lower waves.

Swell waves transmit energy outward from a stormy area in two forms: potential and kinetic. The height of a wave determines the potential energy of position above still-water level. The motion of individual water particles as a wave passes is a measure of the kinetic energy of the wave (Fig. 19-2). The potential energy of the waves moves forward with the wave form, but the kinetic energy of each moving water particle is expended in the nearly circular orbit of the particle. The orbital diameters of water particles beneath a wave decrease rapidly with depth in a geometric progression related to the wavelength. The orbital diameter is halved for each

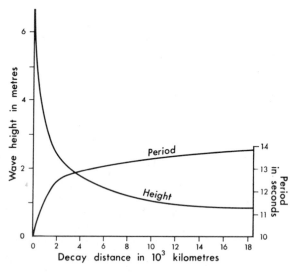

Figure 19-1. The effect of swell decay on the wave period and height of a wave initially 6.1 m high with a period of 10 seconds. (Davies, 1973, Fig. 16.)

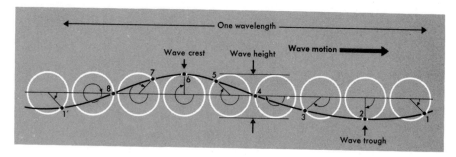

Figure 19-2.   Motion of surface water particles during the passage of a wave. The orbital diameter decreases rapidly with depth beneath the wave.

increase in depth equal to one-ninth of the wavelength. Thus, at a depth equal to the wavelength, the orbital diameter is only about $\frac{1}{512}$, or 0.2 percent, of the surface diameter. This rapid decrease in water motion with depth explains why a submarine can ride out the most severe storm by submerging 50 m or more. A wave is assumed to be moving in deep water when the water depth exceeds one-half the wavelength. At that depth, the drag of the wave on the bottom is negligible. Swell waves are not deflected by the Coriolis effect but move across deep water in a direction perpendicular to their crests, along great-circle tracks (Komar, 1976, p. 99). This tendency, plus the persistence of swell waves, means that energy for coastal erosion can be generated by storms in the opposite hemisphere as much as 15,000 km away.

Breakers and Surf

As deep-water swell approaches a shore, the form of the waves and the method of water and energy transport change. The shoaling sea floor distorts the orbital motion beneath the waves from a circle to an ellipse, then to a to-and-fro linear motion. When the orbital velocity exceeds the threshold of erosion (Fig. 9-8), sediment on the ocean floor is moved back and forth by the waves and absorbs energy from the moving water. At this depth, the conversion of wave energy to geomorphic work begins.

The depth at which waves begin to stir bottom sediment is determined by wave height, wavelength, and period, plus the grain size and sorting of the sediment. Although very long-period swell waves may stir water and bottom sediment to great depths, ordinary waves do not significantly affect bottom-sediment transport or erosion at a depth greater than about 10 m. This depth, which is called **wave base**, or **surf base** (Dietz, 1963), is a convenient reference level, although we should keep in mind that the work of waves does not abruptly cease at a certain depth.

When a swell wave begins to "feel" a shoaling bottom and become a geomorphic agent, the wave velocity is decreased by bottom friction. The wavelength is correspondingly shortened. The wave height also increases. As the crest of each wave approaches, the water moves rapidly forward as well as upward. Ultimately, the forward motion of the water mass in the wave equals the decreasing forward motion of the wave front. The wave assumes a steep face, and then its crest collapses forward into the trough. The deep-water swell becomes **surf**, or lines of **breakers**. Breakers form in the surf zone where the bottom depth is about one-third greater than the breaker height. On gently sloping coasts, the first line of breakers may be as much as a kilometer offshore, and the incoming water may reform into lower waves that break again at successively shallower depths. Wave energy is expended throughout the zone of bottom friction.

Swell can be *reflected* by a vertical seawall or steep cliff with almost no transfer of energy from the waves to the structure, because the mass of water in each wave is not moving forward significantly more than the diameter of the particle orbit. However, when a breaking wave hits a cliff or seawall, thousands of tons of water are in motion against the structure. The greatest pressures are exerted by breaking waves that curl over at their crest and trap air between the wave face and a steep wall of cliff so that the air is compressed; shock pressures of 608 kPa (12,700 lb/ft^2) have been recorded against seawalls. The duration of such great pressure is less than one-hundredth of a second, but large blocks of rock can be pried loose and moved by the repeated assault of the waves.

The wave energy against the coral reef of Bikini Atoll was carefully calculated by Munk and Sargent (1954). For a typical average breaker height of 7 ft (2.1 m), the waves' total power against the reef is 3.73×10^5 kW (500,000 horsepower). On the sector most exposed to the trade winds, the power reaches 20 kW/m (8 horsepower per linear foot) of reef front.

As a deep-sea swell crosses an irregular shoaling coastal bottom, wave *refraction* occurs, analogous to the bending of light rays on passing through a lens. The segment of each wave in the shallowest water is slowed so that in plan view the wave crest becomes concave forward. The effect is best illustrated by constructing a series of orthogonals perpendicular to wave crests (Fig. 19-3). A shallow submarine ridge thus focuses wave energy against its flanks, whereas a submarine valley or basin causes wave orthogonals to diverge and wave crests to be attenuated. Because of the postglacial rise of sea level, most offshore submarine ridges and valleys are continuations of the present subaerial topography, so that modern headlands are usually fronted by submarine ridges, and bays by submarine valleys. The effect is to cause wave energy to converge on headlands and diverge into bays.

Wave refraction is the basis for two important generalizations about the evolution of coasts. First, an irregular initial shoreline becomes more regular as initial seaward protrusions caused by sloping ridges tend to be eroded faster than adjacent shoreline segments in coves or bays at the heads of submarine valleys. Second, wave refraction generates currents flowing along the shore from headlands where the breakers are focused to adjacent coves where the water level is lower. These wave-generated **longshore** or **littoral currents**, along with tidal and other currents, transport the sediment eroded from headlands into adjacent coves where beaches are built. The simplification of a shoreline by refracted waves thus involves the filling of bays as well as the erosion of headlands.

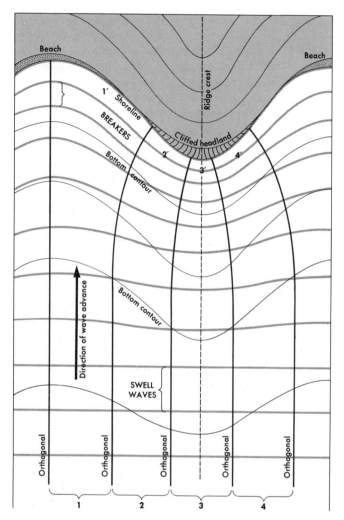

Figure 19-3. Wave refraction over an irregular shoaling bottom. Orthogonals are drawn perpendicular to the wave crests, equally spaced along the swell so that segments 1–4 have equal amounts of energy. Note the convergence of energy on the headland and divergence into the coves.

Waves are *diffracted* when a segment of wave crest passes between headlands or artificial constrictions into open water such as a river mouth or bay. The littoral currents generated within bays and harbors by diffracted waves erode and transport shore sediment until the bay shoreline "fits" the diffraction pattern closely. In theory, shorelines shaped by refracted and diffracted waves should assume gentle curvatures parallel to the wave fronts of the dominant swell. In fact, headlands are commonly composed of more resistant rock than the shores of adjacent coves, and even though wave attack is concentrated on the headlands, the more resistant rocks continue to form structurally controlled seaward projections as the entire shoreline retreats. Straight shorelines are characteristic only of coasts on terranes of uniform erodibility.

Storm Surges and Seismic Sea Waves

As a cyclonic (low atmospheric pressure) storm approaches a coast, a series of processes combine to endanger coastal installations and promote rapid geomorphic change. A drop of 100 Pa (1 mb) in atmospheric pressure causes the sea surface to rise 1 cm (1 bar $\simeq$ 10 m sea water) by the "inverted barometer" effect. Hurricanes or typhoons frequently have central atmospheric pressure 100 mb below normal, so the general rise of the sea surface under the storm may approach 1 m. Even more important is the shear stress of high, persistent winds on the sea surface, actually driving masses of water forward and raising mean sea level several meters along several hundred kilometers of coast. This "wind set-up" combined with the "inverted barometer" effect produces a **storm surge**, with shore-zone water depths several times as great as the highest astronomic tides. Waves, by far the most effective product of such storms, are even more effective because they move shoreward through water several meters deeper than normal and break against land that is poorly shaped to absorb wave energy (Hayes, 1967).

A *seismic sea wave*, often called by the Japanese name **tsunami**, is entirely distinct from wind waves. Sudden motion of the sea floor during an earthquake (Fig. 2-5) creates a solitary wave, or at most two or three waves, of very low amplitude that moves rapidly over the deep ocean. Successive arrival times prove that a tsunami may move with a velocity of over 800 km/hr. Crossing a shoaling bottom, the wavelength shortens, and the wave height greatly increases to 15 m or more, with disastrous consequences when it breaks. Tsunamis are not rare in the Pacific Ocean; their recurrence interval for Hawaii is about 25–50 years.

Tides

Tides are a unique form of energy for geomorphic change. Power is dissipated by the tidal drag of the moon and sun on the rotating earth at a rate of about 3×10^9 kW (Fig. 5-2). The complex periods and heights of tides are driven by lunar and solar gravity but are affected by the size and depth of the various ocean basins, the shape of their shorelines, and the latitudes of the basins, among many other variables (Neumann and Pierson, 1966, pp. 298–325). Seasonal winds and atmospheric pressure patterns also influence the tides. Each ocean, gulf, and sea has its own tidal pattern. The energy of tides is primarily dissipated by shallow seas such as the Bering Sea, the Sea of Okhotsk, the Timor Sea, the Patagonian shelf, and the Hudson Straits. These five shallow seas alone account for nearly one-third of the total energy conversion (Hendershott, 1973, p. 76). We can justifiably conclude that the coastal and submarine geomorphology of shallow seas are strongly shaped by tidal energy.

Tidal ranges are high only where the tides enter a semi-enclosed sea or gulf that has a shape and depth to produce a natural period of oscillation of water that is in resonance with some period of the tide. The greatest tidal range on earth occurs in the Bay of Fundy, Canada, where sea level may vary 16 m between high and low water. The Bay of Fundy has a natural period of oscillation of slightly over 6 hours, almost in phase with the ideal semidiurnal tide (Garrett, 1972).

In addition to their primary geomorphic role of raising and lowering the level of wave attack against the land, tides also generate strong currents. The most rapid currents develop in narrow channels that connect basins with different tidal periods. Hell Gate in the East River at New York City is one such place. The high tide in Long Island Sound takes almost 3 hours to move westward through the Sound. In that time, the tide in the less restricted lower New York Harbor has crested and is half-way down to low water again. Because of the difference in level, the water in the Sound ebbs south through Hell Gate at speeds of 9 km/hr (1 km/hr = 28 cm/s). About 6 hours later, the tide in the west end of the Sound is low, but the water in New York Harbor is nearing high-tide level, and the current floods north into the Sound almost as fast as it previously ebbed. The name of the strait is an obvious reference to the violent and changeable tidal currents.

Occasionally, where a tide enters a large river mouth, it begins to drag the bottom like any long wave and steepen its forward slope. In extreme examples, the river mouth develops a **tidal bore**, a single wave that races upstream at 15–30 km/hr as a vertical wall of water as much as 3 m high. Tidal currents and bores are obviously capable of extensive erosion and transportation of sediment (Ludwick, 1974). The current velocities just cited could erode and transport any sediment from compact clay to boulders (Fig. 9-8). If abrasive tools are available, deep channels can be eroded in bedrock as well.

Wind

Wind is the key link in the conversion of solar energy to wave energy (Fig. 5-2). In the shore zone, wind also acts directly as a geomorphic agent. Low tides expose extensive areas of wave-transported sand for several hours twice each day. Onshore winds dry the sand and blow it into dunes or transport it inland beyond the reach of waves. On some sandy shores, wind accounts for the major loss of sand from the shore system. If wave energy and sediment sources are adequate, the sand is replaced from offshore or along-shore. If not, beaches can be stripped of sand by wind.

The limited travel distance of saltation-transported sand grains (p. 334) keeps most of the windblown sand within the coastal region. Sand dunes are an inherent component of sandy coasts. They are usually of the transverse or parabolic type (Fig. 14-9), because sand supply is renewed by waves, winds are irregular or seasonal, and vegetation restricts dune expansion. Some desert coasts are backed by fields of barchan dunes migrating inland from the beach (Fig. 14-11).

Shore-Zone Weathering and Mass-Wasting

On oceanic shores, repeated wetting and drying by saline water is an effective weathering agent. In addition, a remarkable variety of shore-zone plants and animals bore, scrape, and dissolve rocks, providing biological energy for shore-zone processes. At our present stage of knowledge, it is not practical to attempt a detailed classification of shore-zone biologic, chemical, and mechanical weathering processes, because all three are complexly interrelated.

Intertidal life zones and weathering processes are strongly affected by the length of the alternating emergent and submergent intervals. As each zone has a diagnostic flora and fauna, so it has an associated set of weathering processes (Stephenson and Stephenson, 1972). The spray zone, just above normal high tides, is a zone of especially active chemical weathering. Lower and higher zones, which remain either wet or dry for longer intervals, are less affected. On some rocky shores, the lower intertidal zone is dominated by algal or molluscan encrusting deposits, whereas on other shores, burrowing organisms like sea urchins abrade craters into the rock. The net effect is a set of shore-zone landforms with very pronounced vertical zonation.

The interaction between subaerial and coastal geomorphic processes is illustrated no better than by reviewing mass-wasting (Chap. 8). As waves erode the base of a cliff, debris slides or falls, and a talus may protect the cliff from further erosion for some time. But at the base of the cliff, the talus provides tools for the breakers to use in abrading the shore platform. No widening of shore platforms can occur unless cliff retreat by the normal processes of subaerial mass-wasting equals or exceeds the rate of shoreline erosion. Rarely, major submarine landslides remove an entire section of coast, as when two-thirds of the pirate city of Port Royal, Jamaica, slumped into the sea during an earthquake in 1692 (Link, 1960).

EROSIONAL SHORE-ZONE LANDFORMS

Sea level has risen close to its present level only within the last 6000 years with the melting of the large ice sheets of the Wisconsin glaciation. Considerable uncertainty remains about the last 6000 years, during which sea level is believed by some to have oscillated a few meters above and below present (Fairbridge, 1976), by others to have been essentially stable (Coleman and Smith, 1964; Thom and Chappell, 1975), and by others to have risen slowly toward present level (Shepard and Curray, 1967; Bloom, 1970). The details are complicated by isostatic and other tectonic movements, which have deformed most, if not all, shorelines (pp. 23 and 406). Nevertheless, there are scarcely any coasts of resistant rock that show significant Holocene erosional shore-zone landforms. Those that do have inherited the basic landforms from interglacial ages, the latest of which was about 120,000–140,000 years ago, when sea level was a few meters higher than now (p. 408). If an inherited shore platform or wave-cut cliff base has been reoccupied by the sea for even a few thousand years, it is likely to appear deceptively well developed (Fig. 19-4). Only careful search and good fortune can lead to the discovery of uneroded remnants of post-cutting sedimentary cover on ancient platforms by which their great age can be proved.

The Holocene is a particularly complex time to study shore-zone landforms if so many are relict or exhumed. However, enough weak-rock coasts have modern erosional forms, and enough relict forms are preserved, to give us a good idea of what shore-zone erosion could do if enough time were available (Cotton, 1955).

Figure 19-4. The chalk cliffs at Sangat, France, are fronted by a prominent shore platform. However, a Paleolithic site on the platform near the low-water line proves that the surface was cut during the last interglacial or earlier, was a periglacial hillside terrace during the last glacial, and has only recently been reoccupied by the sea.

Shore Platforms

Where the postglacial rise of sea level has created a shoreline on a former hill slope, shore-zone processes cut a cliff and bench. The cliff is controlled by mass-wasting and structural factors. The eroded bench or platform, approximately intertidal, is given the nongenetic descriptive name **shore platform**. Several processes contribute to the erosion of shore platforms (Wentworth, 1938, 1939).

Processes of Shore-Platform Erosion. The hydraulic pressure and turbulence of breaking waves can easily erode weathered and joint-bounded fresh bedrock (Fig. 19-5). The initial stage of shore-platform erosion is probably the direct removal of rock by waves, which by analogy with river and glacial work is called *wave plucking* or *quarrying*. Even though wave plucking is restricted to a narrow vertical zone, high cliffs can result if sliderock is exported from the talus at the cliff base. **Sea caves** or a **notch** may be cut at the base of a cliff.

As plucking deepens a shore-zone notch or erodes back a sea cliff, the shore platform widens. It is in the surf zone, and sediment from wave plucking and cliff

Figure 19-5. Betty Moody's Cave, Star Island, New Hampshire. Waves have quarried a columnar basalt dike from between granite walls.

Figure 19-6. An abrasion ramp cut across dipping strata, Bay of Fundy coast near Joggins, Nova Scotia.

retreat is dragged back and forth across the platform to shape an **abrasion ramp** (Fig. 19-6). Wave abrasion requires constant removal of plucked or mass-wasted sediment from the abrasion ramp lest the surface be buried. Thus a seaward *slope of transportation* develops on an eroding surface thinly veneered with rock debris in active transport. The abrasion ramp extends seaward until the debris is either abraded to fine sediment, dissolved, or settles into water too deep for further wave movement. From careful profiles across modern abrasion ramps and emerged coastal terraces, Bradley (1958) concluded that their typical seaward gradient is about 1° (about 2 percent). On a tideless shore, such a gradient would extend below the 10-m depth of ordinary wave erosion on a bench no wider than 500 m. If a tide range of 5 m is assumed, the abrasion ramp could still be no wider than 800 m. Shore platforms wider than that cannot develop by abrasion at a fixed sea level, although they could form during progressive submergence.

The term *wave-cut bench* has been commonly used as a synonym for shore platform, and older geomorphic literature proposed continental-scale marine planation as an alternative to subaerial peneplanation. However, the necessity for a seaward slope on an abrasion ramp ensures that even a small island could not be planed off to sea level by abrasion alone. Wide shore platforms, or those with gentle gradients, must be cut by processes other than abrasion (Bradley and Griggs, 1976).

Water-level weathering (Wentworth, 1938) seems to be the dominant process that widens and flattens shore platforms on noncarbonate rocks. Intertidal pools and salt spray on abrasion ramps or in wave-plucked notches promote weathering in the zone just above permanent saturation. Wave erosion during high tides carries off the weathered detritus (Fig. 19-7). On exposed platforms, all stages can be seen from

(a)   (b)

Figure 19-7.   A "tesselated pavement," an intertidal platform shaped by water-level weathering on horizontal, jointed sandstone south of Hobart, Tasmania. Blocks have either raised rims (*a*), or raised centers (*b*).

initial ponding in joint-bounded basins to coalescing flat-floored pools separated by serrate ridges. Bedding planes and joints are frequently etched in high relief.

Shore platforms are developed by water-level weathering at various heights relative to tide level, depending on structural factors such as permeability and fractures and also on wave energy, tidal range, and climate. On Hawaiian coasts exposed to vigorous surf, platforms as much as 7 m above sea level develop by weathering at the edges of spray-filled pools, although better-developed platforms are at 3–4 m (Wentworth, 1938, p. 20). These high platforms on headlands slope landward into bays, where they are only about 1 m above sea level. Such a range in contemporary water-level weathering platforms warns of the danger in attempting to establish the heights of ancient sea levels from the evidence of relict shore platforms and abrasion ramps.

Water-level weathering puts the finishing touches on surfaces prepared by wave abrasion and plucking. Yet excessive abrasion and plucking destroy the evidence of water-level weathering. Water-level weathering platforms become well developed in volcanic rocks, especially weakly indurated, pyroclastic sediments. Their abundance and high perfection on oceanic islands and the peripheral archipelagos of the Pacific Ocean is in part related to the abundance of Cenozoic volcanic rocks in those regions.

On carbonate rocks in the shore zone, solution is so common that a special class of **solution platforms** is distinguished (Wentworth, 1939). These are the most complex shore platforms to describe, because their origin involves both chemical and biologic solution and deposition. Solution platforms are generally intertidal. Their surfaces are flat and smooth but not abraded (Fig. 19-8). Small pools alternate with low domes and ridges, often formed by living carbonate-secreting algae. At the landward edge of a solution platform may be a *solution notch* as deep as 3 m. Because sea water is generally saturated with calcium bicarbonate, it has been argued that a solution platform must form at the level of fresh-water saturation, or water table. However, good solution platforms are also common on desert coral islands that have no fresh ground water. Diurnal changes in the pH of intertidal pools by photosynthesis and temperature change offers one explanation for alternating solution and deposition by sea water (Revelle and Emery, 1957).

Even more effective but complex is the role of biochemical weathering and erosion in developing solution platforms and notches. Many intertidal plants, especially algae, penetrate their substrate with microscopic strands of organic tissue (p. 139). *Lithophagic* (rock-destroying) animals graze on the limestone surface and abrade or dissolve the perforated rock surface to obtain the algae. The net chemical and mechanical effect is rapid erosion, as much as 1 mm/yr by repeated measurements on the exposed parts of pins driven into rock (Hodgkin, 1964). An estimated 154 g/yr of rock was abraded from a square meter area in Barbados by one species of lithophagic gastropod (McLean, 1967, 1972). Submarine erosion rates of 2–10 mm/yr have been estimated for calcareous mudstones in California (Warme and Marshall, 1969).

Considering the complex nature of the processes that shape them, solution platforms may be misnamed. However, the complexity of intertidal carbonate geochemistry is so great and so poorly understood that "solution" is an adequate word for the

Figure 19-8. A solution platform surrounding residual karst blocks of calcareous eolianite near Robe, South Australia.

present. On coral reef coasts, the geomorphology is even more complicated, because solution platforms of the sort described here grade seaward into constructional reef flats and shoreward into abrasion ramps. In the low tidal range common on tropical coral islands, the entire complex may form a flat surface more than a kilometer in width with a relief of only 1–2 m.

Theoretical Equilibrium Form of Shore Platforms. If we assume a maximum depth of wave abrasion of about 10 m, we can deduce an equilibrium profile for abrasion platforms. When wave energy is expended at a uniform rate across the platform, it will have an equilibrium profile that can migrate shoreward but retain the same form. Zenkovich (1967, pp. 151–53) deduced that the stable profile of an abrasion ramp would be one on which the "effective" wave, or the wave form that combines optimum frequency with optimum abrasive energy, would not break. The profile should be convex upward, with no horizontal segments and the minimum slope at the shoreline. The deduced profile should be longer and more gently sloping for larger effective waves. Some shore platforms in the Black Sea may approach the ideal form, but only because they are eroded on limestone by abrasive siliceous beach sediment. Otherwise, the forms are unknown on modern shores. The lack of known examples testifies to the general disequilibrium of the shore zone to present sea level.

Most modern water-level weathering platforms are probably inherited from relict shore platforms that were slightly higher. Old abrasion ramps or other shore platforms, eroded during the last interglacial age when sea level was a few meters above present level, form ideal surfaces on which water-level weathering and solution can operate. Because there is no definite tidal level for the modern forms, it is not possible to assert by altitude alone whether a present-day shore platform is totally relict, an inherited form modified by present processes, or entirely modern. Only where platforms are being exhumed from cover deposits of the last glacial age (Fig. 19-4), or where they grade into constructional reefs that can be radiometrically dated as of last interglacial age, can their relict character be demonstrated.

Coastal Cliffs

The cliffs that rise from the back of shore platforms are commonly called *wave-cut* cliffs, but, as with any subaerial escarpment, mass-wasting is the dominant process of cliff retreat. Wave quarrying at the cliff base keeps it fresh at the appropriate angle of repose. With stable sea level, abrasion ramps eventually widen until wave attack becomes infrequent on the cliff face, when it becomes a vegetated, subdued hillside. The many bold, cliffed coasts of the world are another reminder of the essential disequilibrium between coastal landforms and present sea level.

On some cliffed coasts, the cliffs are so steep that wave-plucked debris falls into water too deep to develop a wave-cut notch or abrasion ramp (Fig. 19-9). These steep coasts are especially common on volcanic terranes, fault scarps, and glaciated coasts. The cliffs are said to *plunge* into deep water (Cotton, 1952). A striking feature of plunging cliffs is that swell waves do not break against them but are reflected back out to sea (p. 440). If rock is strong enough to form a steep cliff, and if that cliff happens to rise from water more than about 10 m deep, it is practically immune from wave attack. Only when a submarine talus forms at the base of a cliff to create breakers can the full energy of waves be impressed on the cliff.

Structural Control of Shore-Zone Landforms

Factors of rock strength, lithology, and solubility have been repeatedly mentioned as contributors to shore-zone erosional morphology. As in the subaerial environment, some rocks are more resistant to weathering and erosion than others. Resistant rocks on coasts tend to protrude seaward as well as to maintain higher elevations. Even though wave refraction focuses more energy onto headlands of resistant rock, they continue to form headlands as erosion progresses. Wave refraction around a protruding headland can be so extreme that only a narrow neck of land is left, connecting a peninsula to the mainland. In suitable structures, especially flat-lying sedimentary rocks, wave erosion may quarry through the neck, creating a **natural arch**. With further erosion, arches collapse leaving isolated pinnacles or **sea stacks** standing on the shore platform (Fig. 19-10).

Shore platforms on coastal terraces of nearly flat-lying sedimentary rocks are especially difficult to interpret. It was noted that water-level weathering takes advantage of pre-existing benches. In horizontal strata, structural benches readily form and may hold enough water to promote water-level weathering at several different intertidal and supertidal levels simultaneously. Even more commonly, slight structural dips are eliminated by water-level weathering to form horizontal platforms that truncate structure but are nevertheless basically controlled by structure, not by shore-zone processes.

Figure 19-9. Plunging cliffs on the drowned margins of a basalt dome, Banks Peninsula, Canterbury, New Zealand. (Photo: D. L. Homer, N. Z. Geological Survey.)

Figure 19-10. The Twelve Apostles, sea stacks eroded from Miocene limestone, Port Campbell National Park, Victoria, Australia (Photo: J. F. Bliney, contributed by B. G. Thom.)

CONSTRUCTIONAL SHORE-ZONE LANDFORMS

Organic Reefs

A reef is by mariners' tradition an obstruction or menace to navigation. By definition then, **reefs** are shallow-water submarine landforms. The term is usually applied to solid, rocky structures rather than to sand bars. Most reefs are constructed by marine organisms, although some are structurally controlled, shallow, submerged ledges such as hogbacks, dikes, or lava flows.

Ecology of Organic Reefs.   Throughout Phanerozoic time, since organisms with hard parts evolved, reef building has been a repeated and persistent form of marine growth, especially for colonial organisms. In shallow oceans of the Cenozoic Era, the Scleractinian colonial corals and associated calcareous algae are the most important builders of organic reefs, although molluscs, sponges, and a variety of other organisms build reefs in special situations. Reefs are built up in shallow water to take advantage of sunlight for photosynthesis, to shed or stay above smothering detrital mud, and to provide a large surface area for continued growth. Waves breaking over and refracting around reefs create currents that sweep nutrients and oxygenated water over the sedentary colonial organisms. Reefs are usually well adapted to local wave and tidal energy so that they are nourished by the energy expenditure rather than being destroyed. Larvae have a free-swimming stage before they settle and attach to the bottom, so reef-builders disperse widely.

Coral and coralline algae reefs are widespread in tropical waters. They require shallow (usually less than 100 m) and clear water for photosynthesis. Reef growth is best in water that does not cool below about 18°C during winter months. Salinity should be close to normal. Fresh water from rivers or torrential rains is especially damaging. A hard substrate is necessary for a colony to become established, but small patch reefs can grow on soft sediments and gradually fuse to form a suitable base. Loose sand is poor for coral growth, as is mud. In spite of all these restrictions,

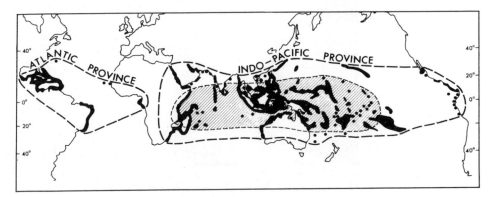

Figure 19-11.   Distribution of Holocene coral reefs. Some form only thin veneers on older structures. Shaded region includes the areas of most prolific reef growth and almost all the atolls. (Davies, 1973, Fig. 47.)

the coastal area of coral reefs is very large (Fig. 19-11). One of the most fundamental subdivisions of coasts could be between those that have corals and all others (Davies, 1973, p. 5). Organic reefs other than coral are trivial landforms and are subsequently ignored.

Coral Reef Morphology.   A coral reef is an enormous mass of limestone that is organically built. The seaward edge of the reef, especially on a windward coast, is where wave-resistant growth forms thrive in the most nutrient-rich environment and provide coralline detritus in the largest quantities. Some detritus breaks off and forms a **fore-reef talus** on the ocean side of the reef. Other, often finer, detritus is swept into the sheltered region behind the reef. In the surf zone on the extreme seaward rim of many Indo-Pacific reefs, calcareous algae build smooth, rounded mounds or rims. The algal limestone is more dense than coral limestone and is a vital contributor to reef stability. More delicate coral species grow in the lee of the seaward reef, adding more detrital sediment. Either a sandy lagoon floor or an intertidal **reef flat** may extend from the lee of the reef (Fig. 19-12).

Coral reefs are generally limited in upward growth to mean low-tide level. The extreme productivity of the reef provides sediment for many subaerial shore-zone landforms, however. Large storms and tsunamis tear away tons of reef-front coralline debris and hurl it over the algal rim onto the reef flat behind (Maragos, et al., 1973). Most reef flats are dotted with large erratic blocks of storm-tossed coral rock, often deeply notched by solution. On the reef flat, islands of coralline sand and gravel may build up high enough to support a fresh ground-water lens and vegetation. The carbonate sand and gravel recements readily especially if exposed to rain and fresh ground water. Major storms can probably remove any part of a coral-reef island that happens

Figure 19-12.   Fringing reef with a low algal rim in the surf zone, Tongatapu, Kingdom of Tonga. Note solution notches on talus blocks and cliffs.

Figure 19-13. High-tide platform of recemented coral debris about 1.5 m above modern reef flat, Turtle Islands, northern Queensland, Australia.

to be struck, but if a mass of debris is not disturbed for a few decades or centuries, it becomes relatively indurated and resistant to all but the worst hurricanes or typhoons. Cementation is generally close to the level of mean high tide in the zone of saturation. Because the tide range in tropical oceans is generally less than 2 meters, low platforms of cemented reef rubble are exposed on most eroding shores (Fig. 19-13). These cemented high-tide platforms have been frequently mistaken for emerged reef flats by which a Holocene fall of sea level has been widely inferred.

Some coral islands become the nesting grounds for sea birds. The phosphate-rich guano reacts with limestone to form phosphate rock, much sought for fertilizer manufacture. Phosphatic islands have a rugged karst surface of lapies and karren (Fig. 7-2) under a thin cover of weathered coral sand and gravel.

Classification of Coral Reefs. The traditional mariners' terms for reefs included **fringing reef**, which is attached to and extended seaward from a coast, and **barrier reef**, which is separated from the coast by a lagoon. Later, the word **atoll**, from the Maldive Islands, was adopted into the scientific literature for an annular reef that encloses a central lagoon. Darwin (1837) conceived of these three reef forms as a genetic sequence by which a fringing reef around a submerging central island would progressively become a barrier reef and then an atoll. Darwin's deduction required subsidence of the ocean floor over vast regions (Fig. 2-3) and sparked a spirited scientific controversy in the late nineteenth century. A series of deep boreholes on atolls have subsequently demonstrated that they are underlain by 1000 m or more of shallow-water reef limestone on basaltic basement, generally validating Darwin's theory for the origin of atolls in the deep ocean basins. The abundance of guyots (p. 491) adds further support.

Darwin's theory does not explain **shelf atolls**, the similar forms on shallow submerged continental shelves. Fairbridge (1950) suggested that shelf atolls could form by crescentic reefs extending their "horns" downwind until they closed into a ring. A simple genetic classification of coral reefs is not possible, because most of the shape is probably inherited from alternating periods of growth and subaerial erosion during Quaternary sea-level oscillations. A simple descriptive classification is given by Table 19-1.

TABLE 19-1. DESCRIPTIVE SCHEME FOR
CLASSIFYING ORGANIC REEFS

Adjacent to coasts or separated by a shallow channel		Fringing reefs
Separated from coast by deep channel		Barrier reefs
Forming islands — without central lagoon — large		Platform reefs
— small		Patch reefs
— very small		Coral pinnacles
— with central lagoon — deep lagoon		Oceanic atolls
— shallow lagoon — large		Shelf atolls
— small		Faros*

SOURCE: Davies, 1973, p. 72.

*Small, elongate atolls with narrow lagoons along the rim of large reefs (Fig. 19-14). Their origin may involve multiple periods of growth separated by subaerial karst development (MacNeil, 1954).

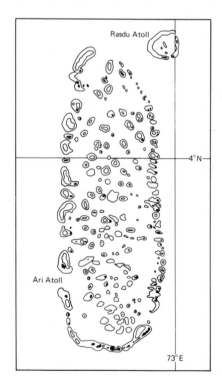

Figure 19-14. Ari Atoll, Maldive Islands. A huge atoll made of numerous smaller, elongate atolls on its rim (faros). Land areas (in black) are very small.

Inheritance and Reef Morphology.   Coral reefs are very strictly controlled by sea-level datum. Emergence or submergence of even a few meters completely displaces all the depositional and erosional zones of the reef and creates relict forms. Quaternary sea-level fluctuations (Figs. 17-14, 18-4) have left a particularly complex record on coral reefs. When sea level was low, reefs were exposed as flat-topped or hill-rimmed islands that were subject to karst solution and peripheral wave erosion or reef growth. Drowned karst topography (p. 146) is well shown on the floors of some atoll lagoons (Shepard, 1970). It is hard to avoid the conclusion that most reef morphology is inherited, in part from constructional forms that date from the last interglacial when sea level was several meters higher than now, and in part from destructional karst landforms that formed during glacial-age low sea levels (Bloom, 1974; Purdy, 1974). Until sophisticated radiometric dating techniques became available, there was no way to evaluate, or often even to detect, the glacial-age unconformities in coral reef stratigraphy. Now, uplifted reefs of known age can be used to measure rates of tectonic uplift, absolute rates of denudation, and the history of Quaternary sea-level oscillations (Figs. 2-6, 17-14).

Sandy Shores

Beaches.   If some wave energy were not absorbed in the work of moving sediment along coasts, far more erosional work would be accomplished. The sediment in motion along a shore is the **beach**. In function, it is analogous to flood-plain alluvium in river valleys; it absorbs energy and moves during times of storm and thereby stabilizes the rate of wave-energy conversion into geomorphic work. A continuous beach along a coast is as good a criterion of the graded condition as is a continuous flood plain on the floor of a river valley or a continuous sheet of sediment on a semiarid pediment.

The postglacial rise of sea level has submerged a variety of sediments, including glacial drift, sand dunes, older marine deposits, and alluvium. As waves move shoreward over the loose sediments, silt and clay are stirred into suspension and carried long distances, usually coming to rest either on tidal mud flats or in water too deep for further wave action. On a shoaling bottom, sand sizes are the first to be stirred by waves (Fig. 9-8), but the sand is moved by saltation, not in suspension, so it stays within the zone of wave action. Seaward of a **null line**, or *null point* on a bottom profile, sand in the offshore zone is either not moved or moved seaward. Landward of the null line, there is a net shoreward movement of sand by waves (Curray, 1969; Swift, 1969; Komar, 1976, pp. 310–14). The rate of landward sand movement is a function of grain size and wave energy. Nearshore sand becomes very well sorted as it is moved shoreward, permitting the sandy bottom to respond to small changes in wave energy. Gravel beaches are also common but move only at much higher energy levels.

Although the null line and the beginning of net onshore sand transport are well seaward of the breaker zone, breakers cause the most dramatic sediment transport. Sand is thrown into suspension by each breaking wave, and before it can settle to the bottom again, it can be moved many meters shoreward or laterally by turbulent currents. In cross profile (Fig. 19-15), two zones of erosion are seen; the most important zone is under the breakers, but another zone of intense sand erosion is in the **swash** zone, where small waves run up the beach and collide with backwash. Between these two eroding zones, when the waves are approaching at an angle, a powerful **littoral current** is generated, moving parallel to the shore. The surf zone is comparable to a shallow alluvial river. When waves break nearly parallel to the shore, only weak littoral currents are produced, and sand movement in the surf zone is nearly perpendicular to the beach. With a breaker angle and height that generate a littoral current of 30 cm/s, sand moves along the shore but with an outward component toward the breaker line. At littoral current velocities of 80–100 cm/s, a strong movement of sand parallel to the shore occurs, like a river with one bank on the beach and the other in the breaker zone (Ingle, 1966). In the surf zone, a ribbon of sand 100 m or more in width and 5–10 cm in thickness moves through a cross section of the littoral zone at a rate of 1000 m³/day or more (Das, 1971). Under such dynamic conditions, sandy beaches can adjust to changing wave and tidal energy within hours, and the nearshore profile is constantly regrading to changing conditions. New sand is added to the littoral system from offshore, from river mouths, and from eroding sea cliffs. Sand is lost from the system by wind transport inland, by tidal-current deposition in sheltered lagoons, by *rip currents* into deep water, and rarely by being trapped in the heads of submarine canyons that lead it out to deep-sea basins (Inman and Brush, 1973).

Wave refraction causes mobile beaches to move readily until the plan view of the shoreline conforms closely to the shape of the refracted wave fronts. On coasts with alternating headlands and beaches and a dominant swell that approaches obliquely, beaches assume a "half-heart," or **zetaform**, curvature (Fig. 19-16). The outline of these headland-bay beaches has been shown to closely approximate a logarithmic spiral (Yasso, 1965; Davies, 1973). Each zetaform beach segment seems to be offset relative to its neighbors with the straight segment of beach rotated from the mean trend of the coast toward the dominant swell direction. Grain size, sorting, and beach profiles all change systematically along each zetaform beach. Each is nearly a closed cell of sand supply and movement, although some sand probably drifts around the down-wave headland to nourish the next beach. The pattern of offset zetaform beaches dominates the high-altitude views of many coasts and is easily recognizable on coastal maps and charts.

Barriers.   On many coasts, the postglacial submergence was across surfaces of low relief and abundant sediment supply. During the later stages of the submergence, the marine transgression created growing ridges of sand that eventually become too massive to be moved further shoreward. The sea flooded the area behind them to form **lagoons**, but the massive sand ridges extend for hundreds of kilometers along many coasts as **barriers**, only rarely tied to headlands (Hoyt, 1967, 1968). Barriers and lagoons are said to extend along 13 percent of the world's coasts (Zenkovich, 1967, pp. 288 and 390).

WATER MOTION	Oscillatory Waves	Wave Collapse	Waves of Translation (bores): Longshore Currents: Seaward Return Flow; Rip Currents	Collision	Swash; Backwash	Wind
DYNAMIC ZONE	Offshore	Breaker	Surf	Transition	Swash	Berm Crest
PROFILE	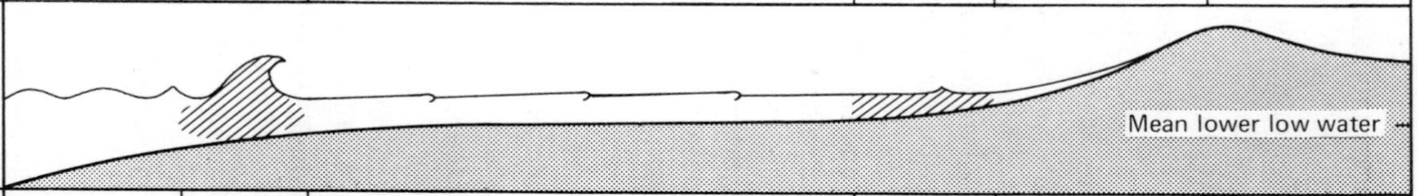				Mean lower low water	
SEDIMENT SIZE TRENDS	← Coarser	Coarsest Grains	← Coarser	Bi-modal Lag deposit	Coarser →	Wind-winnowed lag deposit
PREDOMINANT ACTION	Accretion	Erosion	Transportation	Erosion	Accretion and Erosion	
SORTING	← Better	Poor	Mixed	Poor	Better →	
ENERGY	← Increase	High	Gradient →	High		

Figure 19-15. Major dynamic zones in the nearshore zone of a sandy coast. Two regions of maximum sand suspension and transport are shaded. (Ingle, 1966, Fig. 116.)

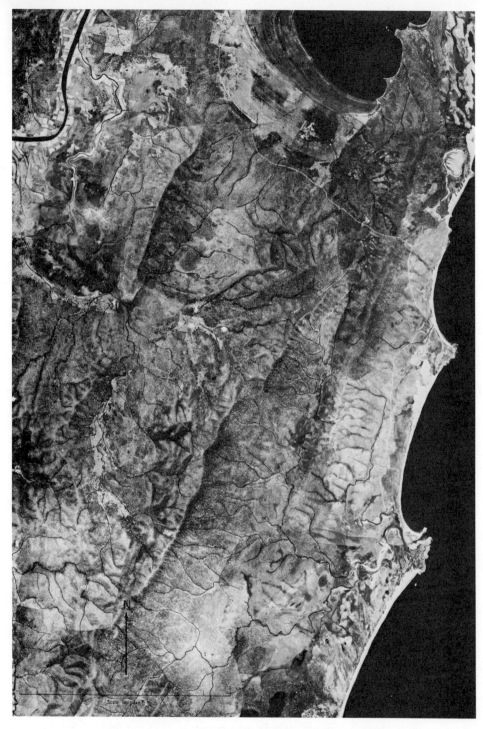

Figure 19-16. Zetaform coast of northern New South Wales, Australia. Longest beaches face the dominant swell from the southeast and trend obliquely to a line connecting the headlands. (Photo courtesy B. G. Thom.)

The origin of barriers was one of the durable controversies of geomorphology. As long as the submarine offshore topography was poorly known, and shore-zone sediment transport unstudied, barriers were thought to represent submarine sand bars that had grown very large and then emerged by tectonic uplift or a drop of sea level. Some authorities argued that wave action actually built barriers above sea level, aided by high storm levels and wind transport. Others maintained that barriers were built along shallow coasts by littoral drift and were not driven landward but grew in place.

The identification of barriers with coastal emergence has now been thoroughly disproved. Numerous boreholes in barriers and their associated lagoons prove that many were built in the last 6000 years, during postglacial submergence (Curray, 1969, p. JC-II, 5). Most of them rest on older foundations, perhaps dating from the last interglacial interval (Otvos, 1972). As with any beach, barriers are fed sand both from offshore and by littoral drifting. Many seem to have grown more rapidly prior to a few thousand years ago and are now eroding (Langford-Smith and Thom, 1969). Possibly sand supply was more abundant as the sea steadily transgressed new terrain, whereas with nearly stable sea level today, sand is being lost into dunes, lagoons, and the offshore zone and is not being replaced.

Barriers are large constructional landforms as much as 1 km wide and 100 m high to the crests of the highest dunes. Unless underlain by a relict landform, the high relief is eolian. Sand moving onshore or along shore is blown off beaches at low tide. If a plentiful supply of sand is available, it is replaced, and the barrier grows. If sand supply is inadequate, the barrier is eroded or driven landward by washover during major storms (Fig. 19-17). Extensive reclamation and beach stabilization projects actually endanger barriers, because unless sand is free to move with changing wave conditions, erosion results (Dolan, et al., 1973).

Most barriers are segmented at intervals by **tidal inlets** that allow tidal currents to ebb and flow into the lagoons and marshes behind the barriers and allow river runoff to escape. Inlets are systematically spaced to provide adequate tidal circulation. They are deep channels, often cutting through the entire cross section of the barrier and into the substrate. As a consequence, if a tidal inlet is forced by littoral sand movement to migrate downcurrent, it completely removes the barrier on the eroding side and builds a new barrier on the depositional side. Historical records show that when inlets migrate too far in this way, a new inlet breaks through the barrier somewhere upcurrent and in turn migrates in the direction of sand transport (Fisher, 1968). On most barrier coasts, the tidal inlets are depositional traps rather than sources for nearshore sediment. On coasts that consist of segmented barrier islands, some sand moves laterally across the mouths of tidal inlets, but some is permanently lost into the inlets. Unless this sand is replaced by headland erosion and river deposition upcurrent or by onshore transport by waves, the barrier will migrate landward over its lagoon or marsh, or grow thinner, or both. Modern coastal conservation techniques include calculations of the sand budgets for each section of a barrier coast. To correct negative budgets, additional sand is provided at the upcurrent end of each littoral cell. The quantities, measured in thousands of cubic meters per

Figure 19-17.   Core Banks, Cape Lookout National Seashore, North Carolina. A barrier in its natural state, broad and low, subject to washover during storms. (Photo: P. J. Godfrey, U.S. National Park Service.)

day (Das, 1971; Komar, 1976, p. 218), can only be provided by dredging sand from the ocean floor well to seaward of the null line. Fortunately, the postglacial rise of sea level seems to have frequently generated incipient barriers that were subsequently overtopped and bypassed and are now relict submarine sand ridges (Fig. 21-3). If these ridges are truly relict and not part of an as-yet-unknown modern sediment-transport system (p. 485), the sand in them can safely be used to restore and maintain modern barriers and other eroding coasts.

Other Landforms of Sandy Shores.   The ready readjustment of beaches and barriers to wave, wind, and tidal energy gives rise to many dynamic landforms. Terminology is extensive and complex but can be found in many books on coastal geomorphology (Guilcher, 1958; Zenkovich, 1967; King, 1972; Davies, 1973; Shepard, 1973; Komar, 1976). Many words, such as **spit** (a sand ridge attached at one end but free on the other) and **bar** (a submerged or intertidal sand ridge), are old mariners' terms. **Tombolo**, a beach that ties an island to the mainland or another island, is of Italian origin. The beach crest, just above high-tide wave action, is called a **berm** by analogy with the curved cross profile of a well-graded roadway. Dunes, blowouts, and other eolian forms (Chap. 14) are common on beaches. On prograded coasts, multiple beach ridges numbering into the hundreds (Fig. 19-18) record the seaward migration of the shoreline by deposition, or **progradation** (Curray, et al., 1969).

Figure 19-18. Multiple beach ridges on a prograding coastal plain, Nayarit, Mexico. (Photo: F. B. Phleger, courtesy J. R. Curray).

Muddy Shores

Intertidal Mud Flats. Clay-size and silt-size particles settle slowly from suspension and are carried by littoral current as suspended load. But in sea water, unlike in rivers, fine sediment *flocculates* into loose, large aggregates and settles out of suspension in quiet water as poorly sorted mud. Mud in suspension that is carried out beyond the null line is lost to the deep ocean floor, but much of the mud stirred up in the shore zone, or carried there by rivers, moves landward into estuaries, bays, lagoons, and other shallow, quiet water. Tidal ebb and flow, which might be thought to be comparable to a river that symmetrically reverses direction every 6 hours, actually cause a net shoreward transfer of mud (van Straaten and Kuenen, 1958; Postma, 1967). During the slack-water intervals when each tide reverses, mud begins to settle. Mud on the shoreward, shallow edge of the flood tide reaches the bottom and coheres, resisting current erosion on the ebb, or outward, tide. However, at the end of the ebb cycle, mud that moves seaward and settles an equal distance has not yet reached the bottom, so it is carried shoreward again by the next flood tide. This tendency to move mud into shallow, quiet water, while waves stir and winnow all the mud out of the surf zone, explains why beaches of clean, well sorted sand are so commonly backed by intertidal mud flats, tidal marshes, or muddy lagoons.

Accretion on mud flats is not only a mechanical process but a biological one as well. Many animals burrow in the mud and feed by filtering suspended organic debris out of the overlying water during tidal submersion. The waste products, including mud, are secreted as fecal pellets that are hydraulically similar to sand grains. These pellets of agglutinated mud account for a significant proportion of the total mud-flat accumulation.

Tidal Marshes.   When mud flats aggrade until they are exposed for about one-half of the tidal cycle, they can be colonized by **halophytic** (salt-tolerant) plants. Algae are the first colonizers, but several genera of grasses, sedges, and trees are uniquely adapted to tolerate immersion in salt water. Once vegetation is established on the mud flats, suspended mud is more effectively filtered out of the tidal waters, and **tidal marshes** quickly accrete to about mean high tide. An equilibrium is reached when the reduced amount of tidal flooding decreases the rate of accretion. The marsh buildup is supplemented by organic accumulation but countered by decay, compaction, and fire.

Tidal marshes are as rigorously controlled by high-tide flooding as coral reefs are by low-tide exposure. For that reason, studies of tidal marsh stratigraphy have been useful to document the history of the postglacial rise of sea level. That history is described more fully in Chapter 20. Marshes, once built to high-tide level, cannot be built higher by marine processes. However, plant succession can permit upland plants to build fresh-water peat seaward over the tidal marshes and convert them to new, although very low, coastal plains. The process is slow and on most coasts has been countered by a continuing slow submergence.

Among the most remarkable landforms of the world are the tropical **mangrove swamps** (Fig. 19-19). A few species of trees from several families have convergently

Figure 19-19. Prop roots of *Rhizophora* mangroves in the intertidal mud of Three Mile Creek, near Townsville, northern Queensland, Australia. (Photo: A. P. Spenceley.)

evolved to survive intertidal saltwater immersion (Lugo and Snedaker, 1974). The adaptations involve specialized root structures and buttressed trunks to support trees in soft mud, pneumatophores to permit oxygen to reach the roots buried in strongly anoxic mud, and viviparous seedlings that remain attached to the parent tree until they have developed leaves and a protoroot. When the seedlings drop, they float right-side up until they wash aground on a suitable bottom and take root.

Mangroves are tropical plants with a distribution almost coincident with coral reefs (Fig. 19-11). Tropical deltaic coasts are vast mangrove swamps with meandering distributaries often completely bridged by mangroves. Huge areas of delta coasts in Africa, India, Malaysia, Indonesia, New Guinea, and Australia, and on the Caribbean coasts of Central and South America, are mangrove swamps. Whether mangroves expand onto exposed mud flats and initiate deposition or simply exploit previous mud deposits (Bird, 1972), they add great regions of semiflooded landscapes to tropical coasts. Progradation rates of 50–100 m/yr in recent centuries have been suggested for the Sumatra coast on the Malacca Straits.

REFERENCES

Bird, E. C. F., 1972, Mangroves and coastal morphology in Cairns Bay, North Queensland: *Jour. Tropical Geog.*, v. 35, pp. 11–16.

Bloom, A. L., 1970, Paludal stratigraphy of Truk, Ponape, and Kusaie, Eastern Caroline Islands: *Geol. Soc. America Bull.*, v. 81, pp. 1895–1904.

———, 1974, Geomorphology of reef complexes, *in* Laporte, L. F., ed., Reefs in time and space: *Soc. Econ. Paleontologists and Mineralogists Spec. Pub. 18*, pp. 1–8.

Bradley, W. C., 1958, Submarine abrasion and wave-cut platforms: *Geol. Soc. America Bull.*, v. 69, pp. 967–74.

Bradley, W. C., and Griggs, G. B., 1976, Form, genesis, and deformation of central California wave-cut platforms: *Geol. Soc. America Bull.*, v. 87, pp. 433–49.

Coleman, J. M., and Smith, W. G., 1964, Late recent rise of sea level: *Geol. Soc. America Bull.*, v. 75, pp. 833–40.

Cotton, C. A., 1952, Criteria for the classification of coasts: *Internat. Geog. Cong., 17th, Washington 1952, Proc.*, pp. 315–19 [1957]. (Reprinted 1974 in *Bold coasts:* A. H. & A. W. Reed, Wellington, N.Z., pp. 118–25.)

———, 1955, Theory of secular marine planation: *Am. Jour. Sci.*, v. 253, pp. 580–89. (Reprinted 1974 in *Bold coasts:* A. H. & A. W. Reed, Wellington, N.Z., pp. 164–74).

Curray, J. R., 1969, Shore zone sand bodies: Barriers, cheniers, and beach ridges, *in* New concepts of continental margin sedimentation: *AGI Short Course Notes*, Am. Geol. Inst., Washington, D.C., pp. JC-II, 1–18.

Curray, J. R., Emmel, F. J., and Crampton, P. J., 1969, Holocene history of a strand plain lagoonal coast, Nayarit, Mexico: *Internat. Sympos. Coastal Lagoons, Mexico City 1967, Proc.*, pp. 63–100.

Darwin, C., 1837, On certain areas of areas of elevation and subsidence in the Pacific and Indian Oceans, as deduced from the study of coral formations: *Geol. Soc. London Proc.*, v. 2, pp. 552–54.

Das, M. M., 1971, Longshore sediment transport rates: A compilation of data: *U.S. Army Coastal Engineering Res. Center, Misc. Paper 1–71*, 81 pp.

DAVIES, J. L., 1973, *Geographical variation in coastal development:* Hafner Publishing Company, New York, 204 pp.

DIETZ, R. S., 1963, Wave-base, marine profile of equilibrium, and wave-built terraces: A critical appraisal: *Geol. Soc. America Bull.*, v. 74, pp. 971–90.

DOLAN, R., GODFREY, P. J., and ODUM, W. E., 1973, Man's impact on the barrier islands of North Carolina: *Am. Scientist*, v. 61, pp. 152–62.

FAIRBRIDGE, R. W., 1950, Recent and Pleistocene coral reefs of Australia: *Jour. Geology*, v. 58, pp. 330–401.

———, 1976, Shellfish-eating Preceramic Indians in coastal Brazil: *Science*, v. 191, pp. 353–59.

FISHER, J. J., 1968, Barrier island formation: Discussion: *Geol. Soc. America Bull.*, v. 79, pp. 1421–26.

GARRETT, C. J. R., 1972, Tidal resonance in the Bay of Fundy and Gulf of Maine: *Nature*, v. 238, pp. 441–43.

GUILCHER, A., 1958, *Coastal and submarine morphology* (transl. by B. W. Sparks and R.H.W. Kneese): Methuen & Co., Ltd., London, 274 pp.

HAYES, M. O., 1967, Hurricanes as geological agents: Case studies of hurricanes Carla, 1961, and Cindy, 1963: *Texas Bureau of Econ. Geology, Rept. of Invest. no. 61*, 56 pp.

HENDERSHOTT, M. C., 1973, Ocean tides: *EOS* (*Am. Geophys. Un., Trans.*), v. 54, pp. 76–86.

HODGKIN, E. P., 1964, Rate of erosion of intertidal limestone: *Zeitschr. für Geomorph.*, v. 8, pp. 385–92.

HOYT, J. H., 1967, Barrier island formation: *Geol. Soc. America Bull.*, v. 78, pp. 1125–36.

———, 1968, Barrier island formation: Reply: *Geol. Soc. America Bull.*, v. 79, pp. 1427–30.

INGLE, J. C., 1966, The movement of beach sand: *Advances in sedimentology*, v. 5, Elsevier Publishing Co., New York, 221 pp.

INMAN, D. L., and BRUSH, B. M., 1973, The coastal challenge: *Science*, v. 181, pp. 20–32.

KING, C. A. M., 1972, *Beaches and coasts*, 2d ed.: Edward Arnold, London, 570 pp.

KINSMAN, B., 1965, *Wind waves: Their generation and propagation on the ocean surface:* Prentice-Hall, Inc., Englewood Cliffs, N.J., 676 pp.

KOMAR, P. D., 1976, *Beach processes and sedimentation:* Prentice-Hall, Inc., Englewood Cliffs, N.J., 429 pp.

LANGFORD-SMITH, T., and THOM, B. G., 1969, New South Wales coastal morphology: *Geol. Soc. Australia Jour.*, v. 16, pp. 572–80.

LINK, M. C., 1960, Exploring the drowned city of Port Royal: *Natl. Geographic*, v. 117, pp. 151–83.

LUDWICK, J. C., 1974, Tidal currents and zig-zag sand shoals in a wide estuary entrance: *Geol. Soc. America Bull.*, v. 85, pp. 717–26.

LUGO, A. E., and SNEDAKER, S. C., 1974, Ecology of mangroves: *Ann. Rev. Ecology and Systematics*, v. 5, pp. 39–64.

MACNEIL, F. S., 1954, Shape of atolls: An inheritance from subaerial erosion forms: *Am. Jour. Sci.*, v. 252, pp. 402–27.

MARAGOS, J. E., BAINES, G. B. K., and BEVERIDGE, P. J., 1973, Tropical cyclone Bebe creates a new land formation on Funafuti Atoll: *Science*, v. 181, pp. 1161–64.

MCLEAN, R. F., 1967, Measurements of beachrock erosion by some tropical marine gastropods: *Bull. Marine Sci.*, v. 17, pp. 551–61.

———, 1972, Nomenclature for rock-destroying organisms: *Nature*, v. 240, p. 490.

MUNK, W. H., and SARGENT, M. C., 1954, Adjustment of Bikini Atoll to ocean waves: *U.S. Geol. Survey Prof. Paper 260-C*, pp. 274–80.

NEUMANN, G., and PIERSON, W. J., JR., 1966, *Principles of physical oceanography:* Prentice-Hall, Inc., Englewood Cliffs, N.J., 545 pp.

OTVOS, E. G., 1972, Mississippi Gulf Coast Pleistocene beach barriers and the age problem of the Atlantic-Gulf Coast "Pamlico"–"Ingleside" beach ridge system: *Southeastern Geology*, v. 14, pp. 241–50.

POSTMA, H., 1967, Sediment transport and sedimentation in the estuarine environment, *in* Lauff, G. H., ed., Estuaries: *Am. Assoc. Adv. Sci. Pub. no. 83*, pp. 158–79.

PURDY, E. G., 1974, Reef configurations: Cause and effect, *in* Laporte, L. F., ed., Reefs in time and space: *Soc. Econ. Paleontologists and Mineralogists Spec. Pub. 18*, pp. 9–76.

REVELLE, R., and EMERY, K. O., 1957, Chemical erosion of beach rock and exposed reef rock: *U.S. Geol. Survey Prof. Paper 260-T*, pp. 699–709.

SHEPARD, F. P., 1970, Lagoonal topography of Caroline and Marshall Islands: *Geol. Soc. America Bull.*, v. 81, pp. 1905–14.

———, 1973, *Submarine geology*, 3d ed.: Harper & Row, Publishers, New York, 517 pp.

SHEPARD, F. P., and CURRAY, J. R., 1967, Carbon-14 determination of sea level changes in stable areas: *Progress in Oceanography*, v. 4, Pergamon Press, Oxford and New York, pp. 283–91.

STEPHENSON, T. A., and STEPHENSON, S., 1972, *Life between tidemarks on rocky shores:* W. H. Freeman and Company, San Francisco, 425 pp.

VAN STRAATEN, L. M. J. U., and KUENEN, P. H., 1958, Tidal action as a cause of clay accumulation: *Jour. Sed. Petrol.*, v. 28, pp. 406–13.

SWIFT, D. J. P., 1969, Inner shelf sedimentation: Processes and products, *in* New concepts of continental margin sedimentation: *AGI Short Course Notes*, Am. Geol. Inst., Washington, D.C., pp. DS-4, 1–46.

THOM, B. G., and CHAPPELL, J., 1975, Holocene sea levels relative to Australia: *Search*, v. 6, pp. 90–93.

WARME, J. E., and MARSHALL, N. F., 1969, Marine borers in calcareous terrigenous rocks of the Pacific coast: *Am. Zoologist*, v. 9, pp. 765–74.

WENTWORTH, C. K., 1938, Marine bench-forming processes: Water-level weathering: *Jour. Geomorph.*, v. 1, pp. 6–32.

———, 1939, Marine bench-forming processes: II, solution benching: *Jour. Geomorph.*, v. 2, pp. 3–25.

YASSO, W. E., 1965, Plan geometry of headland-bay beaches: *Jour. Geology*, v. 73, pp. 702–14.

ZENKOVICH, V. P., 1967, *Processes of coastal development* (transl. by D. G. Fry): Wiley-Interscience, New York, 738 pp.

CHAPTER 20

Explanatory Description of Coasts

Among the earliest historical observations of geomorphic change were reports of coastal erosion and deposition. Long before scholars could accept ideas that subaerial landscapes evolve, they readily noted the retreat of sea cliffs, progressive shoaling of harbors, and forward growth of deltas. Attempts to develop rational classifications and systematic descriptions of coastal landscapes continued in parallel with similar attempts to describe and classify subaerial landscapes. However, even in the modern era, coasts are generally treated as a separate topic of geomorphology, in acknowledgment of the fundamentally different energy transfer mechanisms for coastal change and the narrow vertical range of shore processes.

Systematic coastal geomorphology has paralleled other kinds of systematic geomorphology in recognizing the fundamental roles of structure, process, and time in coastal landscape evolution. However, the three components of geomorphic explanatory description are not applied to coastal studies in quite the same way as to subaerial landscapes. Too much of the coastal landscape is relict, inherited either from earlier shore processes now inoperative or from subaerial or submarine processes. No generally accepted classification of coasts has emerged from two centuries of geomorphic research. The role of structure, process, and time in coastal evolution can be recognized, however. In this chapter, these three component parts of explanatory description are reviewed in succession to illustrate both the similarities and contrasts of coastal and subaerial geomorphic systems.

STRUCTURAL FACTORS OF COASTAL DESCRIPTION

Tectonic Coasts

The narrow vertical range of shore processes ensures that any tectonic movement that affects the relative level of the land and sea will be recorded by relict or modified coastal landforms. However, the postglacial eustatic rise of sea level of 80–140 m has drowned all but the most strongly uplifted coasts, so that many recent tectonic movements are totally masked by late-glacial drowning. The distinction between coasts that have been deformed or displaced by tectonic movements and those that have been passively drowned by the postglacial rise of sea level was made the basis of a coastal classification by Cotton (1952). He recognized the following categories of coasts:

1. *Coasts of stable regions*   These have all been affected by recent submergence:
   a. Coasts dominated by features produced by recent submergence.
   b. Coasts dominated by features inherited from earlier episodes of emergence.
   c. Miscellaneous coasts (fiord, volcanic, etc.).
2. *Coasts of mobile regions*   These have been affected by uplift or depression of the land as well as by recent submergence:
   a. Coasts on which the effects of marine submergence have not been counteracted by uplift of the land.
   b. Coasts on which recent uplift of the land has led to emergence.
   c. Fold and fault coasts.
   d. Miscellaneous coasts (fiord, volcanic, etc.)

Cotton also attempted to classify mobile coasts by the style and orientation of the tectonic movement (Table 20-1). He deduced a series of initial coasts that result from tectonic motion parallel to or transverse to the shoreline, plus some constructional and erosional forms, and then deduced a series of possible subsequent movements, erosion, or deposition. Of interest is the distinction between coasts that are deformed parallel to the shoreline and those that are deformed tranverse to it. In the case of tilting parallel to the shore, the coast might either emerge or submerge depending on the location of the tilt axis. In the case of transverse tilting, the coast might consist of alternating emerged and submerged segments. In New Zealand, where Cotton tested his classification, as well as elsewhere on the rim of the Pacific Basin and in the Mediterranean and Caribbean Seas, a descriptive classification based on the nature of recent or continuing tectonic movements has great usefulness. Nevertheless, in all the examples cited by Cotton, tectonic movement has only modified, but not negated, the effects of glacially controlled eustatic sea-level fluctuation.

TABLE 20-1. CLASSIFICATION OF INITIAL COASTS

	Initial Coasts Resulting from			
I	*II*	*III*	*IV*	*V*
Eustatic change of level, regional movement, or gentle marginal warping on a hingeline parallel to the coast	Faulting or strong mono-clinal flexure ("steep" coasts)	Transverse deformation	Volcanic accumulation	Glacial erosion below sea level

Movement at shoreline causes

                          Afterwards

A	*B*	*A*	*B*
Submergence	Emergence	Drowned	Emergent

Afterwards

(1)   (2)
Retrograded   Prograded by

(a)   (b)
Alluviation   Wave action

SOURCE: Cotton, 1942.

A more fundamental tectonic classification of coasts has grown out of the plate-tectonic theory (Inman and Nordstrom, 1971). If the surface of earth is a series of lithospheric plates on which continents and ocean basins are of only minor significance, we should be able to recognize only four tectonic categories of coasts: (1) those at the edge of active diverging plates, (2) those at zones of plate convergence, (3) those along major transform faults (Fig. 3-10), and (4) those on continents embedded within large plates, far from any zones of convergence, spreading, or translatory plate motion. In category (1) could be placed block-faulted coasts and rift coasts such as the Red Sea coasts of Africa and Arabia. Category (2) includes island-arc systems, such as Indonesia, Japan, and other parts of the western Pacific, and mountainous continental coasts such as the west side of South America and the north side of the Mediterranean Sea. In category (3) are steep transform-fault coasts such as southern California, Venezuela, and the north side of New Guinea. In category (4) are most of the Atlantic Ocean coasts, as well as the Indian Ocean coasts of India, Africa, and Australia. Some of these coasts have shallow, broad continental shelves, but others have steep escarpments and narrow shelves. The difference is whether rivers flow toward or away from the particular coast (Audley-Charles, et al., 1977). If the opposite side of the continent is high, as in South America, a broad, low coast faces the Atlantic, and rivers have built a broad shelf. If river systems drain inland away from coastal escarpments, as in Africa, there is little sediment to build broad shelves except at major deltas.

In many ways, the tectonic categories of coasts are reminiscent of the suggestion of Suess (1904, p. 5) that coasts could be grouped into an Atlantic type, in which continental structural trends are truncated at the coast, and a Pacific type, in which island arcs or mountain ranges parallel the coast. The Atlantic coasts are primarily "trailing edges" of continents, whereas the Pacific coasts are "leading edges," either collision or transform-fault coasts.

The tectonic component of coastal classification includes such issues as the direction and rate of past and present vertical crustal movement, the amount of volcanic activity at the coast or offshore, and the trend of deformational axes relative to the coast.

Structurally Controlled Coasts

As in the case of subaerial landforms (Chaps. 3 and 11), we must distinguish between coasts that have a history of continuing tectonic movements and those that have inherited zones of unequal rock strength from ancient depositional and tectonic events, now inactive. The latter are *structurally controlled* in the sense that inherited structural factors rather than contemporary movements determine the plan view and profile of the coastal region.

Much of the structural control exhibited by coasts is inherited from subaerial landscape evolution. Rocks resistant to subaerial erosion form regions of high relief, and where the sea intersects these regions, cliffed coasts result. Weak-rock lowlands are drowned by the sea to form low coastal plains, perhaps with low islands offshore. Alluvial plains are very likely to develop barriers, because the initial gentle gradients and abundant sediment supply promote constructional shore-zone landforms.

Bold Coasts. In coastal regions of resistant rocks, marine erosion and deposition are quick to exploit structural weaknesses. Joints and faults are the loci for active wave plucking or quarrying. Dikes, especially with columnar joints, are especially subject to selective wave erosion and form deep clefts when they are quarried from between walls of more resistant rocks (Fig. 19-5). Pocket beaches, often of gravel rather than sand, accumulate in coves along shattered zones in otherwise massive rock (Fig. 20-1). In plan view, many coasts on metamorphic terrains show intersecting linear patterns of fiords or estuaries, marking former glacial or fluvial erosion along structural weaknesses (Fig. 20-2). Glacial erosion has been especially effective in etching out structural weaknesses in crystalline rocks prior to their drowning by the postglacial rise of sea level.

Figure 20-1. Broad Cove beach, Appledore Island, Maine. Granite abrasion ramp in foreground has only a single layer of boulders and cobbles overlying it.

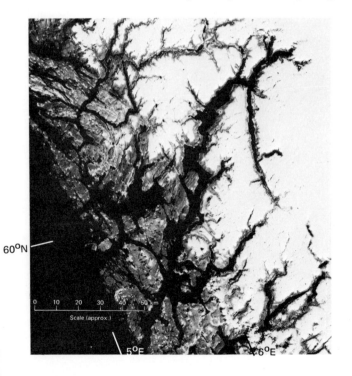

Figure 20-2. Hardanger Fiord, Norway. The structural grain of the coastal metamorphic rocks is clearly shown. Fault lines and joints also influence the dissection. (NASA ERTS E-1299-10211.)

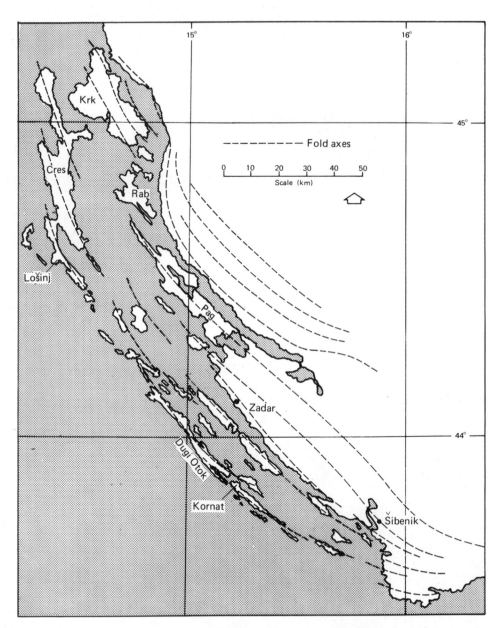

Figure 20-3. The Dalmatian coast of Yugoslavia, the type locality of an embayed coast formed by submergence of a fold belt that parallels the coast. Compare with Figure 11-7.

Where metamorphic or folded sedimentary rocks of contrasting erodibility have been drowned, the typical valley-and-ridge subaerial landscape takes on an even more impressive aspect. Two variants are known. If the drowned fold belt is parallel to the coast, the landscape is called the **Dalmatian type**, from the coastal region of Dalmatia, Yugoslavia (Baulig, 1930, Cotton, 1956). The drowned valleys parallel the coast but are cross-connected by channels. The regional landscape is best appreciated from map study (Fig. 20-3). If the structurally controlled valleys are perpendicular to the regional coastline, a typical **ria** coast develops (Fig. 11-7). Along much of the coasts of western Europe and northeastern North America, Paleozoic fold belts truncated by the opening of the Atlantic basin form deeply indented ria coasts.

Fault-line scarps and monoclines also form bold coasts. As with subaerial forms, some care must be taken to distinguish tectonic coastal scarps from those eroded along old structural trends. On some steep coasts, such as the southeastern Australia coast south of Sydney and the west side of the Indian peninsula, eroded escarpments rise abruptly from the back of narrow coastal plains. No structural control for the scarps is apparent, although they may have retreated many kilometers from initial structural-controlled features that are now submerged.

Low Coasts.   Piedmont alluvial plains (p. 245), deltas, and pediments form substantial portions of the coasts of the world. In fact, "coastal plain" has become synonymous with a variety of weakly indurated sedimentary aprons in which stratification broadly parallels the surface slope and dips gently seaward. Many such coasts are prograding today, with new sedimentary layers accumulating in the offshore zone that may subsequently emerge. Little structural control is shown by these low coasts, exemplified by the Gulf coast of the United States. On some low coasts, resistant layers form low cuestas on the plain, but these rarely affect coastal form. Barriers and lagoons are characteristic.

Low coasts have been strongly affected by Quaternary sea-level fluctuations. Because the weak sediments are easily eroded, rivers entrenched deep valleys across coastal plains during times of low sea levels. Subsequently, the valleys have drowned to form estuaries. On the south and east coasts of the United States, several episodes of drowning and estuarine sedimentation have alternated with sea levels higher than present relative to the land when beach ridges and wave-cut cliffs formed as high as 50 m above present sea level. The ages and even the number of the submergence episodes are widely debated subjects (Cooke, 1930; Flint, 1971, pp. 334–39). It is not known whether the southeastern coasts of the United States are now emerging or submerging, although the coastal plain is made primarily of emerged shallow marine Tertiary sedimentary rocks and has been emerging during Cenozoic time. The episodes of alternating emergence and submergence that built the shoreline features inland and eroded the estuaries below present sea level were at least in part glacially controlled. Some large estuaries, such as the Chesapeake and Delaware Bays, are obviously the drowned lower segments of very large river valleys, but the age of their erosion may date from one or more of the early Quaternary glaciations.

Figure 20-4.   The Norwegian strandflat in the Brønnøysund area (65°N). View is north from an isolated hill 260 m high toward a dissected plateau 500 m high. The intervening low terrain is cut on a variety of metamorphic rocks and has a total submerged and emerged relief of less than 60 m except for fiords that cross it. (Photo: Just Gjessing.)

A few low coasts are anomalous in that they are on resistant rocks. The **strandflat** of Norway is the classic example (Fig. 20-4). It is a low coastal region of crystalline rocks crossed by deep fiords but consisting mainly of low islands and shallow drowned terrain of low relief. The strandflat has been thought to be downfaulted, wave-abraded, or ice-eroded, or a cryoplanation surface. Its origin is still not known (Holtedahl, 1958; Gjessing, 1967).

PROCESS FACTORS IN COASTAL DESCRIPTION

Sea-Level Fluctuations

Worldwide, or *eustatic*, changes of sea level have been the major factor in the development of Quaternary coastal landscapes (pp. 406 and 444). The magnitude of the glacier-controlled fluctuations far exceeded the vertical range of shore processes, and their frequency exceeded the rate of development of erosional equilibrium forms. Depositional coasts, especially coral reefs and sandy barriers, have been able to develop equilibrium landforms even during the brief times, perhaps lasting only 10,000–20,000 years, of interglacial sea-level stability. However, the reefs and sand barriers that together form such a large part of our present coastal landscape are not the result of Holocene events alone: Almost without exception, the modern forms have been shown to be thin veneers resting on ancient but genetically similar foundations dating from the last interglacial or even earlier.

Either by eustatic fluctuation or by tectonic movements, some coasts have *emerged* from the sea, and others have *submerged*. These useful terms refer only to the relative movement of land and sea, or the algebraic sum of all vertical movements at the shoreline. Emergence is by geomorphic convention a net positive movement of the land; submergence is considered a net negative movement. By contrast, stratigraphers, especially those who study marine strata, prefer the opposite connotation of submergence as a positive movement (a marine *transgression*) and emergence as a negative, or *regressive*, phenomenon.

The dichotomy of submergence and emergence has been the basis for most coastal classifications since it was formally proposed by Gulliver (1899) and rigidly codefied by Johnson (1919). The supplemental categories of *neutral coasts* and *compound coasts* were added to cover coasts that were the result of progradation, as by a delta, with neither emergence or submergence a significant factor, or coasts that show evidence of both emergence and submergence. Although extremely influential, the descriptive system of submerged, emerged, neutral, and compound coasts has lost much of its utility with the growing knowledge of multiple Quaternary sea-level fluctuations. With few exceptions, all coasts must be classified as compound in this classification.

Shepard (1937, 1973) recognized that most coasts are now submerged, because we live in a time of shrunken glaciers and high sea levels equivalent to the multiple interglacials of the Pleistocene Epoch (Fig. 18-4). Therefore, his classification, although recognizing that emergence could be found on tectonic coasts, was based on the presence or absence of shoreline modification subsequent to the most recent submergence. He distinguished *primary* coasts, those drowned but otherwise unchanged, and *secondary* coasts, those modified by shoreline erosion or deposition (see p. 478). This system has found considerable use as an alternative to Johnson's scheme.

Valentin (1952, 1970) devised an ingenious coordinate system (Fig. 20-5) that gave submergence and emergence equivalent status with shoreline erosion (retro-

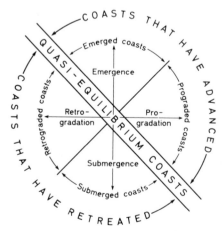

Figure 20-5.   Principal genetic types of coastal configuration. Coasts can be categorized as "emerged and prograded," "submerged and prograded," etc. A balance of vertical and horizontal components can also keep a coast at quasi-equilibrium. (Valentin, 1970.)

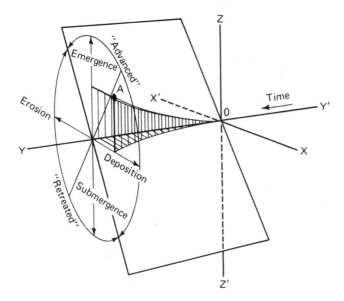

Figure 20-6. Valentin's classification of coasts (Fig. 20-5) simplified and with a time axis added. The history of a coast can be graphed on these axes. For instance, point "A" is shown as the result of emergence and deposition through time. (Bloom, 1965.)

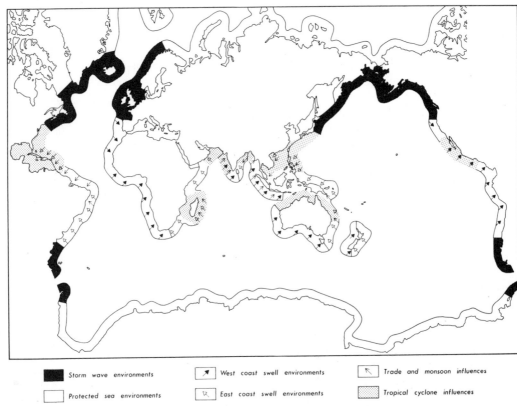

| ■ Storm wave environments | ↗ West coast swell environments | ↖ Trade and monsoon influences |
| □ Protected sea environments | ↙ East coast swell environments | ⬚ Tropical cyclone influences |

Figure 20-7. Major world wave environments. Similar maps of storm frequency and intensity, tidal heights, winds, and currents could potentially be combined as an energy classification of coasts. (Davies, 1973, Fig. 27.)

gradation) and progradation in determining whether a coast has *advanced* seaward or *retreated* landward. The addition of a time axis to this scheme enables the history of a coast to be graphically portrayed (Fig. 20-6) and suggests additional categories for classifying coasts, although the modification is intended primarily for explanatory description rather than classification (Bloom, 1965).

Wave Energy; Coastal Exposure

The rate of energy expenditure at a shoreline has some potential as a basis for classifying coastal landscapes (Tanner, 1960; Davies, 1973). *High-energy coasts* are characterized by bold erosional forms or massive deposits of rather coarse sediment. *Low-energy coasts*, by contrast, are of low relief, often with salt marshes or mangrove swamps. Terrestrial processes, such as deltaic progradation, dominate over shoreline processes.

Shoreline energy is not a random effect of exposure or local coastal configuration. Certain regions, such as coasts in latitudes 45°–60°, and east-facing coasts in the low-latitude trade-wind regions, are generally subject to higher than average storm-induced waves (Fig. 20-7). Equatorial coasts and subarctic coasts generally have low tides and low waves, although the causes for the low energy differ. Subtropical east-facing coasts are especially subjected to cyclonic storms (hurricanes or typhoons) that occur with disastrous regularity every year or two.

The geometric factor of fetch, or exposure, the climatic factors of wind speed, duration, and storm frequency, and geologic factors of offshore depth and sediment supply can be combined to determine the rate at which energy is fed into a coastal segment. Regional patterns of energy distribution can be combined with structural factors of regional coastal rock type to deduce sets of possible landforms. The energy approach is especially useful for explanatory description of shore-zone landforms. The broader coastal landscape is more likely to show the additional effects of tectonic or eustatic changes of level or climatic change.

Climatic Factors

In addition to their fundamental role in determining wave energy input, climatic variables offer other possibilities for explanatory description of coasts. Certain shore processes, ice-push and drift ice on high-latitude coasts for instance, are climate-determined. The new adjective "glaciel" was proposed by L. E. Hamelin in 1959 to describe the geomorphic work of floating ice (Dionne, 1974). Coral-reef and mangrove coasts are uniquely tropical. Coasts in the humid tropics are generally lacking in dune fields, because tropical vegetation quickly stabilizes new beach sand during progradation (Bird and Hopley, 1969). Arid coasts have salt flats, prominent barrier systems built of the abundant sand-size sediment provided from flash floods, and spectacular coastal dunes. A climatic morphogenetic system (p. 307) could be erected for coastal landforms, as for subaerial forms.

TIME AS A FACTOR IN COASTAL GEOMORPHOLOGY

The Concept of Maturity

To geomorphologists of the late nineteenth and early twentieth centuries, initial structures and active processes interacted in a deduced series of stages, analogous to the youth, maturity, and old age of organisms. Time was a subjective variable, marked by the attainment of certain diagnostic landforms rather than by the passage of years or millenniums of real time. It was inevitable that the concept of "stages" in the "life cycle" of subaerial landscapes would be applied to coastal landscapes as well. However, the scheme was never fully adopted for explanatory description of coasts. W. M. Davis, the great champion of the geomorphic cycle concept, cautioned (1905, p. 159):

> One significant peculiarity of the development of the shoreline [as compared to the development of the subaerial landscape] is its immediate recognition of changes of level or interruptions. . . . This contrast leads to the inference that the product of long-continued work of the shore-line forces on a fixed level or on a uniformly changing level is less likely to the found than the product of long-continued work on an inland region, where a series of small and frequent interruptions (elevations and depressions) might hardly make themselves felt.

In spite of this early recognition of the vital importance of emergence and submergence on coasts, subsequent geomorphologists continued to deduce sequential stages in the erosional evolution of coasts on the assumption of sea-level stability. As an abstract logical exercise, no fault can be found with the deductions. The difficulty came in trying to fit real coastal landscapes into the deduced schemes. By analogy with fluvial landscapes, maturity of a coast meant a graded condition, both in plan view and in cross section perpendicular to the shoreline. Cliffs were to be cut back to alignment with prograded beaches until sediment could move freely along shore. The nearshore profile was to be "graded," but insufficient coastal work had been done to define that condition. To D. W. Johnson (1925) and others of his time, the youthful stage of coastal erosion implied irregularities in both plan and profile. "Maturity" was reached only when the entire shoreline had been eroded back beyond the heads of any initial bays. The analogy was to the complete removal of waterfalls in the "mature" river profile, but the real time scale was much different. For that reason, few examples of "mature" coasts on resistant rocks could be cited (Johnson, 1919, 1925). Although widely hailed as a triumph of deductive geomorphic reasoning, the "cycles of erosion" on submerged, emerged, neutral, and compound coasts were not very useful in explanatory description.

Primary and Secondary Coasts

As an alternative to the deductive scheme of submerged and emerged coasts with subsequent temporal stages of erosion, Shepard (1937, 1973) subdivided his dominant class of submerged coasts into *primary* and *secondary* categories (p. 475).

No sequential development of secondary forms was deduced. Neither was any provision made for inherited forms, in spite of the fact that many coasts have landforms relict from the last interglacial, now several meters above sea level.

It is easy to criticize Shepard's subdivision on the grounds that the present degree of marine influence on coastal configuration is dominated by structure and wave energy rather than by a temporal sequence. To a first approximation, the latest eustatic rise of sea level affected all coasts simultaneously, and rather than passing sequentially from the primary to the secondary categories, some coasts have continuously responded to marine influence, either because of weak structures or high energy input, while others have retained the imprint of tectonic forms or subaerial processes and have passively drowned. More rarely, coasts have emerged because of strong tectonic uplift. However, Shepard's dichotomy was not necessarily meant to imply an evolutionary change but rather to provide convenient "pigeonholes" for coastal classification. Attempts to compare it with the earlier sequential scheme distort its intended purpose.

Tempo of Erosion

Rather than treating time as a direct variable in coastal classification, either as measured in years or as in stages of sequential changes, the rate, or *tempo*, of coastal landscape evolution can be used (Cotton, 1952). Some coasts have evolved at a rapid tempo in postglacial time; others have been negligibly modified by marine processes. In a sense, coasts with a rapid tempo of evolution are similar to Shepard's secondary coasts, and slowly changing coasts are the equivalent of Shepard's primary class. There are important differences, however. If steep cliffs were cut by the sea at a lower level, they might now "plunge" (Fig. 19-9) and reflect rather than refract incoming waves. Thus plunging cliffs might presently change only at a very slow tempo, even though they were shaped by marine erosion at a lower sea level.

The concept of tempo has been applied primarily to erosional coasts but could be extended to prograded coasts as well. Deltas and coral reefs have extended some coasts seaward in spite of postglacial submergence. A rapid tempo of accumulation can offset the effects of submergence (Fig. 20-5) just as a rapid tempo of erosion can offset the effects of tectonic emergence. Tempo is not a very promising base for a geomorphic classification of coasts, because rapid tempo might be caused by high wave energy, or weak structures, or changing levels. Too many variables could have combined to produce the same observed results. The concept of tempo in coastal evolution is important in applied research on coasts, however. Coasts undergoing rapid change, either of erosion or progradation, present many more problems for human utilization.

Inheritance of Coastal Landforms

The area covered by the successive Pleistocene ice sheets was remarkably similar, differing by less than 10 percent. If the thicknesses of the ice sheets were also similar, as glacier-flow theory predicts (Chap. 16), then sea level has probably oscillated through a range of 80–140 m during the later Cenozoic Era (Figs. 17-14, 18-4), repeatedly reoccupying shoreline features cut or built much earlier. Most modern coral reefs seem to be veneers of Holocene limestone about 10 m thick, deposited during the last 8000 years or less on foundations that date from the last interglacial. We cannot say how much of other modern coastal landscapes is relict, because we have no dating techniques that are suitable for nonreef coasts. Many broad shore platforms, barriers, and reefs that appear to be in equilibrium with modern sea level and wave energy may be only slightly modified inherited forms (Fig. 19-4). Perhaps the unsatisfactory status of coastal explanatory description is due to the fact that we have not yet learned to separate active from relict forms.

REFERENCES

AUDLEY-CHARLES, M. G., CURRAY, J. R., and EVANS, G., 1977, Location of major deltas: *Geology*, v. 5, pp. 341–44.

BAULIG, H., 1930, Le littoral Dalmate: *Ann. Géogr.*, no. 219, pp. 305–10.

BIRD, E. C. F., and HOPLEY, D., 1969, Geomorphological features on a humid tropical sector of the Australian coast: *Australian Geog. Studies*, v. 7, pp. 89–108.

BLOOM, A. L., 1965, The explanatory description of coasts: *Zeitschr. für Geomorph.*, v. 9, pp. 422–36.

COOKE, C. W., 1930, Correlation of coastal terraces: *Jour. Geology*, v. 38, pp. 577–89.

COTTON, C. A., 1942, Shorelines of transverse deformation: *Jour. Geomorph.*, v. 5, pp. 45–58. (Reprinted 1974 in *Bold coasts*: A. H. & A. W. Reed, Wellington, N.Z., pp. 34–45.)

———, 1952, Criteria for the classification of coasts: *Internat. Geog. Cong., 17th, Washington 1952, Proc.*, pp. 315–19 [1957]. (Reprinted 1974 in *Bold coasts*: A. H. & A. W. Reed, Wellington, N.Z., pp. 118–25.)

———, 1956, Rias *sensu stricto* and *sensu lato*: *Geog. Jour.*, v. 122, pp. 360–64. (Reprinted 1974 in *Bold coasts*: A. H. & A. W. Reed, Wellington, N.Z., pp. 187–94.)

DAVIES, J. L., 1973, *Geographical variation in coastal development*: Hafner Publishing Company, New York, 204 pp.

DAVIS, W. M., 1905, Complications of the geographical cycle: *Internat. Geog. Cong., 8th, Washington 1904, Repts.*, pp. 150–63. (Reprinted 1954 in *Geographical essays*: Dover Publications, Inc., New York, pp. 279–95).

DIONNE, J.-C., 1974, *Bibliographie annotée sur les aspects géologiques du glaciel* (Annotated bibliography on the geological aspects of drift ice): Can. Centre Rech. For. Laurentides, Ste-Foy, Québec, Rapp. Inf. LAU-X-9, 122 pp.

FLINT, R. F., 1971, *Glacial and Quaternary geology*: John Wiley & Sons, Inc., New York, 892 pp.

GJESSING, J., 1967, Norway's paleic surface: *Norsk Geografisk Tidsskrift*, v. 21, pp. 69–132.

GULLIVER, F. P., 1899, Shoreline topography: *Am. Acad. Arts and Sciences Proc.*, v. 34, pp. 151–258.

HOLTEDAHL, H., 1958, Some remarks on geomorphology of continental shelves off Norway, Labrador, and southeast Alaska: *Jour. Geology*, v. 66, pp. 461–71.

INMAN, D. L., and NORDSTROM, C. E., 1971, On the tectonic and morphologic classification of coasts: *Jour. Geology*, v. 79, pp. 1–21.

JOHNSON, D. W., 1919, *Shore processes and shoreline development:* John Wiley & Sons, Inc., New York, 584 pp.

————, 1925, *New England-Acadian shoreline:* John Wiley & Sons, Inc., New York, 608 pp.

SHEPARD, F. P., 1937, Revised classification of marine shorelines: *Jour. Geology*, v. 45, pp. 602–24.

————, 1973, *Submarine geology*, 3d ed.: Harper & Row, Publishers, New York, 517 pp.

SUESS, E., 1904, *The face of the earth*, v. 1 (transl. by H. B. C. Sollas): Clarendon Press, Oxford, 604 pp.

TANNER, W. F., 1960, Florida coastal classification: *Gulf Coast Assoc. Geol. Soc.*, v. 10, pp. 259–66.

VALENTIN, H., 1952, Die Küsten der Erde: *Petermanns Geogr. Mitt. Ergänzungsheft 246*, Justus Perthes, Gotha, 118 pp.

————, 1970, Principles and problems of a handbook on regional coastal geomorphology of the world: Paper read at the Symposium of the IGU Commission on Coastal Geomorphology, Moscow, 10 pp.

CHAPTER 21

Submarine Geomorphology: A Review

INTRODUCTION

Traditional geomorphology was concerned only with the subaerial part of the earth, with at most only the additional few meters of submarine landscape in the nearshore zone. Even today, the number of people who have seen a submarine landscape is very limited. Perhaps only a few hundred scientists have viewed the ocean floor at a depth of more than 100 m. Yet, 71 percent of the lithosphere is under the sea, and most certainly the submarine landscape is important (Fig. 21-1). Now that geology has become marine oriented, with the dominant theory of plate tectonics based primarily on submarine data, a review of submarine geomorphology is especially timely. As will be immediately apparent, little of submarine geomorphology is genetic. Words such as *submarine valley*, *slope*, *rise*, *ridge*, and *plain* are only descriptive terms. Only the most obvious tectonic and volcanic submarine landforms can be given explanatory descriptions based on the theory of plate tectonics. Lesser forms are described in very general terms by analogy with similar subaerial forms.

If submarine landforms are poorly known, the exogenic geomorphic processes that act upon them are even less known. Sound hypotheses and some observational evidence support the opinion that most of the submarine landscape is constructional by tectonic, volcanic, and depositional processes. Evidence of submarine weathering, mass-wasting, and erosion is abundant, but no hypothesis of destructional submarine landscape evolution has been formulated. Many of the sea-floor tectonic and volcanic landforms probably survive with little alteration from the time of their genesis at a

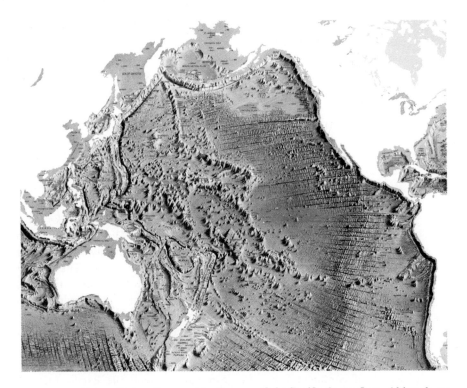

Figure 21-1. Geomorphic diagram of the Pacific Ocean floor. Although not named on this diagram, various geomorphic (or physiographic) provinces are obvious. (Turekian, 1976, Fig. 2-2, based on original art by B. C. Heezen and M. Tharp.)

midocean ridge until they are engulfed at a subduction zone, 10–100 million years later.

In this chapter, the major classes of submarine landforms are first outlined by analogy with the geomorphic or physiographic provinces of the continents. These large, first- and second-order landforms (Chap. 2) are all constructional. Subsequently, smaller-scale erosional landforms are described, and the known or hypothesized geomorphic processes are reviewed. Some time in the future, the submarine landscape will require much more extended consideration.

SUBMARINE GEOMORPHIC PROVINCES

Based primarily on depth and the underlying geologic structure, submarine topography can be subdivided into a relatively small area of *continental margins*, and a much larger area of *deep-sea floor* (Fig. 2-2). Even this subdivision is not simple, nor the boundaries definite. An operational definition could be that the continental margins are structurally part of continents, and either have been subaerially exposed during part of their history or are accumulations of sediment eroded from adjacent subaerial landscapes. The deep-sea floor, by contrast, has oceanic crustal structure and is unlikely ever to have been above sea level.

Continental Margins

On somewhat less than one-half the length of continental margins, thick wedge-shaped bodies of sediment extend seaward as much as several hundred kilometers (Curray, 1969). The upper surface of this **continental terrace** is the **continental shelf**, which slopes gently seaward. At the **shelf break**, the gradient increases sharply as the **continental slope**. The lower part of the margin, extending like an apron onto the oceanic crustal structure, is the **continental rise**. The tempting analog with topset, foreset, and bottomset deltaic structure should be avoided, because continental marginal structure is much more complicated than that. The shelf, slope, and rise form of continental margin is exemplified by the Atlantic margin of the United States and is typical of the **passive margin** of a continent embedded within a lithospheric plate (p. 469).

The other major category of continental margin has either a narrow shelf or none at all, and a deep, linear, marginal depression or **trench** that may be twice as deep as the average ocean floor. Trenches are found along nearly half of the total length of continental margins, especially around the Pacific Ocean (Fig. 21-1). A trench is the most common indication that a continent has a tectonically **active margin**, a subduction zone where an oceanic lithospheric plate is thrusting under the continent. An orogenic zone on the continent is the terrestrial companion of a marginal trench. Other deep-sea trenches occur parallel to *island arcs* at boundaries between two oceanic lithospheric plates (pp. 469, 493).

Other, minor kinds of continental margins include *marginal plateaus, continental borderlands*, and a variety of steep margins that lack trenches (Curray, 1969). The latter are probably *rift margins* that mark fault boundaries between diverging lithospheric plates.

Continental Shelves and Slopes.   About 15 percent of the submarine landscape is continental shelf and slope (Menard and Smith, 1966, table 6). As usually defined (Shepard, 1973, p. 197), the continental shelf is the shallow platform or terrace seaward from a continental coast, terminated by the relatively abrupt increase in gradient called the *shelf break*. The depth of the shelf break ranges from 20–200 m but averages 130 m, very close to the estimate of full-glacial eustatic lowering of sea level (p. 406). The shelf break is almost certainly genetically related to Quaternary sea-level fluctuations, but if so, the present variations in depth must have been caused by isostatic or other tectonic uplift or depression, glacial erosion, and postglacial deposition. The average shelf width is 75 km, and the average seaward gradient is 0°07' (about 2 m/km). By comparison, the average slope of the Great Plains in the central United States is about 1 m/km (p. 246). Hills, valleys, or closed basins with local relief of 20 m or more have been found on 35–60 percent of representative profiles across shelves (Shepard, 1973, p. 277).

The surface morphology and sediments of the continental shelves are generally relict. Fully 70 percent of the sedimentary cover on the shelves was originally deposited at times of lower sea level, probably much of it during the last glacial maximum about 15,000 years ago (Emery, 1968). Most of the surface sediment is fluvial, eolian, or estuarine, proving that much of the shelf morphology also must be either subaerial or subsequently modified during the rapid late-glacial rise of sea level (Fig. 17-14). Many modern estuaries can be traced as submarine valleys out to the shelf break (Fig. 21-2). Sand waves or sand swells on the shelf (Fig. 21-3) may be relict submerged barriers or dune ridges or contemporary, dynamic submarine depositional sand bodies. One hypothesis is that they are the submarine analogs of longitudinal dunes (Uchupi, 1968, p. 17; Hollister, 1973, p. 18). The issue is of considerable economic significance to artificial nourishment of beaches (p. 461). Intertidal and freshwater peat, masses of dead coral, and bones and teeth of terrestrial mammals have been dredged from shelves, all demonstrating the subaerial or coastal origin for most shelf morphology. However, all the shelf sediment, and therefore the morphology, has been affected to some degree by Holocene submarine processes. The term "palimpsest sediment" has been suggested for the reworked shelf deposits (Swift, et al., 1971). As in the case of subaerial palimpsest landscapes (p. 9), the shelf morphology records more than one set of processes that have overlapped to a variable degree.

We can predict that during geologic intervals not affected by glacially controlled sea-level oscillations, the shelf break would be much less prominent, shelf sediments would be arranged in depositional belts parallel to the shoreline, and some kind of shelf-slope equilibrium profile would evolve. Quaternary shelves are not typical of past conditions, and their form and composition are of limited use in interpreting ancient marine environments.

The continental slope (Fig. 21-4) is "by far the steepest, longest and highest topographic feature on the earth's surface [Dietz, 1964]." From the shelf break, rarely deeper than 200 m, the continental slope plunges at least 1 km, and usually 2 or 3 km, down to the top of the continental rise. To a depth of 1800 m, the average gradient is 4°17' (about 75 m/km), more than three times as steep as the westward gradient of the Sierra Nevada down to the Sacramento Valley in California. Off deltas, the gradient is only one-quarter as steep as the world average, but the slope on the west side of the Florida peninsula below a depth of 2 km in the Gulf of Mexico is a spectacular escarpment with an average gradient of 27°.

The steep, escarpmentlike character of the continental slope has been demonstrated to be the result of various kinds of structural control. Some continental slopes are surely fault scarps. Others are reef taluses. Many seem to be the result of a variety of "dams," ridges that have trapped shelf sediments behind them or are to varying degrees buried or draped with sediment (Fig. 21-5). Some of the kinds of dams that have been suggested include horsts and tilted fault blocks, coral reefs, volcanic dike swarms, salt domes, and anticlines (Emery, 1969; Shepard, 1973, p. 280).

CONTOUR INTERVALS 4, 20 and 200 M

BASED ON SOUNDINGS FROM THE U.S. COAST
AND GEODETIC SURVEY SMOOTH SHEETS

Figure 21-2. The Hudson submarine channel on the continental shelf southeast of New York City and the Hudson submarine canyon on the continental slope. (Uchupi, 1970, Plate 1.)

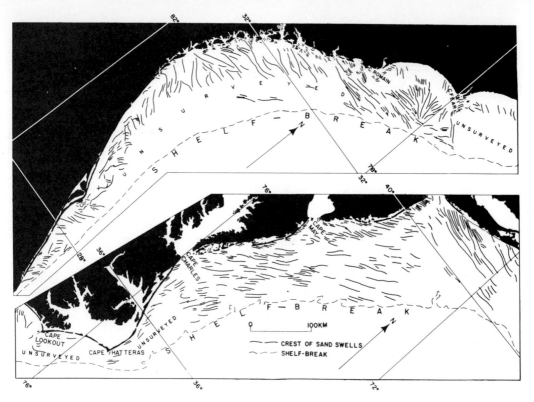

Figure 21-3. Sand swells on the Atlantic continental shelf from New York to Florida. Curved lines are the crests of sand ridges; shelf break is dashed. Figure 21-2 shows some of the sand swells by contour lines. (Uchupi, 1968, Fig. 14.)

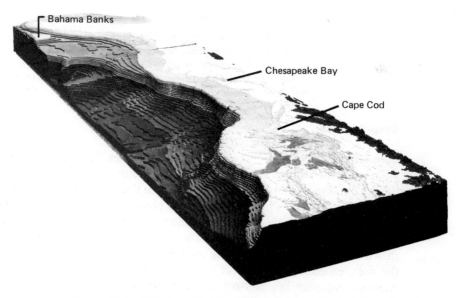

Figure 21-4. Relief model of the Atlantic continental margin from Nova Scotia to Florida. Vertical exaggeration about 20 : 1. Note dominance of the shelf break and continental slope. (Uchupi, 1968, Fig. 1.)

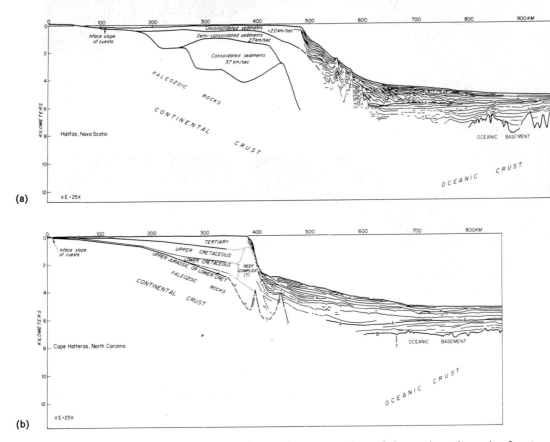

Figure 21-5. Interpretive cross sections of the continental margin of eastern North America at Halifax, Nova Scotia (*a*), and Cape Hatteras, North Carolina (*b*). Vertical exaggeration of 25:1 emphasizes the distinctions between shelf, slope, and rise. The Cape Hatteras section interprets the continental terrace as sediment trapped behind a reef complex. (Emery, et al., 1970, Fig. 38.)

Deep-Sea Floor

Continental Rise. At the base of continental slopes, below water depths of several kilometers, the steep gradients of the slopes decrease to 1° or less (Fig. 21-5). Continuing out into the abyssal hills or plains from this decrease in slope are the undersea morphologic analogs of piedmont alluvial plains (p. 245), the *continental rises*. They comprise about 5 percent of the submarine landscape. Seismic studies show that continental rises overlie oceanic crust and so are part of the oceanic, not continental, structure. On the other hand, the enormous wedge-shaped volume of sediments in the rises, exceeding a thickness of 6 km in some cross sections off eastern North America and extending 300–600 km seaward from the base of the slopes, is clearly derived from the continents and genetically links the rise to the adjacent continental terrace. The continental rises are among the largest sedimentary structures on earth (Emery, et al., 1970).

Continental rises can be subdivided into fan-shaped components that have their apices at the mouths of major submarine canyons in the continental slope. Extending across the fans are meandering, levee-bounded valleys (Stanley, et al.,1971). The sediment of continental rises is a mixture of fine-grained pelagic sediment derived from the overlying water mass and mud, sand, and gravel that were transported down the slope by density currents and submarine mass-wasting. The beds may be highly contorted and slumped. The meandering *fan valleys* probably serve as channels for episodic flows of dense, muddy water or mud that comes down from the adjacent continental slope (Shepard and Dill, 1966, p. 7). The levees along the channels are consistent with this function. Fan valleys often have distributaries but never tributaries. The Monterey fan valley, at the mouth of the Monterey submarine canyon, is more than 36 km long and has levee walls as high as 370 m. The valley floor is of sandy mud with some gravel, clearly derived from the California coastal region at the head of the Monterey Canyon (Shepard and Dill, 1966, p. 84).

Abyssal Hills and Plains.   About 42 percent of the ocean floor, or almost 30 percent of the earth's surface, is as yet classified only as undifferentiated *abyssal hills and plains* (Menard and Smith, 1966). The depth of this vast region ranges from about 3–6 km and is a function of the age of the oceanic crust (Fig. 2-3). The local relief of abyssal hills ranges from a few hundred to 1000 meters. Wherever dredged or photographed, abyssal hills are made of pahoehoe basalt (Bonatti, 1967; Ballard, et al., 1975). Some hills may be volcanic or laccolithic, but most of them are probably fault blocks (Luyendyk, 1970; Ballard and van Andel, 1977). The hilly relief is constructed by volcanic and tectonic processes at midocean rift valleys (Figs. 21-6, 21-7) and is carried laterally away from the ridge crests into abyssal depths by plate motions and thermal contraction (Figs. 2-3, 3-10). With fast spreading rates at ridge crests, the submarine topography is smooth; if the spreading rate is slow, a rugged fault-block topography is generated that persists as ranges of abyssal hills.

Abyssal plains are depositional surfaces of silt and clay of biogenic and terrigenous origin. The sediment thickness is not great, averaging only a few hundred meters. Most of the sediment is very fine brown clay, but beneath areas of highly productive surface waters, siliceous diatoms or calcareous foraminifera form the dominant bottom deposits. Deposition rates in dated deep-sea cores are usually measured in mm/1000 yr, so the entire Quaternary Period is commonly recorded in a few meters of sediment. Quite probably, re-solution of carbonate pelagic sediment has partly destroyed the sedimentary record, especially in the deeper abyssal basins. Seismic reflection profiling suggests that hilly topography forms the surface of the basaltic oceanic crust beneath the depositional abyssal plains, so the geomorphic distinction between abyssal plains and hills is only a matter of sediment thickness.

Figure 21-6. Submarine lava pillows in the median valley of the Mid-Atlantic Ridge. Pillows show effects of repeated surface quenching during expansion. (Photo: R. D. Ballard.)

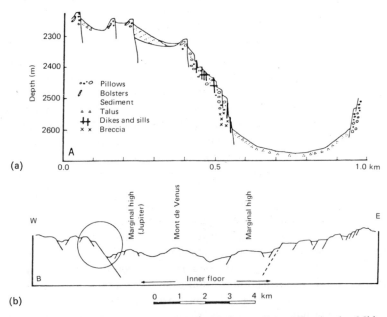

Figure 21-7. Interpretive cross sections of the median valley in the Mid-Atlantic Ridge. (ARCYANA, 1975, Fig. 3.) (*a*) Surface geology observed from a submersible vessel; (*b*) regional cross section showing the location of (*a*).

Oceanic Rises.   Along the midline of the Atlantic Ocean, and continuing as a branching mountain chain through all the oceans, is a rugged volcanic **midocean ridge** or **rise,** generally defined by water depths of less than 4 km but flanked by deeper ocean basins. As arbitrarily defined by depth, the aggregate area of the rises is about one-third of the submarine landscape. Lithospheric plates form new oceanic crust at their accreting margins on the axes of the rises. Rises that show evidence of slow spreading have higher relief, including a characteristic *median trench*, or graben (Fig. 21-7). The northern Mid-Atlantic Ridge is a good example (Ballard and van Andel, 1977). Rapidly spreading rises, like the East Pacific Rise, show broadly convex summit profiles without a median rift. The rises are the youngest parts of the ocean floor and generally have only very thin sediments on a basaltic ocean crust. Linear basalt ridges on the flanks of rises, parallel to the median axis, are horsts and tilted fault blocks of oceanic crust.

Oceanic rises are offset by complexes of faults along major fracture zones that are approximately at right angles to the median axis. These fracture zones are the important *transform faults* (Fig. 3-10) of the plate tectonic theory. Some are striking topographic features, extending for thousands of kilometers as scarps up to 2 km in height.

Volcanoes, Volcanic Ridges, and Poorly Defined Submarine Elevations.   A small percentage of the ocean floor has either bold isolated submarine volcanic peaks called **seamounts**, or volcanic mountain ranges that are not part of the oceanic rises (Fig. 21-8). Some of the conical volcanic peaks rise 3 or 4 km from abyssal depths to flat-topped summits within a few hundred meters below the ocean surface. These unique forms of seamounts were named **guyots** by Hess (1946). Others call them *tablemounts*, with reference to their flat tops. The chance that a random wire-line depth sounding in the open sea would find one of these small summits is very small. Subsequent to the widespread use of continuous seismic depth profiling, at least 2000 have been discovered (Menard, 1964, p. 55). Dredge hauls from the tops of some guyots have shown them to be capped with shallow-water sediment, including beach gravels. A few have low coral banks or reefs as well, ranging in age from mid-Cretaceous (Hamilton, 1956; Matthews, et al., 1974) to Quaternary, but now at depths of 400–2000 m and lightly covered with pelagic sediment. Clearly, guyots developed their flat tops near sea level and have subsequently subsided. All the flattening may not be due to wave erosion or reef growth, however. Volcanic eruptions at oceanic depths are prevented from evolving steam and gas by the great hydrostatic pressure, and they produce smooth-surfaced or pillow lavas. At shallower depths, the pressure release combines with chilling to produce shattered hyaloclastite (Chap. 4). The flat tops of at least some guyots may be a constructional volcanic form, not erosional. Even so, a former shallow depth is implied.

Submarine volcanoes occur in clusters or along linear trends. Some in the tropical oceans acquired coral caps prior to their subsidence and are now the foundations for atolls. Many more sank either too quickly or in a noncoral region and did not acquire reef caps.

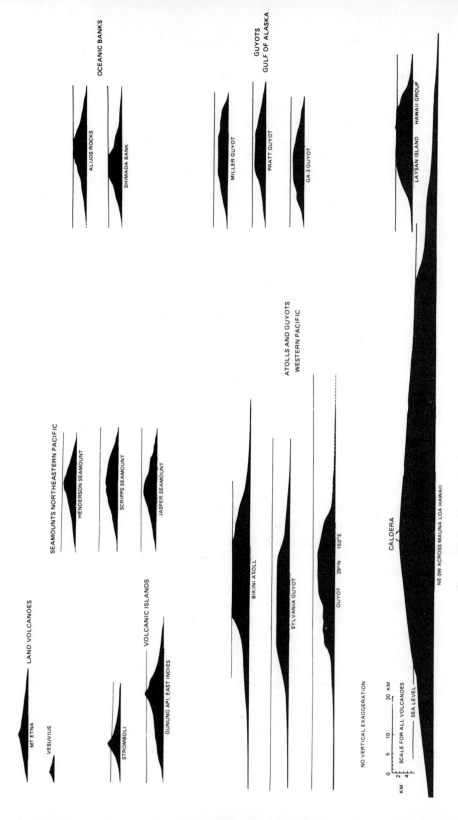

Figure 21-8. Size comparisons for some subaerial volcanoes, volcanic islands, and guyots. No vertical exaggeration. (Menard, 1964, Fig. 4.8.)

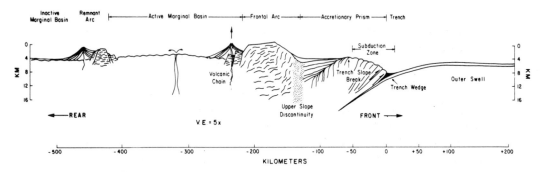

Figure 21-9.   Generalized geologic cross section of a typical island-arc system. Horizontal distances are measured from the trench axis; vertical exaggeration is 5:1. (Karig, 1974.)

Island Arcs and Trenches.   The deepest parts of the oceans are not at their centers but near the margins. About one-half of the continental margins are bounded by trenches twice as deep as the average ocean floor. Other trenches, especially around the western Pacific (Fig. 21-1), are not at continental margins but are associated with arcuate ridges along which volcanic and sedimentary islands stand. The association of trenches and island arcs is well explained by the plate tectonic theory (Fig. 21-9). The trench is the site of subduction where one plate descends beneath another. Volcanoes, often great composite cones of andesitic composition, are the surface expression of remelting along and above the descending slab. Islands of highly deformed deep-water marine sediments form a **frontal arc** by the compression of packets or wedges of sediment literally scraped off the upper part of the descending lithospheric plate (Karig, 1974). Tropical islands on frontal arcs often have flights of emerged coral-reef terraces overlying the deformed sedimentary rocks.

SUBMARINE EROSIONAL PROCESSES AND LANDFORMS

Submarine Weathering

Very little is known about rock weathering in the undersea environment. Volcanic detritus is heavily altered by reaction with sea water. Other nonbiogenic sediments accumulate so slowly on abyssal plains and hills that they have abundant time to become oxidized to brown colors. Montmorillonite clays and zeolite minerals are common alteration products on the deep-sea floor. Manganese and iron in solution are precipitated as oxides at the sea floor, producing "manganese" nodules and crusts on a variety of nuclei. In the deep-sea environment, the crusts grow at extremely slow rates, perhaps slower than 5 mm/1 million yr (Turekian, 1976, p. 81).

Microbial decay, so important in weathering of subaerial landscapes, seems to be almost nonexistent in the deep ocean. Food left unprotected when the research submarine *Alvin* filled with sea water and sank in 1540 m of water off Woods Hole,

Massachusetts, was fresh and palatable when the vessel was raised 10 months later (Jannasch, et al., 1971). The deep ocean is therefore not a good place to dispose of organic waste products from coastal cities.

Calcium carbonate minerals that are stable in surface ocean waters dissolve at an appreciable rate in water depths greater than about 4 km (Turekian, 1976, pp. 67–70). The level at which the solution rate increases rapidly is called the **lysocline**. Calcareous pelagic sediments are differentially dissolved, so the fossils in deep-sea sediments may not record the true proportion of living organisms in the surface waters. This effect has considerable significance to the interpretation of Quaternary climatic change, because we rely on the deep-sea stratigraphic record for relative faunal abundances and oxygen-isotope determinations from which ocean temperatures and continental ice volumes can be inferred (Fig. 18-4).

Where the solution rate below the lysocline equals the deposition rate of calcareous sediment, a "snow line" of calcareous pelagic sediment appears on the submarine landscape. This calcium carbonate **compensation depth** is about 4.5 km in the Pacific Ocean, but it is driven farther below the lysocline under areas of high pelagic carbonate productivity. Above the compensation depth, the submarine landscape is mantled with light-colored calcareous pelagic sediment. Below the compensation depth, calcium carbonate minerals are dissolved, and brown or red clays and siliceous sediments dominate.

Both the lysocline and the compensation depth are the result of water masses moving and mixing in the deep oceans. It is clear that the compensation depth in the equatorial Pacific has risen and fallen through a range of at least 2 km during the Cenozoic Era (van Andel and Moore, 1974). Dramatic Cenozoic changes in oceanic circulation are clearly implied, probably due to the changing positions of continents relative to major ocean currents (p. 407) and to climatic changes (Turekian, 1976, p. 107). When fully understood, the Cenozoic history of the carbonate compensation depth will not only explain patterns of submarine sediment deposition and re-solution but will also establish the chronology and perhaps the causes of the climatic changes under which the modern subaerial landscape evolved.

Submarine Mass-Wasting

Submarine landslides down the continental slopes deliver massive quantities of shallow-water sediment to the continental rises and abyssal plains. Since the early days of submarine cable laying, cable breaks were believed to be associated with submarine slumps and slides. The classic example was the 1929 Grand Banks earthquake off Newfoundland, which broke 12 submarine cables in 28 places. About 15 of the breaks in 6 cables crossing the epicentral region are attributed to earthquake motion or immediate compaction of loose sediment. The explanation of the other breaks, which occurred progressively downslope for as long as 13 hours after the earthquake, has been the subject of controversy. Heezen and Ewing (1952) attributed the progressive breaks to a turbidity current that originated in a mud slide on the slope and flowed 800 km down the continental rise to the abyssal plain (Fig. 21-5a).

The inferred velocity of the turbidity current was 92 km/hr, capable of powerful erosion (Fig. 9-8). Shepard (1973, p. 330) has maintained that many of the cable breaks were caused by successive slumps of an expanding landslide and that the turbidity current did not exceed a velocity of 28 km/hr. Emery, et al. (1970) showed that all of the cables that broke within 59 minutes after the earthquake were within a zone of slumping and landslides on the lower continental slope. Cables that broke more than an hour after the earthquake were well out on the smooth continental rise and were probably broken by a turbidity current with about the velocity calculated by Shepard. The issue has emphasized the role of earthquakes in triggering submarine landslides, but simple sediment loading could just as easily result in similar slope failure.

Divers have observed sand flowing down the axes of submarine valleys off the California coast (Shepard and Dill, 1966). The heads of some California submarine canyons are at shallow depths within a few hundred meters of the shoreline and are major factors in intercepting sand that is drifting alongshore. The sand cascades down the steep submarine valleys, producing visible scouring of the valley walls. Such movement is transitional from mass-wasting to abrasive erosion.

Submarine Erosion

Since their discovery, the large submarine canyons that descend the continental slope (Fig. 21-2) have generated spirited scientific debate. The origin of submarine canyons and the origin of atolls have been the two issues that launched numerous oceanographic studies (Veatch and Smith, 1939). It is significant that the first full-length textbook of submarine geomorphology (Shepard and Dill, 1966) was devoted to the origin of submarine canyons and other sea valleys.

Direct observations are still very limited, but submarine canyons are known to channel sediment-laden waters along their axes. The velocities seem adequate to cause erosion or at least to prevent deposition. Some canyons probably were areas of persistent nondeposition during the Cenozoic accumulation of the shelf-slope sediment pile. During times of Quaternary low sea levels, the canyon heads on the northeastern North American continental slope were near the shoreline at the shelf break and probably acted as sediment chutes, much as the California canyons now function. The canyons on tectonic coasts are more likely to have developed subaerially and have subsided. The great canyons of the Mediterranean Sea have the most spectacular history, eroding to the floor of a dry Miocene tectonic basin that was flooded only when the Straits of Gibralter were breached by the Atlantic Ocean at the beginning of the Pliocene Epoch (Hsü, 1972).

The ocean floors are eroded by currents and waves. Storm surges stir bottom sand on the Gulf of Mexico shelf at least once every five years. A wave-surge velocity of 50 cm/s at a depth of 200 m has been calculated to result from gale-force winds south of England. Swift, et al. (1971, p. 324) reviewed these and other figures and concluded that meteorological, tidal, and other ocean currents can shift sediment and therefore do geomorphic work in the deep ocean.

Internal waves at the boundaries between stratified water masses have a form similar to surface waves, but their wavelengths are on the order of 10 km, and their amplitudes are 5–40 m (Emery and Gunnerson, 1973). There is some evidence that these internal waves "break" and generate a turbulence along the water layers. The role of internal waves in sediment transport is not yet known, but it is conceivable that nondeposition or even erosion of the bottom could result. Internal waves and tidal currents may be responsible for shaping the sand waves on the Atlantic continental shelf of the United States (Fig. 21-3).

REFERENCES

ARCYANA, 1975, Transform fault and rift valley from bathyscaph and diving saucer: *Science*, v. 190, pp. 108–16.

BALLARD, R. D., BRYAN, W. B., HEIRTZLER, J. R., KELLER, G., MOORE, J. G., and VAN ANDEL, T., 1975, Manned submersible observations in the FAMOUS area: Mid-Atlantic Ridge: *Science*, v. 190, pp. 103–8.

BALLARD, R. D., and VAN ANDEL, T. H., 1977, Morphology and tectonics of the inner rift valley at lat 36° 50′ N on the Mid-Atlantic Ridge: *Geol. Soc. America Bull.*, v. 88, pp. 507–30.

BONATTI, E., 1967, Mechanisms of deep-sea volcanism in the South Pacific: *Researches in Geochemistry*, v. 2, John Wiley & Sons, Inc., New York, pp. 453–91.

CURRAY, J. R., ed., 1969, New concepts of continental margin sedimentation: *AGI Short Course Notes*, Am. Geol. Inst., Washington, D.C., 340 pp.

DIETZ, R. S., 1964, Origin of continental slopes: *Am. Scientist*, v. 52, pp. 50–69.

EMERY, K. O., 1968, Relict sediments on continental shelves of the world: *Am. Assoc. Petroleum Geologists Bull.*, v. 52, pp. 445–64.

——, 1969, The continental shelves: *Sci. American*, v. 221, no. 3, pp. 106–22.

EMERY, K. O., and GUNNERSON, C. G., 1973, Internal swash and surf: *Natl. Acad. Sci. Proc.*, v. 70, pp. 2379–80.

EMERY, K. O., UCHUPI, E., PHILLIPS, J. D., BOWIN, C. O., BUNCE, E. T., and KNOTT, S. T., 1970, Continental rise off eastern North America: *Am. Assoc. Petroleum Geologists Bull.*, v. 54, pp. 44–108.

HAMILTON, E. L., 1956, Sunken islands of the Mid-Pacific Mountains: *Geol. Soc. America Mem. 64*, 97 pp.

HEEZEN, B. C., and EWING, M., 1952, Turbidity currents and submarine slumps, and the 1929 Grand Banks earthquake: *Am. Jour. Sci.*, v. 250, pp. 849–73.

HESS, H. H., 1946, Drowned ancient islands of the Pacific basin: *Am. Jour. Sci.*, v. 244, pp. 772–91.

HOLLISTER, C. D., 1973, Atlantic continental shelf and slope of the United States—texture of surface sediments from New Jersey to southern Florida: *U.S. Geol. Survey Prof. Paper 529-M*, 23 pp.

HSÜ, K. J., 1972, When the Mediterranean dried up: *Sci. American*, v. 227, no. 6, pp. 26–36.

JANNASCH, H. W., EIMHJELLEN, K., WIRSEN, C. O., and FARMANFARMAIAN, A., 1971, Microbial degradation of organic matter in the deep sea: *Science*, v. 171, pp. 672–75.

KARIG, D. E., 1974, Evolution of arc systems in the western Pacific: *Ann. Rev. Earth and Planetary Sci.*, v. 2, pp. 51–75.

LUYENDYK, B. P., 1970, Origin and history of abyssal hills in the northeast Pacific Ocean: *Geol. Soc. America Bull.*, v. 81, pp. 2237–60.

MATTHEWS, J. L., HEEZEN, B. C., CATALANO, R., COOGAN, A., THARP, M., NATLAND, J., and RAWSON, M., 1974, Cretaceous drowning of reefs on mid-Pacific and Japanese guyots: *Science*, v. 184, pp. 462–64.

MENARD, H. W., 1964, *Marine geology of the Pacific:* McGraw-Hill Book Company, Inc., New York, 271 pp.

MENARD, H. W., and SMITH, S. M., 1966, Hypsometry of ocean basin provinces: *Jour. Geophys. Res.*, v. 71, pp. 4305–25.

SHEPARD, F. P., 1973, *Submarine geology*, 3d ed.: Harper & Row, Publishers, New York, 517 pp.

SHEPARD, F. P., and DILL, R. F., 1966, *Submarine canyons and other sea valleys:* Rand McNally & Company, Chicago, 367 pp.

STANLEY, D. J., SHENG, H., and PEDRAZA, C. P., 1971, Lower continental rise east of the Middle Atlantic states: Predominant sediment dispersal perpendicular to isobaths: *Geol. Soc. America Bull.*, v. 82, pp. 1831–40.

SWIFT, D. J. P., STANLEY, D. J., and CURRAY, J. R., 1971, Relict sediments on continenal shelves: A reconsideration: *Jour. Geology*, v. 79, pp. 322–46.

TUREKIAN, K. K., 1976, *Oceans*, 2d ed.: Prentice-Hall, Inc., Englewood Cliffs, N.J., 149 pp.

UCHUPI, E., 1968, Atlantic continental shelf and slope of the United States—physiography: *U.S. Geol. Survey Prof. Paper 529-C*, 30 pp.

———, 1970, Atlantic continental shelf and slope of the United States—shallow structure: *U.S. Geol. Survey Prof. Paper 529-I*, 44 pp.

VAN ANDEL, T. H., and MOORE, T. C., 1974, Cenozoic calcium carbonate distribution and calcite compensation depth in the central equatorial Pacific Ocean: *Geology*, v. 2, pp. 87–92.

VEATCH, A. C., and SMITH, P. A., 1939, Atlantic submarine valleys of the United States and the Congo submarine valley: *Geol. Soc. America Spec. Paper 7*, 101 pp.

Author Index

Cvijić, J., 137, 152, 153, 160
Czudek, T., 364, 365

Dale, T. N., 108, 132
Dalrymple, G. B., 283, 304
Dalrymple, J. B., 189, 195
Daly, R. A., 23, 26
Darwin, C., 454, 464
Das, M. M., 457, 461, 464
Da Silva, J. X., 432
Davies, J. L., 438, 452, 453, 455, 457, 461, 465, 476, 477, 480
Davies, W. E., 145, 160
Davis, L. C., 226
Davis, M. B., 421, 433
Davis, W. M., 39, 52, 55, 80, 153, 157, 160, 198, 222, 225, 247, 251, 300, 301, 304, 307, 327, 478, 480
Dawson, J. B., 60, 80
Deevey, E. S., 410, 411
Degens, E. T., 115
Deike, R. G., 143, 160
Deju, R. A., 118
Demek, J., 364, 365
Denny, C. S., 113, 132, 318, 327
Denton, G. H., 407, 411
Dietrich, R. V., 330, 346
Dietz, R. S., 439, 465, 485, 496
Dill, R. F., 146, 160, 161, 489, 495, 497
Dionne, J. C., 477, 480
Dolan, R., 460, 465
Dole, R. B., 284, 304
Domenico, P. A., 140, 160
Donath, F. A., 39, 52
Dorman, J., 25, 26, 101
Dragovich, D., 130, 132
Duennebier, F., 101
Dunbar, C. O., 129, 132
Dunn, J. R., 112, 132
Dunne, K. C., 251
Durloo, L. H., 143, 161
Dury, G. H., 2, 9, 115, 132, 229, 233, 251, 427, 432
Dutton, C. E., 61, 80
Dyson, F. J., 89, 101

Eakin, H. M., 359, 365
Eaton, G. P., 73, 79, 80
Eaton, J. P., 30, 52, 106, 132
Eden, W. J., 177, 195
Edwards, A. R., 327, 412, 433
Eggler, D. H., 255, 276
Ehrlich, P. R., 90, 101
Eimhjellen, K., 496
Embleton, C., 351, 360, 365
Embody, D. R., 195
Emery, K. O., 47, 54, 448, 466, 485, 488, 489, 495, 496
Emmel, F. J., 464
Engel, C., 119, 134
von Engeln, O. D., 17, 26, 32, 52, 188, 195, 263, 276, 302, 304, 398, 411
Ericksen, G. E., 195
Evans, G., 480

Evans, I. S., 110, 111, 112, 133
Everett, D. H., 351, 365
Everitt, B. L., 236, 237, 238, 251
Evernden, J. F., 102
Ewing, G. C., 346
Ewing, M., 26, 101, 494, 496

Fahnestock, R. K., 250
Fairbridge, R. W., 15, 26, 248, 251, 444, 455, 465
Farmanfarmaian, A., 496
Farrand, W. R., 425, 432
Feininger, T., 255, 276, 395, 411
Fenneman, N. M., 17, 19, 26
Ferguson, H. F., 172
Fernold, A. T., 365
Ferrians, O. J., 359
Feth, J. H., 120, 133
Fischer, A. G., 21, 26
Fisher, J. J., 460, 465
Fisher, R. V., 63, 74, 80
Fisk, H. N., 230, 251
Flemal, R. C., 348, 364, 365
Fletcher, E. B., 195
Fletcher, J. E., 185, 195
Flint, R. F., 23, 26, 259, 276, 351, 365, 368, 384, 396, 405, 407, 410, 411, 416, 417, 420, 424, 425, 432, 473, 480
Flohn, H., 88, 91, 101, 416, 432
Folk, R. L., 139, 160
Foose, R. M., 143, 161
Ford, D. C., 142, 147, 148, 161
Foster, H. L., 365
Frakes, L. A., 414, 432
Frederickson, A. F., 118, 133
Friedmann, E. I., 119, 133
de Frietas, M. H., 171, 195
Frye, J. C., 245, 246, 251, 342, 345, 346

Gagliano, S. M., 53
Galloway, R. W., 423, 432
Galvin, C. J., 412
Gaposchkin, E. M., 14, 26
Garfield, D. E., 385
Garrels, R. M., 145, 161, 197, 225
Garrett, C. J. R., 442, 465
Gates, W. L., 408, 411
Geiger, R., 311, 327
Gettings, M. E., 80
Gibbs, R. J., 198, 225, 287, 304
Gilbert, C. M., 39, 53
Gilbert, G. K., 24, 26, 39, 50, 53, 190, 195, 241, 251, 262, 276, 300, 320, 327
Gilluly, J., 260, 276, 287, 304
Gjessing, J., 389, 411, 474, 480
Glen, J. W., 369, 384, 385
Glock, W. S., 297, 304
Goddard, R. H., 366
Godfrey, P. J., 465
Goetze, C., 132
Goldich, S. S., 120, 133
Goldreich, P., 99, 101
Goodlett, J. C., 113, 132, 191, 195
Goody, R. M., 91, 101

Gostin, V. A., 327, 412, 433
Gould, H. R., 49, 53
Gow, A. J., 378, 380, 385
Gramlich, J. W., 283, 304
Gravenor, C. P., 394, 412
Greeley, R., 62, 80
Green, J., 56, 80
Greene, G. W., 365
Gregory, K. J., 201, 225
Griggs, D. T., 109, 133
Griggs, G. B., 447, 464
Grinshpan, Z., 108, 134
Grove, A. T., 430, 433
Guilcher, A., 461, 465
Gulliver, F. P., 475, 480
Gumbel, E. J., 232, 251
Gunnerson, C. G., 496
Gupte, R. B., 74, 80
Gutenberg, B., 390, 412
Guy, H. P., 175, 186, 196

Hack, J. T., 108, 133, 143, 161, 191, 195, 299, 303, 304, 338, 339, 345
Hadley, H., 132
Hadley, R. F., 321, 327
Häfele, W., 90, 101
Hails, J. R., 251
Hajos, M., 327, 412, 433
Halbouty, M. T., 49, 53
Hallam, A., 22, 26
Hamilton, E. L., 491, 496
Hamilton, W., 37, 53
Hamilton, W. L., 99, 101
van der Hammen, T., 421, 432
Hammond, E. H., 17, 26
Hampton, M. A., 327, 412, 433
Harward, M. E., 80
Hawkes, H. K., 50, 53
Hawkes, L., 50, 53
Haxby, W. F., 106, 133
Hayes, M. O., 442, 465
Hays, J., 418, 432
Heezen, B. C., 17, 26, 494, 496, 497
Heirtzler, J. R., 496
Hendershott, M. C., 98, 101, 442, 465
Herak, M., 143, 153, 161
Hesler, J. L., 365
Hess, H. H., 491, 496
Hickok, C. F., Jr., 330, 345
Higashi, A., 173, 195
Hightower, M. L., 53
Hill, R. E., 161
Hills, E. S., 345, 346
Hinkley, K. C., 365
Hjulström, F., 214, 225
Hodgkin, E. P., 448, 465
Holdren, G. R., Jr., 117, 132
Holdren, J. P., 90, 101
Hole, F. D., 132
Holeman, J. N., 285, 304
Holland, H. D., 132
Hollingworth, S. E., 108, 133
Hollister, C. D., 485, 496
Holm, D. A., 341, 346
Holmes, A., 43, 44, 45, 53
Holmes, G. W., 364, 365

499

502

Subject Index

Swash zone, 457
Sweden:
  postglacial uplift, 404
  slope retreat, 193
Swidden agriculture, 131
Switzerland, glacial troughs, 390
Syenite weathering, 104
Syracuse, New York, 146
System:
  closed, 85
  coastal geomorphic, 435–36
  concept in geomorphology, 85–89
  fluvial geomorphic, 197–98, 281
  morphogenetic, 307
  nonpluvial, 423–24
  open, 85–87, 96, 303
  pluvial, 423–24
  steady state, 85–86, 96, 303

Tablemount, 491
Taiwan, mélange terrane, 258
Talus, 75, 188, 193, 315, 350, 359
  submarine, 450
Tarn, 392
Tasman River, New Zealand, 403
Teays Valley, 420
Tectonic mélanges, 258
Temblor Range, California, 36
Tennessee, limestone
  weathering, 123
Tephra, 63–65, 74, 77–78, 407
  size range of, 63
Tephrochronology, 63–65
Terrain, 5
Terrane, 5
Terrestrial thermal gradients, 92–94
Thalweg, 207
Thar desert, India, 431
Thaw lake, 382
Thermal contraction, 109
Thermal expansion, 109
Thermokarst, 142, 269, 364
Tholoid, 76, 77
Through valley, 393
Tiber Delta, Italy, 243
Tidal bore, 443
Tidal inlet, 460
Tidal marsh, 463
Tides, 98–99, 442–43
Tigris-Euphrates, irrigation
  system, 209
Till, 396, 398–400, 402
Till plain, 397
Time, 2, 4, 7–8
  in coastal description, 467, 478–80
  real, 8
  relative, 8
  in soil formation, 123, 125
Timor Sea, 442
Tinajita, 148
Tombolo, 461
Tongatapu, Kingdom of Tonga, 453
Topographic inversion, 22, 228,
  263–64
Tor, 255, 256, 325, 352
Townsville, Australia, 256, 324, 463

Trans-Alaskan oil pipeline, 364
Transcurrent faulting, 35–37
Transform fault, 37, 491
Transitory bar, 235
Trench, submarine, 484, 493
  median, 491
Treppen, 73
Triangular facet, 32, 34
Tributary glacier, 374, 381
Tropical weathering, 128–31
  in karst, 137
  relict in northeastern United
    States, 394
Trough lake, 390
Tsunami, 442, 453
Tufa, 426
Tuff ring, 77–78
Tumulus, 71
Tundra, 364
Turbidity current, 494
Turmkarst, 154
Turtle Islands, Australia, 454
Twelve Apostles, Port Campbell
  National Park, Australia, 451
Typhoon, 95

Uganda, tectonic landforms, 45
Ukranian SSR, loess, 423
Underfit stream, 229
Uniformitarianism, 7, 282, 300
United States soil taxonomy, 125–28
USSR:
  dry regions, 317, 342
  permafrost, 349
Uvala, 152

Vadose zone, 143, 157–58
Vaiont Canyon, Italy, 180, 181
Valley-and-ridge landscape, 264,
  265
  drowned, 473
Valley and Ridge province, 17, 19,
  106, 265
Valley flat, 247
Valley glacier, 390, 393–95, 397,
  410
Valley train, 401, 403
Varved sediment, 396
Veld, 315
  characteristics of, 323
Venezuela coast, 469
Ventifact, 330, 331
Vesuvius, 66
Victoria Island, Northwest
  Territories, Canada, 363
Victoria Nile River, 47
Vietnam, karst, 154
Volcanic neck, 77
Volcanism:
  as an "accident", 55
  dating techniques, 58
  explosion index, 73
Volcano:
  classification, 56, 66–67
  submarine, 491–93
Vulcano, 66

Wadi, 316
Wairarapa fault, New Zealand, 29,
  35
Wang, 153
Wasatch Plateau, Utah, 180
Wash plain, 324, 326
Water, 94–97
  in chemical weathering, 114
  ordered, 112
  properties, 114
  role in mass-wasting, 163–64
Water gap, 275
Water table, 143
  hypothesis of cave
    formation, 157–58
Wave base, 439
Wave-cut bench, 447
Wave-cut cliff, 450
Wave diffraction, 441
Wave energy, 88, 439–40, 443, 448,
  449, 452, 457, 476, 477, 495–96
Wave height factors, 438
Wave motion, 438–39
Wave plucking, 445, 447, 448, 450
Wave quarrying, 445, 446, 450
Wave reflection, 440
Wave refraction, 440, 441, 450, 457
Waves, internal, 496
Weathering, 7, 167, 333
  chemical, 114–20
  mechanical, 105–13
  submarine, 493–94
  water-level, 447–50
Weathering front, 325
Weathering series, 119–20
Welded tuff, 64, 79
Wellington, New Zealand, 42, 43
West Indies, 145
Wetted perimeter, 218
White Mountains, California, 193,
  287–89
Wind-faceted rock, 330
Wind gap, 275
Wine-glass valley, 32
Wisconsin, drumlins, 398
Wyoming Basin, 19

Yangtze River, China, 285
Yardang, 331
Yazoo tributary, 239
Yellowstone National Park region:
  geyser activity, 99
  volcanism, 64, 79–80
Yellowstone River, 211, 275
Yosemite National Park,
  California, 107

Zagros Mountains, Iran, 264, 266
Zambezi River, 44
Zetaform curvature, coasts, 457, 459
Zigzag ridge, 265
Zone of aeration, 143, 145, 157
Zone of saturation, 143